The

Adaptive
Curve
Evolution

Model

for Interest and FX Rates

Matthias Heymann

Cover art (design by Matthias Heymann): The cover photo shows a subway sign in New York City, located on the South East corner of 42nd Street and 8th Avenue, with the lights of New York's Theater District in the background (most notably, the AMC Empire movie theater, Madame Tussauds, and in the far distance the Chrysler Building). Meant to direct tourists and local theater goers to the various available subway lines, this iconic view is just one example of the many times in which the letters A–C–E, which have occupied my mind during my years of working on this book, have randomly revealed themselves to me also in my everyday life.

This particular occurrence appears to further reference the Jackson Five's 1970 signature hit song "A B C," whose chorus *"A B C – It's easy as 1 2 3 – Oh, simple as do re mi – A B C – 1 2 3 – Baby, you and me, girl!"* made the point that *"That's how simple love can be."* Just as— *"A C E – 1 2 3"*—the ACE model can, with the help of this book.

AMS Mathematics Subject Classifications (2010): 91G30, 91G80, 58J90, 83A05

ISBN: 979-8345519370 (hardcover), 979-8345517703 (paperback)
Edition 3.0.1
12/01/2024

published by Matthias Heymann via Kindle Direct Publishing^{TM}
available on www.amazon.com

Chapters 5–9 of this book are
based on a paper by Gregory Pelts.

Preface to Edition 1

Modeling interest rate-related products is an inherently complex task: The underlying set of fundamental instruments—the zero-coupon bonds for all future maturities—is large, correlated in a complex way, and subject to constraints (interest rates are typically bounded below). Furthermore, the family of liquid instruments that the model should be able to calibrate to—caps, floors, and swaptions of all strikes, expirations, and maturities—is large as well, and so global calibration is difficult to achieve.

Over time, various modeling approaches have been developed, first and foremost market and short rate models. Unfortunately, none of those have all the properties that are required of a truly generic and tractable framework. The market model approach (i.e., the BGM model) is high-dimensional and models only finitely many maturities in a consistent manner, which in practice leads to compromises in terms of speed, accuracy, calibration quality, consistency, and time homogeneity. Short rate models are generally fast and numerically tractable, as long as they allow bond prices to be computed via an explicit analytic formula; however, all the popular such models have other downsides: Affine models lack the parametric freedom and dynamical flexibility to achieve true global calibration, while the otherwise powerful Cheyette model class introduces extra state space dimensions to obtain its Markovian structure, thus limiting the number of noise sources that one can use before numerical issues arise.

As a consequence, faced with the problem of achieving global calibration in a numerically feasible way, modelers frequently find themselves having to choose between either ignoring market information, or resorting to using a patchwork of ad hoc methodologies and separate models targeting different types of instruments. The result is an inconsistent modeling environment that can leave its unsuspecting user vulnerable to arbitrage.

The ACE ("Adaptive Curve Evolution") model, in its original form developed by Gregory Pelts in [1] and now carefully rephrased, refined, and made more accessible in the present book, is the first interest rate framework to provide a mathematically accurate solution that does not require any such fixes, while at the same time being computationally extremely effective. It is a fully consistent multi-factor interest rate model that can be robustly calibrated to caps, floors, and swaptions with various strikes, expirations, and maturities, and that can efficiently price Bermudan, accreting, amortizing, and mid-curve swaptions, volatility forwards, spread options, and all other single-currency derivatives.

More precisely, the n-factor ACE model ($n \in \mathbb{N} \setminus \{2\}$) is a short rate process $(r(X_t))_{t \geq 0}$ that is defined via a scalar-valued function $r(\mathbf{x}) > 0$ and an SDE

$$dX_t = \mu(X_t)\,dt + \sigma(X_t, t)\,dW_t, \qquad X_{t=0} = \mathbf{x}_0$$

for its state variable process $(X_t)_{t \geq 0}$, which is of the same dimension n as the Brownian motion $(W_t)_{t \geq 0}$ driving it. Its short rate function $r(\mathbf{x})$ and its drift $\mu(\mathbf{x})$ are specific functions (see Section 2.1 for their exact definitions) that together with the initial state $\mathbf{x}_0$ provide parameters with a total of $2n + 1$ degrees of freedom for calibrating the model to the discount curve; in addition, the scalar-valued noise function $\sigma(\mathbf{x}, t) > 0$ can essentially be chosen freely, which allows the model to be calibrated to the entire caps and swaptions market. Finally, the model provides a closed-form bond price formula that is independent of σ (i.e., it is unspanned), which increases the leverage during the calibration of σ since it allows for bootstrapping by expiration as opposed to maturity. *The ACE model is the first interest rate model to combine all of these properties.*

Our construction of the ACE model relies on a variety of compelling nonstandard techniques. It begins by setting up a general "numeraire-free" mathematical framework for interest rate modeling, in which account values are represented by sections on a line bundle (instead of regular functions), with the no-arbitrage condition expressed by way of a section-based variant of the Kolmogorov backward equation that all sections associated to self-financing accounts have to satisfy. The actual derivation of the model then makes ample use of several tools from theoretical physics that are introduced to the reader along the way, e.g., the Minkowski space (the space used in Einstein's Special Theory of Relativity), the forward light cone, projective spaces, co- and contravariant tensors, pseudo-orthogonal operators, and the d'Alembert operator.

Luckily, the reader who merely wishes to understand the ACE model well enough to use it in practice will not need to learn any of these advanced techniques: After our introduction in Chapter 1, Chapter 2 will lay out all of our main results in detail, along with some straightforward proofs that require only standard knowledge in analysis, stochastic processes, and mathematical finance. The only caveat of this shortcut is that it will leave the reader wondering about the origins of our exact definition of $\mu(\mathbf{x})$, and of our bond price formula, which at that point will appear to be an ingenious guess. It is then left to the reader to decide whether they want to proceed and study also the model's derivation in Chapters 5–8, which will provide invaluable insight into the model's inner workings, and which will equip him with a diverse set of unconventional mathematical tools that may well contain the key also to other financial modeling problems.

I hope that the reader will find this book both useful and inspiring, and that I will succeed in encouraging them to give the ACE model a try. While the path may be steep and rocky at times, I believe that the journey will surely be worth the effort. Fasten your seat belts.

Matthias Heymann

New York, NY, USA
May 2018

Preface to Edition 2

To extend an interest rate model to multiple currencies in a practically meaningful way, it is not enough to simply make multiple copies of the model and correlate their driving Brownian motions: First, this would multiply the state space dimension n by the number of currencies, quickly making the model numerically intractable. Second, this would only allow us to compute the bond prices in their *respective* currencies (i.e., the prices of dollar bonds measured in dollars, the prices of Euro bonds measured in Euros, etc.), and so we would crucially be missing the currencies' exchange rates, which one needs in order to compare the values of these bonds with one another.

Instead, in the newly added Chapter 3 we are seeking to extend our single-currency ACE model without modifying its original state process $(X_t)_{t \geq 0}$ (thus leaving n unchanged), by defining additional deterministic functions for computing the short rates of all the various currencies and the exchange rates between them. More precisely, denoting by C the collection of currencies in scope, e.g., $C := \{\$, €, £\}$, we need to define a short rate function $r^c(\mathbf{x})$ for each $c \in C$, and a spot exchange rate function $a^{c_1/c_2}(\mathbf{x}, t) > 0$ for each pair of currencies $c_1, c_2 \in C$, thus giving rise to the short rate and exchange rate processes

$$(r^c(X_t))_{t \geq 0} \qquad \text{and} \qquad (a^{c_1/c_2}(X_t, t))_{t \geq 0}.$$

We would then need to derive analytic (σ-independent) formulas for the bond prices $B^{c_1}_{c_2, T}(\mathbf{x}, t)$ for $\forall c_1, c_2 \in C$, i.e., for the values measured in units of c_2 of the bond that pays one unit of c_1 at time $T \geq t$. By a simple no-arbitrage argument, this would then in fact give us explicit formulas also for all the forward exchange rates $a^{c_1/c_2}_T(\mathbf{x}, t)$, $T \geq t$.

Unfortunately, upon exploring the various pitfalls that can lead to inconsistencies and arbitrage opportunities, it quickly becomes apparent that finding such an extension to any given single-currency model is a highly non-trivial task that requires r^c, a^{c_1/c_2}, and the SDE's drift μ to interact in seemingly impossible harmony. Even just the basic requisite that restricting the model to any fixed currency c should yield the given single-currency model again is much more complex than one may assume, as it involves changing the measure of $(X_t)_{t \geq 0}$ (to change the numeraire to c), in fact followed by a change of variables.

To the rescue, once again, comes the ACE model's underlying section-based numeraire-free modeling framework, in which the necessity for a change of numeraire elegantly disappears, thus making the construction of a multi-currency extension effortless. As in the single-currency case, after translating the model equations back into the familiar world of $\mathbb{R}^n$, we obtain a fully self-contained $\mathbb{R}^n$-based description of the model that can be understood, verified, and implemented by any skilled Master's student with just a solid training in analysis, stochastic processes, and mathematical finance.

Of course, this necessary detachment of the model from its now inaccessible origins is somewhat regrettable. On the plus side, seeing the many pieces of the multi-currency ACE model's machinery fall into place so perfectly to satisfy its various interconnected analytical requirements (which in the original section-based framework are mere trivialities) will be certain to leave the reader with a feeling of exhilaration akin to witnessing wizardry at work. Wingardium leviosa.

Matthias Heymann   New York, NY, USA
                                                                          March 2020

Preface to Edition 3

The illusion of simplicity is not always easy to come by. Over the course of its lifetime, the ACE model has undergone several transformations to make itself accessible to the widest audience possible. Up until now, this effort had been limited to providing the reader with simple $\mathbb{R}^n$-based model equations and with independent proofs that they work as advertised, leaving the user in the dark about how these formulas had been unearthed in the first place, unless they took it upon themselves to embark with us on our challenging journey through Minkowski space.

As a souvenir for those left behind, this third edition of this book now contains a new chapter with a much shorter alternative "derivation" of the model, based on three comparably simple steps that our model equations have long begged us to explore, in particular since their extension to the multi-currency case. This should certainly be of help to illumine the Fast & Curious.

Why the air quotes, you may ask? Because one can only solve a problem for the first time once, and so in reality, any alternative solution presented later silently builds on all the experiences gained on the original path to success, and may thus be hiding some key insights up the mathemagician's sleeves.

Would this quick derivation be an exception? With our excitement from having traversed the Path of Enlightenment still fresh, it dawns on us that its final passage appears to have relied on some well-hidden stroke of luck. So has it been Fortuna herself who has been holding our hands along our journey? We slowly dare to turn our heads to take a look at the shadow to our right, and then reality is setting in: Yet again, it hasn't been Lady Luck at all, but Gregory Pelts.

Nevertheless, I hope you will enjoy the ride.

Matthias Heymann   New York, NY, USA
                                                         November 2024

x

Summary: Single-Currency ACE Model

Short Rate Process: $(r(X_t))_{t \geq 0}$

Short Rate Function

$$r(\mathbf{x}) := r_\infty + \langle \mathbf{v}, \mathbf{x} \rangle$$

State Process

$$\mathrm{d}X_t = \mu(X_t)\,\mathrm{d}t + \sigma(X_t, t)\,\mathrm{d}W_t, \qquad X_{t=0} = \mathbf{x}_0$$

$$\mu(\mathbf{x}) := (-\lambda I + E)\mathbf{x} + \frac{1}{n-2}\|\mathbf{x}\|^2\mathbf{v} - \frac{2}{n-2}\langle \mathbf{v}, \mathbf{x} \rangle \mathbf{x}$$

Parameters & Constraints

Name	Description	Constraints	DoF
n	state space dimension	$n \geq 3$ ($n = 1$: Sect. 2.9)	–
r_∞	long-term rate	$r_\infty > 0$	1
$\mathbf{v}$	weight vector	$\|\mathbf{v}\| = 1$, (2.42)–(2.43)	$\lceil n/2 \rceil - 1$
$\mathbf{x}_0$	initial state	$r(\mathbf{x}_0) > 0$	n
λ	reversion speed	$\lambda > 0$	1
E	oscillation freq. matrix	$E^T = -E$, (2.41), (2.43)	$\lfloor n/2 \rfloor$
$\sigma(\mathbf{x}, t)$	noise function	$\sigma > 0$, Lemma 2.4	∞

Total DoF: $2n + 1$ discrete parameters, one functional parameter

Other Constraints:	$\delta := r_\infty \lambda - \frac{1}{n-2}r_\infty^2 - \frac{n-2}{4}\|E\mathbf{v}\|^2 \geq 0$
	if $\delta = 0$ then $\lambda < \frac{2}{n-2}r_\infty$

Key Formulas

Bond Prices

$$B_{\$,T}(\mathbf{x}, t) = \mathrm{e}^{-r_\infty \tau}\left\|\|\mathbf{c}_\tau\|\mathbf{x} - \frac{\mathbf{c}_\tau}{\|\mathbf{c}_\tau\|}\right\|^{2-n}$$

$$\mathbf{c}_\tau := \frac{1}{n-2}(E + \lambda I)^{-1}[\mathrm{e}^{-\tau(E+\lambda I)} - I]\mathbf{v}$$

$$\tau := T - t$$

Zero-Noise Process $(\partial_t U_t = \mu)$

$$U_t(\mathbf{x}) = i\left(\mathrm{e}^{t(E+\lambda I)}\left(i(\mathbf{x}) + \hat{\mathbf{v}}\right) - \hat{\mathbf{v}}\right)$$

$$i(\mathbf{x}) := \frac{\mathbf{x}}{\|\mathbf{x}\|^2} \qquad (i(\vec{0}) := \infty, \ i(\infty) := \vec{0})$$

$$\hat{\mathbf{v}} := \frac{1}{n-2}(E + \lambda I)^{-1}\mathbf{v}$$

Forward Rates

$$F(\mathbf{x}, t; s) = r(U_{s-t}(\mathbf{x}))$$

Summary: Multi-Currency ACE Model

Short Rate & Exchange Rate Processes: $(r^c(X_t))_{t\geq 0}$, $(a^{c_1/c_2}(X_t))_{t\geq 0}$

Short Rate & Exchange Rate Functions

$$r^c(\mathbf{x}) := r^c_\infty + \langle \mathbf{v}^c, G_c(\mathbf{x}) \rangle$$

$$a^{c_1/c_2}(\mathbf{x}, t) := \frac{a^{c_1} \cdot e^{-r^{c_1}_\infty t} \cdot \left\| \|\mathbf{c}^{c_2}\| \mathbf{x} - \frac{\mathbf{c}^{c_2}}{\|\mathbf{c}^{c_2}\|} \right\|^{n-2}}{a^{c_2} \cdot e^{-r^{c_2}_\infty t} \cdot \left\| \|\mathbf{c}^{c_1}\| \mathbf{x} - \frac{\mathbf{c}^{c_1}}{\|\mathbf{c}^{c_1}\|} \right\|^{n-2}}$$

Helper Functions

$$\mathbf{c}^c := \frac{1}{n-2}(E + \lambda I)^{-1}(\mathbf{v}^c - \mathbf{v}^{c_0})$$

$$i(\mathbf{x}) := \frac{\mathbf{x}}{\|\mathbf{x}\|^2} \quad (i(\vec{0}) := \infty, \; i(\infty) := \vec{0})$$

$$G_c(\mathbf{x}) := i(i(\mathbf{x}) - \mathbf{c}^c)$$

State Process (c_0-Measure)

$$dX_t = \mu(X_t)\,dt + \sigma(X_t, t)\,dW_t, \qquad X_{t=0} = \mathbf{x}_0$$

$$\mu(\mathbf{x}) := (-\lambda I + E)\mathbf{x} + \frac{1}{n-2}\|\mathbf{x}\|^2 \mathbf{v}^{c_0} - \frac{2}{n-2}\langle \mathbf{v}^{c_0}, \mathbf{x} \rangle \mathbf{x}$$

Parameters & Constraints

Name	Description	Constraints	DoF
C	set of currencies	–	–
c_0	base currency	$c_0 \in C$	–
n	state space dimension	$n \geq 3$	–
r^c_∞	long-term rates	$\forall c \in C: r^c_\infty > 0$	$|C|$
$\mathbf{v}^c$	weight vectors	$\|\mathbf{v}^{c_0}\| = 1$, $\forall c \in C: \mathbf{v}^c \neq \vec{0}$, $\mathbf{v}^{c_0}$ only: (2.42)–(2.43)	$\lceil n/2 \rceil - 1 + (|C|-1)n$
a^c	curr. spot attractors	$a^{c_0} = 1$, $\forall c \in C: a^c > 0$	$|C| - 1$
$\mathbf{x}_0$	initial state	$\forall c \in C: r^c(\mathbf{x}_0) > 0$	n
λ	reversion speed	see below	1
E	oscillation freq. matrix	$E^T = -E$, (2.41), (2.43)	$\lfloor n/2 \rfloor$
$\sigma(\mathbf{x}, t)$	noise function	$\sigma > 0$, Lemmas 3.9+3.11	∞
Total DoF: $n - 1 + |C|(n+2)$ discrete parameters, one functional par.			

Other Constraints:	$\lambda > \max_{c \in C}\left(\frac{r^c_\infty}{n-2} + \frac{n-2}{4r^c_\infty}\frac{\|E\mathbf{v}^c\|^2}{\|\mathbf{v}^c\|^2} \right)$

Key Formulas

General Derivative Pricing

$$V_{c_2}(\mathbf{x}, t) = \frac{1}{a^{c_2/c_0}(\mathbf{x},t)} \, \mathbb{E}\Big[P_{c_1,T} \cdot a^{c_1/c_0}(X_T, T)$$
$$\times \exp\Big(-\int_t^T r^{c_0}(X_s)\, \mathrm{d}s \Big) \Big| X_t = \mathbf{x} \Big]$$

price (in c_2) of a payout of $P_{c_1,T}$ units of c_1 at time T

Bond Prices

$$B_{c_2,T}^{c_1}(\mathbf{x}, t) = \frac{a^{c_1} \cdot \mathrm{e}^{-r_\infty^{c_1} T} \cdot \Big|\big\| \mathbf{c}^{c_2} \big\| \mathbf{x} - \frac{\mathbf{c}^{c_2}}{\|\mathbf{c}^{c_2}\|} \Big|^{n-2}}{a^{c_2} \cdot \mathrm{e}^{-r_\infty^{c_2} t} \cdot \Big|\big\| \mathbf{c}_\tau^{c_1} \big\| \mathbf{x} - \frac{\mathbf{c}_\tau^{c_1}}{\|\mathbf{c}_\tau^{c_1}\|} \Big|^{n-2}}$$

price (in c_2) of the bond paying 1 unit of c_1 at time T

Forward Exchange Rates

$$a_T^{c_1/c_2}(\mathbf{x}, t) = \frac{a^{c_1} \cdot \mathrm{e}^{-r_\infty^{c_1} T} \cdot \Big|\big\| \mathbf{c}_\tau^{c_2} \big\| \mathbf{x} - \frac{\mathbf{c}_\tau^{c_2}}{\|\mathbf{c}_\tau^{c_2}\|} \Big|^{n-2}}{a^{c_2} \cdot \mathrm{e}^{-r_\infty^{c_2} T} \cdot \Big|\big\| \mathbf{c}_\tau^{c_1} \big\| \mathbf{x} - \frac{\mathbf{c}_\tau^{c_1}}{\|\mathbf{c}_\tau^{c_1}\|} \Big|^{n-2}}$$

Zero-Noise Process ($\partial_t U_t = \mu$)

$$U_t(\mathbf{x}) = i\big(\mathrm{e}^{t(E+\lambda I)}\big(i(\mathbf{x}) + \hat{\mathbf{v}} \big) - \hat{\mathbf{v}} \big)$$

Forward Rates

$$F^c(\mathbf{x}, t; s) = r^c(U_{s-t}(\mathbf{x}))$$

Helper Functions

$$\tau := T - t$$
$$\mathbf{c}_\tau^c := \frac{1}{n-2}(E + \lambda I)^{-1}\big[\mathrm{e}^{-\tau(E+\lambda I)} \mathbf{v}^c - \mathbf{v}^{c_0} \big]$$
$$\hat{\mathbf{v}} := \frac{1}{n-2}(E + \lambda I)^{-1} \mathbf{v}^{c_0}$$

Optional: Canonical Parametrization

To simplify these formulas, choose as c_0 an added artificial currency with $r_\infty^{c_0} := 0$, $a^{c_0} := 1$, and $\mathbf{v}^{c_0} := \vec{0}$ (such that $r^{c_0}(\mathbf{x}) \equiv 0$), impose the $\mathbf{v}^{c_0}$- and a^{c_0}-constraints on some other currency, and disregard c_0 in the constraints on $\mathbf{x}_0$ and λ.

This makes μ and thus $U_t(\mathbf{x}) = \mathrm{e}^{t(E-\lambda I)}\mathbf{x}$ linear and simplifies $\mathbf{c}^c$, $\mathbf{c}_\tau^c$, and $V_{c_2}(\mathbf{x}, t)$. Downside: Pricing derivatives in a *real* base currency c_0 removes the need for exchange rates; this is then no longer an option.

Contents

Chapter 1

Introduction

In this introduction we will begin by compiling a comprehensive list of all the features that are desirable of a single-currency interest rate modeling framework. The last of them, unspanned volatility, is probably the hardest to understand, and so we will dedicate a separate section to explaining the main benefit of having this property, namely the ability to bootstrap by expiration. Then we will review the most common approaches to interest rate modeling, discuss their strengths and weaknesses, and demonstrate the advantages of the ACE model over the currently available models. Finally, we will extend our list of desired model features to those of multi-currency interest rate models. The introduction then concludes with brief overviews of the structure of this book and of the history of the ACE model, followed by some credits.

1.1 Desired Features – Single-Currency Models

Focussing on single-currency interest rate models for now (we will get to the multi-currency case later in Section 1.4), we begin our list of desired model features with our prime objective.

✓ **Completeness:** For a fixed given currency (in the following referred to as the *dollar*), the model should allow for the valuation of the set of zero-coupon bonds[1] for *all* future maturities, and of all derived single-currency products.

The remainder of the list is comprised of three categories. The first one contains characteristics that are either directly observed in the market, or that one expects to find under certain idealized market conditions.

---

[1]A maturity-T zero-coupon bond is an instrument that entitles its owner to a one-time payment of one unit of currency (a "dollar") at the future time T. Such bonds typically serve as the set of basis instruments of interest rate models, since all bonds with less trivial payment schedules are equivalent to a collection of zero-coupon bonds with various notionals and maturities.

✓ **Consistency:** The model should be consistent in the sense that the prices it produces are arbitrage free. In particular, this requires all bond price processes to be martingales under the measure associated to the chosen numeraire; e.g., bond price processes expressed in units of a specific maturity-T bond should be martingales under the T-forward measure.

✓ **Positivity of interest rates:** Negative rates, which have been observed in the past for some currencies, do not necessarily present arbitrage opportunities: This apparent disadvantage of investing in bonds over simply holding on to your physical cash can be explained by the inherent risks of this alternative, and by the convenience fees charged by the bank for keeping one's money in an account.

Nevertheless, this can only justify minor forays into the negative rate territory, and so any sensible model should either stay clear of negative rates entirely, or at least keep negative rates tightly bounded. Such a lower bound should be actualized without the use of any brute-force post-simulation flooring, which would compromise consistency.

✓ **Tractable time homogeneity:** By definition, the behavior of a time-homogeneous model at any process time t can only depend on the current state $\mathbf{x}$ of the system, and not explicitly on t. For deterministic smooth dynamics this means that the system can be written in the form $\dot{\mathbf{x}} = \mu(\mathbf{x})$, as opposed to $\dot{\mathbf{x}} = \mu(\mathbf{x}, t)$; for stochastic dynamics we ask that the drift and the noise function of the SDE describing the state dynamics *under the risk-neutral measure* do not depend on t.[2] The parametrization $\mathbf{x}$ in which one wants this property to hold should encode the market information *relative to the time* t at which the state $\mathbf{x}$ is observed, such as the yield curve as a function of the time to maturity (e.g., 10 years, as opposed to the time *of* maturity, e.g., 2028).[3] See Section 5.11 for a more detailed discussion.

In practice, proper calibration may require us to purposely break time homogeneity *in a controlled manner*, by making some of our model parameters time dependent (if the model permits). In the absence of any such manual interference by the user, however, the model should treat all process times equally.

The second category consists of model features that are required to guarantee the numerical feasibility of the model, in particular in regard to simulation, pricing, and calibration.

---

[2]The risk-neutral measure is chosen because its associated numeraire (the dollar) exists for all times and does not single out any specific process time; in contrast, T-forward measures do not have this feature, as their numeraires (the maturity-T bonds) single out the process time $t = T$ at which they cease to exist. To see why one has to decide for a specific measure in the first place, note that when changing to a different numeraire whose dollar value depends on t, the new drift will generally depend on t as well.

[3]In this way, when a state $\mathbf{x}$ is observed at two different times t_1 and t_2, it indeed represents the same view on the respective future.

✓ **Explicit bond prices:** The model should allow for direct computation of all bond prices $B_{\$,T}(\mathbf{x}, t)$ from the current state $\mathbf{x}$ and process time t via a simple analytic formula. Otherwise the calculation of the prices of derivatives such as swaptions (which requires the repeated computation of bond prices) would become expensive, and so the calibration to these instruments (which in turn requires the repeated computation of *their* prices) would become unattainable in practice.

✓ **Low-dimensionality:** The dimension of the model's state space should be small enough to allow for tractable numerics (including backwards induction), but at the same time large enough to allow for realistic yield curve dynamics and robust calibration (see also the item *dynamical flexibility* below). Our own implementation of the ACE model uses the dimension $n = 4$.

Finally, the properties in the third category address the viability of achieving robust global calibration, both in theory and in practice. Aside from standard bonds, the model should be able to calibrate to the most liquid interest rate-related derivatives, including caps, floors,[4] and swaptions[5] of all strikes, expirations, and maturities. Note that since good calibration requires both realistic model behavior and workable numerics, there is some inevitable degree of overlap with the preceding categories.

✓ **Parametric freedom:** The model should provide sufficient parametric freedom to allow for efficient calibration to market prices. In particular, global calibration to the entire spectrum of caps, floors, and swaptions requires at least one *functional* model parameter (i.e., one that depends on both the state variable $\mathbf{x}$ and the process time t). In the ACE model, we will allow the noise coefficient function $\sigma(\mathbf{x}, t)$ in the underlying SDE to be chosen freely (up to some technical conditions).

✓ **Dynamical flexibility:** The range of dynamical behavior that the model can exhibit under its various valid parameter settings should be wide enough to allow the model to mimic any conceivable real-world market behavior with adequate precision under an appropriate choice of parameter values. Otherwise it may not be possible to calibrate the model to any plausible constellation of market-observed instrument prices even if the sheer number of available model parameters is large.

✓ **Unspanned volatility:** To actually find the optimal parameter set *in practice*, it is advantageous if at least some of the model's time-dependent parameters are *unspanned* [5, 6, 7, 8], i.e., if the model's bond price formula

---

[4]A *cap* (*floor*) is a derivative in which the buyer receives payments at the end of each period in which the interest rate exceeds (lies below) the agreed strike value [2].

[5]A *swaption* is an option granting its owner the right but not the obligation to enter into an underlying swap [3]. A *swap* is a derivative in which two counterparties exchange cash flows of one party's financial instrument (e.g., interest rate payments) for those of the other party's financial instrument [4].

happens to not depend on them.[6] The benefits of this technical condition are twofold:

(i) It allows the user to first calibrate all *other* parameters to match the discount curve, and then in a second step to calibrate the *unspanned* parameters to match also the prices of all remaining calibration instruments, without destroying the bond price calibration of the first step.

(ii) It allows the user to calibrate the unspanned parameters by bootstrapping by *expiration* (as opposed to maturity), which provides more leverage, thus generally resulting in a more effective and robust calibration, and in more realistic model behavior. See Section 1.2 for more details on this argument.

We will design the ACE model so that the noise coefficient function $\sigma(\mathbf{x}, t)$ of its SDE is unspanned (i.e., that it does not enter the model's bond price formula).

This concludes our list of desired model features. The ACE ("Adaptive Curve Evolution")[7] model presented in this book is the first interest rate model that fully satisfies all of these requirements, thus positioning itself well within the ranks of the currently available models.

1.2   Bootstrapping by Expiration

The second argument above for adding unspanned volatility to our list of desired model features requires some explanations. Suppose that a time-dependent model parameter, say, $\alpha(t)$, is unspanned, i.e., that the function $B_{\$,T}(\mathbf{x}, t)$ for computing the dollar value of the maturity-T zero-coupon bond from the state $\mathbf{x}$ and the process time t does not depend on α.

Let $(X_t)_{t \geq 0}$ denote the stochastic process representing the state evolution, and suppose that for any given state $\mathbf{x}$ observed at a process time t we can compute the short rate via some function $r(\mathbf{x}, t)$. Then the price at time t of a European call option on a maturity-T zero-coupon bond, with strike $K > 0$ and expiration S (where $t \leq S < T$), is given by

$$C_{S,T,K}(\mathbf{x}, t) = \mathbb{E}\big[\exp\big(-\textstyle\int_t^S r(X_s, s)\,\mathrm{d}s\big)\big(B_{\$,T}(X_S, S) - K\big)^+ \,\big|\, X_t = \mathbf{x}\big], \quad (1.1)$$

where the expected value is taken under the risk-neutral measure.[8] Now observe the following:

---

[6]Of course, models also need to have non-unspanned parameters, since otherwise they cannot be calibrated to the discount curve.

[7]The name of the ACE model was chosen to express that it models the stochastic *evolution* of the yield (or equivalently, the instantaneous forward rate) *curve* in a way that provides enough parametric freedom, dynamic flexibility, and (as a result of being unspanned) numeric means to allow us to *adapt* it to (i.e., to have it reproduce) the market prices of the wide array of liquid interest rate-related calibration instruments.

[8]The arguments to follow will equally apply to more commonly traded instruments like swaptions; we are choosing a bond option as our example merely for simplicity of notation.

- Given that $X_t = \mathbf{x}$, both the exponential discount factor and the random variable X_S in (1.1) only depend on those function values of α on the time interval $[t, S]$.

- Given the state X_S, the bond price $B_{\$,T}(X_S, S)$ can depend only on those values of α on $[S, T]$ (see also (1.3) below).

As a result, $C_{S,T,K}(\mathbf{x}, t)$ can generally only depend on those values of α on $[t, T]$; however, if α is unspanned (i.e., if $B_{\$,T}$ happens to not depend on α) then $C_{S,T,K}(\mathbf{x}, t)$ in fact only depends on those values of α on $[t, S]$. More generally, this statement can analogously be seen to hold in fact for any other (European, American, or Bermudan) option whose payoff only depends on the discount curve up to some time $T > S$, which we call the maturity of the option.

This observation has consequences on the possible strategies that one can choose to calibrate the parameter function $\alpha(t)$. For the calibration of time-dependent parameters one usually utilizes the technique of bootstrapping *by maturity*. In this technique one divides the time line into disjoint consecutive intervals, $[0, \infty) = [0, t_1] \cup (t_1, t_2] \cup \ldots$, and one calibrates the function values of α one interval at a time, by successively attempting to match the prices of only those calibration instruments whose maturities lie within the interval at hand. This procedure works since (as we saw above) option prices can only depend on the values of α up to the time of their maturity, so that each step leaves the (already calibrated) prices of the calibration instruments from all preceding steps untouched. By the same argument, *if α is unspanned then one can—and in fact must—bootstrap by expiration instead*, since in that case (as we saw above) option prices can only depend on the values of α up to the time S of their expiration.

Now why would one prefer bootstrapping by expiration over bootstrapping by maturity? Option prices are generally found to be more affected by changes to the process behavior near the expiration time S than near the maturity T, because changes to earlier stages of the process affect its entire future path. Since bootstrapping by expiration allows us to change α near the expirations of the currently targeted calibration instruments, under this technique any desired price corrections can be achieved with smaller changes to α. This increased leverage has two advantages:

(i) The calibrated values of α will exhibit less fluctuations as a function of the process time t (leading to more realistic model behavior).

(ii) The calibration will be more stable, i.e., α is less affected by day-to-day changes of the market-observed prices of the calibration instruments.

Summary. Unspanned time-dependent parameters (i.e., model parameters that the bond price function $B_{\$,T}(\mathbf{x}, t)$ does not depend on) can be calibrated by bootstrapping by expiration, which provides more leverage, and which therefore generally results in more realistic model behavior and in a more robust calibration.

1.3   Common Modeling Approaches

Let us now discuss the most common approaches to interest rate modeling and highlight their advantages and shortcomings. Please refer to standard textbooks such as [9, 10, 11, 12, 13] for details.

1.3.1   Short Rate Models

In the short rate modeling approach one describes the evolution of the short (i.e., instantaneous spot) rate r_t under the risk-neutral measure, in the form of an SDE

$$\mathrm{d}X_t = \mu(X_t, t)\,\mathrm{d}t + \sigma(X_t, t)\,\mathrm{d}W_t \tag{1.2}$$

for a process $(X_t)_{t\geq 0}$ in $\mathbb{R}^n$ from which the short rate $r_t = r(X_t, t)$ can be computed via some given function $r(\mathbf{x}, t)$. Here $(W_t)_{t\geq 0}$ is a (possibly multidimensional) Brownian motion. A standard example is the Hull–White framework, which includes the Hull–White (HW) model [14], the Cox–Ingersoll–Ross (CIR) model [15], the Black–Derman–Toy (BDT) model [16], and its extension, the Black–Karasinski (BK) model [17].

The most significant problem with this approach is that *in general* the price of the maturity-T zero-coupon bond at time t,

$$B_{\$,T}(\mathbf{x}, t) = \mathbb{E}\big[\exp\big(-\textstyle\int_t^T r(X_s, s)\,\mathrm{d}s\big) \,\big|\, X_t = \mathbf{x}\big], \tag{1.3}$$

cannot be computed analytically, thus requiring the use of either Monte Carlo techniques or of a PDE solver. As a result, calibration and—except in very low dimensions—even pricing becomes computationally unattainable. However, resorting to such low-dimensional cases is not a viable solution, either: One-factor short rate models imply a 100% or -100% instantaneous correlation between the prices of any two instruments[9] (and similarly, between the forward rates of different maturities), which is inconsistent with what is observed in the market. In fact, principle component analyses suggest that in fact at least four factors are necessary to achieve realistic model behavior [18, 19]. Furthermore, *in general* the bond price formula (1.3) depends on all the model parameters that enter the process $(X_t)_{t\geq 0}$ on the time interval $[t, T]$, and so the model is not unspanned, further contributing to the calibration difficulties.

Noteworthy exceptions of solvable models are given by the class of *affine models* [20, 21], which by definition are those models for which $\log(B_{\$,T}(\mathbf{x}, t))$ is affine in $\mathbf{x}$, and which include the HW and the CIR model and some multi-factor generalizations thereof. Every short rate model for which r and all components of μ and $\sigma\sigma^T$ are affine in $\mathbf{x}$ belongs to this class, and in fact for time-homogeneous models the reverse is true under some technical non-degeneracy

---

[9]Indeed, given a one-dimensional process $(X_t)_{t\geq 0}$ following the SDE (1.2) and two instruments with prices $P_1(X_t, t)$ and $P_2(X_t, t)$, by Itô's Lemma the infinitesimal price changes are $\mathrm{d}P_i = (\partial_t + \mu\partial_\mathbf{x} + \frac{1}{2}\sigma^2\partial_\mathbf{x}^2)P_i\,\mathrm{d}t + \sigma\partial_\mathbf{x}P_i\,\mathrm{d}W_t$ for $i = 1, 2$, so that $\mathrm{cov}(\mathrm{d}P_i, \mathrm{d}P_j) = \sigma^2 \cdot (\partial_\mathbf{x}P_i)(\partial_\mathbf{x}P_j)\,\mathrm{d}t$ and thus $\mathrm{corr}(\mathrm{d}P_1, \mathrm{d}P_2) = \mathrm{sign}\big((\partial_\mathbf{x}P_1)(\partial_\mathbf{x}P_2)\big) \in \{-1, 0, 1\}$, with the value 0 being taken only in degenerate situations.

condition [20]. While those affine models with $\mathbf{x}$-independent volatility have short rates that are Gaussian and therefore not bounded below, those with $\mathbf{x}$-dependent volatility successfully address this problem (albeit with $\sigma(\mathbf{x}, t)$ arguably declining too rapidly near states attaining the minimum rate, namely like $\sqrt{r(\mathbf{x}, t) - r_{\min}}$). Either way, the biggest problem shared by affine models is that they do not provide sufficient parametric freedom to achieve global calibration, since (as a result of the specific prescribed $\mathbf{x}$-dependency of all their ingredients) their functional parameters can only depend on t but not on $\mathbf{x}$.

1.3.2 The HJM Framework

In the continuum (complete) market model approach, also known as the *Heath–Jarrow–Morton (HJM) framework* [22], the state $\mathbf{f}$ of the system is the full instantaneous forward rate curve, i.e., the function $\mathbf{f} \colon [t, \infty) \to \mathbb{R}$ from which the prices of the maturity-T zero-coupon bonds can be computed via the formula

$$B_{\$,T}(\mathbf{f}, t) = \exp\left(-\int_t^T \mathbf{f}(s)\,\mathrm{d}s\right). \tag{1.4}$$

The stochastic evolution $(\mathbf{f}_t)_{t \geq 0}$ under this model is described by way of an SDE system of the form

$$\mathrm{d}\mathbf{f}_t(s) = \mu_t(s)\,\mathrm{d}t + \langle \sigma_t(s), \mathrm{d}W_t \rangle, \tag{1.5}$$

where $(W_t)_{t \geq 0}$ is a possibly multidimensional Brownian motion, where $(\sigma_t(s))_{t \in [0,s]}$, for $\forall s \geq 0$, is a stochastic process of the same dimension, and where the drift must be defined as

$$\mu_t(s) := \left\langle \sigma_t(s), \int_t^s \sigma_t(u)\,\mathrm{d}u \right\rangle \tag{1.6}$$

in order to ensure the consistency of the model.

This framework has many advantages: The generality of its model dynamics is essentially limitless. Calibrating the model to the discount curve is as simple as choosing the corresponding forward rate curve as the initial state $\mathbf{f}_0$. Time homogeneity is achieved by having the dynamics of $(\sigma_t(s))_{t \in [0,s]}$ depend on s and t only via their difference $s - t$. Furthermore, while any deterministic choice of this noise process will cause the rates $\mathbf{f}_t(s)$ to be Gaussian and therefore negative with non-vanishing probability, negative rates can be avoided by making σ_t depend on $\mathbf{f}_t$ (i.e., $\sigma_t(s) = \sigma(\mathbf{f}_t, t; s)$), in such a way that $\sigma_t(s) = 0$ whenever $\mathbf{f}_t(s) = 0$. Finally, since the bond price formula (1.4) does not depend on any model parameters, the model is fully unspanned.

The problem with this approach is that the state space is infinite dimensional, and so *as is* it is nothing more than a theoretical construct. In general, any direct numerical implementation will require the discretization of the state $\mathbf{f}$, which will inevitably compromise some of the model's valuable analytical properties, namely consistency and controllable time homogeneity. Furthermore, while in a discretized version of the model bond prices can still be computed comparably quickly using (1.4), accurately pricing more complicated instruments in such a high-dimensional space (and calibrating based on them) becomes computationally near impossible.

1.3.3   The BGM Model

The *Brace–Gątarek–Musiela (BGM) model* [23], also known as the *LIBOR market model (LMM)* or the *lognormal forward-LIBOR model (LFM)*, avoids the numerical problems of the HJM framework by making a tradeoff between numerical feasibility and completeness: It restricts the scope of the model to only a finite set of maturities $T_1, \ldots, T_m$, i.e., the model is agnostic to all other maturities, and its declared goal is to price only those instruments that exclusively depend on this selection of maturities. More precisely, it provides a closed system of SDEs for the forward rates $L(t; T_i, T_{i+1}) := \frac{1}{T_{i+1}-T_i} \int_{T_i}^{T_{i+1}} \mathbf{f}_t(s)\,\mathrm{d}s$, $i = 1, \ldots, m - 1$, in [23] originally derived from the HJM framework with a smart choice of σ_t, but now commonly presented based on the observation that for each i the process $(L(t; T_i, T_{i+1}))_{t \in [0, T_i]}$ is a martingale under the T_{i+1}-forward measure [9, Chapter 6].

Unfortunately, settling for a moderate number of maturities to make the numerics feasible, one is often left with gaps of one or more months between successive properly modeled maturities, and in practice bond prices with other maturities are usually defined via interpolation, thus sacrificing consistency. This problem becomes particularly apparent when pricing spread options[10] (since interpolation methods lead to unrealistic deterministic relations between the spreads within the same interpolation interval), or swaptions with non-standard coupon schedules. In a sense, the model cannot even consistently price the dollar (i.e., the value of the bond maturing now) at every given process time.

Further shortcomings of the BGM model are that it violates time homogeneity (since it singles out specific maturities), and that the approach is inconsistent also in the sense that removing one of the modeled maturities will generally affect the evolution of the remaining ones.

1.3.4   The Cheyette or Ritchken–Sankarasubramanian Approach

Cheyette [25, 26, 27, 28] and independently Ritchken and Sankarasubramanian [29] discovered a class of volatility processes $(\sigma_t(s))_{t \in [0,s]}$ for which one can reduce the HJM model (1.5)–(1.6) to finite dimensions *without any loss of information*. They found that if $\sigma_t = (\sigma_t^1, \ldots, \sigma_t^n)$ is of the form

$$\sigma_t^k(s) = \sum_{i=1}^{N_k} \eta_{k,i}(t) \exp\left(- \int_t^s \lambda_{k,i}(u)\,\mathrm{d}u\right) \quad \text{for} \quad \forall k = 1, \ldots, n, \qquad (1.7)$$

where n is the dimension of the Brownian motion in (1.5), then the forward rate process $(\mathbf{f}_t)_{t \geq 0}$ *recombines*, i.e., it is confined to a finite-dimensional function space at each time t.[11] A process of this class can therefore be reduced to the finite-dimensional Markov process $(X_t)_{t \geq 0}$ that tracks the coefficients X_t of the

---

[10]A *spread option* is a type of option where the payoff is based on the difference in price between two underlying assets [24], for example between two bonds with different maturities.
[11]This subspace may be different for each process time t.

state $\mathbf{f}_t$ w.r.t. a suitable basis of that function space. In practice, this means that one can simply consider the SDE for the process $(X_t)_{t \geq 0}$ and recover the full forward rate curve $\mathbf{f}_t$ (and thus all the bond prices) from the current state X_t via some analytical formula. In particular, there are functions $B_{\$,T}(\mathbf{x}, t)$ and $r(\mathbf{x}, t)$ with which one can compute the bond prices and the short rate from X_t, i.e., one can write any model of this class in the form of a short rate model with an explicitly available formula for the expected value in (1.3).

Furthermore, as it turns out, the basis can be chosen independently of the choice of the $\eta_{k,i}$, which implies that the bond price formula is independent of the $\eta_{k,i}$ as well. Since the $\eta_{k,i}$ end up taking the role of the volatilities of the various components of the resulting SDE for $(X_t)_{t \geq 0}$, this means that all models obtained with this approach will have unspanned volatility. See Appendix A.1 for a simple example that demonstrates this technique.

This powerful recipe gives rise to a wide array of models. Among them are the *Linear Gauss Markov (LGM) models*, which are those models obtained by choosing the $\eta_{k,i}$ as deterministic functions; in particular, for one source of noise (i.e., $n = 1$) the choice $\sigma_t(s) := \eta(t) \cdot e^{-\lambda(s-t)}$ recovers the Hull–White model in a two-dimensional disguise[12] (see Appendix A.1). While the forward rate processes of LGM models are Gaussian (hence the name) and therefore not bounded below, properly making the $\eta_{k,i}$ stochastic by having them depend also on the current state X_t can address this issue. As an added benefit, doing so also gives us the state-dependent unspanned functional parameters $\eta_{k,i}(\mathbf{x}, t)$ that we need to achieve global calibration.

Among the various approaches discussed here, the Cheyette model class comes the closest to satisfying all of our desired properties to full satisfaction; it just has one unfortunate shortcoming, namely that it introduces extra dimensions to its state space. More precisely, its state process $(X_t)_{t \geq 0}$ is of the dimension $\frac{1}{2} \sum_{i=1}^{n} N_k(N_k + 3) \geq 2n$, i.e., it is at least twice the number of Brownian motions that are driving the process; the minimum $2n$ is only achieved in the simple case $N_1 = \cdots = N_n = 1$, and adding terms to the sum in (1.7) increases this dimension quadratically.[13] In practice, this limits the number of noise sources that one can use before running into numerical problems, and thus the framework's dynamical flexibility; in particular, backwards induction (which is only practical in dimensions no higher than four) becomes only feasible for at most two sources of noise.

1.3.5 The Hagan–Woodward Framework

Hagan and Woodward [30] chose a different approach: Rather than trying to reduce the HJM dynamics (1.5)–(1.6) to finite dimensions, they started from

---

[12]Note that in its standard one-dimensional formulation, the Hull–White model is not unspanned; however, as we demonstrate in our calculation in Appendix A.1, one can make it unspanned, by introducing a second (deterministic) state variable that includes all the η-dependencies of the model's bond price function.

[13]While for deterministic $\eta_{k,i}$ this dimension can be reduced, this comes at the cost of losing not only the short rate lower bound, but also the unspannedness of the model.

scratch and developed what can be considered a finite-dimensional analogue of
the HJM framework: Beginning with a general finite-dimensional state process
$(X_t)_{t\geq 0}$ and a function that maps these states to the value of a (not further specified) numeraire, they found conditions under which their framework is consistent
in that it can reproduce any given initial forward rate curve. Their framework
then provides a bond price function and can therefore be written in the form of
a short rate model.

Unfortunately, the consistency condition of this approach is not as explicit as
the condition (1.6) of the HJM framework, since it requires the evaluation of a
certain expected value, which then also enters the bond price and the short rate
function. In practice, this means that it does not provide much improvement
over our original hurdle of having to evaluate (1.3), so that it again restricts
any efforts of creating new analytically tractable models to only simple SDEs
and/or numeraire functions. As a result, while the approach does give rise
to what in [30] is called the class of β–η models, which contains the Hull–
White and the CIR model, these models are all one-dimensional and have a not
generally solvable integral in their bond price functions, and the construction of
any more complex higher-dimensional models seems to be out of reach. Further
problems with this approach are that the models obtained from it are generally
not unspanned, and that it is generally not obvious under which conditions the
model is time homogeneous or restricted to positive rates.

While the Hagan–Woodward framework could not match the practical impact of the other model classes laid out above, it does deserve its spot on our
list because of its noteworthy starting point, which in some sense is similar to
the one that we use in our construction of the ACE model in Chapter 5.

1.3.6   Summary: Model Feature Comparison

Table 1.1 provides an overview over the features and shortcomings of the various
approaches and specific models discussed above. It uses the following symbols:

✓✓	The property is fulfilled to full satisfaction.
✓	The property can easily be fulfilled if desired, by making certain modeling or parameter choices.
(✓✓/✓)	The property is / can be fulfilled, but only with compromises to either this or to other properties.
✕✕	The property is not fulfilled.
✕	The property is not fulfilled *in general*, and it is a non-trivial task to find special cases for which it is fulfilled.

In the *parametric freedom* column we award a ✓ or ✓✓ mark only if the
given approach provides a state-dependent functional parameter. The *dynamic
flexibility* feature is not as black-and-white as the others and clearly leaves room
for discussion; here our personal judgment tries to take into account the given

	market features				numerics		calibration		
	completeness	consistency	positive rates	time homogeneity	explicit bond prices	low-dimensionality	parametric freedom	dynamic flexibility	unspanned volatility
short rate	✓✓	✓✓	✓	✓	✗	✓	✓	✓	✗
affine	✓✓	✓✓	✓	✓	✓✓	✓	✗✗	(✓)	✗✗
HW (1D)	✓✓	✓✓	✗✗	✓	✓✓	✓✓	✗✗	✗✗	✗✗
CIR	✓✓	✓✓	✓✓	✓	✓✓	✓✓	✗✗	✗✗	✗✗
HJM	✓✓	✓✓	✓	✓	✓✓	✗✗	✓	✓	✓✓
discretized	✓✓	(✓✓)	✓	(✓)	✓✓	✗✗	✓	✓	✓✓
BGM	✗✗	✓✓	✓✓	✗✗	✓✓	((✓))	✓✓	✓✓	✓✓
interpolated	✓✓	✗✗	✓✓	✗✗	✓✓	((✓))	✓✓	✓✓	✓✓
Cheyette	✓✓	✓✓	✓	✓	✓✓	(✓)	✓	✓	✓✓
LGM	✓✓	✓✓	✗✗	✓	✓✓	(✓)	✗✗	✗✗	✓✓
HW (2D)	✓✓	✓✓	✗✗	✓	✓✓	✓	✗✗	✗✗	✓✓
Hagan	✓✓	✓✓	✗	✗	✗	✓	✓	✓	✗
β–η	✓✓	✓✓	✓	✓	(✓✓)	✓✓	✗✗	✗✗	✗✗
ACE	✓✓	✓✓	✓✓	✓	✓✓	✓✓	✓✓	✓✓	✓✓

Table 1.1: A comparison of the features and shortcomings of the various approaches to interest rate modeling, and of some specific examples discussed in Sections 1.3.1–1.3.5.

model's dimension, its parametric freedom, its ability to bound rates below, and generally the functional form of its dynamics.

As summarized in the first rows of Table 1.1, *short rate models* are finite dimensional, but in general they do not provide simple analytic bond price functions; all the currently known exceptions that were constructed by solving (1.3) offer only limited parametric freedom and dynamic flexibility, with some models allowing for negative rates. Furthermore, the parameters of short rate models are generally not unspanned, which makes global calibration computationally unattainable.

The *HJM* framework has all of our desired properties, except that its state space is infinite dimensional, and so as is it is no more than a theoretical construct. Brute-force discretization of the state **f** compromises consistency and time homogeneity, while still leaving the state space dimension too high for practical purposes. Two successful techniques of analytically reducing the HJM framework to finite dimensions by considering noise functions of a specific form are the BGM model and the Cheyette (or Ritchken–Sankarasubramanian) approach.

The *BGM* model focusses on only finitely many maturities (i.e., it is incomplete, and as a consequence also time inhomogeneous), and recovering the remaining tenors via interpolation makes the model inconsistent. Either way, the dimension of the model is still too high (one per modeled maturity), leaving us with substantial numerical difficulties and unable to price via backwards induction.

The *Cheyette* model class comes the closest to satisfying all of our desired properties. Unfortunately, for merely analytical reasons these models' state space dimensions are at least twice as high as the number of Brownian motions driving them; in practice, this limits the number of noise sources that one can use before running into numerical problems (in particular with backwards induction) to no more than two. The LGM model subclass (which includes an unspanned two-dimensional model equivalent to the standard one-dimensional Hull–White model) does not take full advantage of the possibilities of the Cheyette approach, with limited parametric freedom and with its rates not bounded below.

Finally, the *Hagan–Woodward* framework requires the evaluation of an expected value to check its consistency condition; as a result, it generally does not provide an explicit bond price formula (with the partially solvable β–η models being only of limited use since they are one-dimensional). Furthermore, models in this framework are generally not unspanned, and it is not obvious how to bound rates below and how to achieve time homogeneity.

The ACE model presented in this book has *all* of our desired model features: It is complete and fully consistent, with an explicit analytic formula to compute the bond prices for all maturities. Its state space dimension coincides with the number of Brownian motions driving it, and it can be chosen to be any value $n \in \mathbb{N} \setminus \{2\}$. Its model dynamics is highly flexible, with its parameter space providing $2n + 1$ degrees of freedom in addition to the freely choosable and unspanned functional volatility parameter $\sigma(\mathbf{x}, t)$. Finally, its rates can be bounded below by any desired value, and it can be made time homogeneous if desired (by choosing σ independent of t).

1.4 Desired Features – Multi-Currency Models

New in this second edition of the book is the added Chapter 3, which introduces the multi-currency generalization of the ACE model. As for the single-currency case discussed in Section 1.1, let us now compile a list of the most desirable features of such a multi-currency interest rate model. We start by augmenting the model's original prime objective.

✓ **Completeness (multi-currency):** Given a collection C of currencies in scope (e.g., $C := \{\$, €, £\}$), the model should allow for the valuation of the set of zero-coupon bonds for all currencies $c \in C$ and all future maturities, for the computation of the spot and forward exchange rates for each pair of currencies $c_1, c_2 \in C$, and for the valuation of all derived single- and cross-currency products.

Next, let us clarify what we consider a well-designed multi-currency extension of a given single-currency interest rate model.

✓ **Extension of the given single-currency model:** Projecting the multi-currency model down to any one of the currencies in scope should yield the short rate process of the given single-currency model again (for a certain currency-specific parameter choice).

✓ **Model symmetry:** All currencies should play equivalent roles in the multi-currency model (aside from their individual parameter settings), with no currency being treated fundamentally differently from the others.

Finally, let us review the desired properties that we had already compiled in Section 1.1 for the single-currency model. While some remain unchanged, others need to be augmented or require some additional commentary.

✓ **Consistency:** The core requirement is unchanged, i.e., the model should still be consistent in the sense that the prices it produces are arbitrage free; in particular, bond price processes for all maturities and currencies should still be martingales under the measures associated to the chosen numeraire.

However, satisfying this requirement is now more involved, for two reasons: First, these numeraires are now specified by a pair consisting of a maturity *and a currency*, i.e., each currency has its own set of T-forward measures. Second, as we will see in Section 3.3, the introduction of the various additional quantities necessary to define a multi-currency model (i.e., the short rate and exchange rate functions for the various currencies) gives rise to further subtle pitfalls that, unless certain relations between these quantities are fulfilled, can make the model inconsistent.

✓ **Positivity of interest rates:** We now require the positivity of the interest rates of *all* currencies. As before, given any model with this property, it is then trivial to modify it in such a way that it imposes any given set of lower rate bounds for the various currencies instead.

✓ **Tractable time homogeneity:** Unchanged.

✓ **Explicit bond prices:** Since now the model has to handle a complete set of bonds for each currency, and since asset prices can now be expressed in terms of each of these currencies, we now need to ask that the model provide simple analytic formulas for the direct computation of all bond prices $B_{c_2,T}^{c_1}(\mathbf{x}, t)$ (for $c_1, c_2 \in C$ and $T \geq t$), i.e., of the value, measured in units of c_2, of the bond that at time T pays out 1 unit of c_1.

Note that clearly it suffices to focus on any numeraire of choice, as long as the model provides spot exchange rate functions $a^{c_1/c_2}(\mathbf{x}, t)$ for each pair of currencies $c_1, c_2 \in C$. Furthermore, by using a simple no-arbitrage argument[14] one can then also obtain formulas for the *forward* exchange rates $a_T^{c_1/c_2}(\mathbf{x}, t)$ for $\forall T \geq t$.

✓ **Low-dimensionality:** Unchanged. To retain numerical feasibility, one cannot allow the multi-currency extension to increase the state space dimension. We will achieve this by defining functions $r^c(\mathbf{x})$ and $a^{c_1/c_2}(\mathbf{x}, t)$ for computing the various interest and exchange rates from the state process $(X_t)_{t \geq 0}$, which will be the same for the single- and the multi-currency case.

✓ **Parametric freedom:** Unchanged. We still only require *one* functional parameter, which in our model will be the noise coefficient function $\sigma(\mathbf{x}, t)$ of the SDE driving the state process $(X_t)_{t \geq 0}$.

✓ **Dynamic flexibility:** Unchanged.

✓ **Unspanned volatility:** We now require that *all* of the model's bond price formulas for the various currencies in scope, as well as the spot and forward exchange rate functions for all currency pairs, are independent of the functional parameter $\sigma(\mathbf{x}, t)$.[15]

This will again allow the user to calibrate the model in two steps, first adjusting only the non-unspanned parameters to have it replicate the various spot exchange rates and the discount curves of the various currencies, and then finding the optimal choice for $\sigma(\mathbf{x}, t)$ by bootstrapping by expiration in order to match the prices of all rate and currency options, which will again increase leverage and stability of the calibration.

The multi-currency ACE model is satisfying all of these requirements completely: It is a fully symmetric extension of the single-currency ACE model that has properly addressed all the potential consistency pitfalls, it produces only positive rates, it is time-homogeneous if desired, and it provides explicit formulas for all bond prices and spot and forward exchange rates. It works in any dimension $n \geq 3$, with no technical need to increase the dimension when further currencies are added, it provides ample parametric freedom (as it provides $(n-1) + |C| \cdot (n+2)$ discrete parameters in addition to the functional parameter $\sigma(\mathbf{x}, t)$), and it has flexible dynamics. Finally, it has unspanned volatility since all its bond price and spot and forward exchange rate formulas are independent of σ.

Since the single-currency ACE model has already been the first to have all of the desired model features listed in Section 1.1, clearly its multi-currency extension is the first to have all of the features listed here in Section 1.4.

---

[14] The reader can find a detailed explanation in the proof of (3.76a) in Lemma 3.12 (iii).

[15] The focus here is really on the set of bond price formulas only: In practice, the spot exchange rate functions will be a part of the model's definition (so that there is no reason why it would depend on σ), and the σ-independence of the forward exchange rates is then a direct consequence of the σ-independence of the bond prices and the general formula (3.76a).

1.5   The Structure of This Book

This book is organized as follows: Part I, which stands on its own, contains a fast track to the ACE model, with Chapter 2 first introducing the single-currency model and then Chapter 3 extending it to multiple currencies. Bypassing its derivation, this part presents the model in the form of a short rate model, illuminates some of its properties, and provides straightforward independent proofs for its bond price, forward rate, and forward exchange rate formulas. The reader who merely wishes to understand the ACE model well enough to use it in practice, and who may have only limited time, mathematical skills, or simply interest in the model's origins, may then decide to not read any further.

Part II contains two independent derivations of the ACE model, i.e., of its dynamics and of its bond price, forward rate, and (spot and forward) exchange rate formulas. First, in Chapter 4 we present a comparably short and non-abstract derivation of both the single- and the multi-currency model, which was devised only after the model equations had already been found with the original, much longer and more abstract derivation presented in Chapters 5–9. See Section 4.3 for a discussion of how this approach relates to the original model derivation, and why that original derivation still has more than just historical value.

The original derivation in Chapters 5–9 is structured as follows: In Chapter 5 we develop a general mathematical framework for interest rate modeling that translates all of our desired model features into concise mathematical properties that our model needs to fulfill, carefully written in a form that facilitates our work in the subsequent chapters. In Chapter 6 we then lay out our road map towards the development of a model that indeed has all these properties, in Chapter 7 we actually carry out these steps with a manifold other than $\mathbb{R}^n$ as our state space, and finally in Chapter 8 we translate the obtained model into the $\mathbb{R}^n$-based form presented earlier in Part I. Finally, in our conclusions in Chapter 9 we look back and outline the various steps of our construction, and we point out some modeling decisions that we had made along the way that may have restricted the generality of our model.

The original derivation of the $\mathbb{R}^n$-based multi-currency extension of the model presented in Chapter 3, which was constructed by extending the interest rate modeling framework in Part II, has not been included in this book. It may be added in a future edition.

1.6   Model History

The ACE model in its original form, along with most of the core techniques presented in this book, was first developed by Gregory Pelts in 2011–2012 [1, 31], and a fully working implementation (including the calibration and pricing logic) was completed by 2013. I began studying his work in late 2013, and over the following months we prepared a series of six firm-internal lectures about his paper.

Realizing the need for a much more comprehensive description of the model that could reach a wider audience, in late 2014 I then started my own write-up, which by May of 2018 had eventually matured into the first edition of the present book, and which focused only on the single-currency case. After some time of rest, between September 2019 and March 2020 I added Chapter 3, which extends the ACE model to multiple currencies. Finally, in October 2024 I decided to add the shorter and less abstract model derivation in the newly inserted Chapter 4.

The main contributions of this book are the following:

- the development of the specific fully $\mathbb{R}^n$-based description of the model presented in Part I, with independent $\mathbb{R}^n$-based proofs of the bond price and evolution map formulas,

- a more rigorous treatment of the parameter constraints that are necessary to guarantee positive rates, in particular of the required boundary behavior of the functional noise parameter $\sigma(\mathbf{x}, t)$,

- the one-dimensional ACE model,

- the multi-currency ACE model,

- the formulation of the general mathematical framework underlying the ACE model's original model derivation (based on Gregory Pelts' original ideas),

- introductions to the various non-standard mathematical techniques used during the model's original derivation, to make the book more accessible to the general reader,

- detailed proofs of all non-obvious claims made during the original model derivation, and

- an independent, much shorter shorter and less abstract, purely $\mathbb{R}^n$-based model derivation.

1.7   Credits

I wish to thank Gregory Pelts for our many discussions that helped me get a better understanding of his original paper [1]. I am also grateful to Huan Yang for his invaluable help in addressing some compliance concerns; without his efforts, I may never have been able to release this book to the public. Finally, I would like to thank Arlie Petters, Amir Aazami, and Jonathan Mattingly for their advice on some logistical questions that arose during the publication process.

Part I

A Fast Track To ACE

In Part I of this book, "A Fast Track To ACE," we will summarize the key equations of our $\mathbb{R}^n$-based parametrization of the ACE model, and we will present short proofs for its bond price, forward rate, and forward exchange rate formulas. This part will aim to serve as a comparably simple window into the *practical* aspects of the model: While it will provide enough information to implement the model in practice, and while its proofs will give the reader the comfort of knowing that our formulas are indeed correct, much of the model's origins will remain in the dark here. In particular, our choices of the second-order terms in the SDE's drift, and later on of the exchange rate function and the reparametrization mappings used in the multi-currency model, will appear as ingenious guesses; furthermore, many proofs (in particular, the proofs of the bond price formula and of the multi-currency model's consistency) will appear to work as if by pure delightful magic, with no indication of how the model's core formulas have been obtained in the first place.

True understanding of the model's inner workings—along with the chance of potentially improving upon it in the future—can only be obtained by putting in the work and studying also the model's original construction laid out in Part II, in its numeraire-free (i.e., section-based) formulation, and with a non-trivial manifold as its state space.

We begin Part I by focusing on the single-currency case in Chapter 2; later in Chapter 3 we will extend the model so that it can handle the interest rates and bond prices of multiple currencies, along with the currencies' spot and forward exchange rates.

Chapter 2

The Single-Currency ACE Model

2.1 Model Description

2.1.1 Model Parameters, State Space

Our $\mathbb{R}^n$-based description of the general n-dimensional ACE model presented in the following is valid for all dimensions

$$n \geq 3. \tag{2.1}$$

As a result of the division by $n-2$ in the drift vector field $\mu(\mathbf{x})$ in (2.11) below, no two-dimensional ACE model exists, while the one-dimensional ACE model does exist but requires some special treatment, as laid out in detail later in Section 2.9.

The model has the following parameters: the *long-term rate*

$$r_\infty > 0, \tag{2.2}$$

the *weight vector* $\mathbf{v} \in \mathbb{R}^n$ with

$$\|\mathbf{v}\| = 1, \tag{2.3}$$

the *oscillation frequency matrix* $E \in \mathbb{R}^{n \times n}$, which we assume to be antisymmetric, i.e.,

$$E^T = -E, \tag{2.4}$$

the *reversion speed*

$$\lambda > 0, \tag{2.5}$$

the *initial state*

$$\mathbf{x}_0 \in M_0 \tag{2.6}$$

in our state space

$$M_0 := \left\{ \mathbf{x} \in \mathbb{R}^n \,\middle|\, \langle \mathbf{v}, \mathbf{x} \rangle > -r_\infty \right\}, \tag{2.7}$$

21

and the continuous *noise function*

$$\sigma \colon M_0 \times [0, \infty) \to (0, \infty), \tag{2.8}$$

which we assume to satisfy the conditions stated in Lemma 2.4, namely that σ has some specific additional regularity, that σ does not grow too fast as $\mathbf{x} \to \infty$, and that it vanishes fast enough as $\mathbf{x}$ approaches the boundary ∂M_0.

In addition to the above, the parameters r_∞, $\mathbf{v}$, E, and λ are subject to the constraints

$$\delta := r_\infty \lambda - \tfrac{1}{n-2} r_\infty^2 - \tfrac{n-2}{4} \| E\mathbf{v} \|^2 \geq 0, \quad \text{and} \tag{2.9a}$$

$$\text{if} \quad \delta = 0 \quad \text{then} \quad \lambda < \tfrac{2}{n-2} r_\infty. \tag{2.9b}$$

As we will learn later in Lemma 2.3 and the discussion following it, δ takes on the role of the *market resiliency*, i.e., the guaranteed amount of force with which the system is pushed back into the positive-rate regime when it approaches a state with a vanishing short rate.

2.1.2 Evolution Equation, Short Rate Function

Given any such set of parameters r_∞, $\mathbf{v}$, E, λ, $\mathbf{x}_0$, and $\sigma(\mathbf{x}, t)$, the ACE model is driven by the n-dimensional process $(X_t)_{t \geq 0}$ satisfying the SDE

$$\mathrm{d}X_t = \mu(X_t)\,\mathrm{d}t + \sigma(X_t, t)\,\mathrm{d}W_t, \qquad X_{t=0} = \mathbf{x}_0, \tag{2.10}$$

where $(W_t)_{t \geq 0}$ is an n-dimensional Brownian motion, and where the drift $\mu \colon \mathbb{R}^n \to \mathbb{R}^n$ is defined as

$$\mu(\mathbf{x}) := (-\lambda I + E)\mathbf{x} + \tfrac{1}{n-2} \| \mathbf{x} \|^2 \mathbf{v} - \tfrac{2}{n-2} \langle \mathbf{v}, \mathbf{x} \rangle \mathbf{x}. \tag{2.11}$$

As we will see later in Lemma 2.3, the constraints on our parameters guarantee that the drift points inwards near the boundary of M_0, and our conditions imposed on σ in Lemma 2.4 then suffice to ensure that the SDE (2.10) has a unique strong solution $(X_t)_{t \geq 0}$ that is confined to M_0 almost surely.

Finally, we define the short rate function $r \colon \mathbb{R}^n \to \mathbb{R}$ as

$$r(\mathbf{x}) := r_\infty + \langle \mathbf{v}, \mathbf{x} \rangle. \tag{2.12}$$

This completes our definition of the ACE model.

2.2 Orthogonal and Antisymmetric Matrices

To prepare for our discussions and calculations to follow, let us briefly review some basic facts about orthogonal and antisymmetric matrices.

Lemma 2.1 (Orthogonal matrices). *Given a matrix $A \in \mathbb{R}^{n \times n}$, the following three statements are equivalent:*

   (i) A is orthogonal, i.e., it is invertible and fulfills $A^{-1} = A^T$.

   (ii) $\forall \mathbf{x}, \mathbf{y} \in \mathbb{R}^n$: $\langle A\mathbf{x}, A\mathbf{y} \rangle = \langle \mathbf{x}, \mathbf{y} \rangle$

   (iii) $\forall \mathbf{x} \in \mathbb{R}^n$: $\|A\mathbf{x}\| = \|\mathbf{x}\|$

Proof. (i) $\Leftrightarrow$ (ii): A is orthogonal according to our definition in (i) if and only if $A^T A - I = 0$, which in turn holds if and only if for $\forall \mathbf{x}, \mathbf{y} \in \mathbb{R}^n$ we have $0 = \langle \mathbf{x}, (A^T A - I)\mathbf{y} \rangle = \langle \mathbf{x}, A^T A \mathbf{y} \rangle - \langle \mathbf{x}, \mathbf{y} \rangle = \langle A\mathbf{x}, A\mathbf{y} \rangle - \langle \mathbf{x}, \mathbf{y} \rangle$.

(ii) $\Leftrightarrow$ (iii): If (ii) holds then (iii) follows by setting $\mathbf{x} = \mathbf{y}$ and taking the square root. Conversely, if (iii) holds then for $\forall \mathbf{x}, \mathbf{y} \in \mathbb{R}^n$ we have $\langle A\mathbf{x}, A\mathbf{y} \rangle = \frac{1}{2}\big(\|A(\mathbf{x}+\mathbf{y})\|^2 - \|A\mathbf{x}\|^2 - \|A\mathbf{y}\|^2\big) = \frac{1}{2}\big(\|\mathbf{x}+\mathbf{y}\|^2 - \|\mathbf{x}\|^2 - \|\mathbf{y}\|^2\big) = \langle \mathbf{x}, \mathbf{y} \rangle$. $\square$

Lemma 2.2 (Antisymmetric matrices). *Given a matrix $E \in \mathbb{R}^{n \times n}$, the following three statements are equivalent:*

   (i) E is antisymmetric, i.e., it fulfills $E^T = -E$.

   (ii) $\forall \mathbf{x} \in \mathbb{R}^n$: $\langle \mathbf{x}, E\mathbf{x} \rangle = 0$

   (iii) The matrices e^{tE}, $\forall t \in \mathbb{R}$, are orthogonal.

Proof. (i) $\Leftrightarrow$ (ii): If $E^T = -E$, then we have $\langle \mathbf{x}, E\mathbf{x} \rangle = \langle E^T\mathbf{x}, \mathbf{x} \rangle = -\langle E\mathbf{x}, \mathbf{x} \rangle = -\langle \mathbf{x}, E\mathbf{x} \rangle$ and thus $\langle \mathbf{x}, E\mathbf{x} \rangle = 0$ for $\forall \mathbf{x} \in \mathbb{R}^n$. Conversely, if (ii) holds then we have $\langle \mathbf{x}, (E + E^T)\mathbf{y} \rangle = \langle \mathbf{x} + \mathbf{y}, E(\mathbf{x} + \mathbf{y}) \rangle - \langle \mathbf{x}, E\mathbf{x} \rangle - \langle \mathbf{y}, E\mathbf{y} \rangle = 0 - 0 - 0 = 0$ for $\forall \mathbf{x}, \mathbf{y} \in \mathbb{R}^n$, and thus $E + E^T = 0$, which is (i).

(i) $\Leftrightarrow$ (iii): If $E^T = -E$, then we have $(e^{tE})^T = e^{tE^T} = e^{-tE} = (e^{tE})^{-1}$ for $\forall t \in \mathbb{R}$, i.e., the matrices e^{tE} are orthogonal. Conversely, if (iii) holds then we have $e^{tE^T} = (e^{tE})^T = (e^{tE})^{-1} = e^{-tE}$ for $\forall t \in \mathbb{R}$, and taking the t-derivative at $t = 0$ yields $E^T = -E$. $\square$

2.3 Model Properties

We begin our discussion of the model and its properties by compiling some quick observations.

Constraints, parametric freedom. The parameter constraints listed in Section 2.1.1 are sufficient to give the model all its desired properties, and to ensure that the bond price and forward rate formulas presented in Section 2.5 are valid. However, as is, one can show that there is still a considerable level of redundancy in the model parameters, with different choices of $(E, \mathbf{v}, \mathbf{x}_0, \sigma)$ leading to the same exact dynamics of the short rate process $(r(X_t))_{t \geq 0}$. This issue will be addressed in detail later in Section 2.7, where we will impose the additional parameter constraints (2.41)–(2.43) to remove this redundancy.

    We will be left with a total of $2n + 1$ degrees of freedom in our non-functional parameter choices, plus the choice of the functional parameter $\sigma(\mathbf{x}, t)$. This gives the model enough parametric freedom to calibrate well to both the discount curve and the swaption market.

The constraint (2.9a), which will the key ingredient in Lemmas 2.3 and 2.4 to ensure that the process $(X_t)_{t\geq 0}$ remains confined to M_0, can be rewritten in several ways, by solving for its various ingredients. One way to think of it is as a stronger lower bound on λ:

$$\lambda \geq \lambda_{\min} := \frac{r_\infty}{n-2} + \frac{n-2}{4r_\infty}\|E\mathbf{v}\|^2. \qquad (2.13)$$

This bound for λ shows more explicitly that large values of r_∞, or values of $\|E\mathbf{v}\|$ that are large compared to $\sqrt{r_\infty}$ (i.e., large long-term rates r_∞ or fast oscillations in the forward rate curves, as discussed below) demand large reversion speeds λ.

Time homogeneity. Since the drift $\mu(\mathbf{x})$ and the short rate function $r(\mathbf{x})$ do not depend on the process time t, the ACE model is time homogeneous if and only if the noise function $\sigma(\mathbf{x},t)$ is in fact chosen to be independent of t, i.e., if $\sigma(\mathbf{x},t) = \sigma(\mathbf{x})$. This gives the modeler full control over the time homogeneity of the model: If time homogeneity is desired then he must choose σ as a function of $\mathbf{x}$ only; if he is willing to sacrifice time homogeneity in order to improve calibration quality, he may allow σ to depend on both $\mathbf{x}$ and t.

Global attractor. The zero-noise process, i.e., the ODE $\dot{\mathbf{x}} = \mu(\mathbf{x})$, can be shown to have the global attractor $\mathbf{x} = \vec{0}$ (global in the sense that the entire state space M_0 is contained in its basin of attraction). Note that we have $\vec{0} \in M_0$ by (2.2) and (2.7).

Indeed, by Lemma 2.2 the antisymmetry assumption (2.4) for E implies that the matrices e^{tE}, $t \in \mathbb{R}$, are orthogonal, so that by Lemma 2.1 the solution $\mathbf{x}_{\mathrm{lin}}(t) = e^{-\lambda t}e^{tE}\mathbf{x}_0$ of the linearized ODE $\dot{\mathbf{x}} = \mu_{\mathrm{lin}}(\mathbf{x}) := (-\lambda I + E)\mathbf{x}$ fulfills $\|\mathbf{x}_{\mathrm{lin}}(t)\| = e^{-\lambda t}\|\mathbf{x}_0\|$, and because of (2.5) this shows that the point $\mathbf{x} = \vec{0}$ is at least a *local* attractor. The fact that this point is indeed a *global* attractor can be shown by actually solving the full zero-noise ODE explicitly. See Lemma 2.7 for our explicit solution of the ODE, and see Lemma 2.13 for a discussion of its limiting behavior.

Short rate reversion, long-term forward rate. As a result, a look at (2.12) shows that in the zero-noise case the short rate process $(r(X_t))_{t\geq 0}$ converges to r_∞ as $t \to \infty$ for any starting point $\mathbf{x}_0 \in M_0$, and so the short rates simulated by the ACE model tend to revert back to this value r_∞ also in the presence of noise. In fact, as we will see later in Corollary 2.2, this also implies that the forward rate curves simulated by the ACE model all have this value as their long-term limit.

Oscillations. The matrix E causes the solution of the deterministic system to rotate parallel to certain hyperplanes as it approaches the attractor $\mathbf{x} = \vec{0}$. More precisely, the matrix e^{tE} in the solution $\mathbf{x}_{\mathrm{lin}}(t)$ of the linearized system is composed of up to $\lfloor n/2 \rfloor$ independent rotations in orthogonal hyperplanes, at different frequencies (see Lemma 2.12 and (2.49) below).

The presence of this matrix therefore adds deterministic oscillations to the short rate dynamics, and thus ultimately to the forward rate curve (see also Lemmas 2.6 and 2.14). This adds some important flexibility to the model, as it allows the model to simulate—and calibrate to—forward rate curves that exhibit such oscillatory behavior, or (by choosing wavelengths smaller than the largest observed maturity) that have a single maximum or minimum. The nonlinear terms in μ add further complexity to this set of possible forward rate curves.

State space, positive rates. By (2.7) and (2.12), the state space M_0 and its boundary ∂M_0 can be written as

$$M_0 = \{\mathbf{x} \in \mathbb{R}^n \mid r(\mathbf{x}) > 0\}, \tag{2.14a}$$

$$\partial M_0 = \{\mathbf{x} \in \mathbb{R}^n \mid r(\mathbf{x}) = 0\}. \tag{2.14b}$$

To see that the short rate process $(r(X_t))_{t \geq 0}$ of the ACE model almost surely remains positive for $\forall t \geq 0$, we therefore need to show that the state process $(X_t)_{t \geq 0}$ remains confined to the supposed domain M_0 defined in (2.7).

Geometric interpretation of $r(\mathbf{x})$. As one can see from (2.12), (2.3), and (2.14b), the short rate function has the geometric interpretation[1]

$$r(\mathbf{x}) = \operatorname{dist}(\mathbf{x}, \partial M_0). \tag{2.15}$$

Flowline diagram. As a first step, let us prove this for the deterministic case $\sigma \equiv 0$, by understanding the behavior of the drift vector field μ near the boundary ∂M_0 and near ∞. The proof for the general stochastic case will then be the content of Section 2.4.

Lemma 2.3. *(i) For $\forall \mathbf{x} \in \partial M_0$ we have*

$$\langle \nabla r(\mathbf{x}), \mu(\mathbf{x}) \rangle = \tfrac{1}{n-2} \|\mathbf{x} - \mathbf{x}_{\min}\|^2 + \delta, \tag{2.16}$$

$$\text{where} \quad \mathbf{x}_{\min} := \tfrac{n-2}{2} E\mathbf{v} - r_\infty \mathbf{v} \in \partial M_0. \tag{2.17}$$

(ii) If $\delta > 0$ then all flowlines of μ emanating from the boundary ∂M_0 are leading into the interior of M_0 at a non-vanishing angle to ∂M_0. If $\delta = 0$ then this statement holds true only on $\partial M_0 \setminus \{\mathbf{x}_{\min}\}$.

(iii) No flowline of μ starting in the state space M_0 can ever leave it or reach ∞ in finite (positive) time. The same statement holds with M_0 replaced by $\overline{M_0}$.

Proof. (i) First note that $\mathbf{x}_{\min} \in \partial M_0$ according to (2.14b), since by (2.12), (2.4) in combination with Lemma 2.2 (ii), and (2.3) we have

$$r(\mathbf{x}_{\min}) = r_\infty + \left\langle \mathbf{v}, \tfrac{n-2}{2} E\mathbf{v} - r_\infty \mathbf{v} \right\rangle = r_\infty + 0 - r_\infty \cdot 1 = 0.$$

---

[1]Indeed, given any $\mathbf{x} \in \mathbb{R}^n$, the point $\hat{\mathbf{x}} := \mathbf{x} - r(\mathbf{x})\mathbf{v}$ fulfills $r(\hat{\mathbf{x}}) = r_\infty + \langle \mathbf{v}, \hat{\mathbf{x}} \rangle = r_\infty + \langle \mathbf{v}, \mathbf{x} \rangle - r(\mathbf{x})\|\mathbf{v}\|^2 = r(\mathbf{x}) - r(\mathbf{x}) \cdot 1 = 0$ and thus $\hat{\mathbf{x}} \in \partial M_0$ by (2.14b), and we have $\mathbf{x} - \hat{\mathbf{x}} = r(\mathbf{x})\mathbf{v} \perp \partial M_0$ and $\|\mathbf{x} - \hat{\mathbf{x}}\| = \|r(\mathbf{x})\mathbf{v}\| = r(\mathbf{x})$.

Using the definitions (2.11)–(2.12), the parameter constraints (2.3)–(2.4), the definition (2.9a), the property in Lemma 2.2 (ii), and finally (2.17), we now find that

$$
\begin{aligned}
\langle \nabla r(\mathbf{x}), &\mu(\mathbf{x})\rangle + r(\mathbf{x})\big(\lambda + \tfrac{2}{n-2}\langle \mathbf{v}, \mathbf{x}\rangle\big) \\
&= \big\langle \mathbf{v}, (-\lambda I + E)\mathbf{x} + \tfrac{1}{n-2}\|\mathbf{x}\|^2\mathbf{v} - \tfrac{2}{n-2}\langle \mathbf{v}, \mathbf{x}\rangle\mathbf{x}\big\rangle + r(\mathbf{x})\big(\lambda + \tfrac{2}{n-2}\langle \mathbf{v}, \mathbf{x}\rangle\big) \\
&= (r(\mathbf{x}) - \langle \mathbf{v}, \mathbf{x}\rangle)\big(\lambda + \tfrac{2}{n-2}\langle \mathbf{v}, \mathbf{x}\rangle\big) + \langle \mathbf{v}, E\mathbf{x}\rangle + \tfrac{1}{n-2}\|\mathbf{x}\|^2\|\mathbf{v}\|^2 \\
&= r_\infty\big(\lambda + \tfrac{2}{n-2}\langle \mathbf{v}, \mathbf{x}\rangle\big) - \langle E\mathbf{v}, \mathbf{x}\rangle + \tfrac{1}{n-2}\|\mathbf{x}\|^2 \\
&= \delta + \tfrac{1}{n-2}\big(r_\infty^2 + \tfrac{(n-2)^2}{4}\|E\mathbf{v}\|^2\big) - \tfrac{2}{n-2}\big\langle -r_\infty \mathbf{v} + \tfrac{n-2}{2}E\mathbf{v}, \mathbf{x}\big\rangle + \tfrac{1}{n-2}\|\mathbf{x}\|^2 \\
&= \delta + \tfrac{1}{n-2}\big\|-r_\infty \mathbf{v} + \tfrac{n-2}{2}E\mathbf{v}\big\|^2 - \tfrac{2}{n-2}\big\langle -r_\infty \mathbf{v} + \tfrac{n-2}{2}E\mathbf{v}, \mathbf{x}\big\rangle + \tfrac{1}{n-2}\|\mathbf{x}\|^2 \\
&= \delta + \tfrac{1}{n-2}\big\|\big(-r_\infty \mathbf{v} + \tfrac{n-2}{2}E\mathbf{v}\big) - \mathbf{x}\big\|^2 \\
&= \delta + \tfrac{1}{n-2}\|\mathbf{x}_{\min} - \mathbf{x}\|^2 \,.
\end{aligned}
$$

This shows that for $\forall \mathbf{x} \in \mathbb{R}^n$ we have

$$
\langle \nabla r(\mathbf{x}), \mu(\mathbf{x})\rangle = \delta + \tfrac{1}{n-2}\|\mathbf{x} - \mathbf{x}_{\min}\|^2 - r(\mathbf{x})\big(\lambda + \tfrac{2}{n-2}\langle \mathbf{v}, \mathbf{x}\rangle\big), \qquad (2.18)
$$

which by (2.14b) implies (2.16).

(ii) In particular, the formula (2.16) shows that we have $\langle \nabla r(\mathbf{x}), \mu(\mathbf{x})\rangle \geq 0$ for $\forall \mathbf{x} \in \partial M_0$, with equality holding only in the case $\delta = 0$ for the point $\mathbf{x} = \mathbf{x}_{\min}$. Since by (2.14a) $\nabla r(\mathbf{x})$ is an inward-pointing normal vector to ∂M_0 at $\mathbf{x}$, this proves (ii).

(iii) For $\delta > 0$ it now follows immediately that no flowline can exit M_0 anywhere on ∂M_0, but for $\delta = 0$ it is not immediately clear whether a flowline can exit M_0 at the exceptional point $\mathbf{x}_{\min}$. However, if such a flowline existed then tracing the flow back from points on ∂M_0 near $\mathbf{x}_{\min}$, we would find a flowline with a nearby starting point that exits M_0 elsewhere, contradicting (ii).

Showing that the flowlines cannot run off to ∞ in finite time requires a change of variables. Defining $\mathbf{y} := g(\mathbf{x}) := (\mathbf{x} - \mathbf{a})/\|\mathbf{x} - \mathbf{a}\|^2$ for some arbitrarily chosen point $\mathbf{a} \in \mathbb{R}^n \setminus \overline{M_0}$, Lemma B.1 in Appendix B.1 shows that $g(M_0)$ is an open ball with $g(\infty) = \vec{0}$ on its boundary, and that $\tilde{\mu} := (\nabla g \cdot \mu) \circ g^{-1}$ (i.e., the vector field whose flowlines are the images under g of the flowlines of μ) has a smooth extension $\tilde{\mu}^+$ to the closed ball $\overline{g(M_0)}$ (and in fact to all of $\mathbb{R}^n$), with $\tilde{\mu}^+(\vec{0})$ pointing towards the ball's center. This shows that in that parametrization no flowline can exit $g(M_0)$ near the point $\vec{0}$, which in our original parametrization just means that no flowline can run off to ∞ in finite time.

The second statement of part (iii) now follows from the fact that the flow under a C^1 vector field is continuous (in fact, C^1) with respect to its starting point. $\square$

Market resiliency. Since according to (2.14b) and (2.16) we have

$$r\big(\mathbf{x} + \mu(\mathbf{x})\,\mathrm{d}t\big) = r(\mathbf{x}) + \langle \nabla r(\mathbf{x}), \mu(\mathbf{x})\rangle\,\mathrm{d}t = 0 + \big(\tfrac{1}{n-2}\|\mathbf{x} - \mathbf{x}_{\min}\|^2 + \delta\big)\,\mathrm{d}t,$$

for $\forall \mathbf{x} \in \partial M_0$, the expression $\tfrac{1}{n-2}\|\mathbf{x} - \mathbf{x}_{\min}\|^2 + \delta$ tells us how fast the drift portion $\mu(X_t)\,\mathrm{d}t$ of our SDE (2.10) drives the short rate back up if it is close to 0. We therefore call this value the *local market resiliency*. The point $\mathbf{x}_{\min}$ is the unique place on the boundary ∂M_0 at which this push-back effect is the smallest; its value δ at this point is called the *(global) market resiliency* of the model.

The constraint (2.9a), which asks that $\delta \geq 0$, was introduced to ensure that the local market resiliency is positive everywhere, except in the degenerate case $\delta = 0$, when this effect vanishes at $\mathbf{x}_{\min}$. In either case, as we have shown in Lemma 2.3 (ii)–(iii) at least for the deterministic case, this suffices to confine the process $(X_t)_{t\geq 0}$ to the set M_0 of all states with positive short rates.

Quadratic drift terms. Before we move on to the general stochastic case, let us use our insights from Lemma 2.3 to understand the quadratic drift terms in (2.11). First observe that they can be written as

$$\mu_{\mathrm{quad}}(\mathbf{x}) = \left(I - 2\frac{\mathbf{x}\mathbf{x}^T}{\|\mathbf{x}\|^2}\right)\big(\tfrac{1}{n-2}\|\mathbf{x}\|^2\mathbf{v}\big),$$

i.e., $\mu_{\mathrm{quad}}(\mathbf{x})$ is the reflection of the vector $\tfrac{1}{n-2}\|\mathbf{x}\|^2\mathbf{v}$ in the hyperplane that contains the origin and is orthogonal to $\mathbf{x}$.

To understand the need for any nonlinear terms, note that unless we are in the degenerate case $E\mathbf{v} = \vec{0}$,[2] the flow under the linear part $\mu_{\mathrm{lin}}(\mathbf{x}) = (-\lambda I + E)\mathbf{x}$ of the drift alone would lead out of our declared state space M_0 at some points on its boundary ∂M_0, and so without any additional terms the process $(X_t)_{t\geq 0}$ would not be confined to M_0. Indeed, the points $\mathbf{x}_c := cE\mathbf{v} - r_\infty\mathbf{v}$, $c \in \mathbb{R}$, for example lie on ∂M_0 by (2.14b) since

$$\begin{aligned}
r(\mathbf{x}_c) &= r_\infty + \langle \mathbf{v}, \mathbf{x}_c\rangle = r_\infty + \langle \mathbf{v}, cE\mathbf{v} - r_\infty\mathbf{v}\rangle \\
&= r_\infty + c\langle \mathbf{v}, E\mathbf{v}\rangle - r_\infty\|\mathbf{v}\|^2 = r_\infty + c\cdot 0 - r_\infty\cdot 1 = 0
\end{aligned}$$

by (2.12), Lemma 2.2 (ii), and (2.3), and if $E\mathbf{v} \neq \vec{0}$ then for sufficiently large $c > 0$ we have

$$\begin{aligned}
\langle \mu_{\mathrm{lin}}(\mathbf{x}_c), \nabla r(\mathbf{x}_c)\rangle &= \big\langle(-\lambda I + E)(cE\mathbf{v} - r_\infty\mathbf{v}), \mathbf{v}\big\rangle \\
&= -\lambda c\langle E\mathbf{v}, \mathbf{v}\rangle + \lambda r_\infty\|\mathbf{v}\|^2 + c\langle E^2\mathbf{v}, \mathbf{v}\rangle - r_\infty\langle E\mathbf{v}, \mathbf{v}\rangle \\
&= 0 + \lambda r_\infty - c\|E\mathbf{v}\|^2 - 0 < 0
\end{aligned}$$

by Lemma 2.2 (ii), (2.3), and (2.4), i.e., $\mu_{\mathrm{lin}}(\mathbf{x}_c)$ points away from M_0 by (2.14a).

---

[2]The unproblematic case $E\mathbf{v} = \vec{0}$ is unfortunately just the one that does not allow for any deterministic interest rate oscillations, since in this case the e^{tE}-rotation of any point $\mathbf{x}$ leaves its $\mathbf{v}$-component unaffected: $\langle \mathrm{e}^{tE}\mathbf{x}, \mathbf{v}\rangle = \langle \mathbf{x}, \mathrm{e}^{tE^T}\mathbf{v}\rangle = \langle \mathbf{x}, \mathrm{e}^{-tE}\mathbf{v}\rangle = \langle \mathbf{x}, \mathbf{v}\rangle$.

Our additional quadratic drift terms fulfill

$$\langle\mu_{\text{quad}}(\mathbf{x}), \nabla r(\mathbf{x})\rangle = \tfrac{1}{n-2}\big\langle\|\mathbf{x}\|^2\mathbf{v} - 2\langle\mathbf{v},\mathbf{x}\rangle\mathbf{x}, \mathbf{v}\big\rangle = \tfrac{1}{n-2}\big(\|\mathbf{x}\|^2\|\mathbf{v}\|^2 - 2\langle\mathbf{v},\mathbf{x}\rangle^2\big)$$

$$= \tfrac{1}{n-2}\|\mathbf{x}\|^2 - \tfrac{2}{n-2}r_\infty^2$$

for $\forall\mathbf{x} \in \partial M_0$ by (2.3), (2.14b), and (2.12), which shows that at least for large $\|\mathbf{x}\|$ they add a component pointing into the interior of M_0 that is sufficiently large to counteract the potential outwards drift from the linear part $\mu_{\text{lin}}(\mathbf{x})$; the proof of the fact that this actually suffices on all of ∂M_0 was the content of Lemma 2.3 (i).

As the reader might have noticed, the second of these two quadratic drift terms is actually counterproductive for this purpose. The reason for introducing it is a purely technical one: It can be seen to make the solution $\mathbf{x}_{\text{quad}}(t) = \frac{\mathbf{x}_0/\|\mathbf{x}_0\|^2 + t\mathbf{v}/(n-2)}{\|\mathbf{x}_0/\|\mathbf{x}_0\|^2 + t\mathbf{v}/(n-2)\|^2}$ for the purely quadratic ODE $\dot{\mathbf{x}} = \mu_{\text{quad}}(\mathbf{x})$ *conformal* in $\mathbf{x}_0$, i.e., the flow μ_{quad} *preserves angles*,[3] and since μ_{lin} has this property as well,[4] so does $\mu = \mu_{\text{lin}} + \mu_{\text{quad}}$.[5] As we will see only later in Chapters 5–8, the conformality of μ is one of the cornerstones of our construction's strategy for obtaining an explicit bond price formula.

2.4   Existence and Uniqueness, Positive Rates

We will now extend our results from Lemma 2.3 (iii) to the general stochastic case, and prove that under certain conditions on our noise function $\sigma(\mathbf{x}, t)$ the ACE SDE (2.10) has a unique strong solution $(X_t)_{t\geq0}$ that remains in M_0 for $\forall t \geq 0$ almost surely. In particular, by (2.14a) this means that the short rates $r(X_t)$ simulated by the ACE model are positive for $\forall t \geq 0$ almost surely, as desired.[6] We want to stress that this result is not only of theoretical significance: The conditions that Lemma 2.4 imposes on σ are important for the practitioner who must choose this function during the calibration of the model. To help the reader internalize them, we will therefore discuss them in detail further below.

To understand the need for these conditions, observe that the existence of a solution $(X_t)_{t\geq0}$ requires that the process can neither run off to ∞ in finite time (which would mean that it is not actually defined for all times $t \geq 0$) nor reach the boundary ∂M_0; to avoid such behavior, one therefore has to impose two kinds of constraints on the noise function $\sigma(\mathbf{x}, t)$: one asking that σ does not increase too fast as $\mathbf{x} \to \infty$ (to allow the drift to pull the process back towards the attractor $\vec{0}$), and one asking that σ dies off sufficiently fast near ∂M_0 (so that the drift can push it away from ∂M_0). This is the purpose of the

---

[3] Indeed, $\mathbf{x}_{\text{quad}}(t)$ is the composition of an inversion $\mathbf{x}_0 \mapsto \mathbf{x}_0/\|\mathbf{x}_0\|^2$, a shift by $t\mathbf{v}/(n-2)$, and an inversion again, all of which are angle-preserving maps.

[4] Recall that $\mathbf{x}_{\text{lin}}(t) = e^{-\lambda t}e^{Et}\mathbf{x}_0$ is the composition of a scaling and a rotation, both of which preserve angles.

[5] To see this, observe that following the combined flow μ for some infinitesimal time dt is equivalent to following first μ_{lin} and then μ_{quad} for the time dt each.

[6] As we will see later in Section 2.5.2, this also implies that in fact the entire forward rate curve is positive for $\forall t \geq 0$ almost surely.

two conditions (2.19) and (2.20) of Lemma 2.4, respectively. The third requirement, the Lipschitz continuity of σ, is inherited from the standard existence and uniqueness result in SDE theory[7] (see, e.g., [32, Chapter 5.2] or [33, Chapter 5, Theorem 2.9]) that our proof is based on.

Finally, recall that having a *strong* solution means that for each *given* Brownian motion $(W_t)_{t \geq 0}$ one can construct a solution $(X_t)_{t \geq 0}$ of the SDE (2.10); in contrast, the existence of a *weak* solution would only mean that one can construct a pair of a Brownian motion $(W_t)_{t \geq 0}$ and a solution $(X_t)_{t \geq 0}$ simultaneously so that the SDE (2.10) is satisfied.[8] Given our eventual goal of numerically simulating $(X_t)_{t \geq 0}$, which is typically done by first simulating the infinitesimal increments of a Brownian motion $(W_t)_{t \geq 0}$ and then deriving from it the associated increments of $(X_t)_{t \geq 0}$ via (2.10), for our practical purposes it is more natural to ask for the existence of a strong solution.

In the following statement, please recall the geometric interpretation (2.15) of our short rate function $r(\mathbf{x})$.

Lemma 2.4 (Existence and uniqueness of the ACE process). *Suppose that the noise function σ is continuous in $(\mathbf{x}, t)$, and that it is chosen such that for some $T > 0$ the following three conditions are met: (i) $\sigma(\mathbf{x}, t)$ is locally Lipschitz continuous in $\mathbf{x}$ uniformly in $t \in [0, T]$,[9] (ii) we have*

$$\limsup_{\mathbf{x} \to \infty} \frac{\sup_{t \in [0,T]} \sigma^2(\mathbf{x}, t)}{r(\mathbf{x}) \|\mathbf{x}\|^2} \cdot \left(1 + 2(n-2) \frac{r(\mathbf{x})}{\|\mathbf{x}\|} \right) < \frac{2}{n-2} ,$$

(growth condition on σ) (2.19)

and (iii) there exist $\alpha \in (0, 1)$, $\varepsilon > 0$, and an open set $N \supset \partial M_0$ such that for $\forall \mathbf{x} \in M_0 \cap N$ we have

$$\sup_{t \in [0,T]} \sigma^2(\mathbf{x}, t) \leq 2\alpha \Big[r(\mathbf{x}) \big(\delta + \tfrac{1}{n-2} \|\mathbf{x} - \mathbf{x}_{\min}\|^2 \big)$$
$$+ \mathbb{1}_{\delta = 0, \|\mathbf{x} - \mathbf{x}_{\min}\| < \varepsilon} \, r(\mathbf{x})^2 \big(\tfrac{2}{n-2} r_\infty - \lambda \big) \Big].$$

(Feller condition) (2.20)

Then the ACE SDE (2.10) has a unique strong solution $(X_t)_{t \in [0,T]}$ that remains in M_0 for $\forall t \in [0, T]$ almost surely.

---

[7]The result states that if two functions $\mu \colon \mathbb{R}^n \times [0, T] \to \mathbb{R}^n$ and $\sigma \colon \mathbb{R}^n \times [0, T] \to \mathbb{R}^{n \times n}$ are continuous in $(\mathbf{x}, t)$ and fulfill the growth condition $\|\mu(\mathbf{x}, t)\| + \|\sigma(\mathbf{x}, t)\| \leq K_1(1 + \|\mathbf{x}\|)$ and the Lipschitz continuity condition $\|\mu(\mathbf{x}, t) - \mu(\mathbf{y}, t)\| + \|\sigma(\mathbf{x}, t) - \sigma(\mathbf{y}, t)\| \leq K_2 \|\mathbf{x} - \mathbf{y}\|$ for $\forall \mathbf{x}, \mathbf{y} \in \mathbb{R}^n$ and $\forall t \in [0, T]$ and for some $K_1, K_2 > 0$, then the SDE $\mathrm{d}X_t = \mu(X_t, t)\, \mathrm{d}t + \sigma(X_t, t)\, \mathrm{d}W_t$, $X_{t=0} = \mathbf{x}_0$, has a unique strong solution $(X_t)_{t \in [0,T]}$.

[8]The difference between these two notions of a solution becomes more apparent if a model includes additional processes that are correlated to $(W_t)_{t \geq 0}$ as well: In this situation, a weak solution (i.e., the construction of a pair of processes $(X_t)_{t \geq 0}$ and $(W_t)_{t \geq 0}$) can generally not take the further dependencies of $(W_t)_{t \geq 0}$ into account, whereas a strong solution can build on a given Brownian motion $(W_t)_{t \geq 0}$ that has previously been constructed with all its additional dependencies.

[9]I.e., for $\forall \mathbf{x} \in M_0$ there $\exists \varepsilon, K > 0$ such that for $\forall \mathbf{x}_1, \mathbf{x}_2 \in M_0$ with $\|\mathbf{x}_1 - \mathbf{x}\| < \varepsilon$ and $\|\mathbf{x}_2 - \mathbf{x}\| < \varepsilon$ and for $\forall t \in [0, T]$ we have $|\sigma(\mathbf{x}_1, t) - \sigma(\mathbf{x}_2, t)| \leq K \|\mathbf{x}_1 - \mathbf{x}_2\|$.

If these conditions hold for $\forall T > 0$, then (2.10) has a unique strong solution $(X_t)_{t \geq 0}$ that remains in M_0 for $\forall t \geq 0$ almost surely.

Proof. In Appendix B.2 we will first prove a variant of the Feller condition for SDEs on the open unit ball B in $\mathbb{R}^n$, which for a general given SDE on B whose drift points inwards near the ball's boundary ∂B, tells us that if the SDE's noise vanishes near ∂B at some specific rate or faster, then the SDE has a unique strong solution that never reaches ∂B.

Our proof of Lemma 2.4, which can then be found in Appendix B.3, works by reducing our problem to just this case, by constructing a bijection $\bar{g} \colon M_0 \to B$ (based on the function g that we had already used in our proof of Lemma 2.3 (iii)) and then considering the image of the ACE SDE (2.10) under $\bar{g}$, obtained via Itô's Lemma. Since $\bar{g}^{-1}(\partial B) = \partial M_0 \cup \{\infty\}$, our Feller condition near ∂B then translates into *two* conditions for the noise function σ of our original ACE SDE: the Feller condition (2.20) near ∂M_0, and the growth condition (2.19) near ∞. $\qquad\square$

Corollary 2.1. *Under the conditions of Lemma 2.4 the short rates $r(X_t)$ simulated by the ACE model are positive for $\forall t \in [0, T]$ almost surely (or for $\forall t \geq 0$ if the conditions are fulfilled for $\forall T > 0$).*

Proof. This now follows from (2.14a). $\qquad\square$

Let us take a closer look at the two conditions (2.19) and (2.20) on our noise function $\sigma(\mathbf{x}, t)$. Although at first glance they may look intimidating, as we will see, all their individual terms actually play a well-defined role that can be properly understood and taken advantage of with just a little effort.[10] The goal of this lemma is to provide the strongest results (i.e., the weakest conditions) possible, but if one is willing to accept only slightly more stringent conditions instead then they can be simplified significantly.

The growth condition as $\mathbf{x} \to \infty$. The condition (2.19) asks that σ does not increase too fast as $\mathbf{x} \to \infty$, which is necessary to ensure that the noise cannot drive the process $(X_t)_{t \geq 0}$ to ∞ in finite time. Since we have $\frac{r(\mathbf{x})}{\|\mathbf{x}\|} = \frac{r_\infty}{\|\mathbf{x}\|} + \left\langle \mathbf{v}, \frac{\mathbf{x}}{\|\mathbf{x}\|} \right\rangle$ and thus

$$\limsup_{\mathbf{x} \to \infty} \left(1 + 2(n-2) \frac{r(\mathbf{x})}{\|\mathbf{x}\|} \right) = 1 + 2(n-2) \cdot 1 = 2n - 3,$$

this condition is certainly fulfilled if the stronger condition

$$\limsup_{\mathbf{x} \to \infty} \frac{\sup_{t \in [0,T]} \sigma^2(\mathbf{x}, t)}{r(\mathbf{x}) \|\mathbf{x}\|^2} < \frac{2}{(n-2)(2n-3)} \qquad (2.21)$$

---

[10]Nevertheless, the impatient reader who is getting stuck during the discussion to follow will be forgiven for skipping ahead to Section 2.5, as long as they promise to return and give it another try before attempting to implement the model.

holds, i.e., if $\exists \alpha \in (0,1)$ such that for all sufficiently large $\mathbf{x} \in M_0$ and $\forall t \in [0,T]$ we have

$$\sigma^2(\mathbf{x},t) \leq \frac{2\alpha}{(n-2)(2n-3)} \, r(\mathbf{x}) \|\mathbf{x}\|^2. \tag{2.22}$$

The second factor in (2.19) weakens this growth condition slightly, namely to

$$\sigma^2(\mathbf{x},t) \leq \frac{2\alpha}{(n-2)\left(1 + 2(n-2)\frac{r(\mathbf{x})}{\|\mathbf{x}\|}\right)} \, r(\mathbf{x}) \|\mathbf{x}\|^2, \tag{2.23}$$

by making the prefactor dependent on $\frac{r(\mathbf{x})}{\|\mathbf{x}\|} \approx \langle \mathbf{v}, \frac{\mathbf{x}}{\|\mathbf{x}\|} \rangle$, i.e., on the direction in which $\mathbf{x}$ goes to ∞: If $\mathbf{x}$ moves into the direction given by $\mathbf{v}$, i.e., perpendicular to ∂M_0, then we have $\frac{r(\mathbf{x})}{\|\mathbf{x}\|} \to 1$, and the prefactor in (2.23) indeed turns into the one in (2.22); however, if $\mathbf{x}$ moves parallelly to ∂M_0 (i.e., if $r(\mathbf{x}) \equiv \text{const} > 0$), then we have $\frac{r(\mathbf{x})}{\|\mathbf{x}\|} \to 0$, and the prefactor in (2.23) becomes $\frac{2\alpha}{n-2} > \frac{2\alpha}{(n-2)(2n-3)}$, thus relaxing our simplified condition (2.22). Finally, rewriting (2.23) as

$$\sigma^2(\mathbf{x},t) \leq \frac{2\alpha}{(n-2)\left(\frac{\|\mathbf{x}\|}{r(\mathbf{x})} + 2(n-2)\right)} \, \|\mathbf{x}\|^3 \tag{2.24}$$

shows that our upper bound (i.e., the *entire* right-hand side of (2.23)) grows the fastest in the direction of $\mathbf{v}$ (where $\frac{r(\mathbf{x})}{\|\mathbf{x}\|}$ is the largest.)

Note that the readily available standard existence and uniqueness result for SDEs in [32, Chapter 5.2] or [33, Chapter 5, Theorem 2.9] could not be applied to our SDE as is, not only because our domain is not all of $\mathbb{R}^n$, but also because that criterion requires both the drift and the noise to grow at most like $O(\|\mathbf{x}\|)$ as $\mathbf{x} \to \infty$, whereas our drift μ defined in (2.11) is of the order $O(\|\mathbf{x}\|^2)$. However, in our case we know that all the flowlines of μ point inwards (they all lead to the attractor $\vec{0}$), and so its strong growth should actually *help* us avoid a noise-driven blow-up. It is for this reason that our result can afford to allow for a stronger growth of our noise function σ, namely of the order $O\big(r(\mathbf{x})^{1/2}\|\mathbf{x}\|\big)$, where our definition (2.12) of $r(\mathbf{x})$ can be interpreted geometrically as the distance of $\mathbf{x}$ to ∂M_0.

The Feller condition near ∂M_0. The condition (2.20) asks that σ vanishes sufficiently fast as $\mathbf{x}$ approaches the boundary ∂M_0 of the domain M_0. (Indeed, by (2.14a) and (2.9a)–(2.9b) all terms on the right of (2.20) are non-negative on M_0, and by (2.14b) the right-hand side vanishes on ∂M_0.) More precisely, if $\delta > 0$, or in the case $\delta = 0$ near all points on ∂M_0 other than $\mathbf{x}_{\min}$, (2.20) asks that $\sigma(\mathbf{x},t)$ must be at most of the order $O\big(r(\mathbf{x})^{1/2}\big)$ (with a prefactor smaller than some bound that depends on the location on ∂M_0), while in the remaining degenerate case it must be at most of the order $O(r(\mathbf{x}))$ (again with some bound on the prefactor).

By (2.16), on ∂M_0 the term $\delta + \frac{1}{n-2}\|\mathbf{x} - \mathbf{x}_{\min}\|^2$ represents the component of the drift μ in the direction of the inner normal to ∂M_0. Writing the first line of the condition (2.20) as

$$\frac{1}{2}\left(\frac{\sup_{t\in[0,T]}\sigma(\mathbf{x},t)}{\sqrt{r(\mathbf{x})}}\right)^2 \leq \alpha\big(\delta + \tfrac{1}{n-2}\|\mathbf{x}-\mathbf{x}_{\min}\|^2\big)$$

therefore shows its analogy to the classical Feller condition

$$\tfrac{1}{2}\sigma_0^2 < b(0)$$

under which the simple one-dimensional CIR model [15] given by the SDE

$$\mathrm{d}R_t = b(R_t)\,\mathrm{d}t + \sigma(R_t)\,\mathrm{d}W_t\,, \qquad R_{t=0} = \mathbf{r}_0,$$

where $b(\mathbf{r}) := -\nu(\mathbf{r}-\mathbf{r}_\infty)$ and $\sigma(\mathbf{r}) := \sigma_0\sqrt{\mathbf{r}}$ for some parameters $\mathbf{r}_0, \mathbf{r}_\infty, \nu, \sigma_0 > 0$, is guaranteed to never reach the point $\mathbf{r} = 0$.

In the degenerate case $\delta = 0$ this drift vanishes at the point $\mathbf{x}_{\min} \in \partial M_0$, and so the term $r(\mathbf{x})\big(\delta + \tfrac{1}{n-2}\|\mathbf{x}-\mathbf{x}_{\min}\|^2\big)$ alone would impose quite a tight bound on σ near that point (as it decreases like $\|\mathbf{x} - \mathbf{x}_{\min}\|^3$ in each direction as $\mathbf{x} \to \mathbf{x}_{\min}$). The term in the second line of (2.20) (which originated from the third term in (2.18), evaluated at $\mathbf{x} = \mathbf{x}_{\min}$) provides some relief, by adding a quadratic (and thus larger) function in a neighborhood of $\mathbf{x}_{\min}$.

2.5   Explicit Formulas

Our key result is that the dollar value of the maturity-T zero-coupon bond,

$$B_{\$,T}(\mathbf{x},t) := \mathbb{E}\big[\exp\big(-\textstyle\int_t^T r(X_s)\,\mathrm{d}s\big)\,\big|\,X_t = \mathbf{x}\big], \qquad (2.25)$$

can for $\forall(\mathbf{x},t) \in M_0 \times [0,T]$ be computed via an analytic formula, i.e., that we can evaluate the expected value in (2.25) explicitly. The bond price formula will be stated and briefly discussed in Section 2.5.1, and it will be proven later in Section 2.6.2.

Naturally, as a result we also obtain an explicit formula for the (absolute) forward rate function $F(\mathbf{x},t;s)$, where $\mathbf{x} \in M_0$ and $0 \leq t \leq s$. Recall that this function is defined via the relation

$$B_{\$,T}(\mathbf{x},t) = \exp\big(-\textstyle\int_t^T F(\mathbf{x},t;s)\,\mathrm{d}s\big) \qquad (2.26)$$

for $\forall(\mathbf{x},t) \in M_0 \times [0,T]$, or equivalently, as

$$F(\mathbf{x},t;s) = -\partial_s \log B_{\$,s}(\mathbf{x},t). \qquad (2.27)$$

However, instead of deriving our forward rate formula from our bond price formula via (2.27), in Section 2.5.2 we will find that (as a result of the ACE model being unspanned) one can alternatively derive it by relating it to the solution of the zero-noise ODE $\dot{\mathbf{x}} = \mu(\mathbf{x})$.

We will therefore conclude Section 2.5 by providing an explicit formula also for the solution of this zero-noise ODE. Besides leading us to our forward rate formula, this may also be useful when simulating the SDE (2.10) numerically, since it will allow us to eliminate all higher-order errors caused by the linear approximation $\mu(X_t)\,\mathrm{d}t$ of its deterministic component.

2.5.1 Bond Prices

We begin by stating and discussing the bond price formula of the ACE model.

Theorem 2.1 (Bond price formula). *The bond prices defined in (2.25) can for $\forall (\mathbf{x}, t) \in M_0 \times [0, T]$ be computed explicitly via the formula*

$$B_{\$,T}(\mathbf{x}, t) = e^{-r_\infty \tau} \left(1 - 2\langle \mathbf{c}_\tau, \mathbf{x} \rangle + \|\mathbf{c}_\tau\|^2 \|\mathbf{x}\|^2 \right)^{1-n/2} \tag{2.28a}$$

$$= \begin{cases} e^{-r_\infty \tau} \left\| \|\mathbf{c}_\tau\| \, \mathbf{x} - \dfrac{\mathbf{c}_\tau}{\|\mathbf{c}_\tau\|} \right\|^{2-n} & \text{for } t \in [0, T), \\ 1 & \text{for } t = T, \end{cases} \tag{2.28b}$$

where $\tau := T - t$, and where the (forward) currency vectors $\mathbf{c}_\tau$, $\tau \geq 0$, are defined as

$$\mathbf{c}_\tau := \tfrac{1}{n-2}(E + \lambda I)^{-1}\left(e^{-\tau(E+\lambda I)} - I\right)\mathbf{v}. \tag{2.29}$$

The expressions inside the parentheses in (2.28a) and inside the norm in (2.28b) do not vanish for any $\mathbf{x} \in M_0$ and for any τ for which they are used.

Proof. The proof is postponed to Section 2.6.2. $\qquad \square$

Before we discuss this formula in detail, let us quickly take a closer look at the currency vectors $\mathbf{c}_\tau$; they will in fact play an even more vital role in the multi-currency case in Chapter 3 (hence the name).

Lemma 2.5. *The currency vectors $\mathbf{c}_\tau$ have the following properties:*

(i) The matrix $E + \lambda I$ is invertible, and so $\mathbf{c}_\tau$ is well-defined for $\forall \tau \geq 0$.

(ii) We have $\mathbf{c}_0 = \vec{0}$ and $\forall \tau > 0$: $\mathbf{c}_\tau \neq \vec{0}$.

(iii) For $\forall \tau \geq 0$ we have

$$\partial_\tau \mathbf{c}_\tau = -\tfrac{1}{n-2} e^{-\tau(E+\lambda I)} \mathbf{v} \tag{2.30a}$$

$$= -\tfrac{1}{n-2}\mathbf{v} - (E + \lambda I)\mathbf{c}_\tau. \tag{2.30b}$$

Proof. (i) By Lemma 2.2 (ii) we have $\langle \mathbf{x}, (E + \lambda I)\mathbf{x} \rangle = \lambda \|\mathbf{x}\|^2$ for $\forall \mathbf{x} \in \mathbb{R}^n$, which shows that $(E + \lambda I)\mathbf{x} = \vec{0}$ implies $\mathbf{x} = \vec{0}$.

(ii) The first part is obvious from the definition (2.29). The second part follows from the fact that for $\forall \tau > 0$ the matrix $e^{-\tau(E+\lambda I)} - I$ is invertible as well; indeed, for $\forall \mathbf{x} \in \mathbb{R}^n$ and we have

$$\left\| (e^{-\tau(E+\lambda I)} - I)\mathbf{x} \right\| \geq \left| \|e^{-\tau(E+\lambda I)}\mathbf{x}\| - \|\mathbf{x}\| \right| = \underbrace{\left| e^{-\lambda \tau} - 1 \right|}_{>0} \|\mathbf{x}\|.$$

(iii) Both relations are direct consequences of the definition (2.29). $\qquad \square$

The bond price formula (2.28) provides two equivalent expressions: (2.28a) is preferable for numerical purposes, as it avoids division by $\|\mathbf{c}_\tau\|$, which by Lemma 2.5 (ii) is zero for $\tau = 0$ (i.e., $t = T$), and it shows that there is no discontinuity at $t = T$. The expression (2.28b) is more useful for our proofs because its dependency on $\mathbf{x}$ is simpler; in particular, for the trained eye it makes it easy to see that $B_{\$,T}(\mathbf{x}, t)$ is harmonic in $\mathbf{x}$ (Lemma 2.10).

Next, observe the decisive fact that our model parameter $\sigma(\mathbf{x}, t)$ does not enter our bond price formula (2.28), despite the fact that it does enter the original definition (2.25) through the process $(X_t)_{t \geq 0}$. This shows that *the ACE model is unspanned with respect to its functional parameter $\sigma(\mathbf{x}, t)$*. Besides leading to the advantages discussed in Sections 1.1–1.2, this has the side effect that although the process $(X_t)_{t \geq 0}$ is not time homogeneous in general due to the t-dependence of $\sigma(\mathbf{x}, t)$, our bond price formula (2.28) depends on T and t only via the tenor $\tau := T - t$ (since it must be the same as for the specific time-homogeneous case of a noise function $\sigma(\mathbf{x})$).

The rate r_∞ in the exponent is the constant term of our short rate function $r(\mathbf{x})$ defined in (2.12). A look at how it enters the definition (2.25) thus fully explains the presence of the exponential prefactor in our bond price formula (2.28).

The remaining part of the formula (2.28b) is a negative (recall (2.1)) power of the norm of some vector. While this immediately implies that the formula can only return positive values (as by (2.25) it should), the negative exponent also requires us to ensure that the expression inside the norm does not vanish (see Lemma 2.9 (ii) below).

The norm is the distance between two vectors, both of which depend on $\mathbf{c}_\tau$, but only one of which actually depends on our state $\mathbf{x}$. The information contained in $\mathbf{c}_\tau$ (which is the only expression through which the parameters λ, E and $\mathbf{v}$ enter our formula) is split into two parts: its length, which enters the first vector as a prefactor of the state $\mathbf{x}$, and its direction, which is used as the second vector.

Finally, we see that our formula returns the value 1 whenever $t = T$, as is expected from (2.25), and from the fact that at the time of maturity a bond is worth \$1.

2.5.2　Forward Rates, Zero-Noise Process

Let us now move on to our forward rate formula, which—as a consequence of the ACE model being unspanned—manages to link the forward rates to the zero-noise process.

Lemma 2.6. *The forward rates defined by the relation (2.26) depend on t and s only via the difference $s - t$, i.e., it is of the form*

$$F(\mathbf{x}, t; s) = F_{\text{rel}}(\mathbf{x}; s - t) \qquad (2.31)$$

for $\forall \mathbf{x} \in M_0$ and for $0 \leq t \leq s$, where F_{rel} is called the relative forward rate *function. This function can be obtained directly from the short rate function*

$r(\mathbf{x})$ *defined in* (2.12) *via the relation*

$$F_{\mathrm{rel}}(\mathbf{x};\tau) = r(U_\tau(\mathbf{x})) \qquad (2.32)$$

for $\forall \mathbf{x} \in M_0$ *and* $\forall \tau \geq 0$, *where the semigroup* $(U_t)_{t\geq 0}$ *of maps* $U_t\colon \overline{M}_0 \to \overline{M}_0$ *is defined as the solution to the zero-noise system, i.e., of the ODE*

$$\forall t \geq 0 \ \forall \mathbf{x} \in \overline{M}_0\colon \qquad \partial_t U_t(\mathbf{x}) = \mu(U_t(\mathbf{x})), \qquad (2.33\mathrm{a})$$

$$\forall \mathbf{x} \in \overline{M}_0\colon \qquad U_{t=0}(\mathbf{x}) = \mathbf{x}. \qquad (2.33\mathrm{b})$$

(Recall that by Lemma 2.3 (iii) these maps are well defined and indeed lead into $\overline{M}_0$ *for* $\forall t \geq 0$.)

Proof. By (2.28) the bond prices $B_{\$,T}(\mathbf{x}, t)$ in the ACE model do not depend on σ. Letting $\sigma \to 0$ in (2.25), we therefore find that

$$B_{\$,T}(\mathbf{x}, t) = \exp\left(-\int_t^T r(U_{s-t}(\mathbf{x}))\, \mathrm{d}s\right),$$

and so (2.27) leads us to (2.32). □

Note the remarkable fact that the forward rate function F is of the form (2.31) even if our process (X_t) is *not* time homogeneous, i.e., if the noise function σ depends on t; for general time-inhomogeneous models this is not the case. By (2.27) this property is in fact equivalent to the bond price functions $B_{\$,T}(\mathbf{x}, t)$ in (2.28) being functions of $\mathbf{x}$ and $\tau = T - t$ only, which in our discussion in Section 2.5.1 we had attributed to the fact that the ACE model is unspanned.

Also observe that since the maps U_t lead into $\overline{M}_0$, (2.32) together with (2.14a) implies that in the ACE model not only the short rates, but in fact the entire forward rate curve is positive.

In order to allow us to utilize the forward rate formula (2.32) in practice, the next lemma provides an explicit solution to the zero-noise ODE (2.33). While it is not immediately obvious that the expression on the right of (2.34) indeed fulfills the semigroup property, this will become more apparent after a slight reformulation, as demonstrated in detail later in Section 2.8.1.

Lemma 2.7. *The ODE* (2.33) *has the following solution for* $\forall (\mathbf{x}, t) \in \overline{M}_0 \times [0, \infty)$:

$$U_t(\mathbf{x}) = \begin{cases} \dfrac{\mathrm{e}^{t(E+\lambda I)}\left(\frac{\mathbf{x}}{\|\mathbf{x}\|^2} - \mathbf{c}_t\right)}{\left\|\mathrm{e}^{t(E+\lambda I)}\left(\frac{\mathbf{x}}{\|\mathbf{x}\|^2} - \mathbf{c}_t\right)\right\|^2} & \text{for } \mathbf{x} \neq \vec{0}, \\[6pt] \vec{0} & \text{for } \mathbf{x} = \vec{0}. \end{cases} \qquad (2.34)$$

The denominator does not vanish for any $\mathbf{x} \in \overline{M}_0 \setminus \{\vec{0}\}$ *and any* $t \geq 0$.

Proof. The proof is postponed to Section 2.6.1. □

2.6   Proofs

We will now present the proofs of the bond price formula (2.28) and the zero-noise evolution formula (2.34). Both proofs are straightforward in that they boil down to checking that the two given functions fulfill certain differential equations.

2.6.1   Proof of the Zero-Noise Evolution Formula

We begin with the proof of the zero-noise evolution formula (2.34). It is easy to see that this formula fulfills the initial condition (2.33b). To see that it also solves the differential equation (2.33a), first note that the case $\mathbf{x} = \vec{0}$ is trivial, since $\mu(\vec{0}) = \vec{0}$. If $\mathbf{x} \neq \vec{0}$, let us abbreviate

$$\mathbf{u}_t := e^{t(E+\lambda I)}\left(\tfrac{\mathbf{x}}{\|\mathbf{x}\|^2} - \mathbf{c}_t\right),$$

so that our zero-noise evolution formula (2.34) can be written as $U_t(\mathbf{x}) = \frac{\mathbf{u}_t}{\|\mathbf{u}_t\|^2}$. Then by (2.30a) we have

$$\partial_t \mathbf{u}_t = (E + \lambda I)\mathbf{u}_t + \tfrac{1}{n-2}\mathbf{v},$$

and so using Lemma 2.2 (ii), we find that

$$
\partial_t\left(\frac{\mathbf{u}_t}{\|\mathbf{u}_t\|^2}\right) = \frac{1}{\|\mathbf{u}_t\|^2}\left(I - 2\frac{\mathbf{u}_t\mathbf{u}_t^T}{\|\mathbf{u}_t\|^2}\right)\left((E + \lambda I)\mathbf{u}_t + \tfrac{1}{n-2}\mathbf{v}\right)
$$

$$
= \frac{E\mathbf{u}_t}{\|\mathbf{u}_t\|^2} - \frac{\lambda\mathbf{u}_t}{\|\mathbf{u}_t\|^2} + \frac{1}{n-2}\frac{\mathbf{v}}{\|\mathbf{u}_t\|^2} - \frac{2}{n-2}\frac{\langle \mathbf{u}_t, \mathbf{v}\rangle\mathbf{u}_t}{\|\mathbf{u}_t\|^4}
$$

$$
= (E - \lambda I)\frac{\mathbf{u}_t}{\|\mathbf{u}_t\|^2} + \frac{1}{n-2}\left\|\frac{\mathbf{u}_t}{\|\mathbf{u}_t\|^2}\right\|^2\mathbf{v} - \frac{2}{n-2}\left\langle \mathbf{v}, \frac{\mathbf{u}_t}{\|\mathbf{u}_t\|^2}\right\rangle\frac{\mathbf{u}_t}{\|\mathbf{u}_t\|^2}
$$

$$
= \mu\left(\frac{\mathbf{u}_t}{\|\mathbf{u}_t\|^2}\right)
$$

for every t with $\mathbf{u}_t \neq 0$.

Since $\mathbf{u}_{t=0} = \mathbf{x}/\|\mathbf{x}\|^2 \neq \vec{0}$, this shows that the function $t \mapsto U_t(\mathbf{x})$ given in (2.34) solves the ODE (2.33) at least until the first time $t > 0$ at which $\mathbf{u}_t = \vec{0}$. Now if we had $\mathbf{u}_{t_0} = \vec{0}$ for some $t_0 > 0$ then we would have $\lim_{t \nearrow t_0} U_t(\mathbf{x}) = \lim_{t \nearrow t_0} \frac{\mathbf{u}_t}{\|\mathbf{u}_t\|^2} = \infty$, i.e., the flowline of μ emanating from $\mathbf{x}$ would run off to ∞ in finite time. But this would contradict our statement in Lemma 2.3 (iii).

2.6.2   Proof of the Bond Price Formula

We are now ready to prove the bond price formula (2.28), based on the well-known Feynman–Kac theorem, whose uniqueness statement (specialized to our specific payoff function[11]) is the following:

---

[11] See Section 5.10.4 for a description of the general case.

Lemma 2.8 (Feynman–Kac). *Suppose that $B_{\$,T}$ is a function that is continuous and bounded on $M_0 \times [0,T]$ and that satisfies the PDE*

$$\forall t \in [0,T) \;\forall \mathbf{x} \in M_0: \; \big(\partial_t + \langle \mu(\mathbf{x}), \nabla \rangle + \tfrac{1}{2}\sigma^2(\mathbf{x},t)\Delta - r(\mathbf{x})\big) B_{\$,T}(\mathbf{x},t) = 0,$$
$$\forall \mathbf{x} \in M_0: \; B_{\$,T}(\mathbf{x},T) = 1. \tag{2.35}$$

Then it must have the representation (2.25).

Proof. See Appendix B.4. □

In order to prove that the expected value (2.25) is given by the explicit expression (2.28), it therefore suffices to show that the functions $B_{\$,T}$ given by (2.28) are continuous and bounded on $M_0 \times [0,T]$, and that they fulfill the Feynman–Kac PDE. We begin with the technical part.

Lemma 2.9. *(i) For any $\tau > 0$ we have*

$$\frac{\mathbf{c}_\tau}{\|\mathbf{c}_\tau\|^2} \notin \overline{M_0}. \tag{2.36}$$

(ii) The norm in (2.28b) is bounded below by some positive constant, uniformly for $\forall (\mathbf{x},\tau) \in \overline{M_0} \times (0,T]$. In particular, this implies that the functions $B_{\$,T}(\mathbf{x},t)$ given in (2.28) are continuous and bounded on $\overline{M_0} \times [0,T]$.

Proof. (i) Suppose that there were a $\tau > 0$ such that $\mathbf{x} := \frac{\mathbf{c}_\tau}{\|\mathbf{c}_\tau\|^2} \in \overline{M_0}$. Then we would have $\frac{\mathbf{x}}{\|\mathbf{x}\|^2} = \mathbf{c}_\tau$, contradicting the second statement in Lemma 2.7.

(ii) Abbreviating expression inside the norm in (2.28b) by $(\dots)$, first note that by (2.3) and (2.7) we have

$$\|\dots\| \geq \langle \dots, \mathbf{v} \rangle$$
$$= \|\mathbf{c}_\tau\| \langle \mathbf{x}, \mathbf{v} \rangle - \langle \mathbf{c}_\tau, \mathbf{v} \rangle / \|\mathbf{c}_\tau\|$$
$$\geq -\|\mathbf{c}_\tau\| \, r_\infty - \langle \mathbf{c}_\tau, \mathbf{v} \rangle / \|\mathbf{c}_\tau\|$$

for all $(\mathbf{x},\tau) \in \overline{M_0} \times (0,T]$. Since by Lemma 2.5 (ii)–(iii) we have $\mathbf{c}_{\tau=0} = \vec{0}$ and $\partial_\tau \mathbf{c}_\tau|_{\tau=0} = -\frac{1}{n-2}\mathbf{v}$, we can use l'Hospital's rule and (2.3) to see that the last ($\mathbf{x}$-independent) expression converges to 1 as $\tau \to 0$. This shows that there $\exists \varepsilon > 0$ such that $\forall (\mathbf{x},\tau) \in \overline{M_0} \times (0,\varepsilon)$: $\|\dots\| \geq \frac{1}{2}$. For $(\mathbf{x},\tau) \in \overline{M_0} \times [\varepsilon,T]$ we can estimate

$$\|\dots\| = \|\mathbf{c}_\tau\| \times \big\|\mathbf{x} - \tfrac{\mathbf{c}_\tau}{\|\mathbf{c}_\tau\|^2}\big\|$$
$$\geq \Big(\min_{\tau \in [\varepsilon,T]} \|\mathbf{c}_\tau\|\Big) \times \mathrm{dist}\Big(\overline{M_0}, \big\{\tfrac{\mathbf{c}_\tau}{\|\mathbf{c}_\tau\|^2} \,\big|\, \tau \in [\varepsilon,T]\big\}\Big),$$

and using Lemma 2.5 (ii) and the continuity of the path $\tau \mapsto \mathbf{c}_\tau$, the first factor (the "min") is positive, while the second one (the "dist") is positive by part (i) of the present lemma (as the distance between a closed and a compact set with empty intersection). □

Moving on to proving that our function satisfies the Feynman–Kac PDE (2.35), first note that—as already mentioned in our discussion in Section 2.5.1— the boundary condition is clearly fulfilled; we can therefore focus on the differential equation itself. Here we will show even more, namely that our proposed solution simultaneously fulfills the two separate equations

$$\Delta B_{\$,T}(\mathbf{x},t) = 0 \qquad \text{and} \qquad \left(\partial_t + \langle \mu(\mathbf{x}), \nabla \rangle - r(\mathbf{x})\right) B_{\$,T}(\mathbf{x},t) = 0 \qquad (2.37)$$

for $\forall t \in [0,T)$ and $\forall \mathbf{x} \in M_0$, which then clearly implies (2.35).

In fact, this split-up of the Feynman–Kac PDE is a necessary consequence of our goal to make our model unspanned with respect to the functional parameter $\sigma(\mathbf{x},t)$: Indeed, note that any single σ-independent function $B_{\$,T}(\mathbf{x},t)$ that fulfills (2.35) for any choice of σ must fulfill the two separate equations in (2.37), as can be seen by simply letting $\sigma \to 0$ in (2.35).

The proofs of the two equations in (2.37) are given in Lemmas 2.10 and 2.11 below; these two lemmas will therefore complete our proof of Theorem 2.1.

Lemma 2.10. *The functions $B_{\$,T}$ defined in (2.28) fulfill the equation*

$$\Delta B_{\$,T}(\mathbf{x},t) = 0 \qquad (2.38)$$

for $\forall t \in [0,T)$ and $\forall \mathbf{x} \in M_0$, i.e., they are harmonic in $\mathbf{x}$.

Proof. For any fixed $t \in [0,T)$ our bond price formula is just a shifted and rescaled version of the function $\mathbf{x} \mapsto \|\mathbf{x}\|^{2-n}$, and so it suffices to show that $\Delta \|\mathbf{x}\|^{2-n} = 0$ for $\forall \mathbf{x} \neq \vec{0}$. This in turn is a well-known fact; we shall show the calculation here for completeness:

$$\partial_{\mathbf{x}_i} \|\mathbf{x}\|^{2-n} = (2-n)\|\mathbf{x}\|^{1-n} \cdot \frac{\mathbf{x}_i}{\|\mathbf{x}\|} = (2-n)\|\mathbf{x}\|^{-n}\mathbf{x}_i$$

$$\partial^2_{\mathbf{x}_i} \|\mathbf{x}\|^{2-n} = (2-n)\left[(\partial_{\mathbf{x}_i}\|\mathbf{x}\|^{-n})\mathbf{x}_i + \|\mathbf{x}\|^{-n}(\partial_{\mathbf{x}_i}\mathbf{x}_i)\right]$$

$$= (2-n)\left[(-n\|\mathbf{x}\|^{-n-2}\mathbf{x}_i)\mathbf{x}_i + \|\mathbf{x}\|^{-n}\right]$$

$$= (2-n)\|\mathbf{x}\|^{-n-2}(-n\mathbf{x}_i^2 + \|\mathbf{x}\|^2)$$

$$\Delta \|\mathbf{x}\|^{2-n} = \sum_{i=1}^{n} \partial^2_{\mathbf{x}_i} \|\mathbf{x}\|^{2-n} = (2-n)\|\mathbf{x}\|^{-n-2}(-n\|\mathbf{x}\|^2 + n\|\mathbf{x}\|^2) = 0$$

$\square$

Lemma 2.11. *The functions $B_{\$,T}$ defined in (2.28) fulfill the equation*

$$\left(\partial_t + \langle \mu(\mathbf{x}), \nabla \rangle - r(\mathbf{x})\right) B_{\$,T}(\mathbf{x},t) = 0 \qquad (2.39)$$

for $\forall t \in [0,T)$ and $\forall \mathbf{x} \in M_0$.

Proof. Throughout this proof let us abbreviate

$$(\ldots) := \|\mathbf{c}_{T-t}\|\mathbf{x} - \frac{\mathbf{c}_{T-t}}{\|\mathbf{c}_{T-t}\|}$$

for $t \in [0, T)$, so that our bond price formula (2.28b) can be written as

$$B_{\$,T}(\mathbf{x},t) = e^{r_\infty(t-T)} \| \ldots \|^{2-n}.$$

(Recall that $(\ldots) \neq \vec{0}$ by Lemma 2.9 (ii).) Then we have

$$\big(\partial_t + \langle \mu(\mathbf{x}), \nabla \rangle\big) B_{\$,T}(\mathbf{x},t)$$
$$= e^{r_\infty(t-T)} \Big[r_\infty \| \ldots \|^{2-n} + (2-n) \| \ldots \|^{1-n} \Big\langle \tfrac{(\ldots)}{\|\ldots\|}, \big(\partial_t + \langle \mu(\mathbf{x}), \nabla \rangle\big)(\ldots) \Big\rangle \Big]$$
$$= \Big[r_\infty + \tfrac{2-n}{\|\ldots\|^2} \big\langle (\ldots), \big(\partial_t + \langle \mu(\mathbf{x}), \nabla \rangle\big)(\ldots) \big\rangle \Big] B_{\$,T}(\mathbf{x},t). \tag{2.40}$$

Since by (2.30b) we have $\partial_t \mathbf{c}_{T-t} = \frac{1}{n-2}\mathbf{v} + (E + \lambda I)\mathbf{c}_{T-t}$, we find that

$$\big(\partial_t + \langle \mu(\mathbf{x}), \nabla \rangle\big)(\ldots)$$
$$= \nabla_{\mathbf{c}}\Big[\|\mathbf{c}\|\mathbf{x} - \tfrac{\mathbf{c}}{\|\mathbf{c}\|} \Big]_{\mathbf{c}=\mathbf{c}_{T-t}} \cdot \partial_t \mathbf{c}_{T-t} + \nabla_{\mathbf{x}}\Big[\|\mathbf{c}_{T-t}\|\mathbf{x} - \tfrac{\mathbf{c}_{T-t}}{\|\mathbf{c}_{T-t}\|} \Big] \cdot \mu(\mathbf{x})$$
$$= \Big[\tfrac{\mathbf{x}\mathbf{c}^T}{\|\mathbf{c}\|} - \tfrac{1}{\|\mathbf{c}\|}\Big(I - \tfrac{\mathbf{c}\mathbf{c}^T}{\|\mathbf{c}\|^2} \Big) \Big]_{\mathbf{c}=\mathbf{c}_{T-t}} \cdot \Big(\tfrac{1}{n-2}\mathbf{v} + (E + \lambda I)\mathbf{c}_{T-t} \Big)$$
$$\qquad\qquad + \|\mathbf{c}_{T-t}\| I \cdot \big((-\lambda I + E)\mathbf{x} + \tfrac{1}{n-2}\|\mathbf{x}\|^2 \mathbf{v} - \tfrac{2}{n-2}\langle \mathbf{v}, \mathbf{x} \rangle \mathbf{x} \big)$$
$$= E\big(\|\mathbf{c}\|\mathbf{x} - \tfrac{\mathbf{c}}{\|\mathbf{c}\|} \big)$$
$$\qquad + \tfrac{1}{n-2}\Big(\tfrac{\mathbf{x}\mathbf{c}^T}{\|\mathbf{c}\|} - \tfrac{1}{\|\mathbf{c}\|}\Big(I - \tfrac{\mathbf{c}\mathbf{c}^T}{\|\mathbf{c}\|^2} \Big) + \|\mathbf{c}\|\|\mathbf{x}\|^2 I - 2\|\mathbf{c}\|\mathbf{x}\mathbf{x}^T \Big)\mathbf{v} \Big|_{\mathbf{c}=\mathbf{c}_{T-t}},$$

where in the last step the terms containing λ canceled out, and two terms containing E vanished because of Lemma 2.2 (ii). Now computing the inner product in (2.40) and abbreviating $\mathbf{c} := \mathbf{c}_{T-t}$, the remaining term containing E vanishes again by Lemma 2.2 (ii), and we obtain

$$\big\langle (\ldots), \big(\partial_t + \langle \mu(\mathbf{x}), \nabla \rangle\big)(\ldots) \big\rangle$$
$$= \tfrac{1}{n-2}\Big\langle \|\mathbf{c}\|\mathbf{x} - \tfrac{\mathbf{c}}{\|\mathbf{c}\|}, \Big(\tfrac{\mathbf{x}\mathbf{c}^T}{\|\mathbf{c}\|} - \tfrac{1}{\|\mathbf{c}\|}\Big(I - \tfrac{\mathbf{c}\mathbf{c}^T}{\|\mathbf{c}\|^2} \Big) + \|\mathbf{c}\|\|\mathbf{x}\|^2 I - 2\|\mathbf{c}\|\mathbf{x}\mathbf{x}^T \Big)\mathbf{v} \Big\rangle$$
$$= \tfrac{1}{n-2}\Big\langle \Big(\tfrac{\mathbf{c}\mathbf{x}^T}{\|\mathbf{c}\|} - \tfrac{1}{\|\mathbf{c}\|}\Big(I - \tfrac{\mathbf{c}\mathbf{c}^T}{\|\mathbf{c}\|^2} \Big) + \|\mathbf{c}\|\|\mathbf{x}\|^2 I - 2\|\mathbf{c}\|\mathbf{x}\mathbf{x}^T \Big)\big(\|\mathbf{c}\|\mathbf{x} - \tfrac{\mathbf{c}}{\|\mathbf{c}\|} \big), \mathbf{v} \Big\rangle$$
$$= \tfrac{1}{n-2}\Big\langle \|\mathbf{x}\|^2 \mathbf{c} - \Big(\mathbf{x} - \tfrac{\langle \mathbf{c}, \mathbf{x} \rangle}{\|\mathbf{c}\|^2}\mathbf{c} \Big) + \|\mathbf{c}\|^2\|\mathbf{x}\|^2 \mathbf{x} - 2\|\mathbf{c}\|^2\|\mathbf{x}\|^2 \mathbf{x}$$
$$\qquad\qquad - \tfrac{\langle \mathbf{x}, \mathbf{c} \rangle}{\|\mathbf{c}\|^2}\mathbf{c} + \tfrac{\mathbf{c}}{\|\mathbf{c}\|^2} - \tfrac{\mathbf{c}}{\|\mathbf{c}\|^2} - \|\mathbf{x}\|^2 \mathbf{c} + 2\langle \mathbf{x}, \mathbf{c} \rangle \mathbf{x}, \mathbf{v} \Big\rangle$$
$$= -\tfrac{1}{n-2}\big(\|\mathbf{c}\|^2\|\mathbf{x}\|^2 - 2\langle \mathbf{x}, \mathbf{c} \rangle + 1 \big)\langle \mathbf{x}, \mathbf{v} \rangle$$
$$= -\tfrac{1}{n-2}\Big\| \|\mathbf{c}\|\mathbf{x} - \tfrac{\mathbf{c}}{\|\mathbf{c}\|} \Big\|^2 \langle \mathbf{x}, \mathbf{v} \rangle$$
$$= \tfrac{1}{2-n}\| \ldots \|^2 \langle \mathbf{v}, \mathbf{x} \rangle.$$

Finally, plugging this back into (2.40), we find that

$$\big(\partial_t + \langle \mu(\mathbf{x}), \nabla \rangle\big) B_{\$,T}(\mathbf{x},t) = \big(r_\infty + \langle \mathbf{v}, \mathbf{x} \rangle \big) B_{\$,T}(\mathbf{x},t) = r(\mathbf{x}) B_{\$,T}(\mathbf{x},t),$$

which is (2.39). $\qquad\square$

2.7   Additional Parameter Constraints

2.7.1   Gauge Fixing

The parameter constraints listed in Section 2.1.1 were sufficient to give the model all its desired properties, and to ensure that the bond price and forward rate formulas presented in Section 2.5 are valid. However, as is, there is still a considerable level of redundancy in the model parameters, with different choices of our parameters leading to the same exact short rate process $(r(X_t))_{t \geq 0}$.

To see this, observe that the value of a state $\mathbf{x}$ is never used *by itself*; we exclusively use it as an argument to the short rate function $r(\mathbf{x})$, from which then all instrument prices of interest can be obtained, such as the bond prices (2.25), and the prices of all dependent derivatives. As a result, given any short rate model such as ours, and given any invertible function $g \colon \mathbb{R}^n \to \mathbb{R}^n$, we could instead consider the process $\tilde{X}_t := g(X_t)$ (with an SDE derived using Itô's Lemma) and the short rate function $\tilde{r}(\tilde{\mathbf{x}}) := r(g^{-1}(\tilde{\mathbf{x}}))$; since $\tilde{r}(\tilde{X}_t) \equiv r(X_t)$, we would then arrive at a model that is identical for all practical purposes.[12]

Multiplication by orthogonal matrices. In our case, the reparametrization $g(\mathbf{x}) := A\mathbf{x}$, for any orthogonal matrix $A \in \mathbb{R}^{n \times n}$, leads us to the short rate model with

$$
\begin{aligned}
\tilde{\mu}(\tilde{\mathbf{x}}) &= A\mu(\mathbf{x}) = A\mu(A^T \tilde{\mathbf{x}}) \\
&= A\big[(-\lambda I + E)A^T \tilde{\mathbf{x}} + \tfrac{1}{n-2}\|A^T \tilde{\mathbf{x}}\|^2 \mathbf{v} - \tfrac{2}{n-2}\langle \mathbf{v}, A^T \tilde{\mathbf{x}}\rangle A^T \tilde{\mathbf{x}}\big] \\
&= (-\lambda I + AEA^T)\tilde{\mathbf{x}} + \tfrac{1}{n-2}\|\tilde{\mathbf{x}}\|^2 A\mathbf{v} - \tfrac{2}{n-2}\langle A\mathbf{v}, \tilde{\mathbf{x}}\rangle \tilde{\mathbf{x}}, \\
\tilde{\sigma}(\tilde{\mathbf{x}}, t) &= \sigma(\mathbf{x}, t) = \sigma(A^T \tilde{\mathbf{x}}, t), \\
\tilde{r}(\tilde{\mathbf{x}}) &= r(A^T \tilde{\mathbf{x}}) = r_\infty + \langle \mathbf{v}, A^T \tilde{\mathbf{x}}\rangle = r_\infty + \langle A\mathbf{v}, \tilde{\mathbf{x}}\rangle,
\end{aligned}
$$

i.e., it leads us back to the ACE model with the modified parameters

$$
E^A := AEA^T, \quad \mathbf{v}^A := A\mathbf{v}, \quad \mathbf{x}_0^A := A\mathbf{x}_0, \quad \text{and} \quad \sigma^A(\mathbf{x}, t) := \sigma(A^T\mathbf{x}, t).
$$

This shows that the parameter sets $(r_\infty, E^A, \mathbf{v}^A, \lambda, \mathbf{x}_0^A, \sigma^A)$ all correspond to the same short rate process.

Parametric redundancies like this one can generally lead to instabilities in the calibration code, since that code does not have a *unique* optimal parameter set to converge to. In order to achieve stable calibration, it is therefore essential to break all symmetries in the parameter space, by imposing additional parameter constraints that within each parameter set equivalence class are only fulfilled by exactly one member. When designing these constraints, one can take advantage of the freedom to choose this member, by picking the one that simplifies the model's analytical expressions and/or its numerics the most. This process is called *gauge fixing*.

---

[12]The bond price function of the reparametrized model will then given by $\tilde{B}_{\$,T}(\tilde{\mathbf{x}}, t) = B_{\$,T}(g^{-1}(\tilde{\mathbf{x}}), t)$.

In the ACE model, one smart way of breaking the symmetry we exposed above can be derived from the following lemma.

Lemma 2.12. *Given any antisymmetric matrix $E \in \mathbb{R}^{n \times n}$ and any vector $\mathbf{v} \in \mathbb{R}^n$, there exists an orthogonal matrix $A \in \mathbb{R}^{n \times n}$ such that $E^A := AEA^T$ and $\mathbf{v}^A := A\mathbf{v}$ are of the form*

$$E^A = \begin{pmatrix} 0 & \eta_1 & & & \\ -\eta_1 & 0 & & \mathbf{0} & \\ & & \ddots & & \\ \mathbf{0} & & & 0 & \eta_{\lfloor n/2 \rfloor} \\ & & & -\eta_{\lfloor n/2 \rfloor} & 0 \end{pmatrix} \quad or \quad \begin{pmatrix} 0 & \eta_1 & & & & \\ -\eta_1 & 0 & & & \mathbf{0} & \\ & & \ddots & & & \\ & & & 0 & \eta_{\lfloor n/2 \rfloor} & \\ \mathbf{0} & & & -\eta_{\lfloor n/2 \rfloor} & 0 & \\ & & & & & 0 \end{pmatrix}$$

(2.41)

for even or odd values of n, respectively,[13] and

$$\mathbf{v}^A = \begin{cases} (v_1, 0, v_2, 0, \ldots, v_{\lceil n/2 \rceil}, 0) & \textit{for even } n, \\ (v_1, 0, v_2, 0, \ldots, v_{\lceil n/2 \rceil}) & \textit{for odd } n, \end{cases}$$

(2.42)

for some scalars

$$\eta_1 \geq \cdots \geq \eta_{\lfloor n/2 \rfloor} \geq 0 \qquad and \qquad v_1, \ldots, v_{\lceil n/2 \rceil} \geq 0. \tag{2.43}$$

Proof. The proof is based on the well-known diagonalization theorem for Hermitian matrices, applied to the matrix iE. See Appendix B.5 for details. □

According to this lemma, within each parameter set equivalence class one can find one member for which E and $\mathbf{v}$ are of the form (2.41)–(2.43), in addition to satisfying the constraint (2.3), i.e.,

$$v_1^2 + \cdots + v_{\lceil n/2 \rceil}^2 = 1. \tag{2.44}$$

In fact, except in degenerate cases the proof of Lemma 2.12 implies that this member is unique: Only if some values η_i coincide then to achieve uniqueness one would also have to enforce an order on the corresponding values v_i; if those coincide as well or if some value v_i vanishes, then one would have to impose additional constraints on $\mathbf{x}_0$.

This shows that we can assume our model parameters E and $\mathbf{v}$ to be of the form (2.41)–(2.44) without losing any generality in the model, thereby removing all its parametric redundancies except for some degenerate cases. This reduces the parameter space formed by all valid choices of r_∞, E, $\mathbf{v}$, λ, and $\mathbf{x}_0$ to

$$\# \text{ degrees of freedom} = 1 + \lfloor n/2 \rfloor + (\lceil n/2 \rceil - 1) + 1 + n$$
$$= 2n + 1 \tag{2.45}$$

---

[13]The matrices in (2.41) have (2×2)-matrices on the diagonal, with all other elements being zero; in other words, every other entry on the two off-diagonals vanishes. In particular, we see that the action on $\mathbb{R}^n$ of any antisymmetric matrix $E \in \mathbb{R}^{n \times n}$ decouples into its actions on certain two-dimensional subspaces of $\mathbb{R}^n$.

degrees of freedom that can be used to calibrate the discount curve; the additional functional parameter $\sigma(\mathbf{x}, t)$ then allows us to also calibrate to the swaption market.

Multiplication by a scalar. In fact, the model as we have defined it in Section 2.1 has already been the result of gauge fixing of a different kind: Suppose that we had defined our state space process $(X_t)_{t \geq 0}$ and our short rate function $r(\mathbf{x})$ exactly as before, via (2.10)–(2.12), but that we had more generally allowed for any weight vector $\mathbf{v} \in \mathbb{R}^n \setminus \{\vec{0}\}$, without the constraint (2.3). (This scenario will in fact be relevant later in Chapter 3 when we discuss the ACE model's multi-currency generalization.)

Then the resulting model with parameters r_∞, E, λ, $\mathbf{v}$, $\mathbf{x}_0$ and $\sigma(\mathbf{x}, t)$ can be seen to be equivalent to the ACE model for another parameter set that does satisfy the constraint (2.3), namely to the one with the same parameters r_∞, E, and λ, but with

$$\tilde{\mathbf{v}} := \frac{\mathbf{v}}{\|\mathbf{v}\|}, \quad \tilde{\mathbf{x}}_0 := \|\mathbf{v}\| \cdot \mathbf{x}_0, \quad \text{and} \quad \tilde{\sigma}(\tilde{\mathbf{x}}, t) := \|\mathbf{v}\| \cdot \sigma(\|\mathbf{v}\|^{-1}\tilde{\mathbf{x}}, t); \quad (2.46)$$

it was for the purpose of eliminating this parametric redundancy that we had introduced the constraint (2.3).

Indeed, given this generalized ACE process $(X_t)_{t \geq 0}$, the process $\tilde{X}_t := g(X_t)$ for $g(\mathbf{x}) := \|\mathbf{v}\| \cdot \mathbf{x}$ would have the starting point $\tilde{\mathbf{x}}_0$, and denoting $\tilde{\mathbf{x}} := \|\mathbf{v}\|\mathbf{x}$, by Itô's formula it would satisfy the SDE with drift and noise function

$$\tilde{\mu}(\tilde{\mathbf{x}}) = \nabla g(\mathbf{x}) \cdot \mu(\mathbf{x}) = \|\mathbf{v}\|\mu(\mathbf{x}) \quad (2.47)$$
$$= \|\mathbf{v}\|\left[(-\lambda I + E)\mathbf{x} + \tfrac{1}{n-2}\|\mathbf{x}\|^2\mathbf{v} - \tfrac{2}{n-2}\langle\mathbf{v}, \mathbf{x}\rangle\mathbf{x}\right]$$
$$= (-\lambda I + E)\tilde{\mathbf{x}} + \tfrac{1}{n-2}\|\tilde{\mathbf{x}}\|^2\tilde{\mathbf{v}} - \tfrac{2}{n-2}\langle\tilde{\mathbf{v}}, \tilde{\mathbf{x}}\rangle\tilde{\mathbf{x}},$$
$$\tilde{\sigma}(\tilde{\mathbf{x}}, t) = \nabla g(\mathbf{x}) \cdot \sigma(\mathbf{x}, t) = \|\mathbf{v}\|\sigma(\|\mathbf{v}\|^{-1}\tilde{\mathbf{x}}, t),$$

i.e., it would satisfy the ACE SDE (2.10)–(2.11) with the parameters (2.46); since the ACE short rate function associated to these parameters can be written as

$$\tilde{r}(\tilde{\mathbf{x}}) = r_\infty + \langle\tilde{\mathbf{v}}, \tilde{\mathbf{x}}\rangle = r_\infty + \langle\mathbf{v}, \|\mathbf{v}\|^{-1}\tilde{\mathbf{x}}\rangle = r(\|\mathbf{v}\|^{-1}\tilde{\mathbf{x}}) = r(g^{-1}(\tilde{\mathbf{x}})),$$

we have $r(X_t) = r(g^{-1}(\tilde{X}_t)) = \tilde{r}(\tilde{X}_t)$, i.e., the generalized ACE short rate process $r(X_t)$ indeed coincides with the one associated to the parameters (2.46).

Of course, for the model with such a generalized weight vector $\mathbf{v} \in \mathbb{R}^n \setminus \{\vec{0}\}$ to have all the properties that we need, we would have to ask that its parameters are such that the equivalent rescaled ACE model given by (2.46) satisfies also all the other parameter constraints, namely (2.9a) and the constraints listed in Lemma 2.4. In other words, we would need to ask that

$$\delta := r_\infty\lambda - \tfrac{1}{n-2}r_\infty^2 - \tfrac{n-2}{4}\frac{\|E\mathbf{v}\|^2}{\|\mathbf{v}\|^2} \geq 0, \quad (2.48)$$

and that the function $\tilde{\sigma}(\tilde{\mathbf{x}}, t)$ satisfies the constraints of Lemma 2.4.

The bond price formula (2.28) then still holds for this generalized model as is: Indeed, by (2.29) and (2.46) the currency vectors for the generalized model are

$$\tilde{\mathbf{c}}_\tau = \frac{\mathbf{c}_\tau}{\|\mathbf{v}\|},$$

and so since the bond price formula was shown to hold for the rescaled model with the parameters (2.46), the bond prices for the generalized model are

$$
\begin{aligned}
B_{\$,T}(\mathbf{x},t) &= \mathbb{E}\big[\exp\big(-\textstyle\int_t^T r(X_s)\,\mathrm{d}s\big)\,\big|\, X_t = \mathbf{x}\big] \\
&= \mathbb{E}\big[\exp\big(-\textstyle\int_t^T \tilde{r}(\tilde{X}_s)\,\mathrm{d}s\big)\,\big|\, \tilde{X}_t = \tilde{\mathbf{x}}\big] \\
&= B_{\$,T}^{\text{rescaled}}(\tilde{\mathbf{x}},t) \\
&= \mathrm{e}^{-r_\infty\tau}\big(1 - 2\langle\tilde{\mathbf{c}}_\tau,\tilde{\mathbf{x}}\rangle + \|\tilde{\mathbf{c}}_\tau\|^2\|\tilde{\mathbf{x}}\|^2\big)^{1-n/2} \\
&= \mathrm{e}^{-r_\infty\tau}\big(1 - 2\langle\mathbf{c}_\tau,\mathbf{x}\rangle + \|\mathbf{c}_\tau\|^2\|\mathbf{x}\|^2\big)^{1-n/2},
\end{aligned}
$$

which is the original formula (2.28) again.

Furthermore, the forward rate formula (2.32) and the zero-noise evolution map formula (2.34) hold as is as well, since their respective proofs found in Lemma 2.6 and in Section 2.6.1 do not rely on the constraint $\|\mathbf{v}\| = 1$.

In summary: While one can generalize the model description to allow for models with any weight vector $\mathbf{v} \in \mathbb{R}^n \setminus \{\vec{0}\}$ (without having to modify the bond price, forward rate, and zero-noise evolution map formulas), this would not yield any new short rate processes, and so it would not be of any use in the single-currency case. However, for the multi-currency ACE model defined in Chapter 3 this insight will turn out to be important, as that model will have multiple weight vectors $\mathbf{v}^c$ as parameters (one for each currency c), only one of which one can constrain to have unit length with the arguments laid out above.

2.7.2 Matrix & Currency Vector Formulas

As a result of our specific choice of constraint, the matrices e^{tE} and $(E + \lambda I)^{-1}$ and the currency vectors $\mathbf{c}_\tau$ that occur repeatedly in our formulas in Section 2.5 can be computed analytically, without having to resort to the exponential power series or to any linear equation solver. Indeed, the necessary calculations decouple into those for the individual (2×2)-matrices of the form $E_\eta = \left(\begin{smallmatrix} 0 & \eta \\ -\eta & 0 \end{smallmatrix}\right)$ on the diagonal, and (for the last diagonal element if n is odd) for the (1×1)-matrix $E' = (0)$. For these basic building blocks, the exponential and the resolvent matrix are given by

$$
\mathrm{e}^{tE_\eta} = \begin{pmatrix} \cos(\eta t) & \sin(\eta t) \\ -\sin(\eta t) & \cos(\eta t) \end{pmatrix} \qquad \text{and} \qquad \mathrm{e}^{tE'} = (1), \qquad (2.49)
$$

as well as

$$
(E_\eta + \lambda I)^{-1} = \frac{1}{\lambda^2 + \eta^2}\begin{pmatrix} \lambda & -\eta \\ \eta & \lambda \end{pmatrix} \qquad \text{and} \qquad (E' + \lambda I)^{-1} = \left(\tfrac{1}{\lambda}\right). \qquad (2.50)
$$

Since by (2.42) we also assume that the subvector of $\mathbf{v}$ on which these matrices act in (2.29) are of the form $\binom{v}{0}$ or (v), respectively, the currency vectors are therefore composed of blocks of the form

$$
\mathbf{c}_\tau = \frac{1}{(n-2)(\lambda^2+\eta^2)} \begin{pmatrix} \lambda & -\eta \\ \eta & \lambda \end{pmatrix} \left[e^{-\lambda\tau} \begin{pmatrix} \cos(\eta\tau) & \sin(\eta\tau) \\ -\sin(\eta\tau) & \cos(\eta\tau) \end{pmatrix} - \begin{pmatrix} 1 & 0 \\ 0 & 1 \end{pmatrix} \right] \begin{pmatrix} v \\ 0 \end{pmatrix}
$$

$$
= \frac{v}{(n-2)(\lambda^2+\eta^2)} \begin{pmatrix} \lambda & -\eta \\ \eta & \lambda \end{pmatrix} \begin{pmatrix} e^{-\lambda\tau}\cos(\eta\tau)-1 \\ -e^{-\lambda\tau}\sin(\eta\tau) \end{pmatrix} \quad \text{and} \tag{2.51a}
$$

$$
\mathbf{c}_\tau = \left(\tfrac{(e^{-\lambda\tau}-1)v}{(n-2)\lambda} \right). \tag{2.51b}
$$

2.8   Analysis

Let us now further deepen our understanding of the model by studying the properties of our various analytical expressions.

2.8.1   Group Property

Since our family of maps U_t given by the formula (2.34) describe the flow of the time-independent ODE (2.33), we know that these maps must satisfy the semigroup property

$$
U_{t_1} \circ U_{t_2} = U_{t_1+t_2} \tag{2.52}
$$

on M_0 for $\forall t \geq 0$. In fact, by extending our space to $\bar{\mathbb{R}}^n := \mathbb{R}^n \cup \{\infty\}$, adopting the convention

$$
\frac{\vec{0}}{\|\vec{0}\|^2} = \infty \qquad \text{and} \qquad \frac{\infty}{\|\infty\|^2} = \vec{0},
$$

and in this way interpreting the maps in (2.34) as a functions $U_t \colon \bar{\mathbb{R}}^n \to \bar{\mathbb{R}}^n$, we should even expect this relation to hold on all of $\bar{\mathbb{R}}^n$ and for $\forall t \in \mathbb{R}$. However, this is hard to see from (2.34), whose design is aiming for brevity stability.

To independently confirm the group property for these maps based on this formula alone, we can use (2.29) and abbreviate

$$
\hat{\mathbf{v}} := \tfrac{1}{n-2}(E+\lambda I)^{-1}\mathbf{v} \tag{2.53}
$$

to rewrite (2.34) as

$$
U_t(\mathbf{x}) = \frac{e^{t(E+\lambda I)}\left(\frac{\mathbf{x}}{\|\mathbf{x}\|^2}+\hat{\mathbf{v}}\right)-\hat{\mathbf{v}}}{\left\|e^{t(E+\lambda I)}\left(\frac{\mathbf{x}}{\|\mathbf{x}\|^2}+\hat{\mathbf{v}}\right)-\hat{\mathbf{v}}\right\|^2} . \tag{2.54}
$$

This formulation shows that the maps U_t are of the form

$$
U_t = i^{-1} \circ s^{-1} \circ u_t \circ s \circ i, \tag{2.55}
$$

where

$$
i(\mathbf{x}) := \tfrac{\mathbf{x}}{\|\mathbf{x}\|^2}, \qquad s(\mathbf{x}) := \mathbf{x}+\hat{\mathbf{v}}, \qquad u_t(\mathbf{x}) := e^{t(E+\lambda I)}\mathbf{x}. \tag{2.56}
$$

The group property (2.52) of the maps U_t therefore follows from the group property of the maps u_t.

2.8.2 Limiting Behavior

Evolution maps. Next, let us understand the behavior of the evolution maps U_t as $t \to \pm\infty$, or in other words, let us find the attractors and repellers of the drift vector field μ. It is easiest to see these limits by looking at our alternative formula (2.54).

Lemma 2.13. *Let us define*

$$\mathbf{x}_{-\infty} := -\frac{\hat{\mathbf{v}}}{\|\hat{\mathbf{v}}\|^2} \,. \tag{2.57}$$

Then the following three statements hold:

(i) We have $\mathbf{x}_{-\infty} \in \mathbb{R}^n \setminus M_0$.

(ii) The maps U_t, $t \neq 0$, all have the points $\vec{0}$ and $\mathbf{x}_{-\infty}$ as their only fixed points.

(iii) For all other points[14] $\forall \mathbf{x} \in \bar{\mathbb{R}}^n$ we have

$$\lim_{t \to \infty} U_t(\mathbf{x}) = \vec{0} \qquad and \qquad \lim_{t \to -\infty} U_t(\mathbf{x}) = \mathbf{x}_{-\infty} \,. \tag{2.58}$$

In particular, together with parts (i) and (ii) this shows that

$$\forall \mathbf{x} \in M_0: \quad \lim_{t \to \infty} U_t(\mathbf{x}) = \vec{0}. \tag{2.59}$$

Proof. (i) First, note that we have $\mathbf{x}_{-\infty} \neq \infty$ since $\hat{\mathbf{v}} \neq \vec{0}$ by (2.53) and (2.3). By (2.7) the statement then follows from the estimate

$$\langle \mathbf{v}, \mathbf{x}_{-\infty} \rangle \quad = \quad \left\langle \mathbf{v}, -\frac{\hat{\mathbf{v}}}{\|\hat{\mathbf{v}}\|^2} \right\rangle = -\frac{\langle \mathbf{v}, \hat{\mathbf{v}} \rangle}{\|\hat{\mathbf{v}}\|^2}$$

$$\overset{(2.53)}{=} \quad -\frac{\langle (n-2)(E + \lambda I)\hat{\mathbf{v}}, \hat{\mathbf{v}} \rangle}{\|\hat{\mathbf{v}}\|^2}$$

$$\overset{\text{L.2.2 (ii)}}{=} -(n-2)\lambda \overset{(2.13)}{\leq} -r_\infty \,,$$

(iii) Using that the matrix e^{tE} is orthogonal by Lemma 2.2 (iii), the property Lemma 2.1 (iii) applied to $A = e^{tE}$, and that we have $\frac{\mathbf{x}}{\|\mathbf{x}\|^2} + \hat{\mathbf{v}} = \vec{0}$ if and only if $\mathbf{x} = \mathbf{x}_{-\infty}$, the limits (2.58) can easily be seen from (2.54).

(ii) Using (2.54), it is easy to check that the given points are indeed fixed points. The fact that they are the *only* fixed points now follows from part (iii) and the group property (2.52) of the maps U_t: Since every fixed point $\mathbf{x}$ of a given map U_t, $t \neq 0$, must fulfill $U_{kt}(\mathbf{x}) = (U_t \circ \cdots \circ U_t)(\mathbf{x}) = \mathbf{x}$ for $\forall k \in \mathbb{Z}$, taking $k \to \pm\infty$ shows that $\mathbf{x}$ must coincide with one of the limits found in part (iii). $\square$

---

[14]Naturally, the fixed points from part (ii) fulfill $\lim_{t \to \pm\infty} U_t(\vec{0}) = \vec{0}$ and $\lim_{t \to \pm\infty} U_t(\mathbf{x}_{-\infty}) = \mathbf{x}_{-\infty}$.

Forward rates. As a result of the limit (2.59) we obtain the following formula for the long-term forward rate.

Corollary 2.2. *The long-term forward rate is given by*

$$\lim_{\tau \to \infty} F_{\mathrm{rel}}(\mathbf{x}; \tau) = r_\infty \qquad (2.60)$$

for $\forall \mathbf{x} \in M_0$, and so the forward rate curves simulated by the ACE model have this fixed long-term rate almost surely.

Proof. This limit now follows from (2.32), (2.59), and (2.12):

$$\lim_{\tau \to \infty} F_{\mathrm{rel}}(\mathbf{x}; \tau) = \lim_{\tau \to \infty} r(U_\tau(\mathbf{x})) = r(\vec{0}) = r_\infty. \qquad \square$$

It is remarkable that the long-term forward rate is independent of the state $\mathbf{x}$, i.e., that it is an invariant of the process. This property can be seen as a weakness of the ACE model, as it limits the model's dynamic flexibility. On the positive side, one can exploit this insight during the calibration, by making the expression on the right of (2.60) coincide with the long-term forward rate presently observed in the market. To further add to our understanding of the ACE model's parameters, let us therefore also compute the next higher-order terms.

Lemma 2.14. *To higher-order accuracy, the limiting behavior of the relative forward rate as $\tau \to \infty$ is given by*

$$F_{\mathrm{rel}}(\mathbf{x}; \tau) = \begin{cases} r_\infty & \text{if } \mathbf{x} = \vec{0}, \\ r_\infty + \mathrm{e}^{-\lambda\tau}\langle \mathbf{v}, \mathrm{e}^{\tau E}\mathbf{z}\rangle + O(\mathrm{e}^{-2\lambda\tau}) & \text{if } \mathbf{x} \neq \vec{0}, \end{cases} \qquad (2.61)$$

where

$$\mathbf{z} := \frac{\frac{\mathbf{x}}{\|\mathbf{x}\|^2} + \hat{\mathbf{v}}}{\left\|\frac{\mathbf{x}}{\|\mathbf{x}\|^2} + \hat{\mathbf{v}}\right\|^2}.$$

In particular, this shows that while r_∞ determines the actual long-term forward rate, λ controls the convergence speed of the relative forward rate curve to this limit, and E encodes the frequencies of the oscillating behavior along the way.

Proof. This formula again follows from (2.32) and (2.12), by considering only the leading-order terms of $U_\tau(\mathbf{x})$ in the expression (2.54) for large τ. The case $\mathbf{x} = \vec{0}$ is then trivial since $U_\tau(\vec{0}) = \vec{0}$ for $\forall \tau \in \mathbb{R}$.

For $\mathbf{x} \neq \vec{0}$, as $\tau \to \infty$ the respective first terms in both the numerator and the denominator in (2.54) dominate the respective second term $\hat{\mathbf{v}}$ (recall that e^{tE} is orthogonal), and so we have

$$U_\tau(\mathbf{x}) \approx \frac{\mathrm{e}^{\tau(E+\lambda I)}\left(\frac{\mathbf{x}}{\|\mathbf{x}\|^2} + \hat{\mathbf{v}}\right)}{\left\|\mathrm{e}^{\tau(E+\lambda I)}\left(\frac{\mathbf{x}}{\|\mathbf{x}\|^2} + \hat{\mathbf{v}}\right)\right\|^2} = \mathrm{e}^{-\lambda\tau}\mathrm{e}^{\tau E}\mathbf{z},$$

with the remaining terms being of the order $O(\mathrm{e}^{-2\lambda\tau})$. $\square$

2.9 The One-Dimensional ACE Model

So far we have restricted our description of the ACE model to the dimensions $n \geq 3$. The two-dimensional case has to be excluded because the division by $n - 2$ in (2.11) would make our drift vector field $\mu(\mathbf{x})$ ill defined.

In contrast, the one-dimensional case was only excluded so far since in this case our parameter constraint (2.9a) no longer allows us to ensure that our state process is confined to its intended domain M_0;[15] aside from that, during the proofs of our various formulas in the preceding sections we have not encountered any serious problems with the specific value $n = 1$. Luckily, this problem can easily be addressed with an alternative parameter constraint that is tailored specifically to the one-dimensional case; in fact, confining the process to M_0 becomes even easier in one dimension since the boundary ∂M_0 then only consists of two points.[16]

Let us therefore now go ahead and see what our equations reduce to when we set $n = 1$, and let us then find the right parameter constraints for this case. As we will see, we will arrive at a model that is reminiscent of the classic Vašíček model, but with an additional higher-order drift term, with an arbitrary state- and time-dependent noise function $\sigma(\mathbf{x}, t)$, and with both a lower and (in contrast to the ACE model for $n \geq 3$) also an upper bound imposed on the short rate process. See also the result by Gabaix in [8, Eqn. (30)].

2.9.1 Reduction of the Standard ACE Equations

Model dynamics. In one dimension, the only matrix $E \in \mathbb{R}^{1 \times 1}$ satisfying (2.4) is the zero matrix $E = (0)$, and the only vector $\mathbf{v} \in \mathbb{R}^1$ satisfying the constraints (2.3) and (2.43) is $\mathbf{v} = 1$. The drift μ and the short rate function r in (2.11)–(2.12) therefore simplify to

$$\mu(\mathbf{x}) = -\lambda \mathbf{x} + \mathbf{x}^2 = \mathbf{x}(\mathbf{x} - \lambda), \tag{2.62}$$

$$r(\mathbf{x}) = r_\infty + \mathbf{x}. \tag{2.63}$$

Parameter constraints, state space. To find an alternative for our constraints (2.9a)–(2.9b) on λ and r_∞ (and to rethink also the necessity of the constraints (2.2) and (2.5)), first note that the case $\lambda = 0$ can be excluded, since for this parameter value the drift μ in (2.62) no longer has a stable attractor. Furthermore, the model given by any parameter set $(\lambda, r_\infty, \mathbf{x}_0, \sigma)$ is equivalent to the model for the parameter set $(-\lambda, r_\infty + \lambda, \mathbf{x}_0 - \lambda, \tilde{\sigma})$, where $\tilde{\sigma}(\tilde{\mathbf{x}}, t) := \sigma(\tilde{\mathbf{x}} + \lambda, t)$, since the state processes $(X_t)_{t \geq 0}$ and $(\tilde{X}_t)_{t \geq 0}$ of the two mod-

---

[15]Indeed, the prefactor $\frac{1}{n-2}$ in the formula (2.16) is negative if $n = 1$, which causes our proof of Lemma 2.3 (ii) to break down.

[16]The only other slight issue in one dimension is that our proof of the boundedness of the bond price function in Lemma 2.9 no longer applies since the exponent in (2.28) is no longer negative. However, checking this technical condition will become trivial for $n = 1$.

els are related via $\tilde{X}_t = X_t - \lambda$[17] and therefore fulfill $\tilde{r}(\tilde{X}_t) = (r_\infty + \lambda) + \tilde{X}_t = r_\infty + X_t = r(X_t)$; this shows that we may decide (as before in the higher-dimensional case) to only consider the parameter values

$$\lambda > 0 \tag{2.64}$$

without losing any generality.

Under this constraint, the drift μ has two roots: the attractor $\mathbf{x} = 0$ (with basin of attraction $(-\infty, \lambda)$), and the repeller $\mathbf{x} = \lambda > 0$ that drives all solutions of the zero-noise process starting from a point $\mathbf{x}_0 > \lambda$ to ∞ in finite time.[18] We therefore find that we need to constrain our process to a state space $M_0 = (\mathbf{x}_{\min}, \mathbf{x}_{\max}) \subset (-\infty, \lambda)$, by requiring that[19]

$$-\infty < \mathbf{x}_{\min} < 0 < \mathbf{x}_{\max} \le \lambda, \tag{2.65}$$

and that σ vanishes sufficiently fast at these two boundary points. This effectively confines the short rate process $(r(X_t))_{t \ge 0}$ to the range $r(M_0) = (r(\mathbf{x}_{\min}), r(\mathbf{x}_{\max}))$, showing that *in the one-dimensional ACE model we must impose an upper bound on the short rates.*

Note, however, that $\mu(\mathbf{x})$ is only monotonically decreasing on $\left(-\infty, \frac{1}{2}\lambda\right]$, and so one could argue that one should in fact choose $\mathbf{x}_{\max} \le \frac{1}{2}\lambda$ to obtain realistic model dynamics.

A constraint on r_∞ no longer appears to be necessary in one dimension, i.e., we may allow for all values $r_\infty \in \mathbb{R}$.[20] However, the choice of r_∞ will determine the range $r(M_0) = (r_\infty + \mathbf{x}_{\min}, r_\infty + \mathbf{x}_{\max})$ of possible short rates; in particular, to obtain a short rate lower bound of 0, one needs to choose $r_\infty := -\mathbf{x}_{\min}$.

Bond prices, forward rates, and zero-noise evolution. To reduce our various formulas in Section 2.5 to the one-dimensional case (by setting $n = 1$, $E = (0)$, and $\mathbf{v} = 1$), first note that by (2.51b) our currency vectors reduce to

$$\mathbf{c}_\tau = \frac{(e^{-\lambda\tau} - 1) \cdot 1}{(1-2)\lambda} \ge 0 = \frac{1}{\lambda}(1 - e^{-\lambda\tau}) \ge 0$$

for $\forall \tau \ge 0$. The bond price formula (2.28) therefore becomes

$$\begin{aligned}
B_{\$,T}(\mathbf{x}, t) &= e^{-r_\infty\tau}\left|\tfrac{1}{\lambda}(1 - e^{-\lambda\tau})\mathbf{x} - 1\right|^1 \\
&= e^{-r_\infty\tau}\left(1 - \tfrac{\mathbf{x}}{\lambda}\left(1 - e^{-\lambda\tau}\right)\right) \\
&= \left(1 - \tfrac{\mathbf{x}}{\lambda}\right)e^{-r_\infty\tau} + \tfrac{\mathbf{x}}{\lambda}e^{-(r_\infty+\lambda)\tau}
\end{aligned} \tag{2.66}$$

---

[17]Indeed, given the state process $(X_t)_{t\ge 0}$ of the first model, the process $\tilde{X}_t := X_t - \lambda$ has the starting point $\mathbf{x}_0 - \lambda$ and satisfies the SDE $d\tilde{X}_t = dX_t = \mu(X_t)\,dt + \sigma(X_t, t)\,dW_t = \tilde{\mu}(\tilde{X}_t)\,dt + \tilde{\sigma}(\tilde{X}_t, t)\,dW_t$, where $\tilde{\mu}(\tilde{\mathbf{x}}) := \mu(\tilde{\mathbf{x}} + \lambda) = (\mathbf{x} + \lambda)\mathbf{x} = \mathbf{x}(\mathbf{x} - (-\lambda))$ and $\tilde{\sigma}(\tilde{\mathbf{x}}, t) := \sigma(\tilde{\mathbf{x}} + \lambda, t)$.

[18]This can also be seen from the explicit formula (2.67) below, which for any $\mathbf{x}_0 > \lambda$ says that $U_{\hat{t}}(\mathbf{x}_0) = \infty$ for $\hat{t} := -\frac{1}{\lambda}\log\left(1 - \frac{\lambda}{\mathbf{x}_0}\right) > 0$.

[19]We do not allow for $\mathbf{x}_{\min} = -\infty$ since we want to bound our short rates below.

[20]Recall that by our discussion leading to (2.13) the constraint $r_\infty \ge 0$ had in fact been a consequence of the very constraint (2.9a) that we have just replaced.

for $\forall \mathbf{x} \in M_0$ and $\tau := T - t \geq 0$. (In the second step we could remove the absolute value since under our constraint (2.64) we have $1 - \mathrm{e}^{-\lambda\tau} \in [0, 1)$ and $\frac{\mathbf{x}}{\lambda} < 1$ for $\forall \mathbf{x} \in M_0 \subset (-\infty, \lambda)$.) Similarly, the zero-noise evolution formula (2.34) simplifies to

$$U_t(\mathbf{x}) = \left[\mathrm{e}^{\lambda t} \cdot \left(\tfrac{1}{\mathbf{x}} - \tfrac{1}{\lambda}(1 - \mathrm{e}^{-\lambda t}) \right) \right]^{-1} = \left(\left(\tfrac{1}{\mathbf{x}} - \tfrac{1}{\lambda} \right) \mathrm{e}^{\lambda t} + \tfrac{1}{\lambda} \right)^{-1}$$
$$= \frac{\lambda \mathbf{x}}{(\lambda - \mathbf{x}) \mathrm{e}^{\lambda t} + \mathbf{x}}, \tag{2.67}$$

and so the relative forward rate function becomes

$$F_{\mathrm{rel}}(\mathbf{x}; \tau) \overset{(2.32)}{=} r(U_\tau(\mathbf{x})) \overset{(2.63)}{=} r_\infty + U_\tau(\mathbf{x})$$
$$\overset{(2.67)}{=} r_\infty + \frac{\lambda \mathbf{x}}{(\lambda - \mathbf{x}) \mathrm{e}^{\lambda\tau} + \mathbf{x}}. \tag{2.68}$$

Conversion to the short rate state variable. We can now further simplify our description of the one-dimensional ACE model, by rewriting it based on the short rate $\mathbf{r} := r_\infty + \mathbf{x}$ itself as our state variable (recall (2.63)). Since in this reformulation r_∞ can be seen as an element of the state space as well, we will denote it as $\mathbf{r}_\infty$ from now on. The following lemma summarizes our results and states the Feller condition for our noise function $\sigma(\mathbf{r}, t)$ explicitly.

Lemma 2.15. *The one-dimensional ACE model on the (bounded) state space $M_0 := (\mathbf{r}_{\min}, \mathbf{r}_{\max})$ is the short rate process*

$$\mathrm{d}R_t = \mu(R_t)\,\mathrm{d}t + \sigma(R_t, t)\,\mathrm{d}W_t, \qquad R_{t=0} = \mathbf{r}_0, \tag{2.69}$$
$$\textit{where} \quad \mu(\mathbf{r}) := -\lambda(\mathbf{r} - \mathbf{r}_\infty) + (\mathbf{r} - \mathbf{r}_\infty)^2, \tag{2.70}$$

for some values $\mathbf{r}_0 \in M_0$ (the initial rate), $\mathbf{r}_\infty \in M_0$ (the long-term rate), and

$$\lambda \geq \mathbf{r}_{\max} - \mathbf{r}_\infty \tag{2.71}$$

(the reversion speed), and for some noise function $\sigma \colon M_0 \times [0, \infty) \to (0, \infty)$.

(i) We have $\mu > 0$ on $(-\infty, \mathbf{r}_\infty)$, $\mu < 0$ on $(\mathbf{r}_\infty, \mathbf{r}_\infty + \lambda)$, and $\mu(\mathbf{r}_\infty) = 0$; in particular, we have $\mu(\mathbf{r}_{\min}) > 0$ and $\mu(\mathbf{r}_{\max}) \leq 0$, with equality in the latter holding only in the degenerate case $\lambda = \mathbf{r}_{\max} - \mathbf{r}_\infty$. As a result, for each $\mathbf{r}_0 \in M_0$ the zero-noise process $t \mapsto U_t(\mathbf{r}_0)$ remains in M_0 for $\forall t \geq 0$ and converges to $\mathbf{r}_\infty$ as $t \to \infty$.

(ii) (Feller condition) Suppose that the noise function σ is continuous, and that it is chosen such that for some $T > 0$ the following two conditions are met: (a) $\sigma(\mathbf{r}, t)$ is locally Lipschitz continuous in $\mathbf{r}$ uniformly in $t \in [0, T]$, and (b) σ vanishes near the boundary points of M_0 so fast that

$$\limsup_{\mathbf{r} \searrow \mathbf{r}_{\min}} \frac{\sup_{t \in [0, T]} \sigma^2(\mathbf{r}, t)}{2(\mathbf{r} - \mathbf{r}_{\min})} < \mu(\mathbf{r}_{\min}) \tag{2.72}$$

at the left end point of M_0 and

$$\limsup_{\mathbf{r} \nearrow \mathbf{r}_{\max}} \frac{\sup_{t \in [0,T]} \sigma^2(\mathbf{r}, t)}{2(\mathbf{r}_{\max} - \mathbf{r})} < -\mu(\mathbf{r}_{\max}) \quad \text{if} \quad \lambda > \mathbf{r}_{\max} - \mathbf{r}_\infty \quad \text{or} \qquad (2.73a)$$

$$\limsup_{\mathbf{r} \nearrow \mathbf{r}_{\max}} \frac{\sup_{t \in [0,T]} \sigma^2(\mathbf{r}, t)}{2(\mathbf{r}_{\max} - \mathbf{r})^2} < \lambda \qquad\qquad \text{if} \quad \lambda = \mathbf{r}_{\max} - \mathbf{r}_\infty \qquad (2.73b)$$

at the right end point of M_0. Then for any starting point $\mathbf{r}_0 \in M_0$ the SDE (2.69) has a unique strong solution $(R_t)_{t \in [0,T]}$ that—just like in fact the entire forward rate curve $\tau \mapsto F_{\mathrm{rel}}(R_t; \tau)$—is confined to M_0 for $\forall t \in [0,T]$ almost surely.

If these conditions hold for $\forall T > 0$, then for any $\mathbf{r}_0 \in M_0$ (2.69) has a unique strong solution $(R_t)_{t \geq 0}$ that—just like the entire forward curve—is confined to M_0 for $\forall t \geq 0$ almost surely.

(iii) The bond dollar prices, the relative forward rates, and the zero-noise evolution are then given by the formulas

$$B_{\$,T}(\mathbf{r}, t) = \left(1 - \tfrac{\mathbf{r} - \mathbf{r}_\infty}{\lambda}\right) e^{-\mathbf{r}_\infty(T-t)} + \tfrac{\mathbf{r} - \mathbf{r}_\infty}{\lambda} e^{-(\mathbf{r}_\infty + \lambda)(T-t)}, \qquad (2.74)$$

$$F_{\mathrm{rel}}(\mathbf{r}; \tau) = \mathbf{r}_\infty + \frac{\lambda(\mathbf{r} - \mathbf{r}_\infty)}{(\lambda + \mathbf{r}_\infty - \mathbf{r}) e^{\lambda \tau} + (\mathbf{r} - \mathbf{r}_\infty)}, \qquad (2.75)$$

$$U_t(\mathbf{r}) = F_{\mathrm{rel}}(\mathbf{r}; t). \qquad (2.76)$$

Proof. The SDE (2.69)–(2.70) is derived from the original SDE (2.10) (with μ given in (2.62)) via a trivial application of Itô's Lemma that shows that the new drift μ and noise σ are just shifted versions of the old ones. The $\mathbf{r}$-based formulas (2.74)–(2.76) for $B_{\$,T}$, F_{rel}, and U_t are then obtained from the original $\mathbf{x}$-based ones in (2.66)–(2.68) via simple shifts as well. By adding $\mathbf{r}_\infty$ to each term in (2.65), we obtain

$$-\infty < \mathbf{r}_{\min} < \mathbf{r}_\infty < \mathbf{r}_{\max} \leq \mathbf{r}_\infty + \lambda, \qquad (2.77)$$

where $\mathbf{r}_{\min} := \mathbf{r}_\infty + \mathbf{x}_{\min}$ and $\mathbf{r}_{\max} := \mathbf{r}_\infty + \mathbf{x}_{\max}$, which leads us to our constraints $\mathbf{r}_\infty \in M_0$ and (2.71).

(i) The first statement in part (i) holds because the roots of the quadratic function μ are $\mathbf{r}_\infty$ and $\mathbf{r}_\infty + \lambda$, and the remark about the signs of $\mu(\mathbf{r}_{\min})$ and $\mu(\mathbf{r}_{\max})$ then follows from (2.77). Similarly, the second statement holds since by (2.77) we have $M_0 \subset (-\infty, \mathbf{r}_\infty + \lambda)$, which is the basin of attraction of $\mathbf{r}_\infty$.

(ii) As a result of our first observation in (i), the statements in (ii)—except for the ones about the forward rates $F_{\mathrm{rel}}(R_t; \tau)$—can be derived by linearly rescaling $(R_t)_{t \geq 0}$ from M_0 to $(-1, 1)$ and invoking our Feller condition on the unit ball in Lemma B.2. See Appendix B.6 for details.

(iii) The fact that the process $(R_t)_{t \geq 0}$—with or without noise—can never leave M_0 in turn has two consequences: First, since the function $B_{\$,T}$ in (2.74) is continuous and therefore bounded on the compact set $\overline{M_0} \times [0,T]$, and since our calculations in Section 2.6.2 imply that it solves the Feynman–Kac PDE on that domain, it allows us to apply Lemma 2.8 and conclude that it has indeed the representation

$$B_{\$,T}(\mathbf{r},t) = \mathbb{E}\big[\exp\big(-\textstyle\int_t^T R_s\,\mathrm{d}s\big)\,\big|\,R_t = \mathbf{r}\big]$$

for $\forall(\mathbf{r},t) \in M_0 \times [0,T]$; the function $B_{\$,T}$ in (2.74) therefore indeed computes the bond prices, and thus the function F_{rel} in (2.75), which is related to it via (2.31) and (2.27), indeed returns the relative forward rates of the model. Second, since in the zero-noise case this statement just means that $\forall \mathbf{r} \in M_0$ $\forall t \geq 0\colon\ U_t(\mathbf{r}) \in M_0$, the relation (2.76) therefore implies that all forward rates $F_{\mathrm{rel}}(R_t;\tau) = U_\tau(R_t)$ are confined to M_0 as well, which was the last missing statement to prove in part (ii).   □

We find that the one-dimensional ACE model, as described by the short rate dynamics (2.69)–(2.70), is a modification of the classic Vašíček model, with an additional second-order drift term, with an arbitrary state- and time-dependent noise function $\sigma(\mathbf{r},t)$, and with both a lower and (in contrast to the ACE model for $n \geq 3$) also an upper bound imposed on the short rate process. See also the result by Gabaix in [8, Eqn. (30)].

Its $2n + 1 = 3$ discrete parameters inherited from the original formulation of the ACE model are $\mathbf{r}_0$, $\mathbf{r}_\infty$, and λ. The values $\mathbf{r}_{\min}$ and $\mathbf{r}_{\max}$ do not affect the zero-noise process but rather the state space M_0 (and thus indirectly the noise σ via the constraint on its boundary behavior); there is no equivalent to them in the standard higher-dimensional ACE model since that model does not impose any upper bound, and since it bounds short rates below by the default value zero (see the paragraph "Choice of the short rate lower bound" in Section 2.10 for how to shift this lower bound to a different value).

As before in the $\mathbf{x}$-based description of the model, note that the drift in (2.70) is monotonically decreasing only on $\big(-\infty, \mathbf{r}_\infty + \tfrac{\lambda}{2}\big]$, and so one can argue that realistic model dynamics are only obtained when choosing $\mathbf{r}_{\max} \leq \mathbf{r}_\infty + \tfrac{\lambda}{2}$.

Taking a closer look at the bond price and forward rate formulas, first observe that (unsurprisingly) this model is still unspanned, as the bond price formula (2.74) does not depend on σ. Further note that setting $\tau = 0$ in the forward rate formula (2.75) indeed recovers our state variable $\mathbf{r}$, confirming its meaning as the short rate of the model. Phrased differently, the short rate function (i.e., the function that maps the state variable to the short rate) is just the identity in this formulation, which shows that the zero-noise evolution formula (2.76) is consistent with (2.32) (which in Lemma 2.6 was shown to hold for any unspanned model).

The reader who is interested in proofs of our formulas (2.74)–(2.76) that are independent of our treatment of the general n-dimensional case in the preceding sections will find the necessary calculations in Appendix B.7.

2.9.2   Generalization to Time-Dependent Drift

Motivated by our results in Lemma 2.15, we will now generalize the one-dimensional ACE model to time-dependent mean reversion speeds $\lambda(t)$ and reversion targets $\theta(t)$. The ansatz (2.83) that we will make to solve the Feynman–Kac equation for this generalized model is rooted in more than just the keen observation that the bond price function (2.74) is linear in $\mathbf{r}$: According to (2.37) and the explanations following these equations, the bond price function in *any* one-dimensional unspanned model must fulfill $0 = \Delta B_{\$,T}(\mathbf{r},t) = \partial_{\mathbf{r}}^2 B_{\$,T}(\mathbf{r},t)$ for $\forall \mathbf{r} \in M_0$ and $\forall t \geq 0$ and must therefore be linear in $\mathbf{r}$.

Lemma 2.16. *The generalized one-dimensional ACE model on the (bounded) state space $M_0 := (\mathbf{r}_{\min}, \mathbf{r}_{\max})$ is the short rate process*

$$\mathrm{d}R_t = \mu(R_t, t)\,\mathrm{d}t + \sigma(R_t, t)\,\mathrm{d}W_t, \qquad R_{t=0} = \mathbf{r}_0, \qquad (2.78)$$

where $\quad \mu(\mathbf{r},t) := -\lambda(t)\big(\mathbf{r} - \theta(t)\big) + \big(\mathbf{r} - \theta(t)\big)^2,$ $\qquad\qquad (2.79)$

for some value $\mathbf{r}_0 \in M_0$ (the initial rate), for some continuous functions $\theta\colon [0,\infty) \to M_0$ (the reversion target) and $\lambda\colon [0,\infty) \to (0,\infty)$ (the reversion speed), which we assume to satisfy[21]

$$\lambda(t) \geq \mathbf{r}_{\max} - \theta(t) \qquad for \qquad \forall t \geq 0, \qquad (2.80)$$

and for some noise function $\sigma\colon M_0 \times [0,\infty) \to (0,\infty)$.

(i) For each $\mathbf{r}_0 \in M_0$ the zero-noise process, i.e., the solution of the ODE $\dot{\mathbf{r}}(t) = \mu(\mathbf{r}(t), t)$, $\mathbf{r}(0) = \mathbf{r}_0$, remains in M_0 for $\forall t \geq 0$.

(ii) (Feller condition) Suppose that the noise function σ is continuous, and that it is chosen such that for some $T > 0$ the following two conditions are met: (a) $\sigma(\mathbf{r},t)$ is locally Lipschitz continuous in $\mathbf{r}$ uniformly in $t \in [0,T]$, and (b) σ vanishes near the boundary points of M_0 so fast that

$$\limsup_{\mathbf{r} \searrow \mathbf{r}_{\min}} \sup_{t \in [0,T]} \left(\frac{\sigma^2(\mathbf{r},t)}{2(\mathbf{r} - \mathbf{r}_{\min})} - \mu(\mathbf{r}_{\min}, t) \right) < 0 \qquad (2.81)$$

at the left end point of M_0 and

$$\limsup_{\mathbf{r} \nearrow \mathbf{r}_{\max}} \sup_{t \in [0,T]} \frac{\sigma^2(\mathbf{r},t)}{2(\mathbf{r}_{\max} - \mathbf{r})(-\mu(\mathbf{r},t))} < 1 \qquad (2.82a)$$

at the right end point of M_0. Note that in the non-degenerate case where $\lambda(t) > \mathbf{r}_{\max} - \theta(t)$ for $\forall t \in [0,T]$, (2.82a) is equivalent to the condition

$$\limsup_{\mathbf{r} \nearrow \mathbf{r}_{\max}} \sup_{t \in [0,T]} \left(\frac{\sigma^2(\mathbf{r},t)}{2(\mathbf{r}_{\max} - \mathbf{r})} + \mu(\mathbf{r}_{\max}, t) \right) < 0. \qquad (2.82b)$$

---

[21] As before, one can argue that realistic model dynamics are only obtained under the stricter constraint $\mathbf{r}_{\max} \leq \theta(t) + \frac{1}{2}\lambda(t)$ for $\forall t \in [0,T]$.

Then for any starting point $\mathbf{r}_0 \in M_0$ *the SDE (2.78) has a unique strong solu-tion* $(R_t)_{t \in [0,T]}$ *that—just like in fact the entire forward rate curve* $s \mapsto F(R_t, t; s)$ *—is confined to* M_0 *for* $\forall t \in [0, T]$ *almost surely.*

If these conditions hold for $\forall T > 0$, *then for any* $\mathbf{r}_0 \in M_0$ *(2.78) has a unique strong solution* $(R_t)_{t \geq 0}$ *that—just like the entire forward curve—is con-fined to* M_0 *for* $\forall t \geq 0$ *almost surely.*

(iii) The bond dollar prices are then given by

$$B_{\$,T}(\mathbf{r}, t) = P_T(t) + Q_T(t)\,\mathbf{r}, \qquad (2.83)$$

where for each fixed $T \geq 0$ *the pair of functions* $P_T, Q_T \colon [0, T] \to \mathbb{R}$ *is the solution of the following system of first-order linear ODEs with given endpoints:*

$$\partial_t \begin{pmatrix} P_T \\ Q_T \end{pmatrix} = \begin{pmatrix} 0 & -(\lambda\theta + \theta^2) \\ 1 & \lambda + 2\theta \end{pmatrix} \begin{pmatrix} P_T \\ Q_T \end{pmatrix}, \qquad \begin{pmatrix} P_T(T) \\ Q_T(T) \end{pmatrix} = \begin{pmatrix} 1 \\ 0 \end{pmatrix}. \qquad (2.84)$$

The absolute forward rates are given as the solution of the ODE

$$\partial_s F(\mathbf{r}, t; s) = \mu(F(\mathbf{r}, t; s), s), \qquad F(\mathbf{r}, t; t) = \mathbf{r}. \qquad (2.85)$$

Proof. (i) Since the roots of the quadratic function $\mu(\cdot, t)$ are $\theta(t)$ and $\theta(t) + \lambda(t)$, and since the lemma assumes that

$$-\infty < \mathbf{r}_{\min} < \theta(t) < \mathbf{r}_{\max} \leq \theta(t) + \lambda(t),$$

for each fixed $t \geq 0$ we have $\mu(\cdot, t) > 0$ just to the right of $\mathbf{r}_{\min}$ and $\mu(\cdot, t) < 0$ just to the left of $\mathbf{r}_{\max}$. This shows that the solution of the zero-noise process cannot escape M_0.

(ii) The statements in part (ii), except for the ones about the forward rates $F(R_t, t; s)$, can be derived analogously to Lemma 2.15 (ii), by linearly rescaling $(R_t)_{t \geq 0}$ from M_0 to $(-1, 1)$ and invoking our Feller condition on the unit ball in Lemma B.2. See Appendix B.8 for details.

(iii) Using the ansatz (2.83) to find the solution of the Feynman–Kac equation associated to this process, we see that it suffices if P_T and Q_T are such that

$$0 = \left(\partial_t + \tfrac{1}{2}\sigma^2 \partial_{\mathbf{r}}^2 + \mu \partial_{\mathbf{r}} - \mathbf{r}\right) B_{\$,T} \qquad (2.86)$$

$$= (\dot{P}_T + \dot{Q}_T\,\mathbf{r}) + 0 + \left(-\lambda(\mathbf{r} - \theta) + (\mathbf{r} - \theta)^2\right) Q_T - \mathbf{r}\left(P_T + Q_T\,\mathbf{r}\right)$$

$$= \left(\dot{P}_T + (\lambda\theta + \theta^2) Q_T\right) + \left(\dot{Q}_T - (\lambda + 2\theta) Q_T - P_T\right)\mathbf{r},$$

i.e., these functions only need to satisfy the ODE system (2.84). The boundary condition of the Feynman–Kac equation, i.e.,

$$1 = B_{\$,T}(\mathbf{r}, T) = P_T(T) + Q_T(T)\,\mathbf{r} \qquad \text{for } \forall \mathbf{r} \in M_0 \text{ and } \forall T \geq 0,$$

is equivalent to the endpoint condition in (2.84). The boundedness of (2.83) on $M_0 \times [0, T]$ simply follows from its continuity on the compact set $[\mathbf{r}_{\min}, \mathbf{r}_{\max}] \times [0, T]$.

Finally, the formula (2.85) is shown just like Lemma 2.6, where the short rate function is now the identity, and where we now need to work with the original *absolute* forward rate function since our drift μ is now time inhomogeneous. See Appendix B.9 for an independent proof that is instead based on the fact that $B_{\$,T}$ satisfies the Feynman–Kac PDE (2.86) with $\sigma = 0$.   □

The reader is encouraged to check that if the functions λ and θ are constant then the functions P_T and Q_T obtained by writing the bond price formula (2.74) in the form (2.83) are indeed the solution of the system (2.84).[22]

2.10   Calibration

Calibrating the parameters of the ACE model to match the market prices of certain liquid interest rate-related instruments is a task that requires both good numerical skills and a fair amount of work experience with the ACE model. There is not a unique way of achieving good calibration, and compromises may not always be avoidable. In the end, the chosen approach may depend on the intended use of the model.

Rather than describing the particular approach that we have chosen in our own implementation, which would bias any efforts that the reader may make on their own, here we will only outline some aspects to consider, and some general tools that are available to address certain issues. Any future publications comparing different calibration strategies for the ACE model would certainly be of great value.

Choice of the dimension n.   When choosing the dimension $n \in \mathbb{N} \setminus \{2\}$, one has to strike a balance between dynamic flexibility (favoring large n) and numerical feasibility (favoring small n). Since the number of independent forward rate oscillations (i.e., the number of (2×2)-matrices on the diagonal of E) is $\lfloor n/2 \rfloor$, increasing n by one leads to more independent oscillations only when an even value of n is reached. This suggests that the tradeoff between flexibility and feasibility is better for even values of n. We have found that the choice $n = 4$ leads to good results.

Choice of calibration instruments.   The ACE model can be calibrated to bonds, caps and floors, and swaptions, with various maturities, expirations and strikes. The user has to decide which exact instruments to calibrate to, based on these instruments' liquidity (and thus on the reliability of their market prices) and on the intended use of the model. For example, if one intends to use the model to price only instruments with no long-term dependency then one

---

[22]In fact, it is a nice exercise to *derive* (2.74) from the general formula in Lemma 2.16, by expressing the solution of (2.84) using a matrix exponential (which is only possible in the time-independent case) and then computing it explicitly by diagonalizing the matrix in the exponent.

may decide to limit the calibration instruments to short-term maturities and expirations as well.

Choice of calibration strategy. Since the ACE model is unspanned w.r.t. its functional noise parameter $\sigma(\mathbf{x}, t)$, one can first calibrate all non-functional parameters to match the discount curve, and then in a second step calibrate σ to match the prices of the remaining instruments, without destroying the discount curve calibration. In the second step one can bootstrap by expiration (as opposed to maturity), i.e., one can assume that $\sigma(\mathbf{x}, t)$ is constant in t on successive time intervals, and then successively calibrate its function values on each time interval to the prices of those instruments whose expirations lie within the given interval. Of course, various modifications of this strategy can be explored to improve the calibration quality.

Choice of objective functions. Once the calibration instruments and a calibration strategy are chosen, one needs to decide upon the objective functions to be minimized. Typically these functions are chosen to be of the form

$$f(r_\infty, E, \mathbf{v}, \lambda, \mathbf{x}_0, \sigma) := \sum_i w_i \times \left(P_i^{\mathrm{market}} - P_i^{\mathrm{ACE}}(r_\infty, E, \mathbf{v}, \lambda, \mathbf{x}_0, \sigma) \right)^2$$

for some weights $w_i > 0$, where P_i^{market} and P_i^{ACE} are the prices[23] of instrument i observed in the market and calculated by the ACE model, respectively. However, depending on the model user's individual preference, other shapes can be explored, in particular based on alternatives to the square function (if one wishes to penalize larger price deviations differently), or on relative instead of absolute errors.

Either way, one must address the problem that market prices of different instruments can be on different orders of magnitude (and in fact near zero), since this may require substantially different weights when using absolute errors, and since it may lead to problems when using relative errors (division by zero).

Choice of numerical optimization method. The numerical minimization can be carried out with a variety of algorithms from readily available software packages. Besides choosing the most efficient algorithm and implementation, one may also experiment with various ways of enforcing our parameter constraints.

Choice of the short rate lower bound. As it is defined here, the short rates $r(X_t)$ simulated by the ACE model (and thus also all forward rates) are positive for $\forall t \geq 0$ almost surely. Imposing a lower bound $r_0 \in \mathbb{R}$ other than 0 (typically slightly negative) can be achieved by leaving the definition of the process $(X_t)_{t \geq 0}$ unchanged but considering the modified short rate function

$$r^{\mathrm{shifted}}(\mathbf{x}) := r(\mathbf{x}) + r_0 \tag{2.87}$$

---

[23]Prices may be expressed in dollars, or for derivatives as implied volatilities.

instead. By (2.25) and (2.27) the bond price and forward rate functions are then given by

$$B_{\$,T}^{\text{shifted}}(\mathbf{x},t) = B_{\$,T}^{\text{ACE}}(\mathbf{x},t) \cdot e^{-r_0(T-t)}, \tag{2.88}$$

$$F^{\text{shifted}}(\mathbf{x},t;s) = F^{\text{ACE}}(\mathbf{x},t;s) + r_0, \tag{2.89}$$

where $B_{\$,T}^{\text{ACE}}(\mathbf{x},t)$ and $F^{\text{ACE}}(\mathbf{x},t;s)$ are those for the standard ACE model.

Fudging the discount curve. Since the bond market is extremely liquid, any compromises on the precision of the calibrated discount curve are usually unacceptable. Remaining errors after calibration can be completely eliminated by sacrificing the time homogeneity and considering a slightly modified version of the ACE model. There are two options for achieving this.

(i) *Shifting:* The straightforward approach is to again consider the shifted short rate function (2.87), but now for some *time-dependent* function $r_0(t)$, i.e.,

$$r^{\text{shifted}}(\mathbf{x},t) := r(\mathbf{x}) + r_0(t), \tag{2.90}$$

which leads to the bond price and forward rate functions

$$B_{\$,T}^{\text{shifted}}(\mathbf{x},t) = B_{\$,T}^{\text{ACE}}(\mathbf{x},t) \cdot e^{-\int_t^T r_0(s)\,ds}, \tag{2.91}$$

$$F^{\text{shifted}}(\mathbf{x},t;s) = F^{\text{ACE}}(\mathbf{x},t;s) + r_0(t). \tag{2.92}$$

By choosing

$$r_0(t) := f^{\text{market}}(s) - F^{\text{ACE}}(\mathbf{x}_0,0;s), \tag{2.93}$$

where f^{market} is the forward rate curve currently observed on the market, and where $F^{\text{ACE}}(\mathbf{x}_0,0;s)$ is the initial forward rate curve of the standard ACE model (after calibrating it as well as possible), this shifted ACE model will reproduce the initial forward rate curve precisely, i.e., we have

$$F^{\text{shifted}}(\mathbf{x}_0,0;s) = f^{\text{market}}(s) \qquad \text{for } \forall s \geq 0. \tag{2.94}$$

Unfortunately, this approach has the downside of making also the short rate lower bound (which is $r_0(t)$) time dependent.

(ii) *Time rescaling:* As an alternative that avoids this usually undesired side effect, one can, for some function $\beta(t) > 0$, consider the time-rescaled process

$$\tilde{X}_t := X_{\tilde{t}(t)}, \qquad \text{where} \quad \tilde{t}(t) := \int_0^t \beta(s)\,ds, \tag{2.95}$$

and the short rate function

$$r^{\text{rescaled}}(\mathbf{x},t) := \beta(t)r(\mathbf{x}) + r_0, \tag{2.96}$$

where $r_0 \in \mathbb{R}$ is the desired (constant) short rate lower bound. A simple change of variables then shows that the bond prices under this fudged ACE model are given by

$$B_{\$,T}^{\text{rescaled}}(\mathbf{x},t) = \mathbb{E}\big[\exp\big(-\int_t^T r^{\text{rescaled}}(\tilde{X}_s,s)\,ds\big)\,\big|\,\tilde{X}_t = \mathbf{x}\big]$$

$$= \mathbb{E}\big[\exp\big(-\textstyle\int_t^T \big(\beta(s)r(X_{\tilde{t}(s)}) + r_0\big)\,ds\big)\,\big|\,X_{\tilde{t}(t)} = \mathbf{x}\big]$$

$$= e^{-r_0(T-t)}\mathbb{E}\big[\exp\big(-\textstyle\int_t^T r(X_{\tilde{t}(s)})\beta(s)\,ds\big)\,\big|\,X_{\tilde{t}(t)} = \mathbf{x}\big]$$

$$= e^{-r_0(T-t)}\mathbb{E}\big[\exp\big(-\textstyle\int_{\tilde{t}(t)}^{\tilde{t}(T)} r(X_u)\,du\big)\,\big|\,X_{\tilde{t}(t)} = \mathbf{x}\big]$$

$$= e^{-r_0(T-t)}B_{\$,\tilde{t}(T)}^{\mathrm{ACE}}\big(\mathbf{x}, \tilde{t}(t)\big), \tag{2.97}$$

and so by (2.27) its forward rates are

$$F^{\mathrm{rescaled}}(\mathbf{x}, t; s) = r_0 + \beta(s)F^{\mathrm{ACE}}\big(\mathbf{x}, \tilde{t}(t); \tilde{t}(s)\big). \tag{2.98}$$

To use this technique in practice, one must first decide for a lower bound $r_0 < \inf\{f^{\mathrm{market}}(s)\,|\,s \geq 0\}$, and calibrate the non-functional parameters of the standard ACE model as well as possible, aiming to achieve that $F^{\mathrm{ACE}} + r_0 = f^{\mathrm{market}}$. To then define the rescaling $\tilde{t}(t)$ (and thus also $\beta(t) = \tilde{t}'(t)$) so that our fudged model reproduces f^{market} exactly, i.e., so that

$$F^{\mathrm{rescaled}}(\mathbf{x}_0, 0; s) = f^{\mathrm{market}}(s) \qquad \text{for } \forall s \geq 0, \tag{2.99}$$

by (2.98) one must simply numerically solve the ODE

$$\tilde{t}'(s) = \frac{f^{\mathrm{market}}(s) - r_0}{F^{\mathrm{ACE}}\big(\mathbf{x}_0, 0; \tilde{t}(s)\big)}, \qquad \tilde{t}(0) = 0. \tag{2.100}$$

Note that by our choice of r_0 this implies that indeed we have $\tilde{t}'(s) > 0$ for $\forall s \geq 0$.

Parametrization of σ. In order to calibrate also the functional parameter $\sigma(\mathbf{x}, t)$, one then needs to decide how to represent this function numerically. Standard grid representations are numerically unfeasible even for small dimensions n, since (i) finding the optimal values at all these grid points numerically would be computationally too expensive, and (ii) the state space M_0 is unbounded. It seems more promising to parameterize σ by assuming some specific functional shape (e.g., the linear combination of some basis functions), and then to calibrate its parameters.

Either way, making full use of $\sigma(\mathbf{x}, t)$ will require some insight into how the shape of this function affects the prices of the calibration instruments (swaptions, caps, and floors), as this would aid the search for both an efficient parametrization of σ, and for an appropriate strategy for optimizing its parameters.

2.11 Model Extensions

Once all options for calibration have been explored, one can consider two further types of generalizations of the basic ACE framework to potentially obtain even better results.

Driving σ by a second process. One may choose to have σ depend on an additional argument **y**, which one can vary over time as a deterministic or stochastic process $(Y_t)_{t\geq 0}$. In other words, we can extend our SDE (2.10) to a system of SDEs for two processes $(X_t)_{t\geq 0}$ and $(Y_t)_{t\geq 0}$, where the SDE for the process $(Y_t)_{t\geq 0}$ cannot depend on X_t, but where the noise prefactor for the ACE SDE for $(X_t)_{t\geq 0}$ may now be any function $\sigma(X_t, Y_t, t)$ (subject to constraints similar to the ones developed in Lemma 2.4).

Since the bond prices in the ACE model are independent of σ and thus of the process $(Y_t)_{t\geq 0}$, our bond price formula (2.28) will remain valid also for this modified process:

$$\mathbb{E}\left[\exp\left(-\int_t^T r(X_s)\,\mathrm{d}s\right) \,\middle|\, X_t = \mathbf{x}\right]$$
$$= \mathbb{E}\left[\mathbb{E}\left[\exp\left(-\int_t^T r(X_s)\,\mathrm{d}s\right) \,\middle|\, X_t = \mathbf{x},\ (Y_s)_{s\geq t}\right]\right]$$
$$= \mathbb{E}\left[B_{\$,T}^{\mathrm{ACE}}(\mathbf{x}, t)\right] = B_{\$,T}^{\mathrm{ACE}}(\mathbf{x}, t).$$

Naturally, this also implies that the forward rate formula will remain unchanged as well.

Adding multiple independent short rate processes. As another generalization, given multiple independent ACE processes $(X_t^i)_{t\geq 0}$ (based on separate parameter sets) and their associated short rate functions r^i, we may consider the short rate process $\left(\sum_i r^i(X_t^i)\right)_{t\geq 0}$ (which is also bounded below by 0). The bond price function of this combined process is then simply the product of those of the individual ACE processes since

$$\mathbb{E}\left[\exp\left(-\int_t^T \left(\sum_i r^i(X_s^i)\right)\mathrm{d}s\right) \,\middle|\, X_t^i = \mathbf{x}^i \text{ for } \forall i\right]$$
$$= \prod_i \mathbb{E}\left[\exp\left(-\int_t^T r^i(X_s^i)\,\mathrm{d}s\right) \,\middle|\, X_t^i = \mathbf{x}^i\right] = \prod_i B_{\$,T}^{\mathrm{ACE}_i}(\mathbf{x}^i, t),$$

and by (2.27) the forward rate function is thus just the sum of the individual forward rate functions.

Note that this technique even allows us to combine the ACE model with other short rate models. The resulting model is only bounded below if all the component models are. In contrast, the resulting model will always be unspanned with respect to the parameter $\sigma(\mathbf{x}, t)$ of the ACE model, even if the added models are not unspanned with respect to any parameter (provided that they do not reuse the function $\sigma(\mathbf{x}, t)$ in their definition).

2.12   Conclusions

Let us conclude by looking back at our achievements, and by pointing out some limitations of the ACE model.

2.12.1 Model Features

We have defined an interest rate model that has all the desired model features listed in Section 1.1:

✓ **Low-dimensionality, dynamic flexibility, parametric freedom:** As the dimension of the model's state space one can choose any value $n \geq 3$, with higher dimensions providing more parametric freedom and dynamic flexibility.[24] Its non-functional parameters r_∞, E, $\mathbf{v}$, λ, and $\mathbf{x}_0$ provide $2n+1$ degrees of freedom that can be used to calibrate the discount curve, while the additional functional parameter $\sigma(\mathbf{x}, t)$ then allows us to also calibrate to the swaption market. The model simulates (and calibrates to) discount curves in a variety of shapes; in particular, the presence of the matrix E in the drift μ allows for oscillations in the associated forward rate curves, with $\lfloor n/2 \rfloor$ independent frequencies and phases, while r_∞ is the long-term forward rate and λ controls the convergence speed of the forward rate curve to this limit (see Lemma 2.14).

✓ **Completeness, consistency, explicit bond prices:** Just like in any short rate model, we define the bond prices for all maturities $T \geq t$ via the formula (2.25), making the model consistent and complete. Our bond price formula (2.28) provides an explicit analytic expression for this expected value, which together with (2.51b) allows us to compute all bond prices accurately and efficiently also in practice.

✓ **Unspanned volatility:** From the formula (2.28) we see that the bond price function $B_{\$,T}(\mathbf{x}, t)$ does not depend on the parameter $\sigma(\mathbf{x}, t)$, despite the fact that this parameter enters the bond price definition (2.25) via the process $(X_t)_{t \geq 0}$. The ACE model is therefore unspanned with respect to its functional model parameter $\sigma(\mathbf{x}, t)$. This opens the door to effective global calibration to the swaption market also in practice, by bootstrapping by expiration as opposed to maturity.

✓ **Time homogeneity:** The ACE model is time homogeneous if and only if the noise function $\sigma(\mathbf{x}, t)$ is chosen to be independent of the process time t, i.e., if $\sigma(\mathbf{x}, t) = \sigma(\mathbf{x})$. This gives the modeler full control over the time homogeneity of the model: If time homogeneity is desirable then one must choose σ as a function of $\mathbf{x}$ only; if one is willing to sacrifice time homogeneity to improve the calibration to the swaption market, then one may allow σ to depend on both $\mathbf{x}$ and t.

✓ **Bounded short rates:** As it is defined here, the short rate process $(r(X_t))_{t \geq 0}$ of the ACE model is positive for $\forall t \geq 0$ almost surely. Imposing a lower bound $r_0 \in \mathbb{R}$ other than 0 on the short rates can be achieved by leaving the definition of $(X_t)_{t \geq 0}$ unchanged but considering the modified short rate function $\tilde{r}(\mathbf{x}) := r(\mathbf{x}) + r_0$ instead. The bond price formula (2.28) must then be multiplied by $e^{-r_0 \tau}$, and in the forward rate formula (2.32) one must replace r by $\tilde{r}$ (effectively shifting the forward rate curve by r_0).

---

[24] In practice, we found that the choice $n = 4$ is a good compromise between parametric freedom and dynamic flexibility on the one side and numerical tractability on the other.

2.12.2   Model Limitations

The model currently has the following limitations:

Discount curve constraints: As with any model that is driven by a finite-dimensional process, for a fixed parameter set the ACE model can only simulate discount curves from a finite-dimensional function space (of the same dimension). While the parameters of the ACE model allow this function space to be equipped with a variety of properties, and thus to adjust the model to the observed market conditions, some discount curve constraints of the model cannot be overcome. In particular, for any fixed parameter set, all discount curves share the same long-term forward rate r_∞, and so in the ACE model this rate can never fluctuate over time.

Analytically/numerically motivated modeling choices: Some modeling choices in our construction were motivated merely by numerical or analytical necessity, rather than by observed (or otherwise theoretically derived) properties of the current market prices or of the historical interest rate evolution. Such modeling decisions may restrict the generality of a model in unforeseen ways, and overcoming the issues they address in alternative ways may lead to a more general model.

Here we will only list two such modeling choices that the reader can already understand now; for a more comprehensive list that includes also those that require understanding of our construction in Chapters 5–8, please see Section 9.2.

- **Unspanned volatility:** Unspanned volatility was added to our list of desired model features to facilitate calibrating $\sigma(\mathbf{x}, t)$ via bootstrapping by expiration. However, leaving aside the issue of finding the optimal choice of σ *in practice*, there is no reason why unspanned models would describe the actual real-world interest rate evolution better than others.

- **Scalar-valued noise function:** In our construction we only allow for scalar-valued (as opposed to matrix-valued) noise functions $\sigma(\mathbf{x}, t)$. While this is not as big a constraint as it may seem at first (see our discussion in Section 6.4.5), it may certainly have led to some loss of generality.

Only one currency: The ACE model presented in this chapter was still restricted to only one currency. Its extension to multiple currencies (i.e., to a model that can simulate the interest rates of multiple currencies, along with the spot and forward exchange rates of the various currency pairs) is the subject of Chapter 3.

Chapter 3

The Multi-Currency ACE Model

3.1 Objective and Outline

Objective. To now extend the ACE model to multiple currencies, it is not enough to simply make multiple copies of the single-currency ACE model and correlate their driving Brownian motions: First, this would multiply the state space dimension n by the number of currencies, quickly making the model numerically untractable. Second, this would only allow us to compute the bond prices in their *respective* currencies (i.e., the prices of dollar bonds measured in dollars, the prices of Euro bonds measured in Euros, etc.), and so we would crucially be missing the currencies' exchange rates, which one needs in order to compare the values of these bonds with one another.

Instead, we are seeking to extend our single-currency ACE model without modifying its original state process $(X_t)_{t \geq 0}$ (thus leaving n unchanged), by defining additional deterministic functions for computing the short rates of all the various currencies and the exchange rates between them. More precisely, denoting by C the collection of currencies in scope, e.g., $C := \{\$, \euro, \pounds\}$, we need to define a short rate function $r^c(\mathbf{x})$ for each $c \in C$, and a spot exchange rate function $a^{c_1/c_2}(\mathbf{x}, t) > 0$ for each pair of currencies $c_1, c_2 \in C$,[1] thus giving rise to the short rate and exchange rate processes

$$(r^c(X_t))_{t \geq 0} \qquad \text{and} \qquad (a^{c_1/c_2}(X_t, t))_{t \geq 0}. \qquad (3.1)$$

To ensure that *all* the rates will remain positive almost surely, we will then have to reduce the state space of the single-currency ACE model by defining it as

$$M_0 := \left\{ \mathbf{x} \in \mathbb{R}^n \,\middle|\, \forall c \in C : r^c(\mathbf{x}) > 0 \right\} \qquad (3.2)$$

---

[1] For example, an exchange rate of $a^{\euro/\$}(\mathbf{x}, t) = 1.10$ would mean that the system is in a state in which $\euro/\$ = 1.10$, i.e., where $\euro 1$ is worth $\$1.10$. In other words, the function a^{c_1/c_2} answers the question, "One unit of currency c_1 is worth how many units of c_2?"

instead, choose the initial state $\mathbf{x}_0$ from this reduced set, and strengthen our parameter constraints to guarantee that the state process $(X_t)_{t \geq 0}$ will be confined to this smaller set M_0 for $\forall t \geq 0$.

The goal is then to find explicit formulas for the bond prices $B_{c_2,T}^{c_1}(\mathbf{x}, t)$ for $\forall c_1, c_2 \in C$, i.e., for the values measured in units of c_2 of the bond that pays one unit of c_1 at time $T \geq t$, given by certain expected values to be discussed in Section 3.2. By a simple no-arbitrage argument (see Lemma 3.12 (iii)), this would then in fact give us explicit formulas also for all the *forward* exchange rates $a_T^{c_1/c_2}(\mathbf{x}, t)$ for $\forall c_1, c_2 \in C$ and $\forall T \geq t$ via the relation

$$a_T^{c_1/c_2}(\mathbf{x}, t) = \frac{B_{c_1,T}^{c_1}(\mathbf{x}, t)}{B_{c_2,T}^{c_2}(\mathbf{x}, t)} \cdot a^{c_1/c_2}(\mathbf{x}, t). \tag{3.3}$$

Of course, since we want the multi-currency ACE model to be fully unspanned (so that one can calibrate it to all bond- or exchange rate-dependent options via bootstrapping by expiration), we ask that *all* of these bond price functions, as well as all the functions $a^{c_1/c_2}(\mathbf{x}, t)$ that we will define, be independent of one's choice of the functional noise parameter $\sigma(\mathbf{x}, t)$.

In summary: The multi-currency ACE model will be based on the state process $(X_t)_{t \geq 0}$ of the single-currency model, confined to a reduced state space M_0, and it will define short rate and exchange rate processes $(r^c(X_t))_{t \geq 0}$ and $(a^{c_1/c_2}(X_t, t))_{t \geq 0}$ for all currencies $c, c_1, c_2 \in C$ by feeding it into certain functions $r^c(\mathbf{x})$ and $a^{c_1/c_2}(\mathbf{x}, t)$. The model will then provide (σ-independent) analytical formulas for the prices (expressed in any currency) of all the bonds associated to the various currencies, and for all the forward exchange rates.

Outline. We begin in Section 3.2 by discussing a problem that had naturally been absent from the single-currency case: Since the stochastic dynamics of the process $(X_t)_{t \geq 0}$ needs to be associated to some fixed numeraire (i.e., to some base currency $c_0 \in C$), in the multi-currency case this introduces an apparent asymmetry to the model, and we need to ensure we can change the numeraire via Girsanov's Theorem as needed, without encountering any inconsistencies.

In Sections 3.3–3.5 we then explore what other technical conditions our model construction has to fulfill in order to define a consistent and properly symmetric model that actually extends the single-currency ACE model as intended. After defining our multi-currency ACE model in Section 3.6, in Sections 3.7–3.8 we then show that this model indeed satisfies all of these conditions, and in Section 3.9 we compile the required constraints of $\sigma(\mathbf{x}, t)$ and discuss the existence and uniqueness of $(X_t)_{t \geq 0}$. In Section 3.10 we derive the model's bond price, forward rate, and forward exchange rate formulas, and in Section 3.11 we interpret these formulas using the notion of numeraire-free values. Finally, in Section 3.12 we present an alternative parametrization of the multi-currency ACE model that leads to a variety of simplifications and practical advantages.

The impatient practitioner who would like to get to the final model equations as quickly as possible, without being held up by learning about the underlying mathematical challenges first, is recommended to focus on Section 3.2 (only the text leading up to Lemma 3.1), Section 3.6, and Section 3.10 on first reading, maybe followed by Section 3.11 to better understand the general pattern in the bond price and (forward) exchange rate functions, and by Section 3.12.

3.2 Base Currency, Change of Numeraire

In contrast to the single-currency ACE model, its multi-currency extension discussed here will have to take into consideration that when one defines the state process $(X_t)_{t \geq 0}$ (again via the SDE (2.10)–(2.11)), one has to declare which of the currencies $c \in C$ we will choose as the numeraire for future calculations of asset values via expected values.

Once a base currency $c_0 \in C$ is chosen, given any asset whose payout[2] at time T is $P_{c,T}$ units of $c \in C$, and supposing that the system is at time $t \leq T$ and in state $\mathbf{x} \in M_0$, the asset's fair value in units of c can be computed via the formula

$$V_c(\mathbf{x}, t) = \frac{1}{a^{c/c_0}(\mathbf{x}, t)} \cdot \mathbb{E}_{c_0} \left[P_{c,T} \cdot a^{c/c_0}(X_T, T) \cdot \exp\left(- \int_t^T r^{c_0}(X_s) \, ds\right) \Big| X_t = \mathbf{x} \right].$$
(3.4)

Indeed, the payout in units of our base currency c_0 is obtained by multiplying $P_{c,T}$ by the exchange rate $a^{c/c_0}(X_T, T)$, and so the *present* c_0-value of this payout at time t is given by the expression inside the expected value; the fair c_0-value of this asset at time t is therefore the expected value above, computed under the risk-neutral measure of $(X_t)_{t \geq 0}$ associated to the numeraire c_0, which as per our agreement is the measure defined via the SDE (2.10)–(2.11). Finally, we have to convert the result back into units of c by dividing by the current exchange rate $a^{c/c_0}(\mathbf{x}, t)$.

Now it would be convenient if we could alternatively carry out this calculation directly, without having to convert the payout into the base currency c_0, namely by taking the expectation under a different measure. In other words, we would like that for some properly chosen measure $\mathbb{P}_c$ the asset value above can be written as

$$V_c(\mathbf{x}, t) = \mathbb{E}_c \left[P_{c,T} \cdot \exp\left(- \int_t^T r^c(X_s) \, ds\right) \Big| X_t = \mathbf{x} \right].$$
(3.5)

Note that here we have to discount the c-based payout using the short rate function r^c associated to c (instead of r^{c_0}).

In particular, this would imply that once these measures $\mathbb{P}_c$ are obtained, we could compute the bond prices $B^c_{c,T}$ based solely on the short rate function r^c associated to the currency c of interest, via the formula

$$B^c_{c,T}(\mathbf{x}, t) = \mathbb{E}_c \left[\exp\left(- \int_t^T r^c(X_s) \, ds\right) \Big| X_t = \mathbf{x} \right].$$
(3.6)

Bond prices measured in other currencies can then easily be derived via the relation

$$B^{c_1}_{c_2,T}(\mathbf{x}, t) = a^{c_1/c_2}(\mathbf{x}, t) \cdot B^{c_1}_{c_1,T}(\mathbf{x}, t).$$
(3.7)

---

[2]In the case of a maturity-T bond, this payout is simply the constant 1; however, in general this payout may depend on all the information available at time T, such as the path of $(X_t)_{t \in [0,T]}$. Technically, if one denotes by $(\mathcal{F}_t)_{t \geq 0}$ the filtration of the model's probability space to which the process $(X_t)_{t \geq 0}$ is adapted, one must ask that $P_{c,T}$ is a random variable that is measurable with respect to the σ-algebra $\mathcal{F}_T$.

Of course, the existence of such a family of risk-neutral measures $\{\mathbb{P}_c \,|\, c \in C\}$ cannot be taken for granted, as it must be contingent on certain relations between the short rate functions r^c and the exchange rate functions a^{c_1/c_2} (in order to prevent contradictions), and on certain technical conditions. The following lemma provides us with sufficient conditions.

Lemma 3.1 (Change of numeraire). *Let $c \in C$, and let us abbreviate*

$$\alpha^c := \tfrac{1}{a^{c/c_0}} (\nabla a^{c/c_0})^T. \tag{3.8}$$

Suppose that the consistency condition

$$\left[\partial_t + \langle \mu(\mathbf{x}), \nabla \rangle + \tfrac{1}{2}\sigma^2(\mathbf{x}, t)\Delta + \left(r^c(\mathbf{x}) - r^{c_0}(\mathbf{x}) \right) \right] a^{c/c_0}(\mathbf{x}, t) = 0 \tag{3.9}$$

holds for $\forall (\mathbf{x}, t) \in M_0 \times [0, \infty)$, and that the Novikoff condition

$$\mathbb{E}_{c_0} \left[\exp \left(\tfrac{1}{2} \int_0^T \| (\sigma \cdot \alpha^c)(X_s, s) \|^2 \, \mathrm{d}s \right) \right] < \infty \tag{3.10}$$

holds for $\forall T > 0$.

Then there exists a measure $\mathbb{P}_c$ such that the expressions in (3.4) and (3.5) indeed coincide for any payout $P_{c,T}$, and the SDE driving $(X_t)_{t \geq 0}$ can be written as

$$\mathrm{d}X_t = \mu^c(X_t, t) \, \mathrm{d}t + \sigma(X_t, t) \, \mathrm{d}\tilde{W}_t^c, \qquad X_{t=0} = \mathbf{x}_0, \quad \text{(3.11a)}$$

where $\qquad \mu^c(\mathbf{x}, t) := \mu(\mathbf{x}) + (\sigma^2 \cdot \alpha^c)(\mathbf{x}, t),$ (3.11b)

and where $\qquad \tilde{W}_t^c := W_t - \int_0^t (\sigma \cdot \alpha^c)(X_s, s) \, \mathrm{d}s$ (3.11c)

is a Brownian motion under $\mathbb{P}_c$.

Proof. The proof, which works by explicitly applying Girsanov's Theorem, can be found in Appendix C.1. The condition (3.9) is needed to write the desired Radon–Nikodym derivative as a Doléans–Dade exponential (thus ensuring that it is a local martingale), while the Novikoff condition (3.10) is imposed to guarantee that it is in fact a proper martingale.

The reader who is not familiar with these techniques will be fine to move on without understanding this proof. However, for those who have long struggled to properly comprehend Girsanov's Theorem, it will be an excellent exercise to give it a try. $\qquad\qquad\qquad\qquad\qquad\qquad\qquad\qquad\qquad\qquad\qquad\qquad$ $\square$

As we define our multi-currency extension, we need the conditions of this lemma to be satisfied for all currencies $c \in C$. The consistency condition (3.9) tells us how the short rate functions $r^c(\mathbf{x})$ and the exchange rate functions $a^{c_1/c_2}(\mathbf{x}, t)$ need to be related. While this lemma introduces this condition only as a sufficient one, it is actually necessary as well; since the lemma's proof buries not only this condition's necessity but also the intuition behind it, we will revisit it again later in Section 3.3.3.

Furthermore, we observe that since we want to define a modeling framework with a freely choosable noise function $\sigma(\mathbf{x}, t)$, the condition (3.9) actually decouples into two separate requirements, much like the Feynman–Kac PDE (2.35) decoupled into the two equations (2.37):

Corollary 3.1. *In a modeling framework with an unspanned noise parameter σ, the condition that (3.9) holds for any choice of σ is equivalent to asking that the following two conditions both be satisfied for $\forall (\mathbf{x}, t) \in M_0 \times [0, \infty)$:*

$$\big[\partial_t + \langle \mu(\mathbf{x}), \nabla \rangle + \big(r^c(\mathbf{x}) - r^{c_0}(\mathbf{x}) \big)\big] a^{c/c_0}(\mathbf{x}, t) = 0, \tag{3.12}$$

$$\Delta a^{c/c_0}(\mathbf{x}, t) = 0. \tag{3.13}$$

Proof. If one asks that (3.9) be satisfied for any choice of σ, then one can take the limit $\sigma \searrow 0$ to find that the condition (3.12) must hold. (Here we used that in an unspanned model, the exchange rate functions a^{c_1/c_2} are not allowed to depend on σ.) Then subtracting (3.12) from (3.9), we also find that (3.13) must hold.

Since on the other hand the conditions (3.12)–(3.13) clearly imply (3.9) for any choice of σ, they are in fact equivalent to it. $\square$

Once the model is defined in a consistent way so that r^c and a^{c_1/c_2} satisfy (3.12)–(3.13), the Novikoff condition (3.10) gives us a technical constraint on the functions $\sigma(\mathbf{x}, t)$ that one can choose. Note that this condition holds in particular if the integrand is uniformly bounded, i.e., if

$$\sup_{\substack{\mathbf{x} \in M_0 \\ t \in [0, T]}} \sigma(\mathbf{x}, t) \, \|\alpha^c(\mathbf{x}, t)\| < \infty \tag{3.14}$$

for $\forall T > 0$. We will revisit this condition later in Section 3.9.1 once we have defined the functions a^{c_1/c_2} of the multi-currency ACE model.

One should, however, keep in mind that this condition is rather conservative (i.e., more restrictive than necessary): First, (3.14) is much stricter than (3.10), which would allow the integrand to go to ∞ near certain critical states $\mathbf{x}_{\mathrm{crit}}$ as long as the probability that the state process approaches these states vanishes sufficiently fast. Second, the Novikoff condition (3.10), which is used to establish that the process $(\mathcal{E}_t)_{t \geq 0}$ in the proof of Lemma 3.1 is a proper martingale, is only sufficient but not necessary to make that desired conclusion. Future efforts may help relax (3.14) and as a result give us more freedom again to choose σ.

3.3 Model Consistency

The introduction of the various functions r^c and a^{c_1/c_2} as the core of our multi-currency model bears the risk of overspecification and unintended contradicions. As an example, we have already learned that we need them to satisfy the condition (3.9) in order for there to be a consistent set of measures $\{\mathbb{P}_c \,|\, c \in C\}$.

Here we will discuss three additional potential pitfalls. The first is obvious and will lead to a further set of conditions on our exchange rate functions; the other two will turn out to already have been addressed in the preceding section, with the last one also providing an intuitive illustration of the necessity of (3.9).

3.3.1 Exchange Rate Triangle Relation

Beginning with the most intuitive relations, we clearly need to require that

$$\forall c \in C: \quad a^{c,c} \equiv 1, \tag{3.15}$$

and under the idealized assumption of no bid–ask spread also that[3]

$$\forall c_1, c_2 \in C: \quad a^{c_1/c_2} = \frac{1}{a^{c_2/c_1}}. \tag{3.16}$$

Indeed, if the left-hand side were larger then one could make a risk-free profit by exchanging one's cash from either of these currencies into the other and then back right away, whereas if the right-hand side were larger then exchanging cash back and forth would lead to a guaranteed loss, which would imply a non-zero bid–ask spread.

Lastly, given any three (not necessarily distinct) currencies $c_1, c_2, c_3 \in C$, the following should lead to identical results: (i) exchanging a unit of c_1 first for the currency c_2 and then exchanging the obtained amount of c_2 further for the currency c_3, and (ii) exchanging the unit of c_1 directly for the equivalent amount of the currency c_3. In other words, we need to ask that

$$\forall c_1, c_2, c_3 \in C: \quad a^{c_1/c_2} \cdot a^{c_2/c_3} = a^{c_1/c_3}. \tag{3.17}$$

Indeed, if this would not hold, then one could create an arbitrage by exchanging a unit of c_1 via the better of these two exchange paths, and by then exchanging the result amount of c_3 back via the other (utilizing (3.16) to compute the rate for the second step).

Note that (3.17) in fact implies (3.15)–(3.16) by considering the cases $c_1 = c_2 = c_3$ and $c_1 = c_3$, respectively, and so we can make (3.17) our only requirement arising from the arguments in this section.

3.3.2 Short Rate Formula

Next, we should quickly convince ourselves that our functions $r^c(\mathbf{x})$ actually take on the meaning of short rate functions, i.e., that they indeed fulfill the relation

$$r^c(\mathbf{x}) = -\partial_T \log B_{c,T}^c(\mathbf{x}, t)\big|_{T=t}. \tag{3.18}$$

Luckily, this relation already holds under the conditions we derived in Section 3.2, as it is a direct consequence of the formula (3.6). This is the result of us introducing these functions into the model via the relations (3.4)–(3.5) based on our understanding of how short rate functions should be used, and then defining our set of measures $\{\mathbb{P}_c \,|\, c \in C\}$ consistent with these relations.

---

[3]In fact, we already quietly used this relation when we converted the outcome of the expected value in (3.4) from units of c_0 back to units of c.

3.3.3 Exchange Rate–Short Rate Consistency

Finally, the short rate functions $r^c(\mathbf{x})$ need to be compatible with the exchange rate functions $a^{c_1/c_2}(\mathbf{x}, t)$ to not allow for obvious arbitrage opportunities.

Indeed, consider a portfolio that at time t has no value, as it is long 1 c_0 worth of c_0-bonds and short 1 c_0 worth of c-bonds (for some fixed $c \in C \setminus \{c_0\}$), with both types of bonds maturing at time $t + dt$. At time $t + dt$ the two bonds mature and pay out their (deterministic) notionals in their respective currencies, and so one needs to use the exchange rate $a^{c/c_0}(X_{t+dt}, t + dt)$ to compute the total value of this portfolio at time $t + dt$ measured in units of c_0. As this exchange rate is based on a future state and thus random, the c_0-value of this portfolio at time $t + dt$ is random as well, and so we need to ensure that the portfolio value's mean is zero, since otherwise (depending on its sign) the creation of either this portfolio or its inverse would be an arbitrage.

The following lemma not only tells us that this type of arbitrage does not exist in our model as long as our previously derived consistency condition (3.9) holds; its simple proof also implies that this condition is in fact necessary, which was not as intuitively clear from our technically challenging proof of Lemma 3.1.

Lemma 3.2. *The strategy described above does not present an arbitrage opportunity if and only if the consistency condition (3.9) of Lemma 3.1 is satisfied.*

Proof. Let us suppose that the system is in state $\mathbf{x}$ at time t. Then the portfolio described above contains $1/B^{c_0}_{c_0,t+dt}(\mathbf{x}, t)$ many bonds of the first type and $-1/B^c_{c_0,t+dt}(\mathbf{x}, t)$ many bonds of the second type, and so the portfolio value at time $t + dt$, measured in units of c_0, is given by

$$\frac{1}{B^{c_0}_{c_0,t+dt}(\mathbf{x}, t)} \cdot 1 - \frac{1}{B^c_{c_0,t+dt}(\mathbf{x}, t)} \cdot a^{c/c_0}(X_{t+dt}, t + dt)$$

$$= \frac{1}{B^{c_0}_{c_0,t+dt}(\mathbf{x}, t)} - \frac{1}{B^c_{c,t+dt}(\mathbf{x}, t) \cdot a^{c/c_0}(\mathbf{x}, t)} \cdot a^{c/c_0}(X_{t+dt}, t + dt)$$

$$\approx \frac{1}{1 - r^{c_0}(\mathbf{x})\, dt} - \frac{1}{(1 - r^c(\mathbf{x})\, dt) \cdot a^{c/c_0}(\mathbf{x}, t)}$$
$$\times \left[a^{c/c_0}(\mathbf{x}, t) + \left(\partial_t + \langle \mu(\mathbf{x}), \nabla \rangle + \tfrac{1}{2}\sigma^2(\mathbf{x}, t)\Delta \right) a^{c/c_0}(\mathbf{x}, t)\, dt + (\ldots)\, dW_t \right]$$

$$\approx (1 + r^{c_0}(\mathbf{x})\, dt) - (1 + r^c(\mathbf{x})\, dt)$$
$$\times \left[1 + \frac{1}{a^{c/c_0}(\mathbf{x}, t)} \left(\partial_t + \langle \mu(\mathbf{x}), \nabla \rangle + \tfrac{1}{2}\sigma^2(\mathbf{x}, t)\Delta \right) a^{c/c_0}(\mathbf{x}, t)\, dt + (\ldots)\, dW_t \right]$$

$$\approx \left[r^{c_0}(\mathbf{x}) - r^c(\mathbf{x}) - \frac{1}{a^{c/c_0}(\mathbf{x}, t)} \left(\partial_t + \langle \mu(\mathbf{x}), \nabla \rangle + \tfrac{1}{2}\sigma^2(\mathbf{x}, t)\Delta \right) a^{c/c_0}(\mathbf{x}, t) \right] dt$$
$$+ (\ldots)\, dW_t \,,$$

where all approximations hold up to order dt. The mean of this expression (i.e., the dt-term) vanishes if and only if (3.9) holds. $\qquad \square$

If the reader is now wondering whether we need to require the condition (3.9) to hold for *all* currency pairs (c_1, c_2) and not only for those of the form

(c, c_0)—of course with μ replaced by the drift μ^{c_2} as defined in (3.11b)—we recommend they challenge themself to show that the conditions for the missing currency pairs in fact follow from the ones involving c_0 and the triangle relation (3.17), which means that they are indeed redundant.

3.4   Single-Currency Projections

Next, we need to ask ourselves under which conditions our multi-currency model is actually an extension of the single-currency ACE model, i.e., when the model's single-currency projections to the various currencies $c \in C$ are each indeed short rate models of the form defined in Chapter 2: While it is obvious that under $\mathbb{P}_{c_0}$ the process $(r^{c_0}(X_t))_{t \geq 0}$ will indeed have the distribution of the single-currency ACE model's short rate process if only we define $(X_t)_{t \geq 0}$ and $r^{c_0}(\mathbf{x})$ as in Chapter 2, we must ask that for *every* $c \in C$, *under the measure* $\mathbb{P}_c$ (!) the process $(r^c(X_t))_{t \geq 0}$ should have a distribution of this class (with certain currency-specific parameters).

Unfortunately, we already know that under $\mathbb{P}_c$ the process $(X_t)_{t \geq 0}$ does *not* have the dynamics of the single-currency ACE model's state process (since the drift μ^c of its SDE (3.11a)–(3.11b) under $\mathbb{P}_c$ depends on σ), and so simply defining the remaining functions r^c as in Chapter 2 won't work. To resolve this issue, we will instead try to find, for each $c \in C$, a reparametrization mapping $G_c \colon M_0 \to \mathbb{R}^n$ such that under $\mathbb{P}_c$ the process

$$\hat{X}_t^c := G_c(X_t) \tag{3.19}$$

has the dynamics of the single-currency ACE model's state process, and such that

$$\hat{r}^c(\hat{\mathbf{x}}) := r^c(G_c^{-1}(\hat{\mathbf{x}})) \tag{3.20}$$

has the form (2.12) of the single-currency ACE model's short rate function (with matching parameters $\mathbf{v}^c$);[4] this would then imply that under $\mathbb{P}_c$ the process $\hat{r}^c(\hat{X}_t^c)$ has the dynamics of the single-currency ACE model's short rate process, and since

$$r^c(X_t) = r^c(G_c^{-1}(\hat{X}_t^c)) = \hat{r}^c(\hat{X}_t^c),$$

this is exactly what we want.

In summary: While $(X_t)_{t \geq 0}$ under $\mathbb{P}_c$ and $r^c(\mathbf{x})$ would *individually* not be as in Chapter 2, the distribution of $(r^c(X_t))_{t \geq 0}$ under $\mathbb{P}_c$ still would, and this is all we want.

Making this construction work and actually finding such reparametrization maps G_c, however, is in fact a lot to ask, and for general models other than the ACE model it may well be that such an extension does not exist: Even just the

---

[4]Note that the challenge here is really only to ensure that (3.19) has the right form; the relation (3.20) can then simply be achieved by defining $r^c(\mathbf{x}) := \hat{r}^c(G_c(\mathbf{x}))$, where $\hat{r}^c$ is of the form (2.12) for some currency-specific parameters r_∞^c and $\mathbf{v}^c$.

requirement that the change of variables (3.19) preserves the multiplicative nature of the noise[5] of the SDE (3.11a)–(3.11b) imposes a very strong constraint on G_c; furthermore, both parts of this two-step transformation (change of measure and change of variables) create terms involving σ in the drift $\hat{\mu}^c$ of $(\hat{X}_t^c)_{t\geq 0}$, and we need to ensure that the combination of these terms vanishes; finally, of course, in addition we have to achieve that the remaining terms of $\hat{\mu}^c$ have the form of the single-currency ACE model's drift again.

The following Lemma will illuminate the situation, by actually carrying out this change of variables with Itô's Lemma, inspecting the resulting SDE for $(\hat{X}_t^c)_{t\geq 0}$, and then explicitly listing all the conditions on the funcions G_c, r^c and a^{c_1/c_2} under which this construction will work.

Lemma 3.3. *Suppose that the consistency conditions (3.12)–(3.13) and the Novikoff condition (3.10) are satisfied. Given any $c \in C \setminus \{c_0\}$ and any C^2-diffeomorphism[6] $G_c\colon M_0 \to \mathbb{R}^n$, further suppose that we apply Itô's Lemma to the SDE (3.11a) to obtain the SDE that describes the dynamics of the reparametrized process $\hat{X}_t^c := G_c(X_t)$ under the measure $\mathbb{P}_c$. Then the following holds true:*

(i) The SDE has multiplicative noise (i.e., its noise term can be written as a scalar function times the increment of a Brownian motion) for all choices[7] of σ if and only if G_c is conformal (i.e., angle-preserving), meaning that for some function $s_c\colon M_0 \to (0,\infty)$ we have

$$(\nabla G_c)^T (\nabla G_c) = s_c^2 \cdot I. \tag{3.21}$$

(ii) The SDE's drift term is independent of σ if and only if $a^{c/c_0} \cdot G_c$ is componentwise harmonic, i.e., if

$$\Delta\big(a^{c/c_0}(\mathbf{x},t) \cdot G_c(\mathbf{x})^i\big) = 0 \qquad for \qquad i = 1, \ldots, n. \tag{3.22}$$

(iii) Under the conditions in (i)–(ii), the SDE reduces to

$$\mathrm{d}\hat{X}_t^c = \hat{\mu}^c(\hat{X}_t^c)\,\mathrm{d}t + \hat{\sigma}^c(\hat{X}_t^c,t)\,\mathrm{d}\hat{W}_t^c, \quad \hat{X}_{t=0}^c = G_c(\mathbf{x}_0), \tag{3.23a}$$

where $$\hat{\mu}^c(\hat{\mathbf{x}}) := \big[\nabla G_c(\mathbf{x})\mu(\mathbf{x})\big]_{\mathbf{x}=G_c^{-1}(\hat{\mathbf{x}})} \tag{3.23b}$$

and $$\hat{\sigma}^c(\hat{\mathbf{x}},t) := \big[s_c(\mathbf{x}) \cdot \sigma(\mathbf{x},t)\big]_{\mathbf{x}=G_c^{-1}(\hat{\mathbf{x}})}, \tag{3.23c}$$

and where $$\hat{W}_t^c := \int_0^t s_c(X_s)^{-1} \nabla G_c(X_s)\,\mathrm{d}\tilde{W}_s^c \tag{3.23d}$$

is a Brownian motion under $\mathbb{P}_c$. Therefore, under $\mathbb{P}_c$ the process $(r^c(X_t))_{t\geq 0}$ has the dynamics of the single-currency ACE model's short rate process if $\hat{\mu}^c$ has the form of the ACE drift (2.11) and if the short rate function

$$\hat{r}^c(\hat{\mathbf{x}}) := r^c(G_c^{-1}(\hat{\mathbf{x}})) \tag{3.24}$$

in this parametrization has the form of the ACE short rate function (2.12), with matching parameters $\mathbf{v}^c$.

Proof. See Appendix C.2. □

---

[5]A process $(X_t)_{t\geq 0}$ is said to have multiplicative noise if the prefactor of the $\mathrm{d}W_t$ term driving its defining SDE is a scalar (as opposed to matrix-valued) function.

[6]A C^2-diffeomorphism is an invertible C^2-mapping whose inverse is C^2 as well.

[7]If σ vanishes in parts of M_0 for $\forall t \geq 0$, then (3.21) need not hold in these regions; however, our functions G_c need to work for *all* choices of σ, in particular strictly positive ones.

3.5　Model Symmetry

Finally, our model definition singles out a specific base currency c_0, apparently breaking the symmetry of the model. Clearly, we would want that the model treats all currencies equally, and that this apparent asymmetry is a mere illusion caused by notational limitations.

To clarify what exactly is causing our concern, recall that the multi-currency ACE model is the joint distribution of the collection of processes

$$(r^c(X_t))_{t\geq 0} \qquad \text{and} \qquad (a^{c_1/c_2}(X_t, t))_{t\geq 0} \tag{3.25}$$

for $c, c_1, c_2 \in C$, which are defined via a set of deterministic functions $r^c(\mathbf{x})$ and $a^{c_1/c_2}(\mathbf{x}, t)$ and a process $(X_t)_{t\geq 0}$ (where a process is really a measure on the space of paths in $\mathbb{R}^n$, often given in the form of an SDE). More precisely, our model actually provides us with a whole *family* of measures $\{\mathbb{P}_c \mid c \in C\}$ for $(X_t)_{t\geq 0}$, which in practical applications we can later choose from depending on which numeraire we want to use in our pricing calculations. The apparent asymmetry of our construction stems from the fact that to define this family, we needed to first define the measure $\mathbb{P}_{c_0}$ for some specific currency c_0 via the SDE (2.10)–(2.11), and then derive the remaining measures via Lemma 3.1.

We would want that the measure $\mathbb{P}_{c_0}$ does in fact *not* play any special role within this family of measures, and that we could equally have defined the joint distribution of the processes (3.25) under any other member $\mathbb{P}_{\bar{c}_0}$ of this family *directly*, by choosing the right parameters of the multi-currency ACE model and using $\bar{c}_0 \in C$ as the base currency.

To ensure that this is indeed the case, we will go one step further than in Lemma 3.3 (iii) and ask that for each $\bar{c}_0 \in C \setminus \{c_0\}$, the reparametrized short rate and exchange rate functions

$$\hat{r}^c_{\bar{c}_0}(\hat{\mathbf{x}}) := r^c(G_{\bar{c}_0}^{-1}(\hat{\mathbf{x}})) \qquad \text{and} \qquad \hat{a}^{c_1/c_2}_{\bar{c}_0}(\hat{\mathbf{x}}, t) := a^{c_1/c_2}(G_{\bar{c}_0}^{-1}(\hat{\mathbf{x}}), t), \tag{3.26}$$

for $c, c_1, c_2 \in C$, have the forms of the functions of the multi-currency ACE model again for certain parameters, but with $\bar{c}_0$ as the base currency; in addition, as before we also ask that under $\mathbb{P}_{\bar{c}_0}$ the process $\hat{X}^{\bar{c}_0}_t = G_{\bar{c}_0}(X_t)$ has the dynamics of the ACE state process (for compatible parameters). In this way, since

$$\hat{r}^c_{\bar{c}_0}(\hat{X}^{\bar{c}_0}_t) = r^c(X_t) \qquad \text{and} \qquad \hat{a}^{c_1/c_2}_{\bar{c}_0}(\hat{X}^{\bar{c}_0}_t, t) = a^{c_1/c_2}(X_t, t),$$

we could then conclude that the joint distribution of the processes (3.25) under $\mathbb{P}_{\bar{c}_0}$ can indeed be defined directly, by considering the multi-currency ACE model with those parameters and with $\bar{c}_0$ as the base currency.

3.6　Model Definition

Now that we have collected all the key properties that our multi-currency extension of the ACE model will have to fulfill, let us go ahead and actually define it. We will begin by introducing all its model parameters and discussing their constraints. Then we will define the state process $(X_t)_{t \geq 0}$ (identically to the one of the single-currency ACE model), the reparametrization maps $G_c(\mathbf{x})$, the short rate functions $r^c(\mathbf{x})$, and the exchange rate functions $a^{c_1/c_2}(\mathbf{x}, t)$. Finally, we will define the state space M_0 and analyze its shape.

Only later in Sections 3.7–3.8 we will then actually show that the model thus defined indeed satisfies all the conditions derived in Sections 3.2–3.5, so that it is really a fully consistent and symmetric multi-currency model that extends the single-currency ACE model, as intended.

3.6.1 Model Parameters

Our $\mathbb{R}^n$-based description of the multi-currency ACE model is again valid for all dimensions
$$n \geq 3.$$
Let us denote by C the finite set of currencies whose rates we intend to model, e.g., $C := \{\$, \euro, \pounds\}$, and let $c_0 \in C$ denote the base currency that we intend to use as the primary numeraire for our pricing calculations.[8]

The model has the following currency-specific parameters for each currency $c \in C$: the *long-term rate*
$$r_\infty^c > 0, \tag{3.27}$$
the *weight vector*[9]
$$\mathbf{v}^c \in \mathbb{R}^n \setminus \{\vec{0}\}, \tag{3.28}$$
and the *initial currency spot value attractor*[10]
$$a^c > 0. \tag{3.29}$$

In addition, the model has the following global parameters: the *mean reversion speed*
$$\lambda > 0, \tag{3.30}$$
the *oscillation frequency matrix* $E \in \mathbb{R}^{n \times n}$, which we assume to be antisymmetric, i.e.,
$$E^T = -E, \tag{3.31}$$
and the *initial state*
$$\mathbf{x}_0 \in M_0 \tag{3.32}$$
in our state space $M_0 \subset \mathbb{R}^n$, which we will define later in Section 3.6.6 (and which will depend on c_0, λ, E, and all the parameters r_∞^c and $\mathbf{v}^c$). Finally, the model allows us to choose the continuous *noise function*
$$\sigma \colon M_0 \times [0, \infty) \to (0, \infty), \tag{3.33}$$
which is subject to certain constraints laid out later in Section 3.6.3.

---

[8]Switching to any other currency as the numeraire will be possible at any time, but that will require carrying out the necessary conversions laid out in the preceding sections.

[9]In the multi-currency ACE model we will *not* ask that *all* vectors $\mathbf{v}^c$ have unit length; see Section 3.6.2 for details.

[10]See Section 3.11 for its interpretation.

3.6.2   Parameter Constraints

To avoid parametric redundancies (i.e., that different choices of parameters lead to equivalent models) we impose the gauge fixing constraints that[11]

$$\|\mathbf{v}^{c_0}\| = 1 \qquad \text{and} \qquad a^{c_0} = 1, \tag{3.34}$$

and (again by the arguments laid out in Section 2.7.1) that E and $\mathbf{v}^{c_0}$ are of the forms (2.41)–(2.43).

In addition, to ensure that the process $(X_t)_{t\geq 0}$ remains confined to M_0, the parameters r_∞^c, $\mathbf{v}^c$, E, and λ are subject to the *resiliency constraint*[12]

$$\delta := \min_{c\in C} \delta_c > 0, \tag{3.35a}$$

$$\text{where} \quad \delta_c := r_\infty^c \lambda - \frac{(r_\infty^c)^2}{n-2} - \frac{n-2}{4}\frac{\|E\mathbf{v}^c\|^2}{\|\mathbf{v}^c\|^2}, \tag{3.35b}$$

or equivalently,

$$\lambda > \lambda_{\min} := \max_{c\in C}\left(\frac{r_\infty^c}{n-2} + \frac{n-2}{4r_\infty^c}\frac{\|E\mathbf{v}^c\|^2}{\|\mathbf{v}^c\|^2}\right), \tag{3.36}$$

and the functional parameter $\sigma(\mathbf{x}, t)$ is assumed to satisfy the conditions stated later in Section 3.9.2, namely that it must have some specific additional regularity, that it must not grow too fast as $\mathbf{x} \to \infty$ (if M_0 is unbounded), and that it must vanish fast enough as $\mathbf{x}$ approaches the boundary ∂M_0.

Taking into account all of these constraints, in addition to the parametric freedom from the functional parameter $\sigma(\mathbf{x}, t)$, the model's discrete parameters r_∞^c, $\mathbf{v}^c$, a^c, λ, E, and $\mathbf{x}_0$ provide a total of

$$\begin{aligned}
\# \text{ degrees of freedom} \\
= |C| + \big[(\lceil n/2\rceil - 1) + (|C| - 1)\cdot n\big] + (|C| - 1) + 1 + \lfloor n/2\rfloor + n \\
= (n-1) + |C|\cdot(n+2) \tag{3.37a} \\
= (2n+1) + (|C| - 1)\cdot(n+2) \tag{3.37b}
\end{aligned}$$

degrees of freedom: $2n + 1$ (as in the single-currency ACE model) from the global parameters and the parameters associated to c_0, plus a further $n + 2$ for every additional currency from the associated parameters r_∞^c, $\mathbf{v}^c$, and a^c.

---

[11] Indeed, just as in the last paragraph of Section 2.7.1, it is an easy exercise to show that for the model that we will define in the following, scaling all the vectors $\mathbf{v}^c$ by a fixed factor can again be counteracted by scaling $\mathbf{x}_0$ and σ as in (2.46) without affecting the short rate processes $(r^c(X_t))_{t\geq 0}$ and the exchange rate processes $(a^{c_1/c_2}(X_t))_{t\geq 0}$; this allows us to ask w.l.o.g. that one of the vectors $\mathbf{v}^c$ have unit length, e.g., $\mathbf{v}^{c_0}$.

The reason why w.l.o.g. one can choose to set any one of the parameters a^c to 1, for example a^{c_0}, is that only their ratios will enter the definition of our model (via (3.51)).

[12] Recall that to allow for general vectors $\mathbf{v}^c \in \mathbb{R}^n \setminus \{\vec{0}\}$, we need to generalize the constraint (2.9a) to (2.48). Furthermore, to facilitate the already very technical proof of our generalized existence and uniqueness criterion in Lemma 3.11, we will now no longer allow for the degenerate case $\delta = 0$.

3.6.3 State Process

Given any such set of currency-specific parameters r_∞^c, $\mathbf{v}^c$, and a^c ($c \in C$), and of global parameters λ, E, $\mathbf{x}_0$, and $\sigma(\mathbf{x}, t)$, the multi-currency ACE model is driven by the n-dimensional process $(X_t)_{t \geq 0}$ satisfying the SDE

$$\mathrm{d}X_t = \mu(X_t)\,\mathrm{d}t + \sigma(X_t, t)\,\mathrm{d}W_t\,, \qquad X_{t=0} = \mathbf{x}_0\,, \tag{3.38}$$

where $(W_t)_{t \geq 0}$ is an n-dimensional Brownian motion, and where the drift $\mu \colon \mathbb{R}^n \to \mathbb{R}^n$ is defined as

$$\mu(\mathbf{x}) := (-\lambda I + E)\mathbf{x} + \tfrac{1}{n-2}\|\mathbf{x}\|^2 \mathbf{v}^{c_0} - \tfrac{2}{n-2}\langle \mathbf{v}^{c_0}, \mathbf{x}\rangle \mathbf{x}. \tag{3.39}$$

In other words, the SDE is identical to the SDE (2.10)–(2.11) for the single-currency case, with the parameter $\mathbf{v}^{c_0}$ associated to the base currency c_0.

As we will see later in Section 3.9.2, the constraints on our parameters guarantee that the drift points inwards near the boundary of our new state space M_0 that we will define soon in Section 3.6.6, and our conditions imposed on σ in Lemma 3.11 will then suffice to ensure that the SDE (3.38) has a unique strong solution $(X_t)_{t \geq 0}$ that is confined to M_0 almost surely.

3.6.4 Reparametrization Maps

Next, we will define the family of reparametrization maps $G_c \colon M_0 \to \mathbb{R}^n$, $c \in C \setminus \{c_0\}$, that our construction relies upon, as motivated in Sections 3.4–3.5. To do so, we begin by defining a group of maps $H_\mathbf{c} \colon \bar{\mathbb{R}}^n \to \bar{\mathbb{R}}^n$, $\mathbf{c} \in \mathbb{R}^n$, from which they will be chosen; these maps $H_\mathbf{c}$ are defined as

$$H_\mathbf{c}(\mathbf{x}) := \frac{\frac{\mathbf{x}}{\|\mathbf{x}\|^2} - \mathbf{c}}{\left\|\frac{\mathbf{x}}{\|\mathbf{x}\|^2} - \mathbf{c}\right\|^2} \tag{3.40}$$

for $\forall \mathbf{c} \in \mathbb{R}^n$ and $\forall \mathbf{x} \in \bar{\mathbb{R}}^n$, where as before we agree on the conventions

$$\frac{\vec{0}}{\|\vec{0}\|^2} = \infty \qquad \text{and} \qquad \frac{\infty}{\|\infty\|^2} = \vec{0}. \tag{3.41}$$

The following lemma will compile their most important properties.

Lemma 3.4. *(i) For each $\mathbf{c} \in \mathbb{R}^n$ we have*

$$H_\mathbf{c}(\vec{0}) = \vec{0}, \qquad H_\mathbf{c}\big(\tfrac{\mathbf{c}}{\|\mathbf{c}\|^2}\big) = \infty, \qquad \text{and} \qquad H_\mathbf{c}(\infty) = -\tfrac{\mathbf{c}}{\|\mathbf{c}\|^2}\,, \tag{3.42}$$

and $H_\mathbf{c}$ is finite on $\bar{\mathbb{R}}^n \setminus \big\{\tfrac{\mathbf{c}}{\|\mathbf{c}\|^2}\big\}$.

(ii) The maps $H_\mathbf{c}$, $\mathbf{c} \in \mathbb{R}^n$, form a group, i.e.,

  (ii.1) the map $H_{\vec{0}}$ is the identity map on $\bar{\mathbb{R}}^n$, and

  (ii.2) for $\forall \mathbf{c}_1, \mathbf{c}_2 \in \mathbb{R}^n$ we have $H_{\mathbf{c}_1} \circ H_{\mathbf{c}_2} = H_{\mathbf{c}_1 + \mathbf{c}_2}$.

In particular, the maps $H_\mathbf{c}$ are invertible, with $H_\mathbf{c}^{-1} = H_{-\mathbf{c}}$.

(iii) For $\forall \mathbf{c} \in \mathbb{R}^n$ and $\forall \mathbf{x} \in \mathbb{R}^n \setminus \{\vec{0}\}$ we have[13]

$$\frac{\|\mathbf{x}\|}{\|H_{\mathbf{c}}(\mathbf{x})\|} = \|\mathbf{x}\| \left\| \frac{\mathbf{x}}{\|\mathbf{x}\|^2} - \mathbf{c} \right\| = \left\| \|\mathbf{c}\|\mathbf{x} - \frac{\mathbf{c}}{\|\mathbf{c}\|} \right\|, \tag{3.43}$$

where here and in the following we use the convention

$$\left\| \|\vec{0}\|\mathbf{x} - \frac{\vec{0}}{\|\vec{0}\|} \right\| := 1 \qquad for \ \forall \mathbf{x} \in \mathbb{R}^n. \tag{3.44}$$

(iv) The maps $H_{\mathbf{c}}$ are C^∞ on $\mathbb{R}^n \setminus \left\{ \frac{\mathbf{c}}{\|\mathbf{c}\|^2} \right\}$, with

$$\nabla H_{\mathbf{c}}(\mathbf{x}) = \begin{cases} \left\| \|\mathbf{c}\|\mathbf{x} - \frac{\mathbf{c}}{\|\mathbf{c}\|} \right\|^{-2} U_{H_{\mathbf{c}}(\mathbf{x})} U_{\mathbf{x}} & for \ \mathbf{x} \neq \vec{0}, \\ I & for \ \mathbf{x} = \vec{0}, \end{cases} \tag{3.45}$$

where $\qquad U_{\mathbf{y}} := I - 2\frac{\mathbf{y} \otimes \mathbf{y}}{\|\mathbf{y}\|^2} \qquad \forall \mathbf{y} \in \mathbb{R}^n \setminus \{\vec{0}\}.$ $\qquad (3.46)$

The matrices $U_{\mathbf{y}}$ fulfill $U_{\mathbf{y}}^T = U_{\mathbf{y}}$ and $U_{\mathbf{y}} U_{\mathbf{y}} = I$, and thus they are orthogonal.

Proof. (i), (ii.1): These parts follow right from the definition of $H_{\mathbf{c}}$ and our conventions (3.41).

(ii.2) Since we can write $H_{\mathbf{c}} = i \circ q_{\mathbf{c}} \circ i$, where $i(\mathbf{x}) := \frac{\mathbf{x}}{\|\mathbf{x}\|^2}$ (which is an involution, i.e., $i \circ i$ is the identity map) and where $q_{\mathbf{c}}(\mathbf{x}) := \mathbf{x} - \mathbf{c}$ (which satisfies the group property $q_{\mathbf{c}_1} \circ q_{\mathbf{c}_2} = q_{\mathbf{c}_1 + \mathbf{c}_2}$), we have

$$H_{\mathbf{c}_1} \circ H_{\mathbf{c}_2} = (i \circ q_{\mathbf{c}_1} \circ i) \circ (i \circ q_{\mathbf{c}_2} \circ i) = i \circ q_{\mathbf{c}_1} \circ (i \circ i) \circ q_{\mathbf{c}_2} \circ i$$
$$= i \circ q_{\mathbf{c}_1} \circ q_{\mathbf{c}_2} \circ i = i \circ q_{\mathbf{c}_1 + \mathbf{c}_2} \circ i = H_{\mathbf{c}_1 + \mathbf{c}_2}.$$

Considering the case $\mathbf{c}_2 = -\mathbf{c}_1$ then also leads to the last statement of part (ii).

(iii) The first equality follows right from the definition of $H_{\mathbf{c}}$ as well; the second one can be seen by computing the squares:

$$\|\mathbf{x}\|^2 \left\| \frac{\mathbf{x}}{\|\mathbf{x}\|^2} - \mathbf{c} \right\|^2 = 1 - 2\langle \mathbf{x}, \mathbf{c} \rangle + \|\mathbf{x}\|^2 \|\mathbf{c}\|^2 = \left\| \|\mathbf{c}\|\mathbf{x} - \frac{\mathbf{c}}{\|\mathbf{c}\|} \right\|^2.$$

(iv) See Appendix C.3. $\qquad\qquad\qquad\qquad\qquad\qquad\qquad\qquad\qquad\qquad\qquad \square$

We now define the *spot currency vectors* $\mathbf{c}^c \in \mathbb{R}^n$, and based on those, the *reparametrization maps* $G_c \colon M_0 \to \mathbb{R}^n$, as

$$\mathbf{c}^c := \frac{1}{n-2}(E + \lambda I)^{-1}(\mathbf{v}^c - \mathbf{v}^{c_0}), \tag{3.47}$$

$$G_c := H_{\mathbf{c}^c} \tag{3.48}$$

for $\forall c \in C$.

Note that under our assumption (3.30) $\lambda I + E$ is invertible, since by Lemma 2.2 (ii) we have $\|(\lambda I + E)\mathbf{x}\|^2 = \|\lambda \mathbf{x}\|^2 + 2\lambda \langle \mathbf{x}, E\mathbf{x} \rangle + \|E\mathbf{x}\|^2 \geq \lambda^2 \|\mathbf{x}\|^2 + 0 + 0$. This ensures that $\mathbf{c}^c$ is indeed well defined.

---

[13] In particular, the expression $\left\| \|\mathbf{c}\|\mathbf{x} - \frac{\mathbf{c}}{\|\mathbf{c}\|} \right\|$ is invariant under the exchange of $\mathbf{x}$ with $\mathbf{c}$.

Also observe that $\mathbf{c}^c$ and thus also G_c implicitly depend on our choice of the base currency c_0. In particular, we have

$$\mathbf{c}^{c_0} = \vec{0} \qquad \text{and thus} \qquad G_{c_0} = id \tag{3.49}$$

by Lemma 3.4 (ii.1).

3.6.5 Short Rate and Exchange Rate Functions

We are now ready to define our short rate functions as

$$r^c(\mathbf{x}) := r_\infty^c + \langle \mathbf{v}^c, G_c(\mathbf{x}) \rangle \tag{3.50}$$

for $\forall c \in C$ and $\forall \mathbf{x} \in \bar{\mathbb{R}}^n \setminus \{\frac{\mathbf{c}^c}{\|\mathbf{c}^c\|^2}\}$, and our exchange rate functions as

$$a^{c_1/c_2}(\mathbf{x}, t) := \frac{a^{c_1} \cdot e^{-r_\infty^{c_1} t} \cdot \left\| \|\mathbf{c}^{c_2}\| \mathbf{x} - \frac{\mathbf{c}^{c_2}}{\|\mathbf{c}^{c_2}\|} \right\|^{n-2}}{a^{c_2} \cdot e^{-r_\infty^{c_2} t} \cdot \left\| \|\mathbf{c}^{c_1}\| \mathbf{x} - \frac{\mathbf{c}^{c_1}}{\|\mathbf{c}^{c_1}\|} \right\|^{n-2}} \tag{3.51}$$

for $\forall c_1, c_2 \in C$ and $\forall \mathbf{x} \in \bar{\mathbb{R}}^n \setminus \{\frac{\mathbf{c}^{c_1}}{\|\mathbf{c}^{c_1}\|^2}\}$. Note that both of these functions implicitly depend on our choice of the base currency c_0, because of their dependence on G_c, $\mathbf{c}^{c_1}$ and $\mathbf{c}^{c_2}$.

3.6.6 State Space

With the short rate functions properly defined, we can now finally define our state space M_0. Since in the multi-currency ACE model we want *all* of the short rates to be positive almost surely, we have to define it as[14]

$$M_0 := \left\{ \mathbf{x} \in \mathbb{R}^n \,\middle|\, \forall c \in C : r^c(\mathbf{x}) > 0 \right\} \tag{3.52a}$$

$$= \bigcap_{c \in C} M^c, \quad \text{where} \quad M^c := \left\{ \mathbf{x} \in \mathbb{R}^n \,\middle|\, r^c(\mathbf{x}) > 0 \right\} \tag{3.52b}$$

and then later in Section 3.9.2 ensure that our state process $(X_t)_{t \geq 0}$ remains confined to this set M_0.

The following lemma helps us understand the shape of M_0. We recommend the reader to not get distracted by the exact formulas in (3.54)–(3.55), as the main takeaway of parts (ii)–(iii) is only the general shape of the sets M^c.

Lemma 3.5. *Consistent with* (3.50), (3.48), *and* (3.42), *for all currencies*[15] $c \in C$ *with* $\mathbf{c}^c \neq \vec{0}$ *let us set*

$$r^c(\infty) := r_\infty^c - \langle \mathbf{v}^c, \frac{\mathbf{c}^c}{\|\mathbf{c}^c\|^2} \rangle. \tag{3.53}$$

---

[14]In both (3.52a) and (3.52b) the condition $r^c(\mathbf{x}) > 0$ is meant to tacitly include the additional requirement that $r^c(\mathbf{x})$ is defined, i.e., that $\mathbf{x} \neq \frac{\mathbf{c}^c}{\|\mathbf{c}^c\|^2}$.

[15]In practice, $r^c(\infty)$ will be defined for $\forall c \in C \setminus \{c_0\}$. Indeed, by (3.47) we have $\mathbf{c}^{c_0} = \vec{0}$, and any other currency $c \in C \setminus \{c_0\}$ with $\mathbf{c}^c = \vec{0}$ would have to fulfill $\mathbf{v}^c = \mathbf{v}^{c_0}$, i.e., by (3.50) its short rate process would be a shifted version of the one of c_0, which is not worth modeling explicitly.

(i) For $\forall c \in C$, the set M^c is open, connected, and contains $\vec{0}$, and so the same holds true for M_0.

(ii) For $\forall c \in C$ such that $\mathbf{c}^c = \vec{0}$ or $r^c(\infty) = 0$, the set M^c is a half space, given by

$$M^c = \left\{ \mathbf{x} \in \mathbb{R}^n \mid \langle \mathbf{v}^c - 2r_\infty^c \mathbf{c}^c, \mathbf{x} \rangle + r_\infty^c > 0 \right\}. \tag{3.54}$$

(iii) For all other $c \in C$, M^c is either the inside (if $r^c(\infty) < 0$) or the outside (if $r^c(\infty) > 0$) of the open ball with

$$center = \frac{r_\infty^c \mathbf{c}^c - \frac{1}{2}\mathbf{v}^c}{\|\mathbf{c}^c\|^2 \, r^c(\infty)} \qquad and \qquad radius = \frac{\frac{1}{2}\|\mathbf{v}^c\|}{\|\mathbf{c}^c\|^2 \, |r^c(\infty)|} \, . \tag{3.55}$$

(iv) In either case, we have[16] $\frac{\mathbf{c}^c}{\|\mathbf{c}^c\|^2} \in \partial M^c$ and thus $\frac{\mathbf{c}^c}{\|\mathbf{c}^c\|^2} \notin M_0$, and so the functions G_c, $c \in C$, are finite on M_0. In any neighborhood of the point $\frac{\mathbf{c}^c}{\|\mathbf{c}^c\|^2}$, $r^c(\mathbf{x})$ takes on every real value, whereas on the remaining part of ∂M^c we have $r^c(\mathbf{x}) = 0$.

Proof. See Section C.4.   □

Note that while the functions r^c are well defined and finite on M_0 by our definition (3.52a) of M_0, the functions a^{c_1/c_2} are well defined and finite on M_0 because by Lemma 3.5 (iv) their only potentially problematic point $\frac{\mathbf{c}^{c_1}}{\|\mathbf{c}^{c_1}\|^2}$ does not belong to M_0.

3.7   Model Consistency and Projections – Proofs

Now that we have defined our model, let us ensure that its short rate, exchange rate, and reparametrization functions indeed satisfy the conditions that we have derived in the preceding sections.

Lemma 3.6. *The short rate functions $r^c(\mathbf{x})$, the exchange rate functions $a^{c_1/c_2}(\mathbf{x}, t)$, and the reparametrization maps $G_c(\mathbf{x})$ defined in Section 3.6 satisfy the consistency conditions (3.12)–(3.13), the exchange rate triangle relation (3.17), and the conditions (3.21)–(3.22) from Lemma 3.3.*

Proof. The mostly straight-forward calculations proving these relations can be found in Appendix C.5. Since it is these calculations that ultimately prove that our multi-currency model works, we strongly recommend the reader to eventually take a closer look.   □

With all these conditions verified, we can now use the formulas derived in Lemma 3.3 (iii) to compute the SDE driving $(\hat{X}_t^c)_{t \geq 0}$ under $\mathbb{P}_c$.

In the statements to come, both in the remainder of this section and later in Section 3.8, note that technically we have yet to prove that the process $(X_t)_{t \geq 0}$ is in fact properly defined (i.e., that the SDE (3.38)–(3.39) has a unique solution),

---

[16]In the case of part (ii) it is understood that the half space ∂M^c contains the point ∞.

and that we can indeed use Girsanov's formula to define the measures $\mathbb{P}_c$ via Lemma 3.1. We will postpone these more technical aspects to Section 3.9.2 to instead focus on the key properties of our model first.

Lemma 3.7. *For each $c \in C$, under the measure $\mathbb{P}_c$ the process $\hat{X}_t^c := G_c(X_t)$ satisfies the ACE SDE*

$$\mathrm{d}\hat{X}_t^c = \hat{\mu}^c(\hat{X}_t^c)\,\mathrm{d}t + \hat{\sigma}^c(\hat{X}_t^c, t)\,\mathrm{d}\hat{W}_t^c, \quad \hat{X}_{t=0}^c = G_c(\mathbf{x}_0), \quad (3.56a)$$

where $\qquad \hat{\mu}^c(\hat{\mathbf{x}}) := (-\lambda I + E)\hat{\mathbf{x}} + \frac{1}{n-2}\|\hat{\mathbf{x}}\|^2 \mathbf{v}^c - \frac{2}{n-2}\langle \mathbf{v}^c, \hat{\mathbf{x}}\rangle \hat{\mathbf{x}}. \qquad (3.56b)$

and $\qquad \hat{\sigma}^c(\hat{\mathbf{x}}, t) := \left\| \|\mathbf{c}^c\|\hat{\mathbf{x}} + \frac{\mathbf{c}^c}{\|\mathbf{c}^c\|} \right\|^2 \cdot \sigma\big(G_c^{-1}(\hat{\mathbf{x}}), t\big), \qquad (3.56c)$

and where $(\hat{W}_t^c)_{t\geq 0}$ is a Brownian motion under $\mathbb{P}_c$ whose increments can be computed from those of $(W_t)_{t\geq 0}$ via the relation

$$\mathrm{d}\hat{W}_t^c = \begin{cases} U_{G_c(X_t)} U_{X_t}\left[\mathrm{d}W_t^c - (n-2)\sigma(X_t, t)\frac{\frac{\mathbf{c}^c}{\|\mathbf{c}^c\|^2} - X_t}{\left\|\frac{\mathbf{c}^c}{\|\mathbf{c}^c\|^2} - X_t\right\|^2}\,\mathrm{d}t \right] & \text{if } X_t \neq \vec{0}, \\ \mathrm{d}W_t^c - (n-2)\sigma(\vec{0}, t)\mathbf{c}^c\,\mathrm{d}t & \text{if } X_t = \vec{0}. \end{cases}$$
$$(3.57)$$

Proof. The calculations can be found in Appendix C.6. □

With these formulas available, we are now ready to conclude that all the single-currency projections of our multi-currency model indeed follow the dynamics of a single-currency ACE model, with currency-specific parameters.

Corollary 3.2 (Single-currency projections). *For each $c \in C$, under the measure $\mathbb{P}_c$ the short rate process $(r^c(X_t))_{t\geq 0}$ follows the dynamics of the single-currency ACE model with parameters λ, E, r_∞^c, $\mathbf{v}^c$, $G_c(\mathbf{x}_0)$ and $\hat{\sigma}^c$.*[17]

Proof. The proof follows our arguments laid out at the beginning of Section 3.4: Under $\mathbb{P}_c$, by Lemma 3.7 the process $(\hat{X}_t^c)_{t\geq 0}$ has the distribution of the ACE state process with the parameters λ, E, $\mathbf{v}^c$, $G_c(\mathbf{x}_0)$ and $\hat{\sigma}^c$; furthermore, the function

$$\hat{r}^c(\hat{\mathbf{x}}) \overset{(3.20)}{=} r^c\big(G_c^{-1}(\hat{\mathbf{x}})\big) \overset{(3.50)}{=} r_\infty^c + \big\langle \mathbf{v}^c, G_c\big(G_c^{-1}(\hat{\mathbf{x}})\big)\big\rangle = r_\infty^c + \langle \mathbf{v}^c, \hat{\mathbf{x}}\rangle$$

is the short rate function of the single-currency ACE model with parameters r_∞^c and $\mathbf{v}^c$. This shows that under $\mathbb{P}_c$ the process $(\hat{r}^c(\hat{X}_t^c))_{t\geq 0}$ has the distribution of the single-currency ACE short rate process with the parameters listed in this corollary, and since by our definitions (3.19)–(3.20) of $\hat{X}_t^c$ and $\hat{r}^c$ we have

$$\hat{r}^c(\hat{X}_t^c) = r^c\big(G_c^{-1}(\hat{X}_t^c)\big) = r^c(X_t), \quad (3.58)$$

this is what we wanted to show. □

---

[17]Note that while these parameters technically do not satisfy the constraints of Chapter 2 since we do not necessarily have $\|\mathbf{v}^c\| = 1$, according to the last paragraph in Section 2.7.1 one can still use them to define a single-currency ACE model, which is equivalent to the one with the rescaled parameters (2.46), which do satisfy all the constraints of Chapter 2.

3.8   Model Symmetry – Proof

Lemma 3.7 will also allow us to conclude that the multi-currency ACE model we defined in Section 3.6 is indeed symmetric in the sense discussed in Section 3.5. To do so, it only remains to compute the functions $\hat{r}^c_{\bar{c}_0}(\hat{\mathbf{x}})$ and $\hat{a}^{c_1/c_2}_{\bar{c}_0}(\hat{\mathbf{x}}, t)$ defined in (3.26).

Lemma 3.8. *Given any $\bar{c}_0 \in C$, the functions*

$$\hat{r}^c_{\bar{c}_0}(\hat{\mathbf{x}}) := r^c\big(G^{-1}_{\bar{c}_0}(\mathbf{x})\big) \qquad and \qquad \hat{a}^{c_1/c_2}_{\bar{c}_0}(\hat{\mathbf{x}}, t) := a^{c_1/c_2}\big(G^{-1}_{\bar{c}_0}(\mathbf{x}), t\big), \qquad (3.59)$$

which are designed to compute the short rates and exchange rates from the variable $\hat{\mathbf{x}} := G_{\bar{c}_0}(\mathbf{x})$, can be written as

$$\hat{r}^c_{\bar{c}_0}(\hat{\mathbf{x}}) = r^c_\infty + \big\langle \mathbf{v}^c, \hat{G}_c(\hat{\mathbf{x}})\big\rangle, \qquad (3.60)$$

$$\hat{a}^{c_1/c_2}_{\bar{c}_0}(\hat{\mathbf{x}}, t) = \frac{a^{c_1} \cdot e^{-r^{c_1}_\infty t} \cdot \left\|\|\hat{\mathbf{c}}^{c_2}\|\hat{\mathbf{x}} - \frac{\hat{\mathbf{c}}^{c_2}}{\|\hat{\mathbf{c}}^{c_2}\|}\right\|^{n-2}}{a^{c_2} \cdot e^{-r^{c_2}_\infty t} \cdot \left\|\|\hat{\mathbf{c}}^{c_1}\|\hat{\mathbf{x}} - \frac{\hat{\mathbf{c}}^{c_1}}{\|\hat{\mathbf{c}}^{c_1}\|}\right\|^{n-2}}, \qquad (3.61)$$

where $\hat{\mathbf{c}}^c$ and $\hat{G}_c$ are the equivalents of $\mathbf{c}^c$ and G_c for $\forall c \in C$, but with $\bar{c}_0$ instead of c_0 as the base currency, i.e.,

$$\hat{\mathbf{c}}^c := \tfrac{1}{n-2}(E + \lambda I)^{-1}(\mathbf{v}^c - \mathbf{v}^{\bar{c}_0}), \qquad (3.62)$$

$$\hat{G}_c := H_{\hat{\mathbf{c}}^c}. \qquad (3.63)$$

Proof. See Appendix C.7.   □

Corollary 3.3 (Model symmetry). *Consider the multi-currency ACE model for some given choice of the base currency c_0 and the parameters λ, E, r^c_∞, $\mathbf{v}^c$, $\mathbf{x}_0$ and σ. For some fixed $\bar{c}_0 \in C$ consider the measure $\mathbb{P}_{\bar{c}_0}$ derived via Lemma 3.1.*

Then the joint distribution of the associated processes $(r^c(X_t))_{t \geq 0}$ and $(a^{c_1/c_2}(X_t, t))_{t \geq 0}$ under $\mathbb{P}_{\bar{c}_0}$ coincides with the one of the multi-currency ACE model with the base currency $\bar{c}_0$ and the parameters λ, E, r^c_∞, $\mathbf{v}^c$, $G_{\bar{c}_0}(\mathbf{x}_0)$ and $\hat{\sigma}^{\bar{c}_0}$, with the latter given by (3.56c).

Proof. As Lemma 3.8 shows, the functions $\hat{r}^c_{\bar{c}_0}(\hat{\mathbf{x}})$ and $\hat{a}^{c_1/c_2}_{\bar{c}_0}(\hat{\mathbf{x}}, t)$ defined in (3.59) coincide with the ACE short rate and exchange rate functions for the base currency $\bar{c}_0$ and for the parameters λ, E, r^c_∞, and $\mathbf{v}^c$. Furthermore, by Lemma 3.7, under the measure $\mathbb{P}_{\bar{c}_0}$ the process $\hat{X}^{\bar{c}_0}_t := G_{\bar{c}_0}(X_t)$ has the dynamics of the ACE state process with parameters λ, E, $\mathbf{v}^{\bar{c}_0}$, $G_{\bar{c}_0}(\mathbf{x}_0)$ and $\hat{\sigma}^{\bar{c}_0}$. The statement of this corollary thus follows from our arguments at the end of Section 3.5.   □

3.9   Constraints on σ

Now that we have shown that all the definitions of our various functions have all the properties that we want, let us investigate what constraints we need to impose on our functional parameter $\sigma(\mathbf{x}, t)$.

In the single-currency ACE model, we only needed to concern ourselves with the question under which conditions the ACE SDE had a unique strong solution that remains confined to the half space M_0 almost surely; in Lemma 2.4 we then found that (aside from a natural regularity condition) this only required us to ask that $\sigma(\mathbf{x}, t)$ vanishes sufficiently fast as $\mathbf{x} \to \partial M_0$ and that it does not grow too fast as $\mathbf{x} \to \infty$.

Now in the multi-currency case, there are two additional aspects to take into account: First, we also need to ensure that our conditions for Lemma 3.1 are fulfiled, so that we can apply Girsanov's Theorem to define our full family of measures $\{\mathbb{P}_c \,|\, c \in C\}$. Second, we need to deal with the more complex shape of M_0, which is now defined in (3.52) as the intersection of the sets M^c, whose shapes are described in Lemma 3.5. Furthermore, throughout our proofs we will need to pay particular attention to the points $\frac{\mathbf{c}^c}{\|\mathbf{c}^c\|^2}$, $c \in C$, which by Lemma 3.5 (iv) lie on ∂M^c and therefore potentially on ∂M_0, since at these points all of the model's core functions (G_c, r^c, and a^{c/c_0}) for their respective associated currency are degenerate.

3.9.1 The Novikoff Condition

We begin by translating the sufficient criterion (3.14) for the Novikoff condition (3.10) of Lemma 3.1 into a constraint on our functional model parameter $\sigma(\mathbf{x}, t)$.

Lemma 3.9 (Novikoff condition)**.** *In order for the Novikoff condition (3.10) to be fulfilled for $\forall c \in C$, it suffices to ask that for $\forall T > 0$ and $\forall c \in C$ with $\mathbf{c}^c \neq \vec{0}$ we have*

$$\sup_{\substack{\mathbf{x} \in M_0 \\ t \in [0,T]}} \frac{\sigma(\mathbf{x}, t)}{\left\| \mathbf{x} - \frac{\mathbf{c}^c}{\|\mathbf{c}^c\|^2} \right\|} < \infty. \qquad (3.64)$$

Proof. Our formula (C.19) for α^c implies that

$$\|\alpha^c(\mathbf{x}, t)\| = \frac{n - 2}{\left\| \mathbf{x} - \frac{\mathbf{c}^c}{\|\mathbf{c}^c\|^2} \right\|}$$

whenever $\mathbf{c}^c \neq \vec{0}$, so that for such $c \in C$ the condition (3.14) turns into (3.64). For currencies $c \in C$ with $\mathbf{c}^c = \vec{0}$, a^{c/c_0} does not depend on $\mathbf{x}$ by (C.12) and (3.44), so that $\alpha^c \equiv \vec{0}$, which means that (3.14) is trivially fulfilled for any choice of σ. $\qquad \square$

In particular, the condition (3.64) imposes two key constraints: First, for any $c \in C$ with $\mathbf{c}^c \neq \vec{0}$ and $\frac{\mathbf{c}^c}{\|\mathbf{c}^c\|^2} \in \partial M_0$ we need to ask that

$$\forall T > 0: \quad \sup_{t \in [0,T]} \sigma(\mathbf{x}, t) = O\left(\left\| \mathbf{x} - \frac{\mathbf{c}^c}{\|\mathbf{c}^c\|^2} \right\| \right) \quad \text{as } \mathbf{x} \to \frac{\mathbf{c}^c}{\|\mathbf{c}^c\|^2}; \qquad (3.65)$$

second, if M_0 is unbounded then we need to require that

$$\forall T > 0: \quad \sup_{t \in [0,T]} \sigma(\mathbf{x}, t) = O(\|\mathbf{x}\|) \qquad \text{as } \mathbf{x} \to \infty. \qquad (3.66)$$

3.9.2   Existence and Uniqueness, Positive Rates

Next, we need to derive criteria under which the state process is well defined in confined to M_0, as desired. We begin by generalizing the statement of Lemma 2.3 about the zero-noise case to the multi-currency model.

Lemma 3.10. *(i) For $\forall c \in C$ and $\forall \mathbf{x} \in \partial M^c \setminus \left\{ \frac{\mathbf{c}^c}{\|\mathbf{c}^c\|^2} \right\}$ we have*

$$\langle \nabla r^c(\mathbf{x}), \mu(\mathbf{x}) \rangle = \tfrac{1}{n-2} \|\mathbf{v}^c\|^2 \big\| G_c(\mathbf{x}) - G_c(\mathbf{x}^c_{\min}) \big\|^2 + \delta_c, \qquad (3.67)$$

$$where \quad \mathbf{x}^c_{\min} := G_c^{-1}\big(\tfrac{n-2}{2} \tfrac{E\mathbf{v}^c}{\|\mathbf{v}^c\|^2} - r^c_\infty \tfrac{\mathbf{v}^c}{\|\mathbf{v}^c\|^2} \big) \in \partial M^c. \qquad (3.68)$$

(ii) For any $c \in C$, all flowlines of μ emanating from the boundary ∂M^c are leading into the interior of M^c at a non-vanishing angle to ∂M^c. As a result, all flowlines of μ emanating from the boundary ∂M_0 are leading into the interior of M_0.

(iii) No flowline of μ starting in the state space M_0 can ever leave it or reach ∞ in finite (positive) time. The same statement holds with M_0 replaced by $\overline{M_0}$.

Proof. The proof, which is a mostly straight-forward generalization of the one of Lemma 2.3, can be found in Appendix C.8.   □

As we had discussed below Lemma 2.3, the expression on the right of (3.67) is the *local market resiliency* for the currency c, and its minimum value δ_c is its *(global) market resiliency*; they measure how fast the drift portion $\mu(X_t)\,dt$ of our SDE (3.38)–(3.39) drives the short rate r^c back up if it is close to 0.

Next, let us move on to the stochastic case, by generalizing Lemma 2.4 to the multi-currency model.

Lemma 3.11 (Existence and uniqueness of the multi-currency ACE process). *Let M_0 refer to the intended domain of the multi-currency ACE model, as defined in (3.52). Suppose that the noise function σ is continuous in $(\mathbf{x}, t)$, and that it is chosen such that for some $T > 0$ the following three conditions are met: (i) $\sigma(\mathbf{x}, t)$ is locally Lipschitz continuous in $\mathbf{x}$ uniformly in $t \in [0, T]$,[18] (ii) if M_0 is unbounded then we have*

$$\lim_{\mathbf{x} \to \infty} \frac{\sup_{t \in [0,T]} \sigma^2(\mathbf{x}, t)}{\mathrm{dist}(\mathbf{x}, \partial M_0) \cdot \|\mathbf{x}\|^2} = 0, \quad \text{(growth condition on } \sigma) \qquad (3.69)$$

and (iii) for $\forall R > 0$ we have[19]

$$\lim_{\substack{\mathbf{x} \to \partial M_0 \\ \|\mathbf{x}\| \le R}} \frac{\sup_{t \in [0,T]} \sigma^2(\mathbf{x}, t)}{\mathrm{dist}(\mathbf{x}, \partial M_0)} = 0. \qquad \text{(Feller condition)} \qquad (3.70)$$

---

[18]I.e., for $\forall \mathbf{x} \in M_0$ there $\exists \varepsilon, K > 0$ such that for $\forall \mathbf{x}_1, \mathbf{x}_2 \in M_0$ with $\|\mathbf{x}_1 - \mathbf{x}\| < \varepsilon$ and $\|\mathbf{x}_2 - \mathbf{x}\| < \varepsilon$ and for $\forall t \in [0, T]$ we have $|\sigma(\mathbf{x}_1, t) - \sigma(\mathbf{x}_2, t)| \le K\|\mathbf{x}_1 - \mathbf{x}_2\|$.

[19]I.e., $\forall R > 0 \; \forall \varepsilon > 0 \; \exists \delta > 0 \; \forall \mathbf{x} \in M_0 \colon \|\mathbf{x}\| \le R \wedge \mathrm{dist}(\mathbf{x}, \partial M_0) < \delta \Rightarrow \sup_{t \in [0,T]} \sigma^2(\mathbf{x}, t) / \mathrm{dist}(\mathbf{x}, \partial M_0) < \varepsilon$.

Then the ACE SDE (3.38) has a unique strong solution $(X_t)_{t \in [0,T]}$ that remains in M_0 for $\forall t \in [0,T]$ almost surely.

If these conditions hold for $\forall T > 0$, then (3.38) has a unique strong solution $(X_t)_{t \geq 0}$ that remains in M_0 for $\forall t \geq 0$ almost surely.

Proof. The proof, which will be added in a future edition of this book, is a generalization of our proof for the single-currency case (Lemma 2.4), based on a generalization of our Feller condition on the unit ball (Lemma B.2 in Appendix B.2) to our specific needs. $\square$

Corollary 3.4. *Under the conditions of Lemma 3.11 the short rates $r^c(X_t)$ simulated by the ACE model are positive for $\forall c \in C$ and $\forall t \in [0,T]$ almost surely (or for $\forall t \geq 0$ if the conditions are fulfilled for $\forall T > 0$).*

Proof. This now follows from (3.52). $\square$

3.9.3 Discussion and Summary

Recalling that in the single-currency model the short rate function has the geometric interpretation (2.15), we see that the conditions listed in Lemma 3.11 are largely analogous to those of our original existence and uniqueness statement in Lemma 2.4; the ones presented here are only slightly stronger, as they ask that the limiting behavior of σ be of the form $o(\ldots)$, as opposed to $O(\ldots)$ with a bounded prefactor. While it would be possible to obtain $O(\ldots)$-criteria also for the present multi-currency case, the more complicated shape of ∂M_0 would make the $\mathbf{x}$-dependent prefactors much more complex than the ones of Lemma 2.4. Since such relatively minor improvements to Lemma 3.11 would likely be useless in practice, they would therefore justify neither the accompanying severe deterioration of the statement's readability, nor the resulting increase in its proof's already ballooning complexity.

To see how the criteria in Lemma 3.11 can be merged with the one of Lemma 3.9, first note that the Feller condition (3.70) implies that the continuous function $\mathbf{x} \mapsto \sup_{t \in [0,T]} \sigma(\mathbf{x}, t)$ must be bounded on any large ball $\{\|\mathbf{x}\| \leq R\}$, and thus that the ratio in (3.64) must be bounded, too, except possibly near $\mathbf{x} = \frac{\mathbf{c}^c}{\|\mathbf{c}^c\|^2}$ and as $\mathbf{x} \to \infty$, which is exactly what the conditions (3.65)–(3.66) (which we have seen to follow from (3.64)) ensure. We therefore find that under the condition (3.70), the Novikoff criterion (3.64) is in fact *equivalent* to (3.65)–(3.66).

Could we potentially even remove also the remaining criteria (3.65)–(3.66) in light of the ones of Lemma 3.11? At present, unfortunately not: Lemma 3.11 only asks that near ∂M_0 we have $\sup_{t \in [0,T]} \sigma(\mathbf{x}, t) = o\big(\mathrm{dist}(\mathbf{x}, \partial M_0)^{1/2}\big)$, and that near ∞ we have $\sup_{t \in [0,T]} \sigma(\mathbf{x}, t) = o\big(\mathrm{dist}(\mathbf{x}, \partial M_0)^{1/2}\|\mathbf{x}\|\big)$; in comparison, the conditions (3.65)–(3.66) require that σ must vanish even faster as $\mathbf{x}$ approaches $\frac{\mathbf{c}^c}{\|\mathbf{c}^c\|^2}$ at a fixed positive angle to ∂M_0, and that σ must grow even slower as $\mathbf{x}$ moves towards ∞ in such a way that $\mathrm{dist}(\mathbf{x}, \partial M_0) \to \infty$.

The fact that our Novikoff criterion in Lemma 3.9 is asking for more than what appears necessary to guarantee that $(X_t)_{t \geq 0}$ is well defined may be in part because, as we had mentioned below (3.14), the criterion (3.14) that Lemma 3.9 is based on is much stricter than the original Novikoff condition (3.10); future work may be able to weaken this condition and thus to potentially remove the conditions (3.65)–(3.66) entirely, given the constraints of Lemma 3.11.

Conclusion: We have found that in order to guarantee both the existence of the family of measures $\mathbb{P}_c$, $c \in C$, and the existence and uniqueness of a solution of the ACE SDE (3.38), it suffices to choose the functional noise parameter σ subject to the regularity condition (i) of Lemma 3.11 and the four constraints (3.65), (3.66), (3.69), and (3.70).

3.10 Bond Prices, Forward Rates, and Forward Exchange Rates

Now that we have confirmed that the multi-currency ACE model is consistent, symmetric, and well defined under appropriate constraints, let us move on to deriving the key formulas for its use in practice, namely the functions for computing its bond prices, forward rates, and forward exchange rates.

At the core of the bond price and forward exchange rate functions will be the *forward currency vectors*, which we define as

$$\mathbf{c}_\tau^c := \tfrac{1}{n-2}(E + \lambda I)^{-1}\big[e^{-\tau(E+\lambda I)}\mathbf{v}^c - \mathbf{v}^{c_0}\big] \tag{3.71}$$

for $\forall c \in C$. Note that by (3.47) we have

$$\mathbf{c}_{\tau=0}^c = \mathbf{c}^c. \tag{3.72}$$

Lemma 3.12. *(i) For any currencies $c_1, c_2 \in C$, any state $\mathbf{x} \in M_0$ and process time $t \geq 0$, and any tenor $\tau \geq 0$, the price—measured in units of c_2—of the bond that at time $t + \tau$ pays out one unit of currency c_1 is given by*

$$B_{c_2, t+\tau}^{c_1}(\mathbf{x}, t) = \frac{a^{c_1} \cdot e^{-r_\infty^{c_1}(t+\tau)} \cdot \left\|\|\mathbf{c}^{c_2}\|\mathbf{x} - \frac{\mathbf{c}^{c_2}}{\|\mathbf{c}^{c_2}\|}\right\|^{n-2}}{a^{c_2} \cdot e^{-r_\infty^{c_2}t} \cdot \left\|\|\mathbf{c}_\tau^{c_1}\|\mathbf{x} - \frac{\mathbf{c}_\tau^{c_1}}{\|\mathbf{c}_\tau^{c_1}\|}\right\|^{n-2}} . \tag{3.73}$$

In particular, the denominator in (3.73) is non-zero for $\forall \mathbf{x} \in M_0$ and $\tau \geq 0$.

(ii) The forward rates $F^c(\mathbf{x}, t; s)$ for $c \in C$, which are defined via the relation

$$B_{c,T}^c(\mathbf{x}, t) =: \exp\big(-\textstyle\int_t^T F^c(\mathbf{x}, t; s)\,\mathrm{d}s\big), \quad i.e., \quad F^c(\mathbf{x}, t; s) := -\partial_s \log B_{c,s}^c(\mathbf{x}, t) \tag{3.74}$$

for $\forall(\mathbf{x}, t) \in M_0 \times [0, \infty)$ and $\forall s \geq t$, can be computed using the formula

$$F^c(\mathbf{x}, t; s) = F_{\mathrm{rel}}^c(\mathbf{x}; s - t) = r^c(U_{s-t}(\mathbf{x})). \tag{3.75}$$

In particular, this implies that all forward rates simulated by the multi-currency ACE model are positive.

Here, as before in Chapter 2, the semigroup $(U_t)_{t\geq 0}$ of maps $U_t \colon \overline{M}_0 \to \overline{M}_0$ is defined as the solution to the zero-noise system (2.33), which can be computed explicitly via (2.34), with $\mathbf{v}$ replaced by $\mathbf{v}^{c_0}$.[20]

(iii) For any currencies $c_1, c_2 \in C$, any state $\mathbf{x} \in M_0$ and process time $t \geq 0$, and any tenor $\tau \geq 0$, the forward exchange rate[21] *for the pair c_1/c_2 is given by*

$$a_{t+\tau}^{c_1/c_2}(\mathbf{x}, t) = \frac{B_{c_1, t+\tau}^{c_1}(\mathbf{x}, t)}{B_{c_1, t+\tau}^{c_2}(\mathbf{x}, t)} \tag{3.76a}$$

$$= \frac{a^{c_1} \cdot e^{-r_\infty^{c_1}(t+\tau)} \cdot \left\| \|\mathbf{c}_\tau^{c_2}\| \mathbf{x} - \frac{\mathbf{c}_\tau^{c_2}}{\|\mathbf{c}_\tau^{c_2}\|} \right\|^{n-2}}{a^{c_2} \cdot e^{-r_\infty^{c_2}(t+\tau)} \cdot \left\| \|\mathbf{c}_\tau^{c_1}\| \mathbf{x} - \frac{\mathbf{c}_\tau^{c_1}}{\|\mathbf{c}_\tau^{c_1}\|} \right\|^{n-2}}. \tag{3.76b}$$

(The formula (3.76a) is valid in any arbitrage-free multi-currency model.)

Note that for bond prices measured in units of the base currency, i.e., for $c_2 = c_0$, the norm in the numerator of (3.73) is equal to 1 by (3.44) (since $\mathbf{c}^{c_0} = \vec{0}$). Also observe that for $c = c_0$ the formula (3.71) for the forward currency vectors reduces to the formula (2.29) for the single-currency case with $\mathbf{v} = \mathbf{v}^{c_0}$. Our multi-currency bond price formula (3.73) for $B_{c_0, t+\tau}^{c_0}(\mathbf{x}, t)$ therefore indeed reduces to the formula (2.28) of the single-currency ACE model.

Finally, note that neither the bond price formula (3.73) nor our exchange rate function defined in (3.51) depend on the functional model parameter $\sigma(\mathbf{x}, t)$, which means that *the multi-currency ACE model is unspanned.*

Proof of Lemma 3.12. (i) Let us first consider the case $c_1 = c_2$. In this case, we can again use our result from Lemma 3.7 that the process $\hat{X}_t^{c_1} := G_{c_1}(X_t)$ follows the dynamics of the ACE state process with parameters λ, E, $\mathbf{v}^{c_1}$, $G_{c_1}(\mathbf{x}_0)$ and $\hat{\sigma}^{c_1}$, and our observation made in the proof of Corollary 3.2 that $\hat{r}^{c_1}$ is the short rate function of the single-currency ACE model with parameters $r_\infty^{c_1}$ and $\mathbf{v}^{c_1}$, to reduce the problem to our bond price formula (2.28b) for the single-currency case:

$$B_{c_1, t+\tau}^{c_1}(\mathbf{x}, t) \stackrel{(3.6)}{=} \mathbb{E}_{c_1}\left[\exp\left(-\int_t^T r^{c_1}(X_s)\,\mathrm{d}s\right) \Big| X_t = \mathbf{x}\right]$$

$$\stackrel{(3.58),(3.19)}{=} \mathbb{E}_{c_1}\left[\exp\left(-\int_t^T \hat{r}^{c_1}(\hat{X}_s^{c_1})\,\mathrm{d}s\right) \Big| \hat{X}_t^{c_1} = G_{c_1}(\mathbf{x})\right]$$

$$\stackrel{(2.25)}{=} B_{\$, t+\tau}(G_{c_1}(\mathbf{x}), t) \tag{3.77}$$

$$\stackrel{(2.28b)}{=} e^{-r_\infty^{c_1}\tau} \left\| \|\tilde{\mathbf{c}}_\tau\| G_{c_1}(\mathbf{x}) + \frac{\tilde{\mathbf{c}}_\tau}{\|\tilde{\mathbf{c}}_\tau\|} \right\|^{2-n}, \tag{3.78}$$

---

[20] Note that according to the last paragraph in Section 2.7.1, the formula (2.34) holds even if we choose not to impose the specific constraint $\|\mathbf{v}^{c_0}\| = 1$ in (3.34) (and instead for example $\|\mathbf{v}^c\| = 1$ for some other fixed $c \in C$).

[21] I.e., the rate for a currency exchange at time $t + \tau$ that at time t can be locked in *at no cost,* by entering a contract that obliges the holder to exchange of one unit of currency c_1 into currency c_2 at that rate at the future time $t + \tau$.

where the function $B_{\$,t+\tau}$ in (3.77) denotes the bond price function for the single-currency ACE model with the parameters listed above,[22] and where we used

$$\tilde{\mathbf{c}}_\tau := \tfrac{1}{n-2}(E+\lambda I)^{-1}(\mathrm{e}^{-\tau(E+\lambda I)} - I)\mathbf{v}^{c_1} \qquad (3.79)$$

to denote the single-currency model's forward currency vectors for the currency c_1. Now using (3.43), (3.48), and the group property of the maps $H_{\mathbf{c}}$, we can further rewrite this expression as

$$
\begin{aligned}
B^{c_1}_{c_1,t+\tau}(\mathbf{x},t) &= \mathrm{e}^{-r^{c_1}_\infty \tau}\left(\frac{\left\|H_{\tilde{\mathbf{c}}_\tau}\left(G_{c_1}(\mathbf{x})\right)\right\|}{\|G_{c_1}(\mathbf{x})\|}\right)^{n-2} \\
&= \mathrm{e}^{-r^{c_1}_\infty \tau}\left(\frac{\left\|H_{\tilde{\mathbf{c}}_\tau}\left(H_{\mathbf{c}^{c_1}}(\mathbf{x})\right)\right\|}{\|H_{\mathbf{c}^{c_1}}(\mathbf{x})\|}\right)^{n-2} \\
&= \mathrm{e}^{-r^{c_1}_\infty \tau}\left(\frac{\left\|H_{\tilde{\mathbf{c}}_\tau + \mathbf{c}^{c_1}}(\mathbf{x})\right\|}{\|H_{\mathbf{c}^{c_1}}(\mathbf{x})\|}\right)^{n-2}.
\end{aligned}
$$

Since by (3.79), (3.47), and (3.71) we have

$$
\begin{aligned}
\tilde{\mathbf{c}}_\tau + \mathbf{c}^{c_1} &= \tfrac{1}{n-2}(E+\lambda I)^{-1}(\mathrm{e}^{-\tau(E+\lambda I)} - I)\mathbf{v}^{c_1} + \tfrac{1}{n-2}(E+\lambda I)^{-1}(\mathbf{v}^{c_1} - \mathbf{v}^{c_0}) \\
&= \tfrac{1}{n-2}(E+\lambda I)^{-1}(\mathrm{e}^{-\tau(E+\lambda I)}\mathbf{v}^{c_1} - \mathbf{v}^{c_0}) \\
&= \mathbf{c}^{c_1}_\tau, \qquad\qquad\qquad\qquad\qquad\qquad\qquad\qquad (3.80)
\end{aligned}
$$

using (3.43) one more time therefore yields

$$
B^{c_1}_{c_1,t+\tau}(\mathbf{x},t) = \mathrm{e}^{-r^{c_1}_\infty \tau}\left(\frac{\|H_{\mathbf{c}^{c_1}_\tau}(\mathbf{x})\|}{\|H_{\mathbf{c}^{c_1}}(\mathbf{x})\|}\right)^{n-2} = \mathrm{e}^{-r^{c_1}_\infty \tau}\cdot\frac{\left\|\|\mathbf{c}^{c_1}\|\mathbf{x} - \frac{\mathbf{c}^{c_1}}{\|\mathbf{c}^{c_1}\|}\right\|^{n-2}}{\left\|\|\mathbf{c}^{c_1}_\tau\|\mathbf{x} - \frac{\mathbf{c}^{c_1}_\tau}{\|\mathbf{c}^{c_1}_\tau\|}\right\|^{n-2}}, \qquad (3.81)
$$

which is (3.73) for $c_1 = c_2$.

To derive (3.73) for general choices of c_1 and c_2, we can now use (3.7) to convert our bond price formula for $B^{c_1}_{c_1,t+\tau}(\mathbf{x},t)$ to units of c_2, simply by multiplying it by the exchange rate $a^{c_1/c_2}(\mathbf{x},t)$ defined in (3.51):

$$
\begin{aligned}
& B^{c_1}_{c_2,t+\tau}(\mathbf{x},t) = a^{c_1/c_2}(\mathbf{x},t) \cdot B^{c_1}_{c_1,t+\tau}(\mathbf{x},t) \\[4pt]
&= \frac{a^{c_1}\cdot \mathrm{e}^{-r^{c_1}_\infty t}\cdot\left\|\|\mathbf{c}^{c_2}\|\mathbf{x} - \frac{\mathbf{c}^{c_2}}{\|\mathbf{c}^{c_2}\|}\right\|^{n-2}}{a^{c_2}\cdot \mathrm{e}^{-r^{c_2}_\infty t}\cdot\left\|\|\mathbf{c}^{c_1}\|\mathbf{x} - \frac{\mathbf{c}^{c_1}}{\|\mathbf{c}^{c_1}\|}\right\|^{n-2}} \times \mathrm{e}^{-r^{c_1}_\infty \tau}\cdot\frac{\left\|\|\mathbf{c}^{c_1}\|\mathbf{x} - \frac{\mathbf{c}^{c_1}}{\|\mathbf{c}^{c_1}\|}\right\|^{n-2}}{\left\|\|\mathbf{c}^{c_1}_\tau\|\mathbf{x} - \frac{\mathbf{c}^{c_1}_\tau}{\|\mathbf{c}^{c_1}_\tau\|}\right\|^{n-2}} \\[6pt]
&= \frac{a^{c_1}\cdot \mathrm{e}^{-r^{c_1}_\infty (t+\tau)}\cdot\left\|\|\mathbf{c}^{c_2}\|\mathbf{x} - \frac{\mathbf{c}^{c_2}}{\|\mathbf{c}^{c_2}\|}\right\|^{n-2}}{a^{c_2}\cdot\ \mathrm{e}^{-r^{c_2}_\infty t}\ \cdot\left\|\|\mathbf{c}^{c_1}_\tau\|\mathbf{x} - \frac{\mathbf{c}^{c_1}_\tau}{\|\mathbf{c}^{c_1}_\tau\|}\right\|^{n-2}}.
\end{aligned}
$$

---

[22] Recall in this context that, as we had shown in the last paragraph in Section 2.7.1, the bond price function (2.28) is still valid for generalized ACE parameter sets such as the ones encountered here that do not satisfy the constraint $\|\mathbf{v}^{c_1}\| = 1$.

It remains to consider potential cases where this entire calculation breaks down due to infinite values or division by zero. First, note that while the calculation leading from (3.78) to (3.81) does not work for $\mathbf{x} = \vec{0}$ (since then we have $G_{c_1}(\mathbf{x}) = \vec{0}$ and therefore cannot use (3.43) to transform (3.78)), in this case the expressions on the right of (3.78) and (3.81) both evaluate to $e^{-r_\infty^{c_1}\tau}$, and so (3.73) holds in this case as well.

Second, observe that in that calculation none of the terms of the form $H_{\mathbf{c}}(\mathbf{x})$ for the various vectors $\mathbf{c}$ (including $G_{c_1}(\mathbf{x})$) evaluate to ∞, either. Indeed, by (3.48) and Lemma 3.5 (iv) our assumption that $\mathbf{x} \in M_0$ implies that

$$H_{\mathbf{c}^{c_1}}(\mathbf{x}) = G_{c_1}(\mathbf{x}) \neq \infty. \tag{3.82}$$

Furthermore, by (3.52b) and (3.50) we have $0 < r^{c_1}(\mathbf{x}) = r_\infty^{c_1} + \langle \mathbf{v}^{c_1}, G_{c_1}(\mathbf{x}) \rangle$, i.e., by (2.14a) and (2.12) the point $G_{c_1}(\mathbf{x})$ lies in the domain of any single-currency ACE model with the parameters $r_\infty^{c_1}$ and $\mathbf{v}^{c_1}$; therefore by Lemma 2.9 (i) and (3.79) we have

$$G_{c_1}(\mathbf{x}) \neq \frac{\tilde{\mathbf{c}}_\tau}{\|\tilde{\mathbf{c}}_\tau\|^2} \qquad \text{for} \qquad \forall \tau > 0,$$

and in fact for $\tau = 0$ this holds as well since this case is just (3.82). By Lemma 3.4 (i) and (3.80) this in turn means that

$$\infty \neq H_{\tilde{\mathbf{c}}_\tau}(G_{c_1}(\mathbf{x})) = H_{\tilde{\mathbf{c}}_\tau}(H_{\mathbf{c}^{c_1}}(\mathbf{x})) = H_{\tilde{\mathbf{c}}_\tau + \mathbf{c}^{c_1}}(\mathbf{x}) = H_{\mathbf{c}_\tau^{c_1}}(\mathbf{x}),$$

i.e., $\mathbf{x} \neq \frac{\mathbf{c}_\tau^{c_1}}{\|\mathbf{c}_\tau^{c_1}\|^2}$.

This shows that the denominator on the right of (3.81) (and thus also the one of (3.73)) is non-zero, and that (3.81) holds for $\forall \mathbf{x} \in M_0$ and $\tau \geq 0$. The generalization to general choices of c_1 and c_2 is unproblematic, since $a^{c_1/c_2}(\mathbf{x}, t)$ is well defined for $\forall \mathbf{x} \in M_0$ by our remark at the end of Section 3.6.6.

(ii) By (3.73) the bond prices $B_{c,T}^c(\mathbf{x}, t)$ in the multi-currency ACE model do not depend on σ. Setting $\sigma \equiv 0$ in (3.6) and recalling that our bond price functions derived in part (i) are valid also for the zero-noise case,[23] we therefore find that

$$B_{c,T}^c(\mathbf{x}, t) = \exp\left(-\int_t^T r^c(U_{s-t}(\mathbf{x}))\,ds\right),$$

and so (3.74) (and the natural definition of F_{rel}^c as in (2.31)) leads us to (3.75).

To see also the second statement of (ii), note that since by Lemma 3.10 (iii) the flowline starting from $\mathbf{x}$—given by the points $U_{s-t}(\mathbf{x})$ for $s \geq t$—is contained in M_0, and since by our definition (3.52a) we have $r^c > 0$ on M_0, the forward rate $F^c(\mathbf{x}, t; s)$ given by (3.75) is always positive for $\forall \mathbf{x} \in M_0$ and $\forall s \geq t \geq 0$.

---

[23] Indeed, our proof in part (i) is based on the formula for the single-currency model, which indeed covers the case $\sigma \equiv 0$ since in that case the Feynman–Kac PDE (2.35) reduces to the PDE (2.39), which in Lemma 2.11 we have shown to hold.

(iii) Suppose the system is in the state $\mathbf{x}$ at time t, and suppose one constructs a portfolio that is short one c_1-bond maturing at time $t + \tau$, and long $w :=$ $B^{c_1}_{c_1,t+\tau}(\mathbf{x},t)/B^{c_2}_{c_1,t+\tau}(\mathbf{x},t)$ many c_2-bonds that are also maturing at time $t + \tau$. At time t, since each c_2-bond is worth $B^{c_2}_{c_1,t+\tau}(\mathbf{x},t)$ units of c_1, the total value of the long position is $B^{c_1}_{c_1,t+\tau}(\mathbf{x},t)$ units of c_1, coinciding with the value of the short position except for the sign. The total value of the portfolio at the time of its construction is therefore zero.

At time $t + \tau$ both bonds will mature, and so the portfolio will then simply hold -1 unit of c_1 and w many units of c_2. This shows that the portfolio replicates a contract that at time $t + \tau$ obliges its owner to exchange 1 unit of c_1 for w many units of c_2, i.e., it replicates a currency future between c_1 and c_2 with rate w. Since the portfolio's initial value was zero, this shows that w must in fact be the forward exchange rate $a^{c_1/c_2}_{t+\tau}(\mathbf{x},t)$,[24] i.e., we have

$$
a^{c_1/c_2}_{t+\tau}(\mathbf{x},t) = w = \frac{B^{c_1}_{c_1,t+\tau}(\mathbf{x},t)}{B^{c_2}_{c_1,t+\tau}(\mathbf{x},t)},
$$

which is (3.76a). Note that these arguments apply to any arbitrage-free model.

Now plugging in the specific bond price formulas (3.73) for our multi-currency ACE model, we obtain

$$
a^{c_1/c_2}_{t+\tau}(\mathbf{x},t) = \frac{e^{-r^{c_1}_\infty \tau} \cdot \dfrac{\left\|\,\|\mathbf{c}^{c_1}\|\,\mathbf{x} - \frac{\mathbf{c}^{c_1}}{\|\mathbf{c}^{c_1}\|}\right\|^{n-2}}{\left\|\,\|\mathbf{c}^{c_1}_\tau\|\,\mathbf{x} - \frac{\mathbf{c}^{c_1}_\tau}{\|\mathbf{c}^{c_1}_\tau\|}\right\|^{n-2}}}{\dfrac{a^{c_2} \cdot e^{-r^{c_2}_\infty (t+\tau)} \cdot \left\|\,\|\mathbf{c}^{c_1}\|\,\mathbf{x} - \frac{\mathbf{c}^{c_1}}{\|\mathbf{c}^{c_1}\|}\right\|^{n-2}}{a^{c_1} \cdot e^{-r^{c_1}_\infty t} \cdot \left\|\,\|\mathbf{c}^{c_2}_\tau\|\,\mathbf{x} - \frac{\mathbf{c}^{c_2}_\tau}{\|\mathbf{c}^{c_2}_\tau\|}\right\|^{n-2}}}
$$

$$
= \frac{a^{c_1} \cdot e^{-r^{c_1}_\infty (t+\tau)} \cdot \left\|\,\|\mathbf{c}^{c_2}_\tau\|\,\mathbf{x} - \frac{\mathbf{c}^{c_2}_\tau}{\|\mathbf{c}^{c_2}_\tau\|}\right\|^{n-2}}{a^{c_2} \cdot e^{-r^{c_2}_\infty (t+\tau)} \cdot \left\|\,\|\mathbf{c}^{c_1}_\tau\|\,\mathbf{x} - \frac{\mathbf{c}^{c_1}_\tau}{\|\mathbf{c}^{c_1}_\tau\|}\right\|^{n-2}},
$$

which is (3.76b). $\square$

3.11   Interpretation: Numeraire-Free Values

The formulas in Lemma 3.12 (i) and (iii) have an intuitive interpretation that can allow us to memorize them more easily:

---

[24]Otherwise, by adding one short position of a currency future with rate $a^{c_1/c_2}_{t+\tau}(\mathbf{x},t)$ to the portfolio (at no cost), we could ensure that at time $t + \tau$ the portfolio value is $w - a^{c_1/c_2}_{t+\tau}(\mathbf{x},t)$ units of c_2. If this were non-zero, then (depending of the sign of this value) either this portfolio or its negative would create a deterministic profit at no cost, i.e., an arbitrage.

Numeraire-free values. At any state $\mathbf{x} \in M_0$ and time $t \geq 0$, we say that a bond that pays out one unit of currency $c \in C$ at the future time $t + \tau$ has a *"numeraire-free value"* given by the formula

$$\mathfrak{B}^c_{t+\tau}(\mathbf{x}, t) := a^c \cdot \mathrm{e}^{-r^c_\infty (t+\tau)} \cdot \left\| \|\mathbf{c}^c_\tau\| \mathbf{x} - \frac{\mathbf{c}^c_\tau}{\|\mathbf{c}^c_\tau\|} \right\|^{2-n}. \tag{3.83}$$

In particular, by setting $\tau = 0$ this implies that the numeraire-free value of the bond that pays out one unit of c right now at time t—in other words, the numeraire-free value of a unit of c itself—is

$$\mathfrak{B}^c_t(\mathbf{x}, t) = a^c \cdot \mathrm{e}^{-r^c_\infty t} \cdot \left\| \|\mathbf{c}^c\| \mathbf{x} - \frac{\mathbf{c}^c}{\|\mathbf{c}^c\|} \right\|^{2-n}. \tag{3.84}$$

We will refer to $\mathfrak{B}^c_{t+\tau}(\mathbf{x}, t)$ as the *(numeraire-free) currency future value* or the *bond value*, and to $\mathfrak{B}^c_t(\mathbf{x}, t)$ as the *(numeraire-free) currency spot value* of c.

Ratios. The general notion of numeraire-free values will be rigorously introduced and discussed in detail in Section 5.1. For now, it suffices to understand that this notion of value carries no meaning by itself, only their ratios do: Indeed, *the ratio of the numeraire-free values associated to any two assets is the value of the first asset expressed in units of the second.*

In our case, there are three combinations, which (consistent with this interpretation of the ratio) lead to the following three quantities: (i) The ratio of any two currency *spot* values is their (spot) exchange rate,

$$a^{c_1/c_2}(\mathbf{x}, t) = \frac{\mathfrak{B}^{c_1}_t(\mathbf{x}, t)}{\mathfrak{B}^{c_2}_t(\mathbf{x}, t)}; \tag{3.85}$$

(ii) the ratio of any two currency *future* values is the forward exchange rate,

$$a^{c_1/c_2}_{t+\tau}(\mathbf{x}, t) = \frac{\mathfrak{B}^{c_1}_{t+\tau}(\mathbf{x}, t)}{\mathfrak{B}^{c_2}_{t+\tau}(\mathbf{x}, t)}; \tag{3.86}$$

and (iii) the ratio of one currency's future (i.e., bond) value and another currency's spot value is the price of the first currency's bond measured in units of the second currency,

$$B^{c_1}_{c_2, t+\tau}(\mathbf{x}, t) = \frac{\mathfrak{B}^{c_1}_{t+\tau}(\mathbf{x}, t)}{\mathfrak{B}^{c_2}_t(\mathbf{x}, t)}. \tag{3.87}$$

Indeed, the reader can easily confirm that plugging the numeraire-free values (3.83)–(3.84) into the ratios in (3.85)–(3.87) does lead to the actual formulas (3.51), (3.76b) and (3.73) of the multi-currency ACE model.

Parametrization. Now that we have identified the numeraire-free currency future value (3.83) as the central quantity in all our key formulas, let us take a closer look at its composition. As we see, the expression in (3.83) is encoded by the two scalars a^c and r^c_∞, and by the forward currency vector $\mathbf{c}^c_\tau$; while the two scalars encode a time-dependent scaling factor, the vector $\mathbf{c}_\tau^c$ fully encodes the function's $\mathbf{x}$-dependency.

The factor a^c is needed to calibrate the initial relationships between the various currencies. Indeed, plugging the process' starting time $t = 0$ and its attractor $\mathbf{x} = \vec{0}$ into the formula's spot version (3.84) yields $\mathfrak{B}_0^c(\vec{0}, 0) = a^c$, which explains why we had introduced a^c in Section 3.6.1 under the name *initial currency spot value attractor*, and which shows that the ratios of the various values a^c, $c \in C$, give us the currency exchange rate attractors at time $t = 0$.

The exponential factor in (3.83) is intuitive, as it is consistent with discounting by the currency's "average" short rate r_∞^c: Indeed, the value of having a unit of c *now* should be exponentially larger than getting it in the future, since having it now would allow us to invest it at an average rate of r_∞^c. At the attractor $\mathbf{x} = \vec{0}$ we therefore have $\mathfrak{B}_t^c(\vec{0}, t) = a^c \cdot \mathrm{e}^{-r_\infty^c t}$, and as a consequence, by (3.85) the exchange rate processes $(a^{c_1/c_2}(X_t, t))_{t \geq 0}$ are not attracted to fixed values, but rather to the moving (i.e., time-dependent) targets $\frac{a^{c_1}}{a^{c_2}} \cdot \mathrm{e}^{(r_\infty^{c_2} - r_\infty^{c_1})t}$.

Finally, the $\mathbf{x}$-dependent component of (3.83) is parametrized by just one vector, namely $\mathbf{c}_\tau^c$. As long as one is only interested in a currency's spot value ($\tau = 0$), one simply needs to consider the vector $\mathbf{c}_{\tau=0}^c = \mathbf{c}^c$ that is directly obtained from the model's parameters via (3.47); the computation of future values ($\tau > 0$) then requires us to simply move this vector from its initial position according to (3.71).

Geometric interpretation. The $\mathbf{x}$-dependent component in (3.83) actually has a geometric interpretation: By (3.43) we can rewrite (3.83) as

$$\mathfrak{B}_{t+\tau}^c(\mathbf{x}, t) = a^c \cdot \mathrm{e}^{-r_\infty^c(t+\tau)} \cdot \left(\frac{\|H_{\mathbf{c}_\tau^c}(\mathbf{x})\|}{\|\mathbf{x}\|} \right)^{n-2} \tag{3.88}$$

for $\forall \mathbf{x} \in M_0 \setminus \{\vec{0}\}$, and the ratio on the right can be interpreted as *the factor by which the map $H_{\mathbf{c}_\tau^c}$ changes the length of the given vector $\mathbf{x}$.*

Since for vectors $\mathbf{c}_\tau^c$ near $\vec{0}$ the map $H_{\mathbf{c}_\tau^c}$ is close to the identity mapping (recall Lemma 3.4 (ii.1)), this shows that in these cases the $\mathbf{x}$-dependent factor will be close to 1. In particular, since $\mathbf{c}^{c_0} = \vec{0}$ by (3.47), $\mathfrak{B}_t^{c_0}(\mathbf{x}, t) = a^{c_0} \cdot \mathrm{e}^{-r_\infty^{c_0} t}$ does not depend on $\mathbf{x}$ at all.

Alternative formulas. Note that since only the ratios of the numeraire-free values $\mathfrak{B}_{t+\tau}^c(\mathbf{x}, t)$ matter, we can multiply their definition by any fixed non-vanishing scalar function of $\mathbf{x}$ and t (but independent of c and τ), as this will not affect any ratios such as the ones in (3.85)–(3.87). Multiplying (3.88) by $\|\mathbf{x}\|^{n-2}$, this shows that we could equally have defined our numeraire-free values in the more compact form

$$\tilde{\mathfrak{B}}_{t+\tau}^c(\mathbf{x}, t) := a^c \cdot \mathrm{e}^{-r_\infty^c(t+\tau)} \cdot \|H_{\mathbf{c}_\tau^c}(\mathbf{x})\|^{n-2}. \tag{3.89}$$

Problematic here is only the point $\mathbf{x} = \vec{0}$: Since by Lemma 3.4 (i) we have $\tilde{\mathfrak{B}}_{t+\tau}^c(\vec{0}, t) = 0$, at $\mathbf{x} = \vec{0}$ one would therefore have to interpret ratios such as (3.85)–(3.87) via the limit $\mathbf{x} \to \vec{0}$, which leads us back to (3.88), i.e., (3.83).

However, there is some insight to be gained from this choice nevertheless: Using (3.40), (3.71), Lemma 2.1 (iii), and Lemma 2.2 (iii) to rewrite (3.89) as

$$\tilde{\mathfrak{B}}_{t+\tau}^c(\mathbf{x}, t) = a^c \cdot e^{-r_\infty^c(t+\tau)} \cdot \left\| \frac{\mathbf{x}}{\|\mathbf{x}\|^2} - \mathbf{c}_\tau^c \right\|^{2-n}$$

$$= a^c \cdot e^{-r_\infty^c(t+\tau)} \cdot \left\| \left(\frac{\mathbf{x}}{\|\mathbf{x}\|^2} + \hat{\mathbf{v}}^{c_0} \right) - e^{-\tau(E+\lambda I)} \hat{\mathbf{v}}^c \right\|^{2-n}$$

$$= a^c \cdot e^{-r_\infty^c t} \cdot e^{-r_\infty^c \tau} \cdot \left\| \mathbf{z} - e^{-\tau(E+\lambda I)} \hat{\mathbf{v}}^c \right\|^{2-n} \tag{3.90a}$$

$$= a^c \cdot e^{-r_\infty^c t} \cdot e^{((n-2)\lambda - r_\infty^c)\tau} \cdot \left\| e^{\tau(E+\lambda I)} \mathbf{z} - \hat{\mathbf{v}}^c \right\|^{2-n}, \tag{3.90b}$$

where we abbreviated

$$\hat{\mathbf{v}}^c := \tfrac{1}{n-2}(E+\lambda I)^{-1}\mathbf{v}^c, \tag{3.91}$$

$$\mathbf{z} := \frac{\mathbf{x}}{\|\mathbf{x}\|^2} + \hat{\mathbf{v}}^{c_0}, \tag{3.92}$$

we see from (3.90a) that in the variable $\mathbf{z}$ these bond sections are harmonic functions (see the proof of Lemma 2.10), and that in (3.90b) we have the conformal maps $\mathbf{z} \mapsto e^{\tau(E+\lambda I)}\mathbf{z}$ applied to $\mathbf{z}$.

Alternative model derivation. For those who have read Chapter 5 already and have then returned here: In the language of that chapter, we therefore could have used this choice of numeraire-free bond values ("bond sections"), without the respective first two factors in (3.90a)–(3.90b) (which are only relevant for the multi-currency model), as the basis of our solution for the problem summarized in Section 5.14, instead of the Minkowski space-based one presented in Chapters 6–8. More specifically, we could have chosen

$$\tilde{D}^{X,nf} := \Delta, \tag{3.93}$$

$$\mathfrak{B}_0^Y(\mathbf{z}, t) := \|\mathbf{z} - \hat{\mathbf{v}}\|^{2-n} \tag{3.94}$$

$$U_\tau(\mathbf{z}) := e^{\tau(E+\lambda I)}\mathbf{z}, \tag{3.95}$$

$$(\hat{U}_\tau \psi)(\mathbf{z}) := e^{((n-2)\lambda - r_\infty)\tau} \cdot \psi(U_{-\tau}(\mathbf{z})) \tag{3.96}$$

as the solution to the problem summarized in Section 5.14. This would have led us to an SDE with a nicer, linear drift

$$\mu(\mathbf{z}) = \partial_\tau U_\tau(\mathbf{z})|_{\tau=0} = (E+\lambda I)\mathbf{z},$$

but with the non-affine short rate function

$$r(\mathbf{z}) = -\partial_\tau \frac{\tilde{\mathfrak{B}}_{t+\tau}(\mathbf{z}, t)}{\tilde{\mathfrak{B}}_t(\mathbf{z}, t)}\Bigg|_{\tau=0}$$

$$= -\partial_\tau \left[e^{-r_\infty \tau} \cdot \frac{\left\| \mathbf{z} - e^{-\tau(E+\lambda I)}\hat{\mathbf{v}} \right\|^{2-n}}{\|\mathbf{z} - \hat{\mathbf{v}}\|^{2-n}} \right]_{\tau=0}$$

$$= r_\infty - (2-n) \frac{\langle \mathbf{z} - \hat{\mathbf{v}}, (E+\lambda I)\hat{\mathbf{v}} \rangle}{\|\mathbf{z} - \hat{\mathbf{v}}\|^2}$$

$$= r_\infty + \left\langle \frac{\mathbf{z} - \hat{\mathbf{v}}}{\|\mathbf{z} - \hat{\mathbf{v}}\|^2}, \mathbf{v} \right\rangle.$$

The goal to simplify r could then have led us to perform the change of variables $\mathbf{x} := \frac{\mathbf{z} - \hat{\mathbf{v}}}{\|\mathbf{z} - \hat{\mathbf{v}}\|^2}$, which inverts (3.92), and therefore to the ACE model equations in the form presented in Chapter 2.

Without a doubt, this would have saved us a lot of work and headache; however, the quickest solution to a problem is rarely discovered first, and there is still more than just historical value in the original model derivation via Gregory Pelts' general interest rate modeling framework.

A quick derivation of the $\mathbb{R}^n$-based ACE model equations based on the choices (3.93)–(3.96) that also cuts out the abstractions in Chapter 5, as well as a discussion of its relation to the original derivation, can be found in Chapter 4.

Adjoint drift vector field. Finally, there is a deeper meaning behind the formula (3.71) for the forward currency vectors $\mathbf{c}_\tau^c$. It is easy to check[25] that these vectors solve the linear ODE

$$\partial_\tau \mathbf{c}_\tau^c = -\mu^*(\mathbf{c}_\tau^c), \qquad \mathbf{c}_{\tau=0}^c = \mathbf{c}^c \tag{3.97}$$

for $\forall c \in C$ and $\forall \tau \geq 0$, where the *adjoint drift vector field*

$$\mu^*(\mathbf{c}) := \frac{1}{n-2}\mathbf{v}^{c_0} + (\lambda I + E)\mathbf{c} \tag{3.98}$$

is the image of the ACE drift μ defined in (3.39) under the inversion mapping $\mathbf{x} \mapsto \frac{\mathbf{x}}{\|\mathbf{x}\|^2}$ (as one can see from Lemma B.1 in Appendix B.1, with $\mathbf{a} = \vec{0}$).

This observation will eventually serve as a generalization of $\mathbf{c}_\tau^c$, and thus of the multi-currency ACE model's bond price and forward exchange rate formulas, to time-dependent parameters $E(t)$ and $\lambda(t)$. This model extension may be added in a future edition of this book.

To get only a first idea why this specific drift would make an appearance, note that in the key ingredient in (3.89), i.e., the expression[26] $\frac{\mathbf{x}}{\|\mathbf{x}\|^2} - \mathbf{c}_\tau^c$, the quantity $\mathbf{c}_\tau^c$ appears side by side with the transformation $\mathbf{x} \mapsto \frac{\mathbf{x}}{\|\mathbf{x}\|^2}$ that led to the definition of μ^*, and that $\tau = T - t$ has a minus sign in front of the t. As a result, the zero-noise evolution of this expression nicely reduces to

$$\partial_t\left(\frac{U_t(\mathbf{x})}{\|U_t(\mathbf{x})\|^2} - \mathbf{c}_{T-t}^c\right) = \mu^*\left(\frac{U_t(\mathbf{x})}{\|U_t(\mathbf{x})\|^2}\right) - \mu^*(\mathbf{c}_{T-t}^c) = (\lambda I + E)\left(\frac{U_t(\mathbf{x})}{\|U_t(\mathbf{x})\|^2} - \mathbf{c}_{T-t}^c\right),$$

revealing yet another powerful cogwheel in the ACE model's machinery.

3.12 The Canonical Parametrization

The description of the multi-currency ACE model presented so far has arisen from the natural way in which one would first start to think about the problem,

---

[25]Indeed, we have $\partial_\tau \mathbf{c}_\tau^c = \partial_\tau\left(\frac{1}{n-2}(E + \lambda I)^{-1}\left[e^{-\tau(E+\lambda I)}\mathbf{v}^c - \mathbf{v}^{c_0}\right]\right) = -\frac{1}{n-2}e^{-\tau(E+\lambda I)}\mathbf{v}^c$
$= -\frac{1}{n-2}\mathbf{v}^{c_0} - (E + \lambda I)\left(\frac{1}{n-2}(E + \lambda I)^{-1}\left[e^{-\tau(E+\lambda I)}\mathbf{v}^c - \mathbf{v}^{c_0}\right]\right) = -\frac{1}{n-2}\mathbf{v}^{c_0} - (E + \lambda I)\mathbf{c}_\tau^c$
$= -\mu^*(\mathbf{c}_\tau^c)$. The initial value $\mathbf{c}_{\tau=0}^c = \mathbf{c}^c$ was already observed in (3.72).
[26]Recall that $\|H_{\mathbf{c}_\tau^c}(\mathbf{x})\| = \left\|\frac{\mathbf{x}}{\|\mathbf{x}\|^2} - \mathbf{c}_\tau^c\right\|^{-1}$ by our definition (3.40).

which required us to choose a base currency c_0 from the set C of currencies that we intend to model; this currency is then used as the numeraire when computing asset values via (3.4), and it is the currency whose single-currency projection was designed to most obviously have the distribution of the single-currency ACE model again. However, this model description has its downsides, both analytically and numerically. In particular, the symmetry of the model is not very apparent (even though we proved it in Section 3.8), and the fact that the parameter $\mathbf{v}^{c_0}$ enters all the vectors $\mathbf{c}^c$ defined in (3.47) and thus all the functions $r^c(\mathbf{x})$ and $a^{c_1/c_2}(\mathbf{x}, t)$ is highly inconvenient during calibration.[27]

We will therefore now introduce an alternative equivalent description of the multi-currency ACE model, called its *canonical parametrization*, that successfully addresses these issues, and that in addition has some other advantages.

Definition. To define the canonical parametrization, suppose we are given a set of parameters E, λ, r_∞^c, $\mathbf{v}^c$ and a^c just as described in Sections 3.6.1–3.6.2, with the gauge fixing constraints (3.34) satisfied for some arbitrary currency c_0. In addition to these *real* currencies, we now add an *artificial* currency that we will now call c_0 instead, thus increasing the set C by this one element. This artificial currency is designed so that its interest rate remains constant at zero, which is achieved by setting

$$r_\infty^{c_0} := 0 \qquad \text{and} \qquad \mathbf{v}^{c_0} := \vec{0}.$$

While technically we had not allowed for this parameter choice in Sections 3.6.1–3.6.2, this was only because we had deemed this case of deterministic rates irrelevant in practice and therefore not worth keeping track of throughout our treatment. However, it is still a valid parameter choice, with all key results remaining intact; the only differences are as follows:

First, since $r^{c_0} \equiv 0$, we do not need to put any effort into bounding this rate below, and so we can simply set

$$M^{c_0} := \mathbb{R}^n$$

and not require the constraint (3.35)–(3.36) (which was introduced to ensure that the zero-noise process remains in M_0) to include this artificial currency.[28]

Second, since we are not calibrating the parameters for c_0, we need to impose the gauge fixing constraints (3.34) and (2.42)–(2.43) on a real currency $c \neq c_0$ instead.

Third, the choice of the parameter a^{c_0} for the artificial currency is inconsequential since we do not care to express any of the model's outputs in this currency, so that this parameter cancels out in all our calculations (such as (3.102) below); we can therefore simply set

$$a^{c_0} := 1$$

---

[27] Indeed, this means that adjusting the parameter $\mathbf{v}^{c_0}$ to improve the calibration of c_0 will inadvertently affect the interest and exchange rates—and thus the calibration—of all the other currencies (whereas this is not the case for any of the other parameters $\mathbf{v}^c$).

[28] In fact, note that δ_{c_0} would be undefined for $\mathbf{v}^{c_0} = \vec{0}$ anyways.

as well.

Choosing this added artificial currency as our new base currency c_0 has several advantages:

Linear state space evolution.   Since we now have $\mathbf{v}^{c_0} = \vec{0}$, the drift $\mu(\mathbf{x})$ of the state space SDE (3.38) turns into the simple linear function

$$\mu(\mathbf{x}) = (-\lambda I + E)\mathbf{x}, \tag{3.99}$$

and as a consequence, the zero-noise evolution (2.34) turns into a linear function as well, namely

$$U_t(\mathbf{x}) = \mathrm{e}^{\tau(-\lambda I + E)}\mathbf{x}. \tag{3.100}$$

Obvious model symmetry.   Similarly, the definitions (3.47) and (3.71) of the spot and forward currency vectors now turn into

$$\mathbf{c}^c := \tfrac{1}{n-2}(E + \lambda I)^{-1}\mathbf{v}^c, \tag{3.101a}$$

$$\mathbf{c}^c_\tau := \tfrac{1}{n-2}(E + \lambda I)^{-1}\mathrm{e}^{-\tau(E+\lambda I)}\mathbf{v}^c \tag{3.101b}$$

for $\forall c \in C$ and $\forall \tau \geq 0$. These vectors therefore now only depend on the global parameters E and λ and on the weight vector $\mathbf{v}^c$ associated to the currency of interest, but no longer on the weight vector of any other currency; as a result, the same applies also to all the functions of our model that are based on these currency vectors, in particular, to the model's core functions $G_c(\mathbf{x})$, $r^c(\mathbf{x})$, and $a^{c_1/c_2}(\mathbf{x}, t)$ (defined in (3.48), (3.50) and (3.51)), and to its derived formulas for $B^{c_1}_{c_2,T}(\mathbf{x}, t)$, $a^{c_1/c_2}_T(\mathbf{x}, t)$, and in fact $\mathfrak{B}^c_T(\mathbf{x}, t)$ (given by (3.73), (3.76b) and (3.83)). This makes it immediately obvious that the model is perfectly symmetric with respect to all the currencies that we intend to model, as none of them is standing out anymore in any of the model's formulas.

Furthermore, according to Corollary 3.3, any multi-currency ACE model that is defined in canonical coordinates with parameters

$$E, \quad \lambda, \quad \mathbf{x}_0, \quad \sigma, \quad r^c_\infty, \quad \mathbf{v}^c \quad \text{and} \quad a^c$$

(for all real currencies $c \in C$) will be equivalent to the multi-currency ACE model as defined in Section 3.6 for any real base currency $\bar{c}_0$ and with parameters

$$E, \quad \lambda, \quad G_{\bar{c}_0}(\mathbf{x}_0), \quad \hat{\sigma}^{\bar{c}_0}, \quad r^c_\infty, \quad \mathbf{v}^c \quad \text{and} \quad a^c$$

(for all real currencies $c \in C$), where $G_{\bar{c}_0}$ and $\hat{\sigma}^{\bar{c}_0}$ are defined via (3.48) and (3.56c), but now based on the spot currency vector $\mathbf{c}^c$ defined in (3.101a) instead of (3.47). This shows that we can indeed write the multi-currency ACE model as defined in Section 3.6 with any given parameters equivalently in canonical parameters, and it illuminates more clearly why the multi-currency ACE model as defined in Section 3.6 is indeed fully symmetric.

Calibration. The fact that the various real currencies $c \in C$ now fully decouple, i.e., that no currency-specific parameter $\mathbf{v}^c$ is affecting the other currencies' short rate and exchange rate processes anymore, makes calibration much more transparent in practice.

Indeed, in our original model description in Section 3.6 the forward rate curves for the various currencies (given in Lemma 3.12 (ii)) implicitly depend on the parameter $\mathbf{v}^{c_0}$ (via both r^c and the maps U_t), and as a result, calibrating $\mathbf{v}^{c_0}$ unintentionally impacts the fit of all the other currencies' curves. In contrast, in the canonical parametrization this is no longer the case, and so during calibration these curves are now only tied together by the global parameters E, λ and $\mathbf{x}_0$, as they should. The same applies to the spot and forward exchange rates for the various currency pairs (given by (3.51) and (3.76b)), which in the original model description implicitly depend on $\mathbf{v}^{c_0}$ as well (via the vectors $\mathbf{c}^c$).

Bounded, convex state space, no growth constraints on $\sigma(\mathbf{x}, t)$. Since by (3.101a), Lemma 2.2 (ii), and (3.35) the quantity $r^c(\infty)$ defined in (3.53) satisfies

$$
\begin{aligned}
r^c(\infty) &= r^c_\infty - \left\langle \mathbf{v}^c, \tfrac{\mathbf{c}^c}{\|\mathbf{c}^c\|^2} \right\rangle \\
&= r^c_\infty - (n-2) \frac{\left\langle \mathbf{v}^c, (E + \lambda I)^{-1} \mathbf{v}^c \right\rangle}{\|(E + \lambda I)^{-1} \mathbf{v}^c\|^2} \\
&= r^c_\infty - (n-2) \frac{\left\langle (E + \lambda I)(E + \lambda I)^{-1} \mathbf{v}^c, (E + \lambda I)^{-1} \mathbf{v}^c \right\rangle}{\|(E + \lambda I)^{-1} \mathbf{v}^c\|^2} \\
&= r^c_\infty - (n-2)(0 + \lambda \cdot 1) \\
&= -\frac{(n-2)\delta_c}{r^c_\infty} - \frac{(n-2)^2}{4r^c_\infty} \frac{\|E \mathbf{v}^c\|^2}{\|\mathbf{v}^c\|^2} \\
&< 0
\end{aligned}
$$

for every real currency $c \in C$, according to Lemma 3.5 (iii) the sets M^c for all those currencies are the insides of open balls in $\mathbb{R}^n$, and so their intersection M_0 is bounded and convex. As a consequence of the boundedness of M_0, the growth constraints (3.66) and (3.69) on $\sigma(\mathbf{x}, t)$ are now moot, which simplifies the calibration of σ; furthermore, the resulting boundedness of both σ and μ limits the size of the state process increments dX_t for any fixed time step dt, which helps to ensure accuracy when simlating the process $(X_t)_{t \geq 0}$.

Asset value calculation without discounting. Finally, since the new base currency's short rate function was designed to be $r^{c_0} \equiv 0$, the discount factor in the asset value formula (3.4) disappears, and so it becomes

$$
V_c(\mathbf{x}, t) = \frac{1}{a^{c/c_0}(\mathbf{x}, t)} \cdot \mathbb{E}_{c_0}\left[P_{c,T} \cdot a^{c/c_0}(X_T, T) \,\middle|\, X_t = \mathbf{x} \right]; \tag{3.102}
$$

as a result, in Monte Carlo simulations one no longer has to compute the discount factor in (3.4) via numerical integration. This only comes at the cost of now having to use the exchange rate function $a^{c/c_0}(\mathbf{x}, t)$ to convert the payouts in *every* currency into units of our artificial currency c_0 and then back, whereas before one could choose one favorite currency c_0 for which the asset valuation via (3.5) would *not* require the use of any exchange rates. As long as the number of different maturities for the various payouts in that favorite currency is small compared to the number of time steps $\mathrm{d}t$, this can be a relevant numerical advantage.

Summary. In summary, the canonical parametrization of the multi-currency ACE model is obtained by choosing as the base currency c_0 an added artificial currency whose interest rates remain constant at zero. It has many advantages over the approach presented before, in which a real currency is chosen: It simplifies the formulas for all the key quantities of the model (for $\mu(\mathbf{x})$, $U_t(\mathbf{x})$, $\mathbf{c}^c$ and $\mathbf{c}^c_\tau$, $G_c(\mathbf{x})$, $r^c(\mathbf{x})$, $a^{c_1/c_2}(\mathbf{x}, t)$, $B^{c_1}_{c_2, T}(\mathbf{x}, t)$, $a^{c_1/c_2}_T(\mathbf{x}, t)$, and $\mathfrak{B}^c_T(\mathbf{x}, t)$), it makes the model's symmetry between all the real currencies immediately obvious, it decouples the calibration of the various currency-specific parameters, it guarantees a bounded and convex state space M_0 (thus eliminating the growth constraints for $\sigma(\mathbf{x}, t)$ and limiting the size of the state space increments $\mathrm{d}X_t$), and it allows for asset valuations without the need for any discount factors (which otherwise would have to be computed via numerical integration).

Given these advantages, the reader may now wonder why we did not choose this parametrization from the beginning, throughout the entire construction in Chapter 3. The answer is that if one looks at it in the right way, then in fact we already did, and more: If one now re-reads Chapter 3 with the additional assumption that c_0 is chosen as an artificial zero-interest currency, and with all the further minor adjustments listed in our definition at the beginning of this section, then the entire sequence of arguments, formulas and proofs in Chapter 3 remains true, with all the various expressions simplifying as discussed above. One should therefore think of our treatment presented in Chapter 3 rather as a *generalization* of the one that we would have obtained if we had confined ourselves to the canonical parametrization from the beginning: a generalization that provides us with the freedom to choose our base currency c_0 either as an artificial currency or as any of the real currencies.

3.13 Recap

We have successfully extended the single-currency ACE model defined in Chapter 2 to multiple currencies, in such a way that our extension has all the desired model features listed in Section 1.4. In particular:

- The projection of the model to any individual currency $c \in C$ (i.e., the distribution of the short rate process $(r^c(X_t))_{t \geq 0}$ under $\mathbb{P}_c$) has the distribution of a single-currency ACE short rate process again, with currency-specific parameters r^c_∞ and $\mathbf{v}^c$.

- The multi-currency model is fully symmetric, i.e., when one compares the joint distribution of the short rate and exchange rate processes under the various measures $\mathbb{P}_c$, no currency stands out, despite the fact that when we originally defined the model we had to focus on a specific numeraire.

- All the currencies' rates are always positive almost surely. (One can always shift them to impose different individual lower bounds instead.)

- The model retains its state space dimension n and thus remains low-dimensional as further currencies are added.

- The model provides $(n-1)+|C|\cdot(n+2)$ discrete parameters, in addition to the unspanned functional volatility parameter $\sigma(\mathbf{x}, t)$, which enters neither the bond price functions nor the exchange rate functions.

To achieve this, we had to define a state process $(X_t)_{t\geq 0}$ (which we chose to coincide with the one of the single-currency ACE model) whose measure $\mathbb{P}_{c_0}$ is understood to be the risk-neutral measure associated to some fixed base currency $c_0 \in C$, as well as carefully constructed short rate functions $r^c(\mathbf{x})$ and exchange rate functions $a^{c_1/c_2}(\mathbf{x}, t)$, for $c, c_1, c_2 \in C$; the state process' risk-neutral measures $\mathbb{P}_c$ under all the remaining currencies $c \in C$ can then be inferred using Girsanov's Theorem. Given these objects, the multi-currency ACE model is then the joint distribution of the short rate and exchange rate processes

$$(r^c(X_t))_{t\geq 0} \qquad \text{and} \qquad (a^{c_1/c_2}(X_t, t))_{t\geq 0}$$

for $c, c_1, c_2 \in C$.

To prepare for our construction, we first started in Section 3.2 by discussing the issue of changing measures and applying Girsanov's Theorem explicitly to obtain the state process' dynamics under all numeraires $c \in C$, which led us to the consistency conditions (3.12)–(3.13) and the Novikoff condition (3.10). Subsequent investigations of further potential model consistency issues in Section 3.3 turned out to only lead to the obvious exchange rate triangle relation (3.17).

Then in Section 3.4 taking a closer look at our goal that all projections to individual currencies should yield the single-currency ACE model again, we quickly realized the need for a family of mappings $G_c \colon M_0 \to \mathbb{R}^n$ such that under $\mathbb{P}_c$ the process $\hat{X}_t^c := G_c(X_t)$ has the dynamics of the single-currency ACE model, and that $\hat{r}^c := r^c \circ G_c^{-1}$ has the form of the single-currency ACE model's short rate function. After applying Itô's formula to actually derive the general dynamics of $(\hat{X}_t^c)_{t\geq 0}$ under $\mathbb{P}_c$, in Lemma 3.3 we obtained explicit analytical conditions on the functions r^c, a^{c_1/c_2} and G_c that are necessary for this construction to work. We then elaborated on this construction based on the maps G_c in Section 3.5, by further discussing our intended notion of model symmetry.

We were then ready to actually define our model (i.e., the functions r^c, a^{c_1/c_2} and G_c, and the process $(X_t)_{t\geq 0}$) in Section 3.6, to prove in Sections 3.7–3.8 that it indeed satisfies all the conditions that we had compiled earlier, and to derive in Section 3.9 the requirements on the behavior of $\sigma(\mathbf{x}, t)$ near ∂M_0 and ∞ that guarantee that the Novikoff condition is satisfied and that the ACE SDE has a unique strong solution $(X_t)_{t \geq 0}$.

In Section 3.10 we then obtained the model's bond price functions (by reducing the problem to the single-currency case) and forward exchange rate functions (by way of a simple no-arbitrage argument), and in Section 3.11 we showed how one can interpret these formulas based on the notion of numeraire-free values, which will be rigorously defined later in Section 5.1. Finally, in Section 3.12 we presented the alternative canonical parametrization of the multi-currency ACE model, which is defined by choosing an artificial zero-interest currency as the base currency c_0, and which leads to a variety of simplifications and practical advantages.

3.14   Closing Statement

This concludes our "Fast Track To ACE." We encourage the reader to now proceed and study also the derivation of the single-currency ACE model laid out in Part II (the derivation of the multi-currency model may be included in a future edition of this book). This will provide valuable insight into the model's inner workings, and it will introduce the reader to a variety of tools and viewpoints that are novel in the world of financial modeling.

Part II

Model Derivation

Chapter 4

Quick Derivation

Historically, the ACE model equations presented in Part I were first derived using Gregory Pelts' interest rate modeling framework, as laid out in Chapters 5–9. As it is so often the case, however, once the solution to our modeling problem had been unearthed, an alternative simpler way could be devised to "find" these equations, which shall be presented in this chapter.

First, in Section 4.1 we will derive the equations of the single-currency ACE model, then in Section 4.2 we will generalize our calculations to the multi-currency case. Finally, in Section 4.3 we will discuss how this approach relates to the original model derivation, and why that original derivation still has more than just historical value.

4.1   Single-Currency Model

The quick derivation consists of three steps:

1. The problem of defining an unspanned interest model, by writing down a Feynman–Kac type PDE and its solution, is simplified by (i) choosing a constant short rate function $r(\mathbf{z}) \equiv \bar{\rho}$ and (ii) ignoring the boundary condition $B_{\$,T}(\mathbf{z}, t = T) \equiv 1$. As we will see, this can be interpreted as a two-currency interest rate model, with an artificial currency $\bar{c}$ as the numeraire that has the constant short rate $\bar{\rho}$, with the bond price function interpreted as the price of the dollar bonds in units of $\bar{c}$, and with the solution's boundary values serving as the exchange rate function between the two currencies.

2. A change of numeraire to the dollar is performed, which eliminates $\bar{c}$ from the model and turns it into a regular non-trivial interest rate model for the dollar. As we will see, the downsides of the model we obtain are that (i) the rate and bond functions contain terms that blow up at a specific point, (ii) the rate function is hard to interpret, and (iii) the functional noise parameter $\sigma(\mathbf{z})$ has bled into the drift $\mu(\mathbf{z})$.

3. A change of variables $\mathbf{z} \mapsto \mathbf{x}(\mathbf{z})$ is carried out that fully resolves all of these three issues.

Upon completion of these three steps, we will arrive at the single-currency ACE model equations introduced in Chapter 2.

Step 1.   Let $\bar{c}$ be an artificial additional currency whose short rate is constant at $\bar{\rho} > 0$, independently of the state $\mathbf{z} \in \mathbb{R}^n$ of the system. Let us denote the currency exchange rate to the dollar (which we assume to be independent of t) as $g(\mathbf{z}) > 0$, i.e., \$1 is worth $g(\mathbf{z})$ many units of $\bar{c}$. Furthermore, let us denote by $B_{\bar{c},T}^{\$}(\mathbf{z},t)$ the value (expressed in units of $\bar{c}$) of the zero-coupon bond that pays out \$1 at time $T \geq t$. It has to satisfy the Feynman–Kac equation with $\bar{c}$ as the numeraire, which is of the form

$$\left(\partial_t + \tfrac{1}{2}\sigma_1^2(\mathbf{z})\Delta + \langle \mu_1(\mathbf{z}), \nabla \rangle - \bar{\rho}\right) B_{\bar{c},T}^{\$}(\mathbf{z},t) = 0, \qquad (4.1\text{a})$$

$$B_{\bar{c},T}^{\$}(\mathbf{z},T) = g(\mathbf{z}). \qquad (4.1\text{b})$$

(The lower indices in σ_1 and μ_1 refer to the fact that we are in Step 1 of our derivation.) Since we want the model to be unspanned (i.e., $B_{\bar{c},T}^{\$}$ is independent of σ_1), letting $\sigma_1 \searrow 0$ in (4.1a) and then subtracting the result from (4.1a) shows that $B_{\bar{c},T}^{\$}$ in fact has to satisfy the two equations

$$\left(\partial_t + \langle \mu_1(\mathbf{z}), \nabla \rangle - \bar{\rho}\right) B_{\bar{c},T}^{\$}(\mathbf{z},t) = 0 \qquad (4.2\text{a})$$

$$\Delta B_{\bar{c},T}^{\$}(\mathbf{z},t) = 0 \qquad (4.2\text{b})$$

individually, together with the boundary condition (4.1b).

Let us start by focusing on (4.2a), which is in fact the Feynman–Kac PDE for a noiseless system. Without noise, the system moves deterministically according to the ODE $\dot{\mathbf{z}} = \mu_1(\mathbf{z})$. Denoting its evolution map by $U_t(\mathbf{z})$, so that

$$\partial_t U_t(\mathbf{z}) = \mu_1(U_t(\mathbf{z})) \qquad \text{and} \qquad U_{t=0}(\mathbf{z}) = \mathbf{z},$$

it is easy to write down the bond price formula for this case: If the system at time t is in state $\mathbf{z}$, then at the time of maturity $T \geq t$ it is in the state $U_{T-t}(\mathbf{z})$, so the bond's \$1 payout at time T is worth $g(U_{T-t}(\mathbf{z}))$ units of $\bar{c}$, and its present value at time t is therefore

$$B_{\bar{c},T}^{\$}(\mathbf{z},t) = \mathrm{e}^{-\bar{\rho}(T-t)} g(U_{T-t}(\mathbf{z})). \qquad (4.3)$$

And indeed, it is easy to check that this formula satisfies both (4.2a) and (4.1b).

Moving on to (4.2b), we now need to ensure that the function in (4.3) is harmonic *for every* $T \geq t$. Considering $T = t$ first, this means that g itself must be harmonic, i.e.,

$$\Delta g = 0. \qquad (4.4)$$

To keep Steps 2–3 analytically manageable, we therefore need to define g as a simple, yet non-trivial, positive harmonic function. The best function that comes to mind[1] is

$$g(\mathbf{z}) := \|\mathbf{z} - \hat{\mathbf{v}}\|^{2-n}, \tag{4.5}$$

for some vector $\hat{\mathbf{v}} \in \mathbb{R}^n$ that we introduce here to inject an additional parameter into our model that can later be used for calibration.

Next, for the function in (4.3) to be harmonic also for $T > t$, the exponential prefactor in (4.3) is not a problem since it is independent of $\mathbf{z}$, but we need to ensure that the evolution map $U_{T-t}(\mathbf{z})$ preserves the harmonic property of g, i.e., that for all $\tau := T - t$ we have[2]

$$
\begin{aligned}
0 &= \Delta[g(U_\tau(\mathbf{z}))] \\
&= \sum_i \partial_i^2[g(U_\tau(\mathbf{z}))] \\
&= \sum_{i,k} \partial_i[(\partial_k g)(U_\tau(\mathbf{z})) \cdot \partial_i U_\tau^k(\mathbf{z})] \\
&= \sum_{i,k,l} (\partial_l \partial_k g)(U_\tau(\mathbf{z})) \cdot \partial_i U_\tau^k(\mathbf{z}) \cdot \partial_i U_\tau^l(\mathbf{z}) + \sum_{i,k} (\partial_k g)(U_\tau(\mathbf{z})) \cdot \partial_i^2 U_\tau^k(\mathbf{z}) \\
&= \operatorname{tr}\big[(\nabla^2 g)(U_\tau(\mathbf{z})) \cdot (\nabla U_\tau)(\mathbf{z}) \cdot (\nabla U_\tau)(\mathbf{z})^T\big] + \sum_k (\partial_k g)(U_\tau(\mathbf{z})) \cdot \Delta U_\tau^k(\mathbf{z}).
\end{aligned}
$$

The second term can be made to vanish by ensuring that U_τ is componentwise harmonic, i.e., that

$$\Delta U_\tau^k = 0 \qquad \text{for every component } k = 1, \dots, n.$$

In particular, this will be satisfied if we choose μ and thus U_τ as linear functions in $\mathbf{z}$, i.e., if

$$\mu_1(\mathbf{z}) := A\mathbf{z} \qquad \text{and thus} \qquad U_\tau(\mathbf{z}) = \mathrm{e}^{\tau A}\mathbf{z} \tag{4.6}$$

for some matrix $A \in \mathbb{R}^{n \times n}$.

Turning our attention to the first term (the trace expression), we observe that if the product of the last two matrices is a multiple of the identity matrix, i.e., if $(\nabla U_\tau)(\nabla U_\tau)^T = \alpha_\tau \cdot I$ for some scalar function $\alpha_\tau(\mathbf{z})$, then this term becomes

$$\operatorname{tr}[\dots] = \alpha_\tau(\mathbf{z}) \cdot \operatorname{tr}[(\nabla^2 g)(U_\tau(\mathbf{z}))] = \alpha_\tau(\mathbf{z}) \cdot (\Delta g)(U_\tau(\mathbf{z})) \overset{(4.4)}{=} 0,$$

as desired. We therefore ask that

$$\alpha_\tau(\mathbf{z}) \cdot I = (\nabla U_\tau)(\mathbf{z}) \cdot (\nabla U_\tau)(\mathbf{z})^T = \mathrm{e}^{\tau A}\mathrm{e}^{\tau A^T}.$$

Now if we didn't have α_τ on the left-hand side, this would just mean that we want $\mathrm{e}^{\tau A}$ to be an orthogonal matrix for every τ, which by Lemma 2.2 is the case if and only if A is anti-symmetric. Since we do allow for this additional scaling factor α_τ, however, we can allow for A to have the more general shape

$$A = E + \lambda I, \qquad \text{where} \qquad E^T = -E \qquad \text{and} \qquad \lambda \in \mathbb{R}; \tag{4.7}$$

---

[1]The fact that this function is harmonic is well known; see also our proof of Lemma 2.10.

[2]We ask the reader to forgive the notational clash that T, which already stands for the time of maturity, will now also denote the operation of matrix transposition.

indeed, we then have

$$e^{\tau A}e^{\tau A^T} = e^{\tau(E+\lambda I)}e^{\tau(-E+\lambda I)} = e^{2\lambda\tau I} = e^{2\lambda\tau}I.$$

Putting everything together, we see that by (4.6) and (4.7) we can choose

$$\mu_1(\mathbf{z}) = (E + \lambda I)\mathbf{z} \qquad \text{and thus} \qquad U_\tau(\mathbf{z}) = e^{\tau(E+\lambda I)}\mathbf{z},$$

so that by (4.3) and (4.5) our bond function becomes

$$B_{\bar{c},T}^{\$}(\mathbf{z}, t) = e^{-\bar{\rho}(T-t)}\big\|e^{(T-t)(E+\lambda I)}\mathbf{z} - \hat{\mathbf{v}}\big\|^{2-n}. \tag{4.8}$$

With all these choices, by construction our bond price function $B_{\bar{c},T}^{\$}(\mathbf{z}, t)$ satisfies both (4.2a) and (4.2b), and thus the Feynman–Kac PDE (4.1a) with $\bar{c}$ as the numeraire, together with the boundary condition (4.1b). Moreover, the function is independent of σ_1.

Note that this function does not satisfy the usual Feynman–Kac boundary condition $B_{\bar{c},T}^{\$}(\mathbf{z}, t = T) = 1$, but rather (4.1b) (since it measures the bond payout in units of $\bar{c}$ and not in dollars), and the rate function in (4.9) is just a constant $\bar{\rho}$; for this reason, at this stage we have not yet completed our task of defining a non-trivial unspanned single-currency short rate model.

Step 2 (Change of numeraire).   Next, let us eliminate $\bar{c}$ from the model by changing the numeraire of our Feynman–Kac PDE to the dollar, thus turning the model into a regular, yet non-trivial, single-currency short rate model for the dollar only. So essentially, we are going to carry out a change of numeraire with Girsanov's formula to obtain the new drift μ_2 (as we know, Girsanov won't affect the noise σ_1, i.e., we have $\sigma_2 = \sigma_1$), and then compute the rate function $r_2(\mathbf{z})$ of the dollar from our bond function.

However, here we will use an equivalent method that achieves both these things in one step that does not require us to have memorized any of these formulas: The Feynman–Kac PDE

$$\big(\partial_t + \tfrac{1}{2}\sigma_1^2(\mathbf{z})\Delta + \langle\mu_1(\mathbf{z}), \nabla\rangle - \bar{\rho}\big)f_{\bar{c}}(\mathbf{z}, t) = 0 \tag{4.9}$$

is fulfilled by any function $f_{\bar{c}}$ representing the value *in units of* $\bar{c}$ of a self-financing account. So what equation will a function $f_{\$}$ that measures the value *in dollars* instead fulfill? Given any such function $f_{\$}$, the function $f_{\bar{c}} := g \cdot f_{\$}$ expresses the value in units of $\bar{c}$, and so it must fulfill (4.9). Dividing by $g(\mathbf{z})$ and using the product rule of differentiation, we therefore have

$$0 = \tfrac{1}{g(\mathbf{z})}\big(\partial_t + \tfrac{1}{2}\sigma_1^2(\mathbf{z})\Delta + \langle\mu_1(\mathbf{z}), \nabla\rangle - \bar{\rho}\big)\big[(g(\mathbf{z})f_{\$}(\mathbf{z}, t)\big] \tag{4.10}$$

$$= \big(\partial_t + \tfrac{1}{2}\sigma_1^2(\mathbf{z})\Delta\big)f_{\$}(\mathbf{z}, t)$$

$$+ \big\langle\mu_1(\mathbf{z}) + \tfrac{1}{g(\mathbf{z})}\sigma_1^2(\mathbf{z})\nabla g(\mathbf{z}), \nabla\big\rangle f_{\$}(\mathbf{z}, t)$$

$$- \big(\bar{\rho} - \tfrac{1}{2g(\mathbf{z})}\sigma_1^2(\mathbf{z})\Delta g(\mathbf{z}) - \tfrac{1}{g(\mathbf{z})}\langle\mu_1(\mathbf{z}), \nabla g(\mathbf{z})\rangle\big)f_{\$}(\mathbf{z}, t). \tag{4.11}$$

This has again the form

$$\left(\partial_t + \tfrac{1}{2}\sigma_1^2(\mathbf{z})\Delta + \langle\mu_2(\mathbf{z}),\nabla\rangle - r_2(\mathbf{z})\right)f_\$(\mathbf{z},t) = 0 \tag{4.12}$$

of a Feynman–Kac PDE, and using

$$\frac{1}{g(\mathbf{z})}\nabla g(\mathbf{z}) = \|\mathbf{z}-\hat{\mathbf{v}}\|^{n-2}\cdot(2-n)\|\mathbf{z}-\hat{\mathbf{v}}\|^{1-n}\cdot\frac{\mathbf{z}-\hat{\mathbf{v}}}{\|\mathbf{z}-\hat{\mathbf{v}}\|}$$

$$= -(n-2)\frac{\mathbf{z}-\hat{\mathbf{v}}}{\|\mathbf{z}-\hat{\mathbf{v}}\|^2},$$

we can write its drift as

$$\mu_2(\mathbf{z}) := \mu_1(\mathbf{z}) + \frac{1}{g(\mathbf{z})}\sigma_1^2(\mathbf{z})\nabla g(\mathbf{z})$$

$$= (E+\lambda I)\mathbf{z} - (n-2)\sigma_1^2(\mathbf{z})\frac{\mathbf{z}-\hat{\mathbf{v}}}{\|\mathbf{z}-\hat{\mathbf{v}}\|^2} \tag{4.13}$$

and its rate function as

$$r_2(\mathbf{z}) := \bar{\rho} - \frac{1}{2g(\mathbf{z})}\sigma^2(\mathbf{z})\Delta g(\mathbf{z}) - \frac{1}{g(\mathbf{z})}\langle\mu_1(\mathbf{z}),\nabla g(\mathbf{z})\rangle$$

$$= \bar{\rho} - 0 + (n-2)\frac{\langle(E+\lambda I)\mathbf{z},\mathbf{z}-\hat{\mathbf{v}}\rangle}{\|\mathbf{z}-\hat{\mathbf{v}}\|^2}$$

$$= \bar{\rho} + (n-2)\frac{\langle(E+\lambda I)(\mathbf{z}-\hat{\mathbf{v}}),\mathbf{z}-\hat{\mathbf{v}}\rangle}{\|\mathbf{z}-\hat{\mathbf{v}}\|^2} + (n-2)\frac{\langle(E+\lambda I)\hat{\mathbf{v}},\mathbf{z}-\hat{\mathbf{v}}\rangle}{\|\mathbf{z}-\hat{\mathbf{v}}\|^2}$$

$$= \bar{\rho} + (n-2)\lambda + (n-2)\big\langle(E+\lambda I)\hat{\mathbf{v}},\frac{\mathbf{z}-\hat{\mathbf{v}}}{\|\mathbf{z}-\hat{\mathbf{v}}\|^2}\big\rangle$$

$$= r_\infty + \big\langle\mathbf{v},\frac{\mathbf{z}-\hat{\mathbf{v}}}{\|\mathbf{z}-\hat{\mathbf{v}}\|^2}\big\rangle, \tag{4.14}$$

where we used (4.4) and Lemma 2.2 (ii) and then defined

$$r_\infty := \bar{\rho} + (n-2)\lambda \qquad \text{and} \qquad \mathbf{v} := (n-2)(E+\lambda I)\hat{\mathbf{v}}.$$

We have therefore obtained the Feynman–Kac equation (4.12) for functions measuring account values in units of dollars instead of $\bar{c}$. In particular, it will be fulfilled by the bond price function that measures the value of the bond in dollars. Using (4.8) and (4.5), Lemmas 2.2 (iii) and 2.1 (iii), and the abbreviations

$$\mathbf{c}_\tau := \left(e^{-\tau(E+\lambda I)} - I\right)\hat{\mathbf{v}}$$

$$= \frac{1}{n-2}(E+\lambda I)^{-1}\left(e^{-\tau(E+\lambda I)} - I\right)\mathbf{v} \tag{4.15}$$

and $\tau := T - t$, it is given by

$$B^\$_{\$,t+\tau}(\mathbf{z},t) = \frac{B^\$_{\bar{c},t+\tau}(\mathbf{z},t)}{g(\mathbf{z})}$$

$$= \frac{e^{-\bar{\rho}\tau}\big\|e^{\tau(E+\lambda I)}\mathbf{z} - \hat{\mathbf{v}}\big\|^{2-n}}{\|\mathbf{z}-\hat{\mathbf{v}}\|^{2-n}} \tag{4.16a}$$

$$= \frac{e^{-\bar{\rho}\tau}e^{(2-n)\lambda\tau}\big\|\mathbf{z} - e^{-\tau(E+\lambda I)}\hat{\mathbf{v}}\big\|^{2-n}}{\|\mathbf{z}-\hat{\mathbf{v}}\|^{2-n}}$$

$$= \frac{e^{-[\bar{\rho}+(n-2)\lambda]\tau}\left\|(\mathbf{z}-\hat{\mathbf{v}}) - \left(e^{-\tau(E+\lambda I)} - I\right)\hat{\mathbf{v}}\right\|^{2-n}}{\|\mathbf{z}-\hat{\mathbf{v}}\|^{2-n}}$$

$$= e^{-r_\infty\tau}\left\|\frac{\mathbf{z}-\hat{\mathbf{v}}}{\|\mathbf{z}-\hat{\mathbf{v}}\|} - \frac{\mathbf{c}_\tau}{\|\mathbf{z}-\hat{\mathbf{v}}\|}\right\|^{2-n}$$

$$= e^{-r_\infty\tau}\left(1 - 2\left\langle\frac{\mathbf{z}-\hat{\mathbf{v}}}{\|\mathbf{z}-\hat{\mathbf{v}}\|^2}, \mathbf{c}_\tau\right\rangle + \|\mathbf{c}_\tau\|^2\|\mathbf{z}-\hat{\mathbf{v}}\|^{-2}\right)^{1-n/2} \tag{4.16b}$$

$$= e^{-r_\infty\tau}\left\|\|\mathbf{c}_\tau\|\frac{\mathbf{z}-\hat{\mathbf{v}}}{\|\mathbf{z}-\hat{\mathbf{v}}\|^2} - \frac{\mathbf{c}_\tau}{\|\mathbf{c}_\tau\|}\right\|^{2-n}. \tag{4.16c}$$

In the last three steps, we successfully managed to lump all occurrences of the term $\mathbf{z}-\hat{\mathbf{v}}$ together to better understand the function's dependency on $\mathbf{z}$, ending up with just one expression of the form $\frac{\mathbf{z}-\hat{\mathbf{v}}}{\|\mathbf{z}-\hat{\mathbf{v}}\|^2}$, exactly as it occurs also in our rate function $r_2(\mathbf{z})$ in (4.14).

Note also that by construction we have $B_{\$,T}^\$(\mathbf{z}, t\!=\!T) \equiv 1$; to check this from our formulas (4.16b)–(4.16c), simply observe that $\mathbf{c}_{\tau=0} = \vec{0}$. At this point, we have therefore succeeded in defining a flexible single-currency interest rate model that is unspanned with respect to the functional parameter $\sigma_1(\mathbf{z})$.

Unfortunately, (i) both the rate function $r_2(\mathbf{z})$ and the bond price function $B_{\$,T}^\$(\mathbf{z}, t)$ contain the expression $\frac{\mathbf{z}-\hat{\mathbf{v}}}{\|\mathbf{z}-\hat{\mathbf{v}}\|^2}$ that blows up at $\hat{\mathbf{v}}$, (ii) the rate function $r_2(\mathbf{z})$ is hard to interpret, and (iii) the functional parameter $\sigma_1(\mathbf{z})$ has bled into the drift $\mu_2(\mathbf{z})$, so that we cannot interpret $\sigma_1(\mathbf{z})$ *solely* as the size of the noise anymore, and that the calibration of σ_1 will change the attractor of the stochastic state process. So next, we will address all three of these issues with a single change of variables.

Step 3 (Change of variables).   A look at both the rate function (4.14) and the bond function (4.16b)–(4.16c) suggests that we should try looking at the model in the variable

$$\mathbf{x} := \frac{\mathbf{z} - \hat{\mathbf{v}}}{\|\mathbf{z}-\hat{\mathbf{v}}\|^2} \tag{4.17}$$

instead. Not only will this simplify these functions significantly:[3]

$$r_3(\mathbf{x}) = r_\infty + \langle\mathbf{v}, \mathbf{x}\rangle, \tag{4.18}$$

$$B_{\$,t+\tau}^\$(\mathbf{x}, t) = e^{-r_\infty\tau}\left(1 - 2\langle\mathbf{x}, \mathbf{c}_\tau\rangle + \|\mathbf{c}_\tau\|^2\|\mathbf{x}\|^2\right)^{1-n/2} \tag{4.19a}$$

$$= e^{-r_\infty\tau}\left\|\|\mathbf{c}_\tau\|\mathbf{x} - \frac{\mathbf{c}_\tau}{\|\mathbf{c}_\tau\|}\right\|^{2-n}; \tag{4.19b}$$

they will also crucially make the bond function in (4.19b) harmonic in $\mathbf{x}$.[4] To see why the latter is important, note the following:[5]

---

[3]For simplicity of notation, we will denote the $\mathbf{x}$-based bond function with the same name again.

[4]To see this, observe that the expression inside the outer norm only applies a simple scale and shift to $\mathbf{x}$, so that the harmonic property of the function $\mathbf{x} \mapsto \|\mathbf{x}\|^{2-n}$ is inherited by (4.19b).

[5]The reader may skip the next two paragraphs if they seem overwhelming on first reading.

As we know, as we will carry out this change of variables, Itô's formula will add an additional term to our drift μ_2 that originates from the noise term and will therefore have σ_1^2 as a prefactor. For a general given change of variables, it would be an extraordinary stroke of luck if that additional term would happen to cancel out the σ_1-dependent second term in (4.13) from our change of numeraire, which is what we want (see issue (iii) above). However, by ensuring that our bond function satisfies $\Delta_{\mathbf{x}} B_{\$,T}^{\$} = 0$, this will intuitively become a near certainty.

That is because just like how we had obtained (4.2a)–(4.2b), we can again let $\sigma_1 \searrow 0$ in our final $\mathbf{x}$- and dollar-based Feynman–Kac PDE to show that $B_{\$,T}^{\$}$ will make its σ_1-dependent part vanish separately. Since $B_{\$,T}^{\$}$ also satisfies the PDE $\Delta_{\mathbf{x}} B_{\$,T}^{\$} = 0$, we can have high hopes that this is because that σ_1-dependent part *is* (a scalar multiple of) $\Delta_{\mathbf{x}}$,[6] which means that (a) the two σ_1-dependent drift-terms (i.e., the part's first-order derivative terms) must actually have canceled each other out, and (b) the change of variables has not affected the multiplicative nature of the noise.

With this motivation and our hopes high, let us see what under the change of variables (4.17) our Feynman–Kac PDE (4.12) turns into. Luckily,[7] we have actually already performed a change of variable of this type on a general SDE in Section B.3, and according to (B.19) we obtain the following new drift and noise functions:

$$\sigma_3(\mathbf{x}) := \frac{\sigma_1(\mathbf{z})}{\|\mathbf{z} - \hat{\mathbf{v}}\|^2} = \|\mathbf{x}\|^2 \sigma_1 \big(\tfrac{\mathbf{x}}{\|\mathbf{x}\|^2} + \hat{\mathbf{v}} \big), \tag{4.20}$$

$$\mu_3(\mathbf{x}) := \frac{1}{\|\mathbf{z} - \hat{\mathbf{v}}\|^2} \left(I - 2\frac{(\mathbf{z} - \hat{\mathbf{v}}) \otimes (\mathbf{z} - \hat{\mathbf{v}})}{\|\mathbf{z} - \hat{\mathbf{v}}\|^2} \right) \mu_2(\mathbf{z}) + \frac{(2-n)\sigma_1^2(\mathbf{z})}{\|\mathbf{z} - \hat{\mathbf{v}}\|^4} (\mathbf{z} - \hat{\mathbf{v}})$$

$$= \|\mathbf{x}\|^2 \left(I - 2\frac{\mathbf{x} \otimes \mathbf{x}}{\|\mathbf{x}\|^2} \right) \mu_2(\mathbf{z}) - (n-2)\|\mathbf{x}\|^2 \sigma_1^2(\mathbf{z})\mathbf{x}$$

$$= \|\mathbf{x}\|^2 \left(I - 2\frac{\mathbf{x} \otimes \mathbf{x}}{\|\mathbf{x}\|^2} \right) \big[(E + \lambda I)\big(\tfrac{\mathbf{x}}{\|\mathbf{x}\|^2} + \hat{\mathbf{v}} \big) - (n-2)\sigma_1^2(\mathbf{z})\mathbf{x} \big]$$

$$\qquad\qquad - (n-2)\|\mathbf{x}\|^2 \sigma_1^2(\mathbf{z}, t)\mathbf{x}$$

$$= (E - \lambda I)\mathbf{x} + \tfrac{1}{n-2}\|\mathbf{x}\|^2 \left(I - 2\frac{\mathbf{x} \otimes \mathbf{x}}{\|\mathbf{x}\|^2} \right) \mathbf{v}$$

$$= (E - \lambda I)\mathbf{x} + \tfrac{1}{n-2}\|\mathbf{x}\|^2 \mathbf{v} - \tfrac{2}{n-2}\langle \mathbf{v}, \mathbf{x}\rangle \mathbf{x}. \tag{4.21}$$

We can therefore conclude that the (σ_3-independent) bond function in (4.19) satisfies the Feynman–Kac PDE

$$\big(\partial_t + \tfrac{1}{2}\sigma_3^2(\mathbf{x})\Delta + \langle \mu_3(\mathbf{x}), \nabla \rangle - r_3(\mathbf{x}) \big) B_{\$,T}^{\$}(\mathbf{x}, t) = 0, \tag{4.22a}$$

$$B_{\$,T}^{\$}(\mathbf{x}, T) = 1, \tag{4.22b}$$

---

[6]Phrased differently: By the argument leading to (4.2), we know that in an unspanned model whose functional parameter $\sigma_1(\mathbf{x})$ does not affect the drift, the bond price function must satisfy $\Delta B_{\$,T}^{\$} = 0$ (i.e., it is a *necessary* condition for issue (iii) to be resolved), and it seems unlikely that this condition is satisfied by luck.

[7]This is in fact no coincidence: In the proof in Section B.3 we performed this change of variables because already then we had realized that it would simplify our SDE.

where σ_3, μ_3, and r_3 are given by (4.20), (4.21), and (4.18), respectively. Here the function $\sigma_3(\mathbf{x})$ can be chosen freely (by choosing $\sigma_1(\mathbf{z})$ correspondingly), and in fact it can be chosen as a time-dependent functions $\sigma_3(\mathbf{x}, t)$.[8] As these formulas coincide with those introduced in Chapter 2, this completes our quick derivation of the single-currency ACE model.

4.2   Multi-Currency Model

To generalize this to multiple currencies, we still introduce one artificial currency, but now with the constant short rate $\bar{\rho} := 0$, and for each of the multiple real currencies $c \in C$ we define the exchange rate as the time-dependent function

$$g_c(\mathbf{z}, t) := a^c e^{-\rho^c t} \|\mathbf{z} - \hat{\mathbf{v}}^c\|^{2-n} \tag{4.23}$$

for some currency-dependent scalars $a^c, \rho^c > 0$ and vectors $\hat{\mathbf{v}}^c$.[9] In Step 1, this would lead us to the bond price functions

$$B^c_{\bar{c}, T}(\mathbf{z}, t) = a^c e^{-\rho^c T} \|e^{(T-t)(E+\lambda I)} \mathbf{z} - \hat{\mathbf{v}}^c\|^{2-n}. \tag{4.24}$$

In Step 2, we would pick one of the real currencies as our base currency (let us call it c_0) and change the numeraire of the Feynman–Kac PDE to that currency, as before. The calculations will now change as follows:

First, in (4.10), which will now have $g_{c_0}(\mathbf{z}, t)$ instead of $g(\mathbf{z})$, the time derivative will now produce an additional term acting on $g_{c_0}(\mathbf{z}, t)$, and so in (4.11) we will have to add ρ^{c_0} inside the parentheses in the last line, essentially replacing the now vanishing $\bar{\rho}$. Note that all other terms are unaffected since our new factor $a^c e^{-\rho^c t}$ passes right by all the $\mathbf{z}$-derivatives and gets absorbed again when dividing by $g_{c_0}(\mathbf{z}, t)$. The extra term ρ^{c_0} then finds its way into $r_2(\mathbf{z})$ in (4.14), which we shall now call

$$r^{c_0}_2(\mathbf{z}) = r^{c_0}_\infty + \left\langle \mathbf{v}^{c_0}, \frac{\mathbf{z} - \hat{\mathbf{v}}^{c_0}}{\|\mathbf{z} - \hat{\mathbf{v}}^{c_0}\|^2} \right\rangle, \tag{4.25}$$

where we define

$$r^c_\infty := \rho^c + (n-2)\lambda \qquad \text{and} \qquad \mathbf{v}^c := (n-2)(E + \lambda I)\hat{\mathbf{v}}^c$$

for $\forall c \in C$. The bond price calculation generalizes to the bonds of any currency c as follows:

$$\begin{aligned} B^c_{c_0, t+\tau}(\mathbf{z}, t) &= \frac{B^c_{\bar{c}, t+\tau}(\mathbf{z}, t)}{g_{c_0}(\mathbf{z}, t)} \\ &= \frac{a^c e^{-\rho^c(t+\tau)} \|e^{\tau(E+\lambda I)} \mathbf{z} - \hat{\mathbf{v}}^c\|^{2-n}}{a^{c_0} e^{-\rho^{c_0} t} \|\mathbf{z} - \hat{\mathbf{v}}^{c_0}\|^{2-n}} \end{aligned}$$

---

[8]Our derivation did not use time-dependent noise merely for simplicity of notation.

[9]We could have used this generalized form in the single-currency case as well, but (i) when dealing with only one currency, the additional factor a^{c_0} would have canceled out in Step 2 and disappeared from our model, and (ii) the approach via $\bar{\rho}$ (which only works for the single-currency model) seems more intuitive than the equivalent approach of making the exchange rate time-dependent.

$$
= \frac{a^c e^{-\rho^c (t+\tau)} e^{(2-n)\lambda \tau} \left\| \mathbf{z} - e^{-\tau(E+\lambda I)} \hat{\mathbf{v}}^c \right\|^{2-n}}{a^{c_0} e^{-\rho^{c_0} t} \left\| \mathbf{z} - \hat{\mathbf{v}}^{c_0} \right\|^{2-n}}
$$

$$
= \frac{a^c e^{-[\rho^c + (n-2)\lambda](t+\tau)} \left\| (\mathbf{z} - \hat{\mathbf{v}}^{c_0}) - \left(e^{-\tau(E+\lambda I)} \hat{\mathbf{v}}^c - \hat{\mathbf{v}}^{c_0} \right) \right\|^{2-n}}{a^{c_0} e^{-[\rho^{c_0} + (n-2)\lambda]t} \left\| \mathbf{z} - \hat{\mathbf{v}}^{c_0} \right\|^{2-n}}
$$

$$
= \frac{a^c e^{-r_\infty^c (t+\tau)}}{a^{c_0} e^{-r_\infty^{c_0} t}} \left\| \frac{\mathbf{z} - \hat{\mathbf{v}}^{c_0}}{\left\| \mathbf{z} - \hat{\mathbf{v}}^{c_0} \right\|} - \frac{\mathbf{c}_\tau^c}{\left\| \mathbf{z} - \hat{\mathbf{v}}^{c_0} \right\|} \right\|^{2-n}
$$

$$
= \frac{a^c e^{-r_\infty^c (t+\tau)}}{a^{c_0} e^{-r_\infty^{c_0} t}} \left(1 - 2 \left\langle \frac{\mathbf{z} - \hat{\mathbf{v}}^{c_0}}{\left\| \mathbf{z} - \hat{\mathbf{v}}^{c_0} \right\|^2}, \mathbf{c}_\tau^c \right\rangle + \left\| \mathbf{c}_\tau^c \right\|^2 \left\| \mathbf{z} - \hat{\mathbf{v}}^{c_0} \right\|^{-2} \right)^{1-n/2}
$$

$$
= \frac{a^c e^{-r_\infty^c (t+\tau)}}{a^{c_0} e^{-r_\infty^{c_0} t}} \left\| \left\| \mathbf{c}_\tau^c \right\| \frac{\mathbf{z} - \hat{\mathbf{v}}^{c_0}}{\left\| \mathbf{z} - \hat{\mathbf{v}}^{c_0} \right\|^2} - \frac{\mathbf{c}_\tau^c}{\left\| \mathbf{c}_\tau^c \right\|} \right\|^{2-n}, \tag{4.26}
$$

where now we defined

$$
\mathbf{c}_\tau^c := e^{-\tau(E+\lambda I)} \hat{\mathbf{v}}^c - \hat{\mathbf{v}}^{c_0}
$$

$$
= \frac{1}{n-2} (E + \lambda I)^{-1} \left(e^{-\tau(E+\lambda I)} \mathbf{v}^c - \mathbf{v}^{c_0} \right). \tag{4.27}
$$

In Step 3, we now perform the change of variables

$$
\mathbf{x} := \frac{\mathbf{z} - \hat{\mathbf{v}}^{c_0}}{\left\| \mathbf{z} - \hat{\mathbf{v}}^{c_0} \right\|^2}, \tag{4.28}
$$

which leads us to the same functions $\sigma_3(\mathbf{x})$ and $\mu_3(\mathbf{x})$ as before, except with $\mathbf{v}^{c_0}$ instead of $\mathbf{v}$, and to the bond functions

$$
B_{c_0, t+\tau}^c(\mathbf{x}, t) = \frac{a^c e^{-r_\infty^c (t+\tau)}}{a^{c_0} e^{-r_\infty^{c_0} t}} \left(1 - 2 \langle \mathbf{x}, \mathbf{c}_\tau^c \rangle + \left\| \mathbf{c}_\tau^c \right\|^2 \left\| \mathbf{x} \right\|^2 \right)^{1-n/2}
$$

$$
= \frac{a^c e^{-r_\infty^c (t+\tau)}}{a^{c_0} e^{-r_\infty^{c_0} t}} \left\| \left\| \mathbf{c}_\tau^c \right\| \mathbf{x} - \frac{\mathbf{c}_\tau^c}{\left\| \mathbf{c}_\tau^c \right\|} \right\|^{2-n}. \tag{4.29}
$$

Using these bond functions, we can then easily derive the formulas for the forward exchange rates in general,

$$
a_{t+\tau}^{c_1/c_2}(\mathbf{x}, t) = B_{c_0, t+\tau}^{c_1}(\mathbf{x}, t) / B_{c_0, t+\tau}^{c_2}(\mathbf{x}, t)
$$

$$
= \frac{a^{c_1} \cdot e^{-r_\infty^{c_1}(t+\tau)} \cdot \left\| \left\| \mathbf{c}_\tau^{c_2} \right\| \mathbf{x} - \frac{\mathbf{c}_\tau^{c_2}}{\left\| \mathbf{c}_\tau^{c_2} \right\|} \right\|^{n-2}}{a^{c_2} \cdot e^{-r_\infty^{c_2}(t+\tau)} \cdot \left\| \left\| \mathbf{c}_\tau^{c_1} \right\| \mathbf{x} - \frac{\mathbf{c}_\tau^{c_1}}{\left\| \mathbf{c}_\tau^{c_1} \right\|} \right\|^{n-2}}, \tag{4.30}
$$

and the currency spot exchange rates in particular,

$$
a^{c_1/c_2}(\mathbf{x}, t) = a_t^{c_1/c_2}(\mathbf{x}, t)
$$

$$
= \frac{a^{c_1} \cdot e^{-r_\infty^{c_1} t} \cdot \left\| \left\| \mathbf{c}^{c_2} \right\| \mathbf{x} - \frac{\mathbf{c}^{c_2}}{\left\| \mathbf{c}^{c_2} \right\|} \right\|^{n-2}}{a^{c_2} \cdot e^{-r_\infty^{c_2} t} \cdot \left\| \left\| \mathbf{c}^{c_1} \right\| \mathbf{x} - \frac{\mathbf{c}^{c_1}}{\left\| \mathbf{c}^{c_1} \right\|} \right\|^{n-2}}, \tag{4.31}
$$

where we defined

$$\mathbf{c}^c := \mathbf{c}^c_{\tau=0} = \hat{\mathbf{v}}^c - \hat{\mathbf{v}}^{c_0}$$
$$= \tfrac{1}{n-2}(E + \lambda I)^{-1}(\mathbf{v}^c - \mathbf{v}^{c_0}) \tag{4.32}$$

for $\forall c \in C$. In particular, since $\mathbf{c}^{c_0} = \vec{0}$, we have to interpret

$$\left\| \|\mathbf{c}^{c_0}\| \mathbf{x} - \tfrac{\mathbf{c}^{c_0}}{\|\mathbf{c}^{c_0}\|} \right\| = \left(\|\mathbf{c}^{c_0}\|^2 \|\mathbf{x}\|^2 - 2\langle \mathbf{c}^{c_0}, \mathbf{x}\rangle + 1 \right)^{1/2} = 1$$

and therefore

$$a^{c_1/c_0}(\mathbf{x}, t) = \tfrac{a^{c_1}}{a^{c_0}} \, e^{(r^{c_0}_\infty - r^{c_1}_\infty)t} \left\| \|\mathbf{c}^{c_1}\| \mathbf{x} - \tfrac{\mathbf{c}^{c_1}}{\|\mathbf{c}^{c_1}\|} \right\|^{2-n}, \tag{4.33}$$

and so we find that the price of a c_1-bond expressed in units of an arbitrary currency c_2 is given by

$$B^{c_1}_{c_2, t+\tau}(\mathbf{x}, t) = \frac{B^{c_1}_{c_0, t+\tau}(\mathbf{x}, t)}{a^{c_2/c_0}(\mathbf{x}, t)} = \frac{\dfrac{a^{c_1} e^{-r^{c_1}_\infty(t+\tau)}}{a^{c_0} e^{-r^{c_0}_\infty t}} \left\| \|\mathbf{c}^{c_1}_\tau\| \mathbf{x} - \tfrac{\mathbf{c}^{c_1}_\tau}{\|\mathbf{c}^{c_1}_\tau\|} \right\|^{2-n}}{\dfrac{a^{c_2}}{a^{c_0}} e^{(r^{c_0}_\infty - r^{c_2}_\infty)t} \left\| \|\mathbf{c}^{c_2}\| \mathbf{x} - \tfrac{\mathbf{c}^{c_2}}{\|\mathbf{c}^{c_2}\|} \right\|^{2-n}}$$

$$= \frac{a^{c_1} \cdot e^{-r^{c_1}_\infty(t+\tau)} \cdot \left\| \|\mathbf{c}^{c_2}\| \mathbf{x} - \tfrac{\mathbf{c}^{c_2}}{\|\mathbf{c}^{c_2}\|} \right\|^{n-2}}{a^{c_2} \cdot e^{-r^{c_2}_\infty t} \cdot \left\| \|\mathbf{c}^{c_1}_\tau\| \mathbf{x} - \tfrac{\mathbf{c}^{c_1}_\tau}{\|\mathbf{c}^{c_1}_\tau\|} \right\|^{n-2}} . \tag{4.34}$$

Finally, since the short rate of any currency is determined merely by its bond price function expressed in units of that same currency, and since in Step 1 we had defined our model completely symmetrically in all real currencies, the short rate function (4.25) must generalize to

$$r^c_2(\mathbf{z}) = r^c_\infty + \left\langle \mathbf{v}^c, \tfrac{\mathbf{z} - \hat{\mathbf{v}}^c}{\|\mathbf{z} - \hat{\mathbf{v}}^c\|^2} \right\rangle$$

for $\forall c \in C$, and therefore to

$$r^c_3(\mathbf{x}) = r^c_\infty + \left\langle \mathbf{v}^c, G_c(\mathbf{x}) \right\rangle, \tag{4.35}$$

where

$$G_c(\mathbf{x}) := \left. \frac{\mathbf{z} - \hat{\mathbf{v}}^c}{\|\mathbf{z} - \hat{\mathbf{v}}^c\|^2} \right|_{\mathbf{z}=\mathbf{z}(\mathbf{x})} = \frac{\left(\tfrac{\mathbf{x}}{\|\mathbf{x}\|^2} + \hat{\mathbf{v}}^{c_0} \right) - \hat{\mathbf{v}}^c}{\left\| \left(\tfrac{\mathbf{x}}{\|\mathbf{x}\|^2} + \hat{\mathbf{v}}^{c_0} \right) - \hat{\mathbf{v}}^c \right\|^2}$$

$$= \frac{\tfrac{\mathbf{x}}{\|\mathbf{x}\|^2} - \mathbf{c}^c}{\left\| \tfrac{\mathbf{x}}{\|\mathbf{x}\|^2} - \mathbf{c}^c \right\|^2} . \tag{4.36}$$

We have thus defined a multi-currency model with all the key formulas introduced in Chapter 3, concluding our quick derivation of also the multi-currency ACE model.

4.3 Discussion

Ideas leading to the quick derivation. Given the model equations presented in Chapters 2–3, what has led us to the strategy behind the quick derivation in the present chapter?

First, as the multi-currency model had to allow for general non-normalized weight vectors $\mathbf{v}^c$, the question of what would happen if $\mathbf{v}^c = \vec{0}$ for some currency c was a natural one to ask, which led to the idea of the optional canonical parametrization proposed in Section 3.12 via the introduction of an artificial currency with a constant short rate. The next natural question to ask was then whether it would make sense to use this idea also in the single-currency model.

This idea would eventually lead to the insight that Pelts' strategy of first modeling "numeraire-free" values of assets and then obtaining the classical values measured in dollars in a final step by taking ratios (see Section 5.1) is really equivalent to using "values expressed in units of an artificial currency" first and then changing the numeraire.

Finally, our formula (2.54)–(2.56) for the zero-noise process U_t had long indicated that a change of variables $\mathbf{z} := \frac{\mathbf{x}}{\|\mathbf{x}\|^2} + \hat{\mathbf{v}}$ would lead to a linear drift, and therefore likely to a more insightful representation of the model that would be simpler at least in some aspects, even though this parametrization of the model would be less useful in practice (since it moves the attractor $\vec{0}$ to ∞ and complicates our rate function $r(\mathbf{x})$).

Comparison to the original derivation. Historically, however, the approach presented in this chapter was not the way in which the ACE model was originally developed, and how its $\mathbb{R}^n$-based model equations were first unearthed. Those who may now wonder whether the strategy outlined above would not have been much more "obvious" than the extremely abstract path on which Pelts had chosen to embark, should keep in mind the following:

First, Pelts' original approach aimed at (and succeeded in) creating an unspanned interest rate modeling framework *with maximal generality*. In Step 1, we make our lives easy by choosing U_t as a semigroup of *linear* conformal maps, well knowing that there are further conformal maps that might work if they only interacted nicely with g. And indeed, Pelts' use of a unified representation of conformal maps as pseudo-orthogonal maps in Minkowski space is guaranteed to cover *all* semigroups of conformal maps, including the maps of the form $H_{\mathbf{c}}(\mathbf{x})$ introduced in Section 3.6.4 that are nonlinear in $\mathbb{R}^n$ but linear in the Minkowski space representation (see Section 8.7). So it was feasible to think that this might have led to a more general model, and it was only after careful gauge fixing and changes of variables that it became clear that the specific semigroup $(U_\tau)_{\tau \geq 0}$ used here actually already provides maximal generality.

Second, one should not fail to appreciate the apparent outrageous luck in our quick derivation that the final change of variables $\mathbf{z} \mapsto \mathbf{x}$ has exactly the properties we need: Not only does it not lead to any regularity issues near $\mathbf{x} = \vec{0}$ as it maps the "attractor" $\mathbf{z} = \infty$ to $\mathbf{x} = \vec{0}$, it also happens to be angle-preserving (except for the sign of the angle) and therefore crucially preserves the multiplicative nature of our SDE's noise, and it perfectly cancels out the σ_1-dependent part of the drift.[10]   Only from Pelts' framework, in which all currencies (including the artificial one) are treated completely symmetrically, do we know that all this in fact wasn't luck at all, since the final step in Chapter 8 gave us a lot of freedom how to map its abstract state space back to $\mathbb{R}^n$ in a conformal manner, with one way leading to the parametrization $\mathbf{z}$ (to be used for $\bar{c}$ as the numeraire)[11] and another way leading to the parametrization $\mathbf{x}$ (to be used for c_0).

Third, what may be considered as unnecessarily difficult and abstract by many mathematicians or financial engineers, may come much more naturally to some trained theoretical physicists like Pelts for whom many of the concepts and techniques in his framework are second nature. Indeed, Pelts' original paper [1] is considerably more compact than our presentation in Chapters 5–9, as it does not feel the need to teach any of these techniques.

Finally, there are a lot of ingenious mathematical techniques to learn from Pelts' general interest rate modeling framework, and several types of powerful abstractions that shed light on what we are *really* doing in this quick derivation; in particular, his use of sections as numeraire-free asset values is a very powerful modeling technique that may well prove useful in other problems. We would therefore definitely advise the reader to invest the time and read on.

---

[10]While, as we had explained at the beginning of Step 3, this can be rationalized by observing that this change of variables makes the bond price function harmonic, that did not need to be the case with the same change of variables that gives $r_3(\mathbf{x})$ its simplest possible form. More so, the fact that the bond price function given by (4.16a) can be made harmonic at all via *any* τ-independent change of variables is not to be taken for granted.

[11]More precisely, the ACE framework would base $\bar{c}$ on $\mathbf{z}' := \mathbf{z}/\|\mathbf{z}\|^2$ instead (which also fixes the sign switch in λ in our proof); we absorbed this inversion in the starting point of our proof to simplify our calculations.

Chapter 5

The Mathematical Framework

The remainder of this book is dedicated to our original derivation of the single-currency ACE model, along with its bond price and forward rate formulas.[1] Let us therefore now step back to the end of Chapter 1 and ignore all the results that we have already given away in Part I.

Our construction is neither based on the short rate modeling approach described in Section 1.3.1 (defining a process $(X_t)_{t\geq 0}$ and a short rate function $r(\mathbf{x}, t)$ and then trying to solve (1.3)), nor on the HJM framework in Section 1.3.2 (trying to define a noise process $\sigma_t(s)$ such that the process $(\mathbf{f}_t)_{t\geq 0}$ defined via (1.5)–(1.6) recombines, as in Cheyette's method in Section 1.3.4).

Instead, our goal will be to define a process $(X_t)_{t\geq 0}$ on some state space $M_0 \subset \mathbb{R}^n$ and a family of bond price functions $B_{\$,T}(\mathbf{x}, t)$, $T \geq 0$, in such a way that the resulting model has all the properties listed in Section 1.1. Note that in order to make the model unspanned w.r.t. the functional parameter $\sigma(\mathbf{x}, t)$ that will control the size of the noise of the process $(X_t)_{t\geq 0}$, the bond functions $B_{\$,T}(\mathbf{x}, t)$ that we choose must not depend on $\sigma(\mathbf{x}, t)$. In a final step, the short rate formulation of the model can then be obtained via the formula

$$r(\mathbf{x}, t) = -\partial_T \log B_{\$,T}(\mathbf{x}, t)|_{T=t}, \tag{5.1}$$

which is the definition of the short rate.[2]

5.1   Numeraire-Free Modeling

We begin with a discussion of our mathematical approach to constructing a *numeraire-free* modeling framework. The underlying fundamental viewpoint touches the very core of financial modeling and applies to problems unrelated to interest rates as well.

---

[1] The derivation of its multi-currency extension may be added in a future edition.
[2] We will construct our model so that r does not depend on t.

5.1.1   Motivation and Mathematical Approach

Motivation.   Choosing the most convenient numeraire for our stochastic evolution equation of the state process $(X_t)_{t\geq 0}$ is a tricky task for which the two standard options both have significant downsides: When measuring asset values in dollars (i.e., under the risk-neutral measure), the Kolmogorov backward equation, which is used to express the consistency condition, contains a multiplicative rate term that is hard to deal with.[3] Choosing some specific maturity-T bond as a numeraire instead (i.e., working under a T-measure) would remove this problematic term, but unfortunately this would also destroy the time symmetry in our equations even for a perfectly time-homogeneous model, since we are singling out the process time $t = T$ at which the bond matures (and thus ceases to exist, which is another problem).

To successfully eliminate both of these problems, we will design our framework in a *numeraire-free* way, as laid out in the following. To do so, we will make use of the concept of *sections on line bundles* (see also [34, 35, 36]). A section, which we will typically denote by $\psi(\mathbf{x})$ or $\mathfrak{B}(\mathbf{x})$ (for *bond*), is a generalized notion of a standard real-valued function (explained in more detail below) that lends itself well to representing the *numeraire-free value* of a given asset at each state $\mathbf{x} \in M_0$.

Numeraire-free asset values as one-dimensional vectors.   What do we mean by "numeraire-free value"? Consider the statement "the asset A is worth \$100," which should be read as "$v_A = 100 \cdot v_\$$," where v_A is the value of the asset A and where $v_\$$ is the value of a one-dollar bill. It shows that asset values are typically expressed relative to the value of some other asset (the numeraire), i.e., it is given in the form of an *exchange rate*, here the number $100 \in \mathbb{R}$. Models usually decide for some numeraire first and then describe the evolution of the corresponding exchange rate. In contrast, *our numeraire-free modeling approach avoids having to choose a numeraire altogether, and instead it models the numeraire-free asset values v themselves.*

But what type of mathematical object should we choose to model numeraire-free values? To decide, let us think what we plan on doing with them. First we realize that the set of asset values forms a vector space V, since one can add any two asset values (e.g., $v_{A_1} + v_{A_2}$ should correspond to the value of an account containing one asset A_1 and one asset A_2), and since one can multiply asset values by any real number (e.g., $5 \cdot v_A$ is the value of 5 assets A, and $(-5) \cdot v_A$ is the value of the obligation to repay 5 assets A). Furthermore, we need to ensure that for any two asset values v_1 and $v_2 \neq \vec{0}$ there is a unique number $r \in \mathbb{R}$ such that $v_1 = r \cdot v_2$ (the exchange rate, denoted by $\frac{v_1}{v_2} := r \in \mathbb{R}$), i.e., that any two vectors in our vector space are linearly dependent; this requires V to be one-dimensional. In short, *we will model asset values as the elements of some fixed one-dimensional vector space V.*

---

[3]More precisely, the problem is that we must choose our bond price functions $B_{\$,T}(\mathbf{x}, t)$ as solutions of the Kolmogorov backward equation, which because of its short rate term is not fully known as long as the bond price functions are unavailable (see (5.1)).

We now see how to properly interpret the equation $v_1 = r \cdot v_2$: Choosing a numeraire (i.e., deciding to measure asset values in units of v_2) means to choose a vector $v_2 \in V \setminus \{\vec{0}\}$ as the basis vector of our one-dimensional vector space V; the exchange rate $r \in \mathbb{R}$ is then the coefficient of the vector v_1 w.r.t. that basis.

Positive and negative values. Finally, we would like to be able to ask whether a given asset has "positive" value, i.e., whether possessing it is better than not possessing it. For example, having a dollar bill is certainly better than not having it (no matter how strong or weak the currency currently is compared to others), and so its value is considered positive; in fact, all assets that one commonly uses as a numeraire (such as bonds) usually have positive value. On the other hand, a debt certificate over one dollar has negative value, since one would rather prefer not to have it. This leads us to decompose V into a set V^+ of "positive" vectors, a set V^- of "negative" vectors, and the zero vector (which represents assets with no value):

$$V = V^+ \cup V^- \cup \{\vec{0}\}.$$

Furthermore, we expect that positive multiples of any asset with positive value should have positive value as well, while debt certificates over any positive-valued asset should have negative value. More precisely, introducing the notation $\operatorname{sign}(v) := \mathbb{1}_{V^+}(v) - \mathbb{1}_{V^-}(v)$, we expect that

$$\forall r \in \mathbb{R} \ \forall v \in V: \quad \operatorname{sign}(rv) = \operatorname{sign}(r) \cdot \operatorname{sign}(v), \tag{5.2}$$

which allows us to conclude that V^+ and V^- must be the two pieces obtained by cutting through the line V at the origin. This shows that we must impose some further structure on V (in addition to its vector space properties): We ask that V is a *signed* one-dimensional vector space, i.e., that one of its two sides of the origin is agreed to be called positive and the other one negative.

State-dependent asset values as sections on the wealth line bundle. To model a *state-dependent* asset value, i.e., a map that assigns to every hypothetical state of the world $\mathbf{x} \in M_0$ the numeraire-free value $\psi(\mathbf{x})$ of some given asset in that state, it now seems natural to think of ψ as a function $\psi \colon M_0 \to V$, for some fixed signed one-dimensional vector space V. However, doing so would impose a very subtle yet decisive unnecessary constraint onto our framework that would in fact render our construction impossible.

To understand this, observe that we do not plan on ever comparing the values $\psi(\mathbf{x}_1)$ and $\psi(\mathbf{x}_2)$ of some asset at two different states $\mathbf{x}_1 \neq \mathbf{x}_2$. For example, there is no such thing as an exchange rate $\frac{\psi(\mathbf{x}_1)}{\psi(\mathbf{x}_2)}$, since a person living in a world in the state $\mathbf{x}_1$ cannot make trades with another person who is living in a world in a different state $\mathbf{x}_2$.[4] Similarly, the calculation $\psi(\mathbf{x}_1) + \psi(\mathbf{x}_2)$ is meaningless

---

[4]The reader may now argue that it should be possible to say whether a certain asset A is worth more in one state or the other. However, this holds true only for exchange rates, and not for numeraire-free values: The asset may be worth more *dollars* in state $\mathbf{x}_1$ than in

since an account cannot contain assets from different states of the world at the same time.

We can therefore safely *allow ψ to map every $\mathbf{x} \in M_0$ into a different signed one-dimensional vector space $V_{\mathbf{x}}$*. Such maps ψ are no longer called functions, but *sections*. The family $\mathcal{V}_0 := (V_{\mathbf{x}})_{\mathbf{x} \in M_0}$ is called the *wealth line bundle* (we say that it is a *signed* line bundle since all the vector spaces $V_{\mathbf{x}}$ are signed), and the space of all sections ψ on that line bundle is denoted by $\mathcal{S}(\mathcal{V}_0)$.

Asset values $\psi(\mathbf{x}, t)$ that also depend on the process time t should be thought of as a family of sections $\psi(\cdot, t) \in \mathcal{S}(\mathcal{V}_0)$, one section for each fixed t. Alternatively, one can think of ψ as a section on a line bundle $(V_{\mathbf{x},t})_{\mathbf{x} \in M_0, t \in \mathbb{R}}$, where for simplicity we set $V_{\mathbf{x},t} := V_{\mathbf{x}}$ for $\forall t \in \mathbb{R}$. Note that since we do not plan on comparing values for different process times t, we could have allowed $V_{\mathbf{x},t}$ to be different for each t; however, in our model we do not make use of this freedom.

Summary.   Numeraire-free modeling is based on the following principal ideas:

(i) There is *one* family $\mathcal{V}_0 := (V_{\mathbf{x}})_{\mathbf{x} \in M_0}$ of one-dimensional signed vector spaces $V_{\mathbf{x}}$, called the wealth line bundle. For each $\mathbf{x}$, one side of the origin of $V_{\mathbf{x}}$ is labeled as "positive" and the other one as "negative," with the two sides denoted as $V_{\mathbf{x}}^+$ and $V_{\mathbf{x}}^-$, respectively.[5]

(ii) State-dependent numeraire-free asset values are modeled as sections ψ on this wealth line bundle (one section for each asset), i.e., by mappings that assign to every state $\mathbf{x} \in M_0$ a vector $\psi(\mathbf{x}) \in V_{\mathbf{x}}$ that represents the numeraire-free value of the asset if the world is in the state $\mathbf{x}$. We denote the space of all sections on $\mathcal{V}_0$ by $\mathcal{S}(\mathcal{V}_0)$.

(iii) Positive-valued assets, i.e., those that are always better to have than not to have (such as bonds or the dollar, and usually any asset used as a numeraire), are modeled as sections that map each state $\mathbf{x}$ into the subset $V_{\mathbf{x}}^+ \subset V_{\mathbf{x}}$ of positive asset values. We denote the subset of all such *positive sections* by $\mathcal{S}_+(\mathcal{V}_0)$. The set of all *non-zero sections* (i.e., sections $\psi \in \mathcal{S}(\mathcal{V}_0)$ such that $\forall \mathbf{x} \in M_0 \colon \psi(\mathbf{x}) \neq \vec{0} \in V_{\mathbf{x}}$) is denoted by $\mathcal{S}_{\varnothing}(\mathcal{V}_0)$.

(iv) Asset values $\psi(\mathbf{x}, t)$ that also depend on the process time t should be thought of as a family of sections $\psi(\cdot, t) \in \mathcal{S}(\mathcal{V}_0)$, one section for each fixed t.

(v) Given any two assets with associated sections $\psi_1 \in \mathcal{S}(\mathcal{V}_0)$ and $\psi_2 \in \mathcal{S}_{\varnothing}(\mathcal{V}_0)$, respectively, the exchange rate between the two is given by the standard function $\frac{\psi_1}{\psi_2} \colon M_0 \to \mathbb{R}$, defined as the unique function f such that $\psi_1(\mathbf{x}) = f(\mathbf{x})\psi_2(\mathbf{x})$ for $\forall \mathbf{x} \in M_0$.[6]

---

state $\mathbf{x}_2$, i.e., we have $\frac{\psi_A(\mathbf{x}_1)}{\psi_\$(\mathbf{x}_1)} > \frac{\psi_A(\mathbf{x}_2)}{\psi_\$(\mathbf{x}_2)}$. This however does not necessarily imply that the asset is also worth more *Euros* in state $\mathbf{x}_1$ than in state $\mathbf{x}_2$, since the dollar-to-euro conversion rates in these two states may be different. This shows that comparing asset values in two different states *without choosing a numeraire* is an ill-defined problem.

[5]The rays $V_{\mathbf{x}}^+$ and $V_{\mathbf{x}}^-$ do not include the origin.

[6]In contrast to our notational convention in the introduction in Chapter 1, throughout

Going forward, we may refer to the "numeraire-free value" of an asset simply as its "value"; in contrast, its price measured in dollars will explicitly be referred to as its "dollar value."

5.1.2 Discussion

Before we proceed, let us discuss a few potential sources of confusion about our strategy of using one-dimensional vectors to model numeraire-free asset values.

How is a one-dimensional vector space any different from $\mathbb{R}$?

The specific vector space $\mathbb{R}$ of real numbers has more structure than what is required of a general vector space: $\mathbb{R}$ is a *field* since it has a multiplication operation $\cdot: \mathbb{R} \times \mathbb{R} \to \mathbb{R}$ defined on it whose neutral element we call 1. In contrast, general vector spaces V are *not* required to have a multiplication operation $\cdot: V \times V \to V$, and so the only element that they single out is the zero vector (i.e., the neutral element w.r.t. vector addition).

While it is certainly possible to map any one-dimensional vector space V to $\mathbb{R}$ (more generally, any n-dimensional vector space can be mapped to $\mathbb{R}^n$), this requires choosing a basis first (which in our one-dimensional case consists of only one vector): Only once a basis vector $v_0 \in V$ is chosen, every other vector $v \in V$ can get mapped to the unique scalar $r \in \mathbb{R}$ such that $v = rv_0$. As we pointed out in Section 5.1.1, the basis vector v_0 should be interpreted as the numeraire-free value of a numeraire, and the real number r is then the value of v in units of that numeraire.

How can one interpret the norm of a vector $v \in V$?

The definition of a general vector space does not require it to have a norm, or even an inner product, and so we do not require our vector space V to be equipped with any such additional structure either. Defining a norm on V would single out two vectors whose length is 1, which we want to avoid: We want that the only way of introducing a scale to our vector space is the choice of a numeraire.

Can't we just use $\mathbb{R}$ as our vector space V anyways? After all, that is a signed one-dimensional vector space.

We could, as long as we ensure that in all of our calculations and considerations only the ratios of values $v_1, v_2 \in \mathbb{R}$ matter, and never their individual sizes. However, this would make us vulnerable to some subtle traps that could lead us to inadvertently violate this rule, and furthermore, the familiarity of working with $\mathbb{R}$ may obstruct our view and prevent us from seeing some possible geometric constructions that may solve our problem at hand. To avoid this, we recommend the reader to truly embrace the idea of numeraire-free modeling, which means to think in terms of the mathematical objects whose properties are closest to what we are trying to model, namely of vector spaces that are *not* equipped with any additional unnecessary structure.

Part II of this book, functions denoted by f will no longer have the meaning of a forward rate function.

Are there any other advantages of not using $\mathbb{R}$ as our vector space?

Yes. Thinking of asset values as vectors makes it easier to keep track of the meaning of the mathematical objects in our calculations: If an expression of interest in our calculations is a vector in V then one can immediately conclude that it must be interpreted as a numeraire-free asset value, whereas exchange rates must be real numbers. If we were to use $\mathbb{R}$ as our vector space then this distinction based on the type of the object alone would not be possible.

5.1.3   Example: Bond Sections and Bond Price Functions

Following our strategy laid out in Section 5.1.1, we will base our numeraire-free interest rate modeling framework on the *bond sections* $\mathfrak{B}_T(\mathbf{x}, t)$, $t \leq T$, i.e., on a family of positive sections that represent the numeraire-free values of the maturity-T zero-coupon bonds.[7],[8] Once we have derived formulas for the bond sections $\mathfrak{B}_T$, we can obtain the bond price functions $B_{\$,T}$ (our actual objects of interest that measure the bond prices in dollars) as follows:

First, for $\forall T_1, T_2 \in \mathbb{R}$ the value of the maturity-T_2 bond in units of maturity-T_1 bonds (i.e., the exchange rate between the two) is given by the function $B_{T_1,T_2}: M_0 \times (-\infty, T_1 \wedge T_2] \to \mathbb{R}$,[9] defined as

$$B_{T_1,T_2}(\mathbf{x}, t) := \frac{\mathfrak{B}_{T_2}(\mathbf{x}, t)}{\mathfrak{B}_{T_1}(\mathbf{x}, t)} \,, \tag{5.3}$$

where the ratio on the right is to be understood as described in Section 5.1.1, item (v) above. Since by the very definition of a bond at each process time t the bond with maturity $T = t$ is worth \$1, the dollar value $B_{\$,T}(\mathbf{x}, t)$ of any maturity-T bond can therefore be computed via the formula

$$B_{\$,T}(\mathbf{x}, t) = B_{t,T}(\mathbf{x}, t) = \frac{\mathfrak{B}_T(\mathbf{x}, t)}{\mathfrak{B}_t(\mathbf{x}, t)} \,. \tag{5.4}$$

5.2   Spot Parametrization

While *any* choice of a family of bond sections $(\mathfrak{B}_T)_{T \in \mathbb{R}}$ and of a process $(X_t)_{t \geq 0}$ can be seen as a valid interest rate model, equipping the model with all of our desired properties is a difficult mathematical problem. Making this problem as tractable as possible requires us to carefully think how to interpret our state space, and how to best write our bond price functions in terms of a few well-chosen key ingredients. This will be the content of Sections 5.2–5.6. Throughout those sections, the only constraints that we will impose on our model are (i) that

---

[7]Throughout this book, all our bonds will be zero-coupon bonds, and so we will simply refer to them as bonds from this point forward. Bonds that do entitle to coupon payments do not require any further thought, as they can be replicated by baskets of zero-coupon bonds.

[8]Note that although we start our state process $(X_t)_{t \geq 0}$ at the (arbitrarily chosen) time $t = 0$, we will consider all maturities $T \in \mathbb{R}$ and allow our bond functions and sections to take also negative process times t as an argument.

[9]We use the notation $a \wedge b := \min\{a, b\}$ and later also $a \vee b := \max\{a, b\}$ for $\forall a, b \in \mathbb{R}$.

the bond price functions $B_{\$,T}(\mathbf{x}, t)$ do not depend on our functional parameter $\sigma(\mathbf{x}, t)$ that controls the size of the noise of $(X_t)_{t \geq 0}$, and (ii) that in the zero-noise case $\sigma(\mathbf{x}, t) \equiv 0$ the model is arbitrage free and time homogeneous.

From this point on we will use the upper index X for functions or sections (e.g., $B_{\$,T}^X$, B_{T_1,T_2}^X, $\mathfrak{B}_T^X$, r^X, etc.) to emphasize that they expect a state in the *spot parametrization* as its argument, which we will introduce in the following and which we will denote by $\mathbf{x}$. In Section 5.3 we will then introduce an alternative parametrization (the *forward parametrization*) whose states we denote by $\mathbf{y}$, and we will analogously use the index Y (e.g., $B_{\$,T}^Y$, B_{T_1,T_2}^Y, $\mathfrak{B}_T^Y$, r^Y, etc.) to emphasize that a function or section expects its argument in that format.

5.2.1 Definition

At any process time $t \in \mathbb{R}$, let us think of the state $\mathbf{x} \in M_0$ as some parametrization of the *relative forward rate curve* $\tau \mapsto F_{\mathrm{rel}}(\mathbf{x}; \tau)$, where $\tau \geq 0$ denotes the tenor relative to the time t at which $\mathbf{x}$ is observed. In other words, we have the relation

$$B_{\$,T}^X(\mathbf{x}, t) = \exp\left(-\int_0^{T-t} F_{\mathrm{rel}}(\mathbf{x}; \tau)\, \mathrm{d}\tau\right) \qquad \forall \mathbf{x} \in M_0 \; \forall t \leq T, \qquad (5.5)$$

which means that if we define the *absolute* forward rate function $F(\mathbf{x}, t; s)$ for $\forall \mathbf{x} \in M_0$ and $\forall s \geq t$ via the relation

$$B_{\$,T}^X(\mathbf{x}, t) = \exp\left(-\int_t^T F(\mathbf{x}, t; s)\, \mathrm{d}s\right) \qquad \forall \mathbf{x} \in M_0 \; \forall t \leq T, \qquad (5.6)$$

we have

$$F_{\mathrm{rel}}(\mathbf{x}; \tau) = F(\mathbf{x}, t; t + \tau). \qquad (5.7)$$

This is called the *spot parametrization*.

5.2.2 Time Homogeneity, Evolution Semigroup

Let us assume that in the zero-noise case $\sigma(\mathbf{x}, t) \equiv 0$ our modeling framework is fully time homogeneous in this parametrization,[10] i.e., that the only natural model parameter that allows the user to infuse time-dependent behavior into the model is the noise parameter $\sigma(\mathbf{x}, t)$.[11] Furthermore, let us make the assumption that the zero-noise process $(X_t^{\sigma=0})_{t \geq 0}$ is continuous.

We therefore assume that $(X_t^{\sigma=0})_{t \geq 0}$ is the solution of an ODE of the form

$$\mathrm{d}X_t^{\sigma=0} = \mu(X_t^{\sigma=0})\, \mathrm{d}t \qquad (5.8)$$

for some C^1 vector field $\mu\colon M_0 \to \mathbb{R}^n$. Our time homogeneity requirement then states that the field μ that controls the dynamics of this process cannot

---

[10] For a discussion about why we chose our framework to be time homogenous in this specific parametrization, see Section 5.11.

[11] In addition, one still has the option of reparametrizing the process time to fudge the discount curve, as described in Section 2.10; this, too, leads to time inhomogeneity.

explicitly depend on the process time t, since it asks that this process behaves in the same way when started from the same state $\mathbf{x}_0$ at different times t. Furthermore, μ must be such that the solutions of (5.8), when started from any point $\mathbf{x}_0 \in M_0$ at $t = 0$, are defined for $\forall t \geq 0$ and do not lead out of M_0.

The solutions of this ODE therefore give rise to an evolution semigroup $(U_s)_{s \geq 0}$ of maps

$$U_s \colon M_0 \to M_0, \qquad s \geq 0; \tag{5.9}$$

in other words, if we denote by $U_s(\mathbf{x})$ the state at time $s \geq 0$ of the process (5.8) started from $X_{s=0}^{\sigma=0} = \mathbf{x} \in M_0$, i.e., if

$$\partial_s U_s(\mathbf{x}) = \mu(U_s(\mathbf{x})), \qquad U_{s=0}(\mathbf{x}) = \mathbf{x} \tag{5.10}$$

for $\forall s \geq 0$ and $\forall \mathbf{x} \in M_0$, then we have

$$\forall s_1, s_2 \geq 0 \quad \forall \mathbf{x} \in M_0 \colon \qquad U_{s_1}(U_{s_2}(\mathbf{x})) = U_{s_1+s_2}(\mathbf{x}). \tag{5.11}$$

Note that setting $s = 0$ in (5.10) yields

$$\forall \mathbf{x} \in M_0 \colon \qquad \partial_s U_s(\mathbf{x})\big|_{s=0} = \mu(\mathbf{x}). \tag{5.12}$$

5.2.3   Consistency, Reduction to the Short Rate Function

In order for our modeling framework to be arbitrage free, in the zero-noise case it must always exhibit the "expected" behavior, since this is the behavior that instrument prices are based on. Intuitively, this should imply that the relative forward rate curve, which encodes this expected future behavior, simply shifts to the left at unit speed as the deterministic process (5.8) evolves. The following lemma makes this statement rigorous.

Lemma 5.1. *A necessary condition for our framework to be arbitrage free in the zero-noise case is that*

$$F_{\mathrm{rel}}(U_s(\mathbf{x}); \tau) = F_{\mathrm{rel}}(\mathbf{x}; s + \tau) \tag{5.13}$$

for $\forall \mathbf{x} \in M_0$ and $\forall s, \tau \geq 0$. (In fact, later in Section 5.8.3 we will show that this condition is also sufficient.)

Proof. Let $s, \tau \geq 0$ be given, and suppose that at $t = 0$ the system is in the state $\mathbf{x} \in M_0$. Then let us consider, for any $T_1, T_2 \geq s$, the trading strategy of buying a_1 maturity-T_1 bonds and selling a_2 maturity-T_2 bonds, where $a_1 := 1/B_{\$,T_1}^X(\mathbf{x}, 0)$ and $a_2 := 1/B_{\$,T_2}^X(\mathbf{x}, 0)$. The total price for initiating this trade is

$$a_1 B_{\$,T_1}^X(\mathbf{x}, 0) - a_2 B_{\$,T_2}^X(\mathbf{x}, 0) = 1 - 1 = 0,$$

and so in an arbitrage-free model the dollar value of this account must remain 0 until the first bond expires.[12] In particular, at the time $t = s$ we must still have

$$a_1 B_{\$,T_1}^X(U_s(\mathbf{x}), s) - a_2 B_{\$,T_2}^X(U_s(\mathbf{x}), s) = 0.$$

---

[12] Indeed, if the account value were positive at any future time, this strategy would lead to guaranteed profit; if it were negative then the reversed strategy of selling a_1 maturity-T_1 bonds and buying a_2 maturity-T_2 bonds would.

Solving for $\frac{a_1}{a_2}$ and applying (5.5), we therefore find that

$$\exp\left(-\int_{T_1}^{T_2} F_{\mathrm{rel}}(\mathbf{x};\tau')\,\mathrm{d}\tau'\right) = \frac{B_{\$,T_2}^X(\mathbf{x},0)}{B_{\$,T_1}^X(\mathbf{x},0)} = \frac{a_1}{a_2} = \frac{B_{\$,T_2}^X(U_s(\mathbf{x}),s)}{B_{\$,T_1}^X(U_s(\mathbf{x}),s)}$$

$$= \exp\left(-\int_{T_1-s}^{T_2-s} F_{\mathrm{rel}}(U_s(\mathbf{x});\tau')\,\mathrm{d}\tau'\right)$$

$$= \exp\left(-\int_{T_1}^{T_2} F_{\mathrm{rel}}(U_s(\mathbf{x});\tau'-s)\,\mathrm{d}\tau'\right),$$

and then taking the log and differentiating by T_2 on both sides, we obtain

$$F_{\mathrm{rel}}(\mathbf{x};T_2) = F_{\mathrm{rel}}(U_s(\mathbf{x}),T_2-s).$$

Choosing $T_2 := s + \tau$ now implies (5.13). $\qquad\square$

As a consequence of Lemma 5.1, we now realize that given the semigroup $(U_s)_{s\geq 0}$, we no longer need the full function F_{rel} in order to compute the complete relative forward rate curve from any state $\mathbf{x}$; indeed, the *short rate function*

$$r^X(\mathbf{x}) = F_{\mathrm{rel}}(\mathbf{x};0)$$

(recall (5.1) and (5.5)) suffices since setting $\tau = 0$ in (5.13) and then renaming s to τ implies that

$$F_{\mathrm{rel}}(\mathbf{x};\tau) = r^X(U_\tau(\mathbf{x})) \qquad \forall \tau \geq 0. \tag{5.14}$$

Note that since by (5.9) the maps U_τ lead into M_0, (5.14) further implies that if we design our model such that its short rates stay within a certain range (i.e., if r^X maps M_0 into that range), then in fact the entire forward curve will.

5.2.4   Bond Prices and Bond Exchange Rates

We can therefore express our bond price functions $B_{\$,T}$ solely based on the function $r^X(\mathbf{x})$ and the semigroup $(U_s)_{s\geq 0}$: Once these two are given, we can obtain the full relative forward rate curve $\tau \mapsto F_{\mathrm{rel}}(\mathbf{x};\tau)$ via (5.14) and then use (5.5) to obtain

$$B_{\$,T}^X(\mathbf{x},t) = \exp\left(-\int_0^{T-t} r^X(U_\tau(\mathbf{x}))\,\mathrm{d}\tau\right) \tag{5.15}$$

for $t \leq T$. As a result, we can also compute the bond exchange rates $B_{T_1,T_2}^X(\mathbf{x},t)$ solely based on these two key ingredients, via the formula

$$B_{T_1,T_2}^X(\mathbf{x},t) = \frac{B_{\$,T_2}^X(\mathbf{x},t)}{B_{\$,T_1}^X(\mathbf{x},t)} = \exp\left(-\int_{T_1-t}^{T_2-t} r^X(U_\tau(\mathbf{x}))\,\mathrm{d}\tau\right) \tag{5.16}$$

for $t \leq T_1 \wedge T_2$.

5.3 Forward Parametrization

Next, we will discuss a second way of parametrizing our model, as an alternative to the spot parametrization introduced in Section 5.2. While we will always be able to switch from one parametrization to the other without affecting the model's effective behavior, certain properties of the model will be more easily understood in one parametrization than in the other. In particular, time homogeneity will best be expressed in the spot parametrization, while our consistency condition will best be stated in the forward parametrization, which we will introduce in the following.

5.3.1 Motivation & Definition

Suppose that our semigroup $(U_s)_{s \geq 0}$ actually extended to a group $(U_s)_{s \in \mathbb{R}}$, i.e., that every operator U_s had an inverse U_{-s}. Then we could rewrite our bond exchange rate functions in (5.16) as

$$
\begin{aligned}
B_{T_1,T_2}^X(\mathbf{x},t) &= \exp\left(-\int_{T_1-t}^{T_2-t} r^X\left(U_\tau(\mathbf{x})\right) d\tau\right) \\
&= \exp\left(-\int_{T_1}^{T_2} r^X\left(U_{\tau-t}(\mathbf{x})\right) d\tau\right) \tag{5.17} \\
&= \exp\left(-\int_{T_1}^{T_2} r^X\left(U_\tau(U_{-t}(\mathbf{x}))\right) d\tau\right) \\
&= B_{T_1,T_2}^Y(U_{-t}(\mathbf{x})), \tag{5.18}
\end{aligned}
$$

where we define

$$
B_{T_1,T_2}^Y(\mathbf{y}) := \exp\left(-\int_{T_1}^{T_2} r^X\left(U_\tau(\mathbf{y})\right) d\tau\right). \tag{5.19}
$$

If we then introduced the time-dependent change of variables

$$
(t,\mathbf{x}) \leftrightarrow (t,\mathbf{y}) \qquad \text{where} \qquad \mathbf{y} = U_{-t}(\mathbf{x}), \tag{5.20}
$$

then we would have
$$
B_{T_1,T_2}^X(\mathbf{x},t) = B_{T_1,T_2}^Y(\mathbf{y}). \tag{5.21}
$$

As a result, if instead of $(X_t)_{t \geq 0}$ we considered the process $(Y_t)_{t \geq 0}$ defined as

$$
Y_t := U_{-t}(X_t) \tag{5.22}
$$

then the bond exchange rates for the state $Y_t = \mathbf{y}$ would be given by the function $B_{T_1,T_2}^Y(\mathbf{y})$, *which is independent of the process time t at which the state $\mathbf{y}$ is observed*. This can be a significant advantage in some of our future calculations that will include the time derivative of these bond exchange rate functions (in particular, in the Kolmogorov backward equation used to describe our consistency requirement). We would therefore like to be able to switch into this $\mathbf{y}$-parametrization for such calculations.

We will call this $\mathbf{y}$-parametrization the *forward parametrization*. Note that by (5.20) at the process time $t = 0$ the spot and the forward parametrizations coincide.

5.3.2 Interpretation

The interpretation of the forward parametrization is as follows: Suppose that at some time t the state in the forward parametrization is $\mathbf{y}$. Then by (5.20) the corresponding state in the spot parametrization is $\mathbf{x} = U_t(\mathbf{y})$, and so the *relative* forward rate curve in that state is given by

$$\tau \mapsto F_{\text{rel}}(\mathbf{x}; \tau) = F_{\text{rel}}(U_t(\mathbf{y}); \tau) \stackrel{(5.13)}{=} F_{\text{rel}}(\mathbf{y}; t + \tau) \qquad (5.23)$$

for $\tau \geq 0$. The *absolute* forward rate curve, whose argument is the absolute time $s \geq t$ instead of the tenor τ, is therefore given by

$$s \mapsto F(\mathbf{x}, t; s) \stackrel{(5.7)}{=} F_{\text{rel}}(\mathbf{x}; s - t) \stackrel{(5.23)}{=} F_{\text{rel}}(\mathbf{y}; s).$$

Observe that this expression does not depend on the process time t.

In other words, we find that while the *spot* parametrization of a given yield curve is the state $\mathbf{x}$ such that $\tau \mapsto F_{\text{rel}}(\mathbf{x}; \tau)$ for $\tau \geq 0$ is the *relative* forward rate curve, its *forward* parametrization is the state $\mathbf{y}$ such that $s \mapsto F_{\text{rel}}(\mathbf{y}; s)$ for $s \geq t$ is the *absolute* forward rate curve. Phrased differently, $\mathbf{x}$ encodes the relative forward rate curve, while $\mathbf{y}$ encodes the absolute forward rate curve; converting one curve into the other requires knowledge of the process time t, which is consistent with our definition (5.20).

This interpretation also shows why (as observed in the preceding section) the bond exchange rates in the forward parametrization can be computed without knowledge of t: Given the state $\mathbf{y}$, i.e., given the absolute forward rate curve, we can easily compute the exchange rate between a T_1- and a T_2-bond, by integrating this absolute forward rate curve from T_1 to T_2 and then negating and exponentiating the result. Given the state $\mathbf{x}$, i.e., given the relative forward rate curve, however, we do not know where on this curve to find the times T_1 and T_2, and thus we do not know which interval to integrate this curve over, unless we are also given the process time t.

5.3.3 Evolution Group

To make all of the above calculations possible, however, and to make the reparametrization (5.20) a bijection, we first need to extend the semigroup $(U_s)_{s \geq 0}$ to a group $(U_s)_{s \in \mathbb{R}}$ in an appropriate way. The most general way of doing this while leaving our assumptions on the spot parametrization practically untouched is as follows.

Let us assume that the drift μ of the zero-noise process (5.8) extends to some larger set $M \supset M_0$ (i.e., to a C^1 vector field $\mu \colon M \to \mathbb{R}^n$), in such a way that the ODE (5.8) starting from any $\mathbf{x} \in M_0$ has solutions *for all times* $t \in \mathbb{R}$ (and not only for $t \geq 0$). W.l.o.g. we may assume that M only consists of flowlines that eventually reach (and then by (5.9) stay in) M_0.

If analogously to Section 5.2.2 we now denote by $U_s(\mathbf{x})$ the state at time $s \in \mathbb{R}$ of the process (5.8) started from $X^{\sigma=0}_{t=0} = \mathbf{x} \in M$, then the operators

$$U_s \colon M \to M, \qquad s \in \mathbb{R},$$

thus defined form a group $(U_s)_{s \in \mathbb{R}}$, and for $\forall s \geq 0$ they extend the operators $U_s \colon M_0 \to M_0$ defined in Section 5.2.2. Furthermore, the properties (5.10)–(5.12) extend to $\forall \mathbf{x} \in M$ and $\forall s \in \mathbb{R}$, i.e., we have

$$\partial_s U_s(\mathbf{x}) = \mu(U_s(\mathbf{x})), \qquad U_{s=0}(\mathbf{x}) = \mathbf{x} \tag{5.24}$$

$$U_{s_1}(U_{s_2}(\mathbf{x})) = U_{s_1+s_2}(\mathbf{x}), \tag{5.25}$$

$$\partial_s U_s(\mathbf{x})\big|_{s=0} = \mu(\mathbf{x}) \tag{5.26}$$

for $\forall \mathbf{x} \in M$ and $\forall s, s_1, s_2 \in \mathbb{R}$, and since (as a result of the group property) the operators U_s are invertible, we have in fact

$$\forall s \in \mathbb{R}: \quad U_s(M) = M. \tag{5.27}$$

5.3.4   State Space Family

While the state space of the process $(X_t)_{t \geq 0}$ in the spot parametrization remains M_0, the associated process $(Y_t)_{t \geq 0}$ in the forward parametrization defined by (5.22) has the time-dependent state space

$$M_t := U_{-t}(M_0) \tag{5.28}$$

for $\forall t \geq 0$, i.e., we have

$$\textit{almost surely} \qquad \forall t \geq 0: \ Y_t \in M_t.$$

For future use we define the sets M_t given by (5.28) for $\forall t \in \mathbb{R}$. Note that these sets fulfill

$$\forall s, t \in \mathbb{R}: \quad U_s(M_t) = M_{t-s}, \tag{5.29}$$

and that the state space family $(M_t)_{t \in \mathbb{R}}$ is increasing, i.e.,

$$M_{t_1} \subseteq M_{t_2} \subseteq M \qquad \text{for} \qquad t_1 \leq t_2, \tag{5.30}$$

since by (5.9) we have

$$M_{t_1} = U_{-t_1}(M_0) = U_{-t_2}(U_{t_2-t_1}(M_0)) \subseteq U_{-t_2}(M_0) = M_{t_2}.$$

In particular, this implies that

$$\forall t_1, t_2 \in \mathbb{R}: \quad M_{t_1} \cap M_{t_2} = M_{t_1 \wedge t_2}, \tag{5.31a}$$

$$M_{t_1} \cup M_{t_2} = M_{t_1 \vee t_2}. \tag{5.31b}$$

Finally, since we assumed that every flowline in M eventually reaches M_0, we have

$$M = \bigcup_{t \geq 0} M_t. \tag{5.32}$$

5.3.5 State–Time Pairs

The state–time pairs $(\mathbf{x}, t)$ and $(\mathbf{y}, t)$ in the spot and forward parametrizations are confined to the sets

$$M_X := M_0 \times \mathbb{R},$$
$$M_Y := \{(\mathbf{y}, t) \mid t \in \mathbb{R}, \ \mathbf{y} \in M_t\},$$

respectively, and with this notation our mapping (5.20) can be written as

$$U \colon M_Y \to M_X, \qquad U(\mathbf{y}, t) = (U_t(\mathbf{y}), t) \ = (\mathbf{x}, t), \qquad (5.33\text{a})$$
$$U^{-1}(\mathbf{x}, t) = (U_{-t}(\mathbf{x}), t) = (\mathbf{y}, t). \qquad (5.33\text{b})$$

Note that for $\forall t \geq 0$ we have

$$U(Y_t, t) = (X_t, t), \qquad (5.34)$$

as well as

$$(X_t, t) \in M_X \qquad \text{and} \qquad (Y_t, t) \in M_Y$$

almost surely. Finally, we denote the subsets of all possible state–time pairs before some fixed time $T \geq 0$ by

$$M_{X,T} := \{(\mathbf{x}, t) \in M_X, \ t \leq T\},$$
$$M_{Y,T} := \{(\mathbf{y}; t) \in M_Y, \ t \leq T\}.$$

5.3.6 Line Bundles, Functions, and Sections

In order to express account values $\psi(\mathbf{y})$ in the forward parametrization, we also need to extend our line bundle $\mathcal{V}_0 = (V_\mathbf{x})_{\mathbf{x} \in M_0}$ to a line bundle $\mathcal{V} = (V_\mathbf{x})_{\mathbf{x} \in M}$. Defining the subbundles

$$\mathcal{V}_T := (V_\mathbf{x})_{\mathbf{x} \in M_T}$$

for $\forall T \in \mathbb{R}$, the set of all sections $\psi(\mathbf{y})$ that are only defined for $\forall \mathbf{y} \in M_T$, for some fixed $T \in \mathbb{R}$, is then given by $\mathcal{S}(\mathcal{V}_T)$.

To allow for time-dependent sections, we again define $V_{\mathbf{x},t} := V_\mathbf{x}$ for $\forall \mathbf{x} \in M$ and $\forall t \in \mathbb{R}$, and we consider the line bundles

$$\mathcal{V}_{X,T} = (V_{\mathbf{x},t})_{(\mathbf{x},t) \in M_{X,T}}, \qquad\qquad \mathcal{V}_X = (V_{\mathbf{x},t})_{(\mathbf{x},t) \in M_X},$$
$$\mathcal{V}_{Y,T} = (V_{\mathbf{y},t})_{(\mathbf{y},t) \in M_{Y,T}}, \qquad\qquad \mathcal{V}_Y = (V_{\mathbf{y},t})_{(\mathbf{y},t) \in M_Y}$$

for $\forall T \in \mathbb{R}$, which are subbundles of

$$\mathcal{V}_{M,T} := (V_{\mathbf{x},t})_{(\mathbf{x},t) \in M \times (-\infty, T]} \qquad (5.35)$$
$$\text{and} \qquad \mathcal{V}_{M \times \mathbb{R}} := (V_{\mathbf{x},t})_{(\mathbf{x},t) \in M \times \mathbb{R}}, \qquad (5.36)$$

respectively. In this way, time-dependent values of instruments that exist only until some time $T \in \mathbb{R}$ are expressed using sections $\psi(\mathbf{x}, t)$ in $\mathcal{S}(\mathcal{V}_{X,T})$ (in the spot parametrization), or using sections $\psi(\mathbf{y}, t)$ in $\mathcal{S}(\mathcal{V}_{Y,T})$ (in the forward parametrization), while time-dependent values of instruments that exist for all times are expressed using sections in $\mathcal{S}(\mathcal{V}_X)$ or $\mathcal{S}(\mathcal{V}_Y)$.

Similarly, if for any set A we denote by $\mathcal{F}(A)$ the set of all real-valued functions on A, exchange rates between any two instruments that exist until times $T_1, T_2 \in \mathbb{R}$ are given by functions in $\mathcal{F}(M_{X,T_1 \wedge T_2})$ or $\mathcal{F}(M_{Y,T_1 \wedge T_2})$, while time-dependent exchange rates between instruments that exist for all times are expressed using sections in $\mathcal{S}(\mathcal{V}_X)$ or $\mathcal{S}(\mathcal{V}_Y)$.

Observe that sections $\psi \in \mathcal{S}(\mathcal{V}_{X,T})$ or $\psi \in \mathcal{S}(\mathcal{V}_{Y,T})$ that happen to not depend on t can be identified with sections in $\mathcal{S}(\mathcal{V}_0)$ and $\mathcal{S}(\mathcal{V}_T)$, respectively. Similarly, functions $f \in \mathcal{F}(\mathcal{V}_{X,T})$ or $f \in \mathcal{F}(\mathcal{V}_{Y,T})$ that happen to not depend on t are really functions in $\mathcal{F}(M_0)$ and $\mathcal{F}(M_T)$, respectively.

5.3.7 Short Rate Function

Our bond exchange rate function $B^Y_{T_1,T_2}(\mathbf{y})$ in (5.19) is in fact already well defined on $M_{T_1 \wedge T_2}$ as is: Indeed, for $\forall \mathbf{y} \in M_{T_1 \wedge T_2}$ and $\forall \tau \geq T_1 \wedge T_2$ we have

$$U_\tau(\mathbf{y}) \in U_\tau(M_{T_1 \wedge T_2}) \overset{(5.28)}{=} U_{\tau - T_1 \wedge T_2}(M_0) \overset{(5.28)}{=} M_{(T_1 \wedge T_2) - \tau} \overset{(5.30)}{\subseteq} M_0\,,$$

which implies that the integrand $r^X(U_\tau(\mathbf{y}))$ in (5.19) is well defined. It is therefore *not* necessary for us to extend the short rate function $r^X : M_0 \to \mathbb{R}$ to the larger set M.

As it should be expected from Section 5.3.2, the function $r^Y(\mathbf{y}, t)$ for computing the short rate in the forward parametrization requires knowledge of the time t at which the state $\mathbf{y}$ is observed. Since $\mathbf{x} = U_t(\mathbf{y})$, it can be obtained from r^X via the relation

$$r^Y(\mathbf{y}, t) = r^X(U_t(\mathbf{y})) \tag{5.37}$$

for $\forall (\mathbf{y}, t) \in M_Y$.

5.3.8 Zero-Noise Process

Besides leading to time-independent bond exchange rate functions, another advantage of the forward parametrization is the simplicity of the zero-noise process: Since in the spot parametrization the zero-noise process starting from a point $\mathbf{x}_0 \in M_0$ is given by $X^{\sigma=0}_t = U_t(\mathbf{x}_0)$ for $\forall t \geq 0$ (by definition of our operators U_t), its forward parametrization

$$Y^{\sigma=0}_t \overset{(5.22)}{=} U_{-t}(X^{\sigma=0}_t) \equiv \mathbf{x}_0$$

is constant for $\forall t \geq 0$. The zero-noise process in the forward parametrization is therefore given by the trivial ODE

$$dY^{\sigma=0}_t = 0, \qquad dY^{\sigma=0}_{t=0} = \mathbf{x}_0\,.$$

5.3.9 Summary: Spot vs. Forward Parametrization

In summary: The forward parametrization has the disadvantage that the state space M_t of our process changes over time, but it has the advantages that (i) the zero-noise process leaves states constant and that (ii) the bond exchange rates are independent of t. While the concept of time homogeneity will be best understood in the spot parametrization, our consistency requirement will be expressed more efficiently in the forward parametrization.

5.4 Bond Sections

As discussed in Section 5.1.3, for our numeraire-free modeling approach we need to define positive-valued sections $\mathfrak{B}_T^X(\mathbf{x}, t)$ and $\mathfrak{B}_T^Y(\mathbf{y}, t)$ (i.e., sections in $\mathcal{S}_+(\mathcal{V}_{X,T})$ and $\mathcal{S}_+(\mathcal{V}_{Y,T})$, respectively) whose ratios are the bond exchange rate functions $B_{T_1,T_2}^X(\mathbf{x}, t)$ and $B_{T_1,T_2}^Y(\mathbf{y})$ from the preceding sections, i.e.,

$$\frac{\mathfrak{B}_{T_2}^X(\mathbf{x}, t)}{\mathfrak{B}_{T_1}^X(\mathbf{x}, t)} = B_{T_1,T_2}^X(\mathbf{x}, t) \qquad \forall (\mathbf{x}, t) \in M_{X,T},$$

$$\frac{\mathfrak{B}_{T_2}^Y(\mathbf{y}, t)}{\mathfrak{B}_{T_1}^Y(\mathbf{y}, t)} = B_{T_1,T_2}^Y(\mathbf{y}) \qquad \forall (\mathbf{y}, t) \in M_{Y,T}.$$

Clearly, this requirement leaves us with some freedom, since multiplying all bond sections by the same function $f(\mathbf{x}, t)$ or $f(\mathbf{y}, t)$ leaves their ratios unchanged. We can make use of this freedom to simplify our framework without affecting the effective model behavior, since our actual interest lies solely in the exchange rates.

5.4.1 Forward Parametrization

In particular, we can choose to make our bond sections $\mathfrak{B}_T^Y$ independent of time, i.e., $\mathfrak{B}_T^Y = \mathfrak{B}_T^Y(\mathbf{y})$. According to our remark at the end of Section 5.3.6, we are therefore looking for sections

$$\mathfrak{B}_T^Y \in \mathcal{S}_+(\mathcal{V}_T) \quad \text{for} \quad \forall T \in \mathbb{R}$$

whose ratios are the desired exchange rates

$$\frac{\mathfrak{B}_{T_2}^Y(\mathbf{y})}{\mathfrak{B}_{T_1}^Y(\mathbf{y})} = B_{T_1,T_2}^Y(\mathbf{y}) = \exp\left(-\int_{T_1}^{T_2} r^X(U_\tau(\mathbf{y})) \, d\tau\right) \tag{5.38}$$

for $\forall T_1, T_2 \in \mathbb{R}$ and $\forall \mathbf{y} \in M_{T_1 \wedge T_2}$. Note that expressions or equations such as this one that involve functions and/or sections that are defined on different domains are always understood to be valid only on the intersection of these domains (here $M_{T_1 \wedge T_2}$ by (5.31a)).

We will postpone our specific construction of $(\mathfrak{B}_T^Y)_{T \in \mathbb{R}}$ to Section 5.6.3; for now, we will find comfort in the following lemma.

Lemma 5.2. *(i) There exists a family $(\mathfrak{B}_T^Y)_{T \in \mathbb{R}}$ of sections $\mathfrak{B}_T^Y \in \mathcal{S}_+(\mathcal{V}_T)$ such that (5.38) holds.*

(ii) Given any two such families $\left(\mathfrak{B}_T^{Y,1}\right)_{T \in \mathbb{R}}$ and $\left(\mathfrak{B}_T^{Y,2}\right)_{T \in \mathbb{R}}$, there $\exists f \in \mathcal{F}(M)$ such that $\forall T \in \mathbb{R}$: $\mathfrak{B}_T^{Y,1} = f \cdot \mathfrak{B}_T^{Y,2}$.

Proof. (i) Let $\psi \in \mathcal{S}_+(\mathcal{V})$ be any positive-valued section, and let $T^\star \colon M \to [0, \infty)$ be any function such that $\forall \mathbf{y} \in M$: $\mathbf{y} \in M_{T^\star(y)}$ (recall (5.32)). Then the section

$$\mathfrak{B}_T^Y(\mathbf{y}) := \exp\left(-\int_{T^\star(\mathbf{y})}^T r(U_\tau(\mathbf{y})) \, d\tau\right) \cdot \psi(\mathbf{y})$$

is well defined for $\forall T \in \mathbb{R}$ and $\forall \mathbf{y} \in M_T$; indeed, for any such $\mathbf{y}$ we have $\mathbf{y} \in M_T \cap M_{T^\star(\mathbf{y})} = M_{T \wedge T^\star(\mathbf{y})}$, i.e., $U_{T \wedge T^\star(\mathbf{y})}(\mathbf{y}) \in M_0$, and thus

$$U_\tau(\mathbf{y}) = U_{\tau - T \wedge T^\star(\mathbf{y})}\big(U_{T \wedge T^\star(\mathbf{y})}(\mathbf{y})\big) \in M_0$$

for $\forall \tau \geq T \wedge T^\star(\mathbf{y})$ by (5.9). It is easy to see that the bond sections thus defined are positive by (5.2), and that they have the desired ratios (5.38).

(ii) Given any two such families and any $T_1, T_2 \in \mathbb{R}$, we have

$$\frac{\mathfrak{B}_{T_2}^{Y,1}}{\mathfrak{B}_{T_1}^{Y,1}} = B_{T_1, T_2}^Y = \frac{\mathfrak{B}_{T_2}^{Y,2}}{\mathfrak{B}_{T_1}^{Y,2}} \qquad \text{and thus} \qquad \frac{\mathfrak{B}_{T_2}^{Y,1}}{\mathfrak{B}_{T_2}^{Y,2}} = \frac{\mathfrak{B}_{T_1}^{Y,1}}{\mathfrak{B}_{T_1}^{Y,2}}$$

on $M_{T_1 \wedge T_2}$. In other words, given any $\mathbf{y} \in M$, the value $f(\mathbf{y}) := \frac{\mathfrak{B}_T^{Y,1}(\mathbf{y})}{\mathfrak{B}_T^{Y,2}(\mathbf{y})}$ is the same for any $T \in \mathbb{R}$ with $\mathbf{y} \in M_T$. This defines the desired function $f \in \mathcal{F}(M)$. $\square$

5.4.2   Spot Parametrization

Now suppose that we have decided upon a specific family $(\mathfrak{B}_T^Y)_{T \in \mathbb{R}}$. Motivated by the observation that for $\forall (\mathbf{x}, t) \in M_{X, T_1 \wedge T_2}$ we have

$$B_{T_1, T_2}^X(\mathbf{x}, t) = B_{T_1 - t, T_2 - t}^Y(\mathbf{x}) \tag{5.39}$$

by (5.16) and (5.19), we can then make further use of our freedom by defining

$$\mathfrak{B}_T^X(\mathbf{x}, t) := \mathfrak{B}_{T-t}^Y(\mathbf{x}). \tag{5.40}$$

(Note that this is well defined since $T - t \geq 0$ and thus $\mathbf{x} \in M_0 \subseteq M_{T-t}$.) The bond sections $\mathfrak{B}_T^X$ thus defined indeed fulfill

$$\frac{\mathfrak{B}_{T_2}^X(\mathbf{x}; t)}{\mathfrak{B}_{T_1}^X(\mathbf{x}, t)} \overset{(5.40)}{=} \frac{\mathfrak{B}_{T_2 - t}^Y(\mathbf{x})}{\mathfrak{B}_{T_1 - t}^Y(\mathbf{x})} \overset{(5.38)}{=} B_{T_1 - t, T_2 - t}^Y(\mathbf{x}) \overset{(5.39)}{=} B_{T_1, T_2}^X(\mathbf{x}, t)$$

for $\forall (\mathbf{x}, t) \in M_{X, T_1 \wedge T_2}$, as required.

5.5 Change of Parametrization

Since the spot and the forward parametrization both have their advantages, we would like to be able to switch between the two at will. To do so, we not only need to convert state–time pairs $(\mathbf{x}, t)$ into $(\mathbf{y}, t)$ and vice versa, as described in (5.33), we also need to convert between exchange rate functions $f_X(\mathbf{x}, t)$ and $f_Y(\mathbf{y}, t)$, and between sections $\psi_X(\mathbf{x}, t)$ and $\psi_Y(\mathbf{y}, t)$. *The conversions of functions and sections should be designed in such a way that one can carry out any calculation involving states, functions, and sections in either parametrization and obtain the same result.*

5.5.1 Functions

First, let us define an operator

$$\bar{U} \colon \mathcal{F}(M_Y) \to \mathcal{F}(M_X), \qquad \bar{U} f_Y = f_X, \tag{5.41}$$

that maps a function $f_Y(\mathbf{y}, t)$ to a function $f_X(\mathbf{x}, t)$ returning the same values, i.e., such that

$$f_X(\mathbf{x}, t) = f_Y(\mathbf{y}, t) \tag{5.42}$$

(see for example (5.21)); in other words, we want this operator to have the property

$$(\bar{U} f_Y)(\mathbf{x}, t) = f_X(\mathbf{x}, t) = f_Y(\mathbf{y}, t) = f_Y(U^{-1}(\mathbf{x}, t)). \tag{5.43}$$

We therefore define it as

$$\bar{U} f_Y := f_Y \circ U^{-1} \qquad \text{for} \qquad \forall f_Y \in \mathcal{F}(M_Y), \tag{5.44a}$$

which has the inverse

$$\bar{U}^{-1} f_X := f_X \circ U \qquad \text{for} \qquad \forall f_X \in \mathcal{F}(M_X). \tag{5.44b}$$

More explicitly, these operators can be written as

$$(\bar{U} f_Y)(\mathbf{x}, t) = f_Y(U_{-t}(\mathbf{x}), t), \tag{5.45a}$$

$$(\bar{U}^{-1} f_X)(\mathbf{y}, t) = f_X(U_t(\mathbf{y}), t) \tag{5.45b}$$

for $\forall(\mathbf{x}, t) \in M_X$ and $\forall(\mathbf{y}, t) \in M_Y$. As an example, we can use $\bar{U}$ to rewrite (5.18) and (5.37) as

$$B_{T_1, T_2}^X = \bar{U} B_{T_1, T_2}^Y, \tag{5.46}$$

$$r^X = \bar{U} r^Y. \tag{5.47}$$

Observe that $\bar{U}$ and $\bar{U}^{-1}$ are linear, and that they have the property

$$\forall f_Y^1, f_Y^2 \in \mathcal{F}(M_Y) \colon \qquad \bar{U}(f_Y^1 \cdot f_Y^2) = (\bar{U} f_Y^1) \cdot (\bar{U} f_Y^2), \tag{5.48a}$$

$$\forall f_X^1, f_X^2 \in \mathcal{F}(M_X) \colon \quad \bar{U}^{-1}(f_X^1 \cdot f_X^2) = (\bar{U}^{-1} f_X^1) \cdot (\bar{U}^{-1} f_X^2). \tag{5.48b}$$

5.5.2   Sections

Similarly, we need to define an operator

$$\hat{U}: \mathcal{S}(\mathcal{V}_Y) \to \mathcal{S}(\mathcal{V}_X), \qquad \hat{U}\psi_Y = \psi_X,$$

that maps a section $\psi_Y(\mathbf{y}, t)$ to a section $\psi_X(\mathbf{x}, t)$. However, we do *not* require that $\psi_X(\mathbf{x}, t) = \psi_Y(\mathbf{y}, t)$; this relation is nonsensical since the two sides of this equation are elements of the two different vector spaces, namely $V_{\mathbf{x}}$ and $V_{\mathbf{y}}$, respectively. Instead, what we require is only that our actual objects of interest, namely the exchange rates computed from these sections, coincide in both parametrizations, i.e., that $\frac{\psi_X^1(\mathbf{x},t)}{\psi_X^2(\mathbf{x},t)} = \frac{\psi_Y^1(\mathbf{y},t)}{\psi_Y^2(\mathbf{y},t)}$. The operator $\hat{U}$ therefore needs to fulfill

$$\left(\frac{\hat{U}\psi_Y^1}{\hat{U}\psi_Y^2}\right)(\mathbf{x}, t) = \frac{(\hat{U}\psi_Y^1)(\mathbf{x}, t)}{(\hat{U}\psi_Y^2)(\mathbf{x}, t)} = \frac{\psi_X^1(\mathbf{x}, t)}{\psi_X^2(\mathbf{x}, t)} = \frac{\psi_Y^1(\mathbf{y}, t)}{\psi_Y^2(\mathbf{y}; t)} = \frac{\psi_Y^1}{\psi_Y^2}(\mathbf{y}, t)$$

$$\overset{(5.43)}{=} \left(\bar{U}\frac{\psi_Y^1}{\psi_Y^2}\right)(\mathbf{x}, t),$$

or in short,

$$\frac{\hat{U}\psi_Y^1}{\hat{U}\psi_Y^2} = \bar{U}\frac{\psi_Y^1}{\psi_Y^2} \tag{5.49}$$

for $\forall\psi_Y^1 \in \mathcal{S}(\mathcal{V}_Y)$ and $\psi_Y^2 \in \mathcal{S}_\varnothing(\mathcal{V}_Y)$. Concerning the denominator on the left, note that clearly we expect that the operator $\hat{U}$ fulfills

$$\forall\psi_Y \in \mathcal{S}_\varnothing(\mathcal{V}_Y): \quad \hat{U}\psi_Y \in \mathcal{S}_\varnothing(\mathcal{V}_X), \tag{5.50a}$$

$$\forall\psi_Y \in \mathcal{S}_+(\mathcal{V}_Y): \quad \hat{U}\psi_Y \in \mathcal{S}_+(\mathcal{V}_X), \tag{5.50b}$$

since the property of an asset to have non-zero or positive value should be preserved when moving from one parametrization to the other.

Now setting $\psi_Y^1 := f_Y \cdot \psi_Y^2$, we find that (5.49) implies that (and is in fact equivalent to)

$$\hat{U}(f_Y \cdot \psi_Y) = (\bar{U}f_Y) \cdot (\hat{U}\psi_Y) \tag{5.51}$$

for $\forall f_Y \in \mathcal{F}(M_Y)$ and $\forall\psi_Y \in \mathcal{S}_\varnothing(\mathcal{V}_Y)$. This relation states that the operation of multiplying a section by a function (i.e., of an asset value by an exchange rate) can be carried out in either parametrization and lead to corresponding results. We will certainly want this relation to hold in general, i.e., for $\forall\psi_Y \in \mathcal{S}(\mathcal{V}_Y)$.

Furthermore, note that our desired property (5.49) and the linearity of $\bar{U}$ imply that also the operator $\hat{U}$ must be linear: Given any $\forall a, b \in \mathbb{R}$ and $\forall\psi_Y^1, \psi_Y^2 \in \mathcal{S}(\mathcal{V}_Y)$, we can fix any arbitrary $\psi_Y^3 \in \mathcal{S}_\varnothing(\mathcal{V}_Y)$ to find that

$$\frac{\hat{U}(a\psi_Y^1 + b\psi_Y^2)}{\hat{U}\psi_Y^3} = \bar{U}\left(\frac{a\psi_Y^1 + b\psi_Y^2}{\psi_Y^3}\right) = \bar{U}\left(a\frac{\psi_Y^1}{\psi_Y^3} + b\frac{\psi_Y^2}{\psi_Y^3}\right)$$

$$= a\bar{U}\frac{\psi_Y^1}{\psi_Y^3} + b\bar{U}\frac{\psi_Y^2}{\psi_Y^3} = a\frac{\hat{U}\psi_Y^1}{\hat{U}\psi_Y^3} + b\frac{\hat{U}\psi_Y^2}{\hat{U}\psi_Y^3}$$

$$= \frac{a\hat{U}\psi_Y^1 + b\hat{U}\psi_Y^2}{\hat{U}\psi_Y^3},$$

which implies that $\hat{U}(a\psi_Y^1 + b\psi_Y^2) = a\hat{U}\psi_Y^1 + b\hat{U}\psi_Y^2$.

In short, we need to define a linear operator $\hat{U}\colon \mathcal{S}(\mathcal{V}_Y) \to \mathcal{S}(\mathcal{V}_X)$ with the property (5.50) such that (5.51) holds for $\forall f_Y \in \mathcal{F}(M_Y)$ and $\forall \psi_Y \in \mathcal{S}(\mathcal{V}_Y)$.

However, in contrast to $\bar{U}$, the operator $\hat{U}$ is not yet fully determined by its desired properties: To satisfy (5.51), we still have the freedom to choose the value $\hat{U}\psi_Y$ for any fixed $\psi_Y \in \mathcal{S}_\varnothing(\mathcal{V}_Y)$ of our choice at will; only then can the images $\hat{U}\psi_Y'$ of all other sections $\psi_Y' \in \mathcal{S}(\mathcal{V}_Y)$ be determined by setting $f_Y := \psi_Y'/\psi_Y$ in (5.51). In addition, if we choose ψ_Y and its image $\hat{U}\psi_Y$ as positive sections, then in fact (5.51) also ensures that (5.50) holds, since any other non-vanishing (positive) section ψ_Y' can be written as the product of a non-vanishing (positive) function f_Y and ψ_Y.

This freedom is used up by our earlier decision to enforce the convenient relation (5.40): In order for $\hat{U}$ to fulfill its purpose of consistently mapping **y**-based sections into **x**-based ones, which in particular requires that

$$\hat{U}\mathfrak{B}_T^Y = \mathfrak{B}_T^X \tag{5.52}$$

for $\forall T \in \mathbb{R}$, while at the same time fulfilling (5.40), our construction needs to satisfy the additional condition

$$(\hat{U}\mathfrak{B}_T^Y)(\mathbf{x}, t) = \mathfrak{B}_{T-t}^Y(\mathbf{x}) \tag{5.53}$$

for $\forall T \in \mathbb{R}$ and $\forall (\mathbf{x}, t) \in M_{X,T}$.

We will continue our construction of $\hat{U}$ later in Section 5.6.3, where we will also discuss the precise meaning of (5.52) and (5.53) (note that $\mathfrak{B}_T^Y$ is only defined on $M_{Y,T}$ and not on all of M_Y). First, however, let us gain some more insight by further inspecting the operator $\bar{U}$ that we have already fully defined.

5.6 Evolution Groups for Functions and Sections

A closer look at the relation (5.45a) shows that the map $\bar{U}$ actually maps every "time slice" $f_Y(\cdot, t)$ individually, via the mapping $f \mapsto f \circ U_{-t}$. In order to prepare for an analogous property for the mapping $\hat{U}$, let us now describe this observation and its consequences rigorously.

5.6.1 Evolution Group for Functions

Let the group $(\bar{U}_s)_{s \in \mathbb{R}}$ of linear operators

$$\bar{U}_s \colon \mathcal{F}(M) \to \mathcal{F}(M)$$

be defined as

$$\bar{U}_s f := f \circ U_{-s} \tag{5.54}$$

for $\forall f \in \mathcal{F}(M)$, i.e., $(\bar{U}_s f)(\mathbf{x}) := f(U_{-s}(\mathbf{x}))$ for $\forall \mathbf{x} \in M$. Note that the operator family $(\bar{U}_s)_{s\in\mathbb{R}}$ encodes the same information as the family $(U_s)_{s\in\mathbb{R}}$, since every operator U_s can be recovered from $\bar{U}_{-s}$ by applying $\bar{U}_{-s}$ to the test functions $f_i(\mathbf{x}) := \mathbf{x}_i$ (the ith component of $\mathbf{x}$).[13]

Using the formula (5.54), these operators $\bar{U}_s$ can consistently be applied also to functions $f \in \mathcal{F}(M_T)$ that are defined only on a smaller domain M_T, $T \in \mathbb{R}$, and we have

$$\forall s, T \in \mathbb{R}: \quad \bar{U}_s: \mathcal{F}(M_T) \to \mathcal{F}(M_{T-s}), \tag{5.55}$$

as well as the consistency[14] property

$$\forall s, T \in \mathbb{R} \ \forall f \in \mathcal{F}(M): \quad \left. (\bar{U}_s f) \right|_{M_{T-s}} = \bar{U}_s(f|_{M_T}). \tag{5.56}$$

Observe that the group property

$$\forall s_1, s_2 \in \mathbb{R}: \quad \bar{U}_{s_1}(\bar{U}_{s_2} f) = \bar{U}_{s_1+s_2} f \tag{5.57}$$

holds not only for functions $f \in \mathcal{F}(M)$, but also in the sense of (5.55) for functions $f \in \mathcal{F}(M_T)$ for any $T \in \mathbb{R}$. Further observe that

$$\bar{U}_s(f_1 \cdot f_2) = \bar{U}_s(f_1) \cdot \bar{U}_s(f_2) \tag{5.58}$$

for $\forall f_1, f_2 \in \mathcal{F}(M)$ (or for $\forall f_1, f_2 \in \mathcal{F}(M_T)$), and that the operators $\bar{U}_s$ preserve the subsets $\mathcal{F}_\varnothing(M)$ and $\mathcal{F}_+(M)$ of non-zero and of positive functions, respectively, i.e.,

$$\forall s \in \mathbb{R} \ \forall f \in \mathcal{F}_\varnothing(M): \quad \bar{U}_s f \in \mathcal{F}_\varnothing(M), \tag{5.59a}$$

$$\forall s \in \mathbb{R} \ \forall f \in \mathcal{F}_+(M): \quad \bar{U}_s f \in \mathcal{F}_+(M). \tag{5.59b}$$

Using these new operators $\bar{U}_s$, we can now express the map $\bar{U}$ as

$$(\bar{U} f_Y)(\mathbf{x}, t) = (\bar{U}_t f_Y(\cdot, t))(\mathbf{x}) \tag{5.60}$$

for $\forall f_Y \in \mathcal{F}(M_Y)$ and $\forall (\mathbf{x}, t) \in M_X$. In particular, for time-independent functions $f_Y \in \mathcal{F}(M_{Y,T})$ (i.e., for functions $f_Y \in \mathcal{F}(M_T)$ by our remark at the end of Section 5.3.6) this means that

$$(\bar{U} f_Y)(\mathbf{x}, t) = (\bar{U}_t f_Y)(\mathbf{x})$$

for $\forall (\mathbf{x}, t) \in M_{X,T}$.

5.6.2   The Time Shift Property for Bond Exchange Rates

We can now write the formula (5.38) for our bond exchange rates as

$$B_{T_1,T_2}^Y = \exp\left(-\int_{T_1}^{T_2} \bar{U}_{-\tau} r^X \, d\tau\right) \in \mathcal{F}(M_{T_1 \wedge T_2}). \tag{5.61}$$

The following lemma lists some of their key properties.

---

[13]The slight notational clash with the initial state $\mathbf{x}_0 \in M_0$ will not be an issue for us.

[14]This is *not* related to the desired consistency property of our model described in Section 1.1.

Lemma 5.3. *(i) The functions* $B^Y_{T_1,T_2}$ *given by* (5.61) *fulfill*

$$r^X = -\partial_\tau B^Y_{0,\tau}\big|_{\tau=0} \in \mathcal{F}(M_0),\,^{15} \tag{5.62}$$

$$\forall T_1, T_2, T_3 \in \mathbb{R}: \quad B^Y_{T_1,T_2} \cdot B^Y_{T_2,T_3} = B^Y_{T_1,T_3}\,, \tag{5.63}$$

$$\forall T_1, T_2, s \in \mathbb{R}: \quad \bar{U}_s B^Y_{T_1,T_2} = B^Y_{T_1-s,T_2-s}\,, \tag{5.64}$$

where (5.63) *is an equation on* $M_{T_1 \wedge T_2 \wedge T_3}$, *and where* (5.64) *is to be understood in the sense of* (5.55), *with* $T = T_1 \wedge T_2$.

(ii) Let $(B^Y_{T_1,T_2})_{T_1,T_2 \in \mathbb{R}}$ *be any family of functions* $B^Y_{T_1,T_2} \in \mathcal{F}_\varnothing(M_{T_1 \wedge T_2})$ *that have the properties* (5.63)–(5.64), *and for which the derivative defining* r^X *via* (5.62) *exists. Then the functions* $B^Y_{T_1,T_2}$ *are given by* (5.61).

Proof. (i) The properties (5.62)–(5.63) are trivial; (5.64) can be shown as follows:

$$\begin{aligned}
\bar{U}_s B^Y_{T_1,T_2} &= \bar{U}_s \exp\Big(-\int_{T_1}^{T_2} \bar{U}_{-\tau} r^X \,\mathrm{d}\tau\Big) = \exp\Big(-\int_{T_1}^{T_2} \bar{U}_{-\tau} \bar{U}_s r^X \,\mathrm{d}\tau\Big) \\
&= \exp\Big(-\int_{T_1}^{T_2} \bar{U}_{-(\tau-s)} r^X \,\mathrm{d}\tau\Big) = \exp\Big(-\int_{T_1-s}^{T_2-s} \bar{U}_{-\tau} r^X \,\mathrm{d}\tau\Big) \\
&= B^Y_{T_1-s,T_2-s}\,.
\end{aligned}$$

(ii) Given any family $(B^Y_{T_1,T_2})_{T_1,T_2 \in \mathbb{R}}$ of functions fulfilling (5.62)–(5.64), the formula (5.61) follows from

$$\partial_{T_2} B^Y_{T_1,T_2} = \partial_\Delta B^Y_{T_1,T_2+\Delta} \stackrel{(5.63)}{=} B^Y_{T_1,T_2} \partial_\Delta B^Y_{T_2,T_2+\Delta}$$
$$\stackrel{(5.64)}{=} B^Y_{T_1,T_2} \partial_\Delta \bar{U}_{-T_2} B^Y_{0,\Delta} = B^Y_{T_1,T_2} \bar{U}_{-T_2} \partial_\Delta B^Y_{0,\Delta} \stackrel{(5.62)}{=} -B^Y_{T_1,T_2} \bar{U}_{-T_2} r^X,$$

where all derivatives are right-sided and all Δ-derivatives are taken at $\Delta = 0$, and from the initial value $B^Y_{T_1,T_1} = 1$, which is derived by setting $T_2 := T_1$ in (5.63). $\qquad\square$

5.6.3 Evolution Group for Sections

Desired properties. To construct our reparametrization map $\hat{U}$ with the required properties (5.50)–(5.52), it now only seems natural to define $\hat{U}$ via a group of operators $\hat{U}_t$ acting on individual time slices as well, i.e.,

$$(\hat{U}\psi_Y)(\mathbf{x},t) := (\hat{U}_t\psi_Y(\cdot,t))(\mathbf{x}) \tag{5.65}$$

for $\forall \psi_Y \in \mathcal{S}(\mathcal{V}_Y)$ and $\forall(\mathbf{x},t) \in M_X$, which in particular for time-independent sections $\psi_Y \in \mathcal{S}(M_{Y,T})$ (i.e., for sections $\psi_Y \in \mathcal{S}(M_T)$) means that

$$(\hat{U}\psi_Y)(\mathbf{x},t) = (\hat{U}_t\psi_Y)(\mathbf{x}) \tag{5.66}$$

for $\forall(\mathbf{x},t) \in M_{X,T}$. This reduces our construction of $\hat{U}$ to the construction of a group $(\hat{U}_s)_{s\in\mathbb{R}}$ of operators

$$\hat{U}_s \colon \mathcal{S}(\mathcal{V}) \to \mathcal{S}(\mathcal{V})$$

---

[15] On the boundary ∂M_0 this derivative only exists as a right-sided derivative.

acting on sections that—in analogy to $(\bar{U}_s)_{s\in\mathbb{R}}$—can consistently be applied also to sections ψ defined on a smaller domain M_T, with

$$\forall s, T \in \mathbb{R}: \quad \hat{U}_s : \mathcal{S}(\mathcal{V}_T) \to \mathcal{S}(\mathcal{V}_{T-s}) \tag{5.67}$$

and the consistency condition

$$\forall \psi \in \mathcal{S}(\mathcal{V}) \ \forall s, T \in \mathbb{R}: \quad (\hat{U}_s\psi)\big|_{M_{T-s}} = \hat{U}_s(\psi|_{M_T}), \tag{5.68}$$

and such that the group property

$$\forall s_1, s_2 \in \mathbb{R}: \quad \hat{U}_{s_1}(\hat{U}_{s_2}\psi) = \hat{U}_{s_1+s_2}\psi \tag{5.69}$$

holds not only for sections $\psi \in \mathcal{S}(\mathcal{V})$, but also in the sense of (5.67) for sections $\psi \in \mathcal{S}(\mathcal{V}_T)$ for any $T \in \mathbb{R}$.

In order for $\hat{U}$ to fulfill (5.50) and (5.51), all we need to require is that

$$\forall s \in \mathbb{R} \ \forall \psi \in \mathcal{S}_\varnothing(\mathcal{V}): \quad \hat{U}_s f \in \mathcal{S}_\varnothing(\mathcal{V}), \tag{5.70a}$$

$$\forall s \in \mathbb{R} \ \forall \psi \in \mathcal{S}_+(\mathcal{V}): \quad \hat{U}_s f \in \mathcal{S}_+(\mathcal{V}), \tag{5.70b}$$

and that

$$\forall f \in \mathcal{F}(M) \ \forall \psi \in \mathcal{S}(\mathcal{V}): \quad \hat{U}_s(f\psi) = (\bar{U}_s f) \cdot (\hat{U}_s\psi); \tag{5.71}$$

the desired relations (5.50) and (5.51) then follow from (5.65) and (5.60).

Observe that the desired property (5.71) of $\hat{U}_s$ alone still leaves us with the freedom to choose the value $\hat{U}_s\psi_0$ for any fixed $\psi_0 \in \mathcal{S}_\varnothing(\mathcal{V})$ of our choice at will; only then can the images $\hat{U}_s\psi'$ of all other sections $\psi' \in \mathcal{S}(\mathcal{V})$ be determined by setting $f := \psi'/\psi_0$ and $\psi := \psi_0$ in (5.71). In addition, if we choose ψ_0 and its image $\hat{U}_s\psi_0$ as positive sections, then in fact (5.71) (together with (5.59) and (5.2)) also ensures that (5.70) holds, since any other non-vanishing (positive) section ψ' can be written as the product of a non-vanishing (positive) function f and ψ_0.

Finally, note that setting $f := \frac{\psi_2}{\psi_1}$ and $\psi = \psi_1$ in (5.71) implies that

$$\forall s \in \mathbb{R} \ \forall \psi_1 \in \mathcal{S}_\varnothing(\mathcal{V}) \ \forall \psi_2 \in \mathcal{S}(\mathcal{V}): \quad \bar{U}_s\left(\frac{\psi_2}{\psi_1}\right) = \frac{\hat{U}_s\psi_2}{\hat{U}_s\psi_1}. \tag{5.72}$$

To summarize, we have shown that to construct our reparametrization map $\hat{U}$ with the required properties (5.50)–(5.51), it suffices to define a group $(\hat{U}_s)_{s\in\mathbb{R}}$ of operators $\hat{U}_s : \mathcal{S}(\mathcal{V}) \to \mathcal{S}(\mathcal{V})$ with the properties (5.67)–(5.71); the map $\hat{U}$ is then defined via (5.65). We will call such a group $(\hat{U}_s)_{s\in\mathbb{R}}$ an **extension** of $(U_s)_{s\in\mathbb{R}}$.

Bond section formula. For our reparametrization $\hat{U}$ to also fulfill the additional convenient condition (5.53), we need to have

$$(\hat{U}_t\mathfrak{B}_T^Y)(\mathbf{x}) \overset{(5.66)}{=} (\hat{U}\mathfrak{B}_T^Y)(\mathbf{x}, t) \overset{(5.53)}{=} \mathfrak{B}_{T-t}^Y(\mathbf{x})$$

for $(\mathbf{x}, t) \in M_{X,T}$, i.e.,

$$\hat{U}_t \mathfrak{B}_T^Y = \mathfrak{B}_{T-t}^Y \tag{5.73}$$

on M_0 for $\forall T \in \mathbb{R}$ and $\forall t \leq T$. In particular, setting $t = T$ and applying $\hat{U}_{-T}$ to both sides implies that

$$\forall T \in \mathbb{R}: \quad \mathfrak{B}_T^Y = \hat{U}_{-T} \mathfrak{B}_0^Y . \tag{5.74}$$

This relation shows that aside from being used to construct our reparametrization map $\hat{U}$, *the group* $(\hat{U}_s)_{s \in \mathbb{R}}$ *will allow us to compute the full family* $(\mathfrak{B}_T^Y)_{T \in \mathbb{R}}$ *of bond sections* $\mathfrak{B}_T^Y \in \mathcal{S}_+(\mathcal{V}_T)$ *from only* $\mathfrak{B}_0^Y$, and thus by (5.40) also the full family $(\mathfrak{B}_T^X)_{T \in \mathbb{R}}$ of bond sections $\mathfrak{B}_T^X \in \mathcal{S}_+(\mathcal{V}_0)$ in the spot parametrization via the formula

$$\forall T \in \mathbb{R} \; \forall t \leq T \; \forall \mathbf{x} \in M_0: \quad \mathfrak{B}_T^X(\mathbf{x}, t) = (\hat{U}_{t-T} \mathfrak{B}_0^Y)(\mathbf{x}). \tag{5.75}$$

Furthermore, together with the group property (5.69) this relation implies that (5.73) in fact holds for $\forall t \in \mathbb{R}$ and on all of M_{T-t}:

$$\hat{U}_t \mathfrak{B}_T^Y \overset{(5.74)}{=} \hat{U}_t \hat{U}_{-T} \mathfrak{B}_0^Y \overset{(5.69)}{=} \hat{U}_{-(T-t)} \mathfrak{B}_0^Y \overset{(5.74)}{=} \mathfrak{B}_{T-t}^Y . \tag{5.76}$$

Existence and uniqueness. The following lemma tells us that the additional desired relation (5.74) is in fact just enough to uniquely specify the group $(\hat{U}_s)_{s \in \mathbb{R}}$ and thus via (5.65) our reparametrization map $\hat{U}$.

Lemma 5.4. *Let* $(\mathfrak{B}_T^Y)_{T \in \mathbb{R}}$ *be any family of sections* $\mathfrak{B}_T^Y \in \mathcal{S}_+(\mathcal{V}_T)$ *whose bond exchange rates* $B_{T_1, T_2}^Y := \mathfrak{B}_{T_2}^Y / \mathfrak{B}_{T_1}^Y$ *are of the form* (5.61). *Then there exists a unique extension* $(\hat{U}_s)_{s \in \mathbb{R}}$ *of* $(U_s)_{s \in \mathbb{R}}$ *fulfilling* (5.74).

Proof. See Appendix D.1. $\square$

In this proof and throughout the rest of this book we will at times use symbols such as $\mathfrak{B}_T^Y$ and $\frac{1}{\mathfrak{B}_T^Y}$ as a short-hand notation for the multiplication and division operators

$$\mathfrak{B}_T^Y : \mathcal{F}(M_T) \to \mathcal{S}(\mathcal{V}_T), \quad f \mapsto f \cdot \mathfrak{B}_T^Y , \tag{5.77a}$$

$$\frac{1}{\mathfrak{B}_T^Y} : \mathcal{S}(\mathcal{V}_T) \to \mathcal{F}(M_T), \quad \psi \mapsto \frac{\psi}{\mathfrak{B}_T^Y} . \tag{5.77b}$$

Similarly, we will use expressions such as B_{T_1, T_2}^Y and $\frac{1}{B_{T_1, T_2}^Y}$ to denote the operators that multiply or divide a given function or section by B_{T_1, T_2}^Y.

5.6.4 Summary: Bond Sections, Reparametrization, Evolution Groups

Given any group $(U_s)_{s\in\mathbb{R}}$ as defined in Sections 5.2.2 and 5.3.3, our model intends to consider all bond exchange rate functions $B^Y_{T_1,T_2}$ of the form (5.61), for any short rate function r^X. By Lemma 5.2 there are many ways to express these exchange rates as the ratios of bond sections $\mathfrak{B}^Y_T(\mathbf{y})$; which way we choose has no impact on our model since we only care about exchange rates.

Furthermore, there are many ways to specify how to map sections $\psi_Y(\mathbf{y},t)$ in the forward parametrization to sections $\psi_X(\mathbf{x},t)$ in the spot parametrization such that (5.51) holds. Again, the specific choice of this reparametrization map $\hat{U}$ has no influence on the actual objects of interest in our model, namely the exchange rates between any two instruments; it solely affects how easy it will be for us to work with our section-based equations.

In Lemma 5.4 we showed that every valid choice of a family $(\mathfrak{B}^Y_T)_{T\in\mathbb{R}}$ uniquely specifies an extension $(\hat{U}_s)_{s\in\mathbb{R}}$ of $(U_s)_{s\in\mathbb{R}}$ such that (5.74) holds, which allows us to compute the full family of bond sections from just $\mathfrak{B}^Y_0$ and $(\hat{U}_s)_{s\in\mathbb{R}}$. The group $(\hat{U}_s)_{s\in\mathbb{R}}$ can then be used to define a reparametrization map $\hat{U}$ via (5.65), in such a way that the convenient relation (5.40) holds (which was motivated by our observation (5.39)).

5.6.5 Representation of the Bond Section Family

The following lemma will show the reverse, namely that we may consider any extension $(\hat{U}_s)_{s\in\mathbb{R}}$ of $(U_s)_{s\in\mathbb{R}}$ and any section $\mathfrak{B}^Y_0 \in \mathcal{S}_+(\mathcal{V}_0)$, and then *define* the full family $(\mathfrak{B}^Y_T)_{T\in\mathbb{R}}$ via (5.74); the lemma states that any family of bond sections constructed in this way leads to exchange rates of the form (5.61), as long as some technical smoothness condition is fulfilled. According to our findings that we summarized in Section 5.6.4, following this strategy does not make us lose generality, since *every* family of exchange rate functions of the form (5.61) can be obtained in this way.

Lemma 5.5. *Let $\mathfrak{B}^Y_0 \in \mathcal{S}_+(\mathcal{V}_0)$ be any positive section and $(\hat{U}_s)_{s\in\mathbb{R}}$ some extension of $(U_s)_{s\in\mathbb{R}}$, and let the family $(\mathfrak{B}^Y_T)_{T\in\mathbb{R}}$ of sections $\mathfrak{B}^Y_T \in \mathcal{S}_+(\mathcal{V}_T)$ be defined via (5.74). Then as long as the bond exchange rates*

$$B^Y_{T_1,T_2} := \mathfrak{B}^Y_{T_2}/\mathfrak{B}^Y_{T_1} \tag{5.78}$$

are smooth enough that the derivative (5.62) is well defined, they are given by the formula (5.61).

Proof. As we showed in (5.76), the relation (5.74) together with the group property (5.69) implies (5.73), and this in turn implies the time shift property (5.64) for the bond exchange rates defined via (5.78):

$$\bar{U}_s B^Y_{T_1,T_2} = \bar{U}_s \frac{\mathfrak{B}^Y_{T_2}}{\mathfrak{B}^Y_{T_1}} \stackrel{(5.72)}{=} \frac{\hat{U}_s\mathfrak{B}^Y_{T_2}}{\hat{U}_s\mathfrak{B}^Y_{T_1}} \stackrel{(5.73)}{=} \frac{\mathfrak{B}^Y_{T_2-s}}{\mathfrak{B}^Y_{T_1-s}} = B^Y_{T_1-s,T_2-s}.$$

Since the bond exchange rates can easily be seen to satisfy (5.63), the desired statement therefore follows from Lemma 5.3 (ii). □

Following this insight, we will consider the group $(U_s)_{s\in\mathbb{R}}$, an extension $(\hat{U}_s)_{s\in\mathbb{R}}$, and a section $\mathfrak{B}_0^Y \in \mathcal{S}_+(\mathcal{V}_0)$ as our basic building blocks for a general zero-noise interest rate model that satisfies our consistency condition and that allows us to conveniently switch between the spot and the forward parametrization.

5.7 Conversion Rules – Overview

The diagram in Figure 5.1, which will accompany us throughout the rest of Section 5, provides a comprehensive overview over all the key objects of our framework, and it illustrates all the necessary rules for converting them when switching between the spot and the forward parametrization, or from one numeraire to another.

Each box in this diagram corresponds to a combination of one of the two parametrizations (*left*=spot, *right*=forward), and of a specific numeraire (as stated on the top of each box): the dollar ("cash"), any maturity-T bond, or none ("numeraire-free"). The four rows in each box list the key objects in our framework that need to be converted (in addition to the state–time pairs themselves):

(i) the functions (or in the numeraire-free case, sections) that are used to express the state- and time-dependent values of accounts (this row first lists the standard notation for a function or section, and then the specific notation for the value of a maturity-T bond),

(ii) the Kolmogorov backward operator,

(iii) its deterministic part, and

(iv) its stochastic part.

The items in rows (ii)–(iv) will be introduced later in Sections 5.8–5.10 and can be ignored for time being.

Each arrow connecting two boxes illustrates the conversion between the associated parametrization–numeraire combinations. To convert a function or section (as found in the first row of the box at the start of the arrow), the map annotating the arrow must be applied to it as is. In contrast, the operators in rows (ii)–(iv) are converted via a similarity transformation, as discussed in more detail in the subsequent sections.

For example, when working in the cash numeraire, the conversion from the forward to the spot parametrization is illustrated by the arrow between the two large boxes on the top of the diagram. Its annotation $\bar{U}$ tells us that the desired

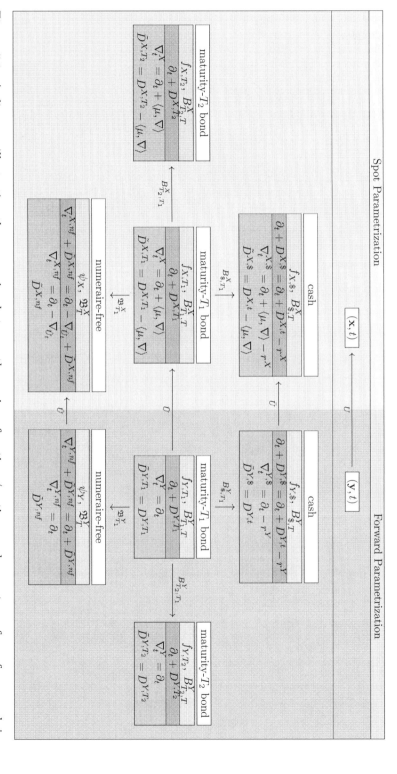

Figure 5.1: A diagram illustrating the conversion between the various functions/sections and operators of our framework in the two parametrizations (*left*=spot, *right*=forward) and in the various numeraires (as stated on the top of each box): the dollar ("cash"), any maturity-T bond, or none ("numeraire-free").

conversion of exchange rate functions and of the Kolmogorov backward operator introduced later is achieved via the formulas

$$f_{X,\$} = \bar{U} f_{Y,\$} \quad \text{and} \quad \partial_t + D^{X,\$} = \bar{U} \circ \left(\partial_t + D^{Y,\$} \right) \circ \bar{U}^{-1},$$

respectively, consistent with (5.41) and (5.132).

As another example, when working in the spot parametrization, switching numeraires from the maturity-T_1 bond to the maturity-T_2 bond (for some $T_1, T_2 \in \mathbb{R}$) requires the conversion illustrated by the arrow pointing at the leftmost large box in the diagram. Its annotation $B^X_{T_2,T_1}$, which is to be understood as a multiplication operator (as explained at the end of Section 5.6.3), tells us that the two analogous conversions are achieved via the formulas

$$f_{X,T_2} = B^X_{T_2,T_1} \cdot f_{X,T_1} \quad \text{and} \quad \partial_t + D^{X,T_2} = B^X_{T_2,T_1} \circ \left(\partial_t + D^{X,T_1} \right) \circ \frac{1}{B^X_{T_2,T_1}},$$

respectively, where the first relation is intuitive and the second one is given in (5.88a).

We encourage the reader to periodically revisit this diagram throughout the remainder of Section 5, and to ensure that by the end of this section they have understood it in its entirety.

5.8   Infinitesimal Generators

So far we have only considered the deterministic process (i.e., the zero-noise case $\sigma \equiv 0$), and for this case we have ensured that our model is time homogeneous and satisfies our consistency condition (i.e., that it is arbitrage free). We will now begin to consider the general stochastic case.

Introducing stochasticity to our framework adds another layer of difficulty: We now not only have *two* equivalent descriptions $(X_t)_{t \geq 0}$ and $(Y_t)_{t \geq 0}$ of the model dynamics (and thus two SDEs); the classical modeling approach also requires us to choose a numeraire, i.e., which T-measure we want to consider.[16] The SDEs corresponding to the various parametrization–numeraire combinations are equivalent in that any one of them can be obtained from any other via Itô's lemma (for switching between the two parametrizations) and Girsanov's formula (for switching between two different T-measures).

Unfortunately, these two formulas are not very easy to work with, in particular in the multidimensional case. Things become much more tractable if one encodes the model dynamics in a more abstract way, namely in the form of the infinitesimal generators of these processes. Not only does this approach make both transformation formulas simpler (and in fact similar), it also has the additional benefit of not limiting our mathematical framework to diffusion processes a priori. In this way, while the specific model that we construct in Chapters 6–8

---

[16] For any $T \geq 0$, the T-measure is the probability measure under which the price evolution $(P_t)_{t \in [0,T]}$ of any self-financing account, expressed in units of the maturity-T bond, is a local martingale.

is a diffusion process, our general mathematical framework supports also other types of processes (such as Markov jump processes), as long as the limiting case $\sigma \equiv 0$ is an ODE of the form (5.8).

The goal of the following sections will be to express our consistency and our time homogeneity requirement in terms of the infinitesimal generator of our process, and to explore how the obtained conditions change when moving from one parametrization or numeraire to the other. Designing a process with these two properties will then boil down to first finding an infinitesimal generator that satisfies these equivalent conditions, and then reading off the dynamics of the process that this generator encodes. To facilitate our search for such a generator in Chapters 6–8, we will need to make this strategy compatible with our numeraire-free framework, by defining a section-based Kolmogorov backward operator.

5.8.1 Definition

Let (Z_t) be a continuous-time Markov process in $\mathbb{R}^n$. Note that we left out the subscript "$t \geq 0$" (as in $(Z_t)_{t \geq 0}$), to indicate that (i) we did not specify any starting time and initial state, i.e., (Z_t) stands symbolically only for the transition rules for propagating the distribution at any time t to $t + \Delta t$, and that (ii) these rules are in fact defined for $\forall t \in \mathbb{R}$.

The *infinitesimal generator* D^Z (or in short simply the *generator*) of (Z_t) is the linear operator

$$(D^Z f)(\mathbf{z}, t) := \lim_{\Delta t \searrow 0} \tfrac{1}{\Delta t} \mathbb{E}[f(Z_{t+\Delta t}, t) - f(\mathbf{z}, t) | Z_t = \mathbf{z}], \tag{5.79}$$

which is defined on the subset of $\mathcal{F}(\mathbb{R}^n \times \mathbb{R})$ consisting of all those real-valued functions $f(\mathbf{z}, t)$ that are sufficiently well behaved for this limit to exist. Observe that under some additional smoothness assumption on f we have

$$((\partial_t + D^Z)f)(\mathbf{z}, t) = \lim_{\Delta t \searrow 0} \tfrac{1}{\Delta t} \mathbb{E}[f(Z_{t+\Delta t}, t + \Delta t) - f(\mathbf{z}, t) | Z_t = \mathbf{z}]. \tag{5.80}$$

We will call the operator $\partial_t + D^Z$ the *Kolmogorov backward operator* (or in short simply the *Kolmogorov operator*[17]).

Although (5.79) and (5.80) are really only defined and consistent for sufficiently well-behaved functions $f(\mathbf{z}, t)$, we choose to not overload our notation and simply write expressions such as

$$D^Z \colon \mathcal{F}(\mathbb{R}^n \times \mathbb{R}) \to \mathcal{F}(\mathbb{R}^n \times \mathbb{R}) \quad \text{or} \quad \partial_t + D^Z \colon \mathcal{F}(\mathbb{R}^n \times \mathbb{R}) \to \mathcal{F}(\mathbb{R}^n \times \mathbb{R}),$$

and we will also say that some formula or calculation holds "for $\forall f \in \mathcal{F}(\mathbb{R}^n \times \mathbb{R})$" even if it holds only for those sufficiently well-behaved functions. This notational convention, which we will also make for the deterministic and stochastic parts of the Kolmogorov operator defined later in Section 5.8.6 and in a similar fashion

---

[17]In this book we will not use the Kolmogorov *forward* operator.

for their section-based counterparts defined in Section 5.9, will become particularly convenient for the latter, since there is no notion of smoothness for sections that one could keep track of.[18]

Our notation that we use to distinguish the Kolmogorov operators for the various parametrization–numeraire combinations can be found in the respective second rows of the large boxes in Figure 5.1. Furthermore, some of these boxes also provide some useful formulas for these Kolmogorov operators that we will derive later on.

Example: SDE. If (Z_t) is described by an SDE

$$\mathrm{d}Z_t = \tilde{\mu}(Z_t, t)\,\mathrm{d}t + \tilde{\sigma}(Z_t, t)\,\mathrm{d}W_t \,, \tag{5.81}$$

where (W_t) is a Brownian motion, then it is well known that D^Z is the second-order differential operator

$$D^Z = \sum_{i=1}^{n} \tilde{\mu}_i(\mathbf{z}, t)\,\partial_{\mathbf{z}_i} + \frac{1}{2}\sum_{i,j=1}^{n}(\tilde{\sigma}\tilde{\sigma}^T)_{ij}(\mathbf{z}, t)\,\partial_{\mathbf{z}_i}\partial_{\mathbf{z}_j} \,, \tag{5.82}$$

and so D^Z is defined on the space of all functions $f(\mathbf{z}, t)$ for which the derivatives $\partial_t f$ and $\partial_{\mathbf{z}_i}\partial_{\mathbf{z}_j} f$ exist and are continuous. For general dynamics that allow for jumps, D^Z will no longer be a *differential* operator.

Key property. The generator can be shown to generally encode the full information contained in (Z_t). E.g., in the SDE case one can recover $\tilde{\mu}$ and $\tilde{\sigma}\tilde{\sigma}^T$ from D^Z by applying this operator to the test functions $f(\mathbf{z}, t) := \mathbf{z}_i$ and $f(\mathbf{z}, t) := \mathbf{z}_i\mathbf{z}_j$, and then recover $\tilde{\sigma}$ by choosing a matrix square root of $\tilde{\sigma}\tilde{\sigma}^T$ (there are multiple such square roots, but they all lead to the same process dynamics (Z_t)). This allows us to first express all our desired process properties as properties of D^Z, then in Chapters 6–8 to find an operator D^Z that fulfills all these properties, and finally to read off the corresponding process dynamics.

5.8.2 Kolmogorov Backward Equation, Self-Financing Accounts

The following lemma is a standard result from the theory of stochastic processes. It transforms the probabilistic question of whether a process of a certain type is a local martingale into a PDE.

Lemma 5.6 (Kolmogorov backward equation). *Given a function $f \in \mathcal{F}(\mathbb{R}^n \times \mathbb{R})$ and a Markov process (Z_t) with generator D^Z, the process $(f(Z_t, t))$ is a local martingale if and only if*

$$(\partial_t + D^Z)f = 0. \tag{5.83}$$

---

[18]Recall that one cannot add or subtract vectors from different vector spaces $V_{\mathbf{x}}$, and so for sections $\psi(\mathbf{x})$ the standard derivative $\partial_{\mathbf{x}}\psi(\mathbf{x})$ is ill-defined.

This theorem becomes particularly important in mathematical finance because of its application to self-financing accounts. A *self-financing account* (or portfolio) is a collection of assets with the properties that (i) all purchases of new assets are financed with the sale of existing ones, and (ii) there is no other investment strategy that with probability 1 will gain value relative to it.

For example, an account with a single bond of some fixed maturity T is self-financing until the time $t = T$, at which point its payoff will have to be reinvested in another instrument. On the other hand, a zero-interest cash account with one dollar in it is not self-financing, since one can expect to make more money by investing it in a bond (whose value in a positive-rate environment will be less than one dollar) and waiting for its payoff at maturity.

Now given some fixed self-financing account that will be used as the numeraire, let us assume that the evolution of some other self-financing accounts' values with respect to that numeraire are given by $(f_i(Z_t, t))$ for some functions $f_i(\mathbf{z}, t)$ and some Markov process (Z_t). Then by the fundamental theorems of asset pricing, in a consistent (i.e., arbitrage-free) model there is an equivalent measure under which the processes $(f_i(Z_t, t))$ are local martingales, and so by Lemma 5.6 the functions f_i must fulfill the Kolmogorov backward equation (5.83). That equivalent measure of course depends on the chosen numeraire, and therefore the generator D^Z used in (5.83), which is the generator of (Z_t) *under that measure*,[19] depends on that numeraire as well.

5.8.3   Consistency

This allows us to rephrase our consistency requirement from Section 1.1 in the form of a PDE: According to Section 5.8.2, for our model to be arbitrage free, the process $(B^X_{T_1,T_2}(X_t, t))_{t \le T_1 \wedge T_2} = (B^Y_{T_1,T_2}(Y_t))_{t \le T_1 \wedge T_2}$ (which represents the value of the maturity-T_2 bond measured in units of the maturity-T_1 bond) must be a local martingale under what is commonly referred to as the T_1-measure, for $\forall T_1, T_2 \in \mathbb{R}$. (The T_1-measure is the measure that corresponds to the choice of the maturity-T_1 bond as the numeraire.) According to Lemma 5.6, this is equivalent to asking that

$$\forall T_1, T_2 \in \mathbb{R}: \ (\partial_t + D^{X,T_1}) B^X_{T_1,T_2} = 0, \tag{5.84a}$$

or equivalently, $$\forall T_1, T_2 \in \mathbb{R}: \ (\partial_t + D^{Y,T_1}) B^Y_{T_1,T_2} = 0, \tag{5.84b}$$

where we denote by $D^{X,T}$ and $D^{Y,T}$ the generators of the processes (X_t) and (Y_t), respectively, under the T-measure. (For simplicity we will assume that—as already in the deterministic case—the dynamics of (X_t) and thus also (Y_t) is defined for all times $t \in \mathbb{R}$; in particular for the time-homogeneous case this is not a constraint.)

Note the decisive fact that the ∂_t-term in (5.84b) vanishes since $B^Y_{T_1,T_2}$ does not depend on t (recall Section 5.3.1); this had been our original motivation for introducing the forward parametrization.

---

[19] The generator *under a certain probability measure* is the operator defined in (5.79), where the given measure is used when computing the expected value $\mathbb{E}$.

Consistency of the deterministic system. As an immediate application of our criteria (5.84), we can now show that for the deterministic system (5.8) the necessary condition (5.13) for our framework to be arbitrage free, which we have already used throughout our construction, is in fact sufficient.

Indeed, setting $\tilde{\sigma} = 0$ in (5.82) to get a formula for the generator D^{X,T_1} in the zero-noise case (5.8), and using the bond exchange rate formula (5.17) that we have obtained as a consequence of (5.13), we obtain

$$(\partial_t + D^{X,T_1}) B^X_{T_1,T_2}(\mathbf{x},t) = (\partial_t + \langle \mu(\mathbf{x}), \nabla \rangle) \exp \left(-\int_{T_1}^{T_2} r^X(U_{\tau-t}(\mathbf{x})) \, \mathrm{d}\tau \right)$$

$$= -\exp(\dots) \int_{T_1}^{T_2} \left(\partial_t + \langle \mu(\mathbf{x}), \nabla \rangle \right) r^X(U_{\tau-t}(\mathbf{x})) \, \mathrm{d}\tau \, .$$

Since by (5.25)–(5.26) we have

$$\partial_t \, r^X(U_{\tau-t}(\mathbf{x})) = \partial_s \, r^X \big(U_{\tau-t}(U_{-s}(\mathbf{x})) \big) \big|_{s=0} = -\langle \mu(\mathbf{x}), \nabla \rangle \big(r^X(U_{\tau-t}(\mathbf{x})) \big),$$

the last integral vanishes, and so we have successfully checked our consistency criterion (5.84a). $\qquad\square$

5.8.4 Time Shift, Change of Parametrization, Change of Numeraire

In this section we will explore how the generator changes under various types of transformations of the given process. We begin with a simple time shift, which will become relevant in Section 5.11 when we discuss time homogeneity.

Lemma 5.7. *For some given process (Z_t) and some fixed $\Delta t \in \mathbb{R}$, let the delayed process $(Z_t^{\Delta t})$ be defined as $Z_t^{\Delta t} := Z_{t-\Delta t}$, and let us denote by $S_{\Delta t}$ the operator that shifts a given function $f(\mathbf{z}, t)$ to the left by the time Δt, i.e., we define $(S_{\Delta t} f)(\mathbf{z}, t) := f(\mathbf{z}, t + \Delta t)$.*

Then the generator $D^{Z^{\Delta t}}$ of the delayed process fulfills

$$\partial_t + D^{Z^{\Delta t}} = S_{-\Delta t} \circ (\partial_t + D^Z) \circ S_{\Delta t}, \tag{5.85}$$

$$\textit{i.e.,} \qquad D^{Z^{\Delta t}} = S_{-\Delta t} \circ D^Z \circ S_{\Delta t}. \tag{5.86}$$

Proof. See Appendix D.2. $\qquad\square$

For our next two types of transformations, recall that the two parametrizations (X_t) and (Y_t) of our model dynamics are related to each other via the reparametrization formula (5.34); and for each of these two parametrizations, the various T-measures are by definition related via a Radon–Nikodym derivative proportional to $B^X_{T_1,T_2}(X_{T_1 \wedge T_2}, T_1 \wedge T_2)$ or $B^Y_{T_1,T_2}(Y_{T_1 \wedge T_2})$, respectively.[20]

---

[20]More precisely, if one starts the process (X_t) from the state $\mathbf{x}_0$ at the time t_0, then for every random variable Z that is measurable with respect to the σ-algebra $\sigma\big((X_t)_{t \in [t_0, T_1 \wedge T_2]}\big)$ we have $\mathbb{E}^{X,T_2} Z = \frac{1}{c} \mathbb{E}^{X,T_1} \big[B^X_{T_1,T_2}(X_{T_1 \wedge T_2}, T_1 \wedge T_2) \cdot Z \big]$, with the normalizing constant $c = B^X_{T_1,T_2}(\mathbf{x}_0, t_0)$. An analogous statement holds in the forward parametrization.

Both types of transformations change the generator of the process under consideration, and the following lemma tells us how.

Lemma 5.8. *(i) The generators $D^{X,T}$ and $D^{Y,T}$ (for the same value T) are related via the formula*

$$\partial_t + D^{Y,T} = \bar{U}^{-1} \circ \left(\partial_t + D^{X,T}\right) \circ \bar{U}, \tag{5.87}$$

with $\bar{U}$ defined as in Section 5.5.1.

(ii) For $\forall T_1, T_2 \in \mathbb{R}$ we have

$$\partial_t + D^{X,T_2} = B^X_{T_2,T_1} \circ \left(\partial_t + D^{X,T_1}\right) \circ B^X_{T_1,T_2}, \tag{5.88a}$$

$$\partial_t + D^{Y,T_2} = B^Y_{T_2,T_1} \circ \left(\partial_t + D^{Y,T_1}\right) \circ B^Y_{T_1,T_2} \tag{5.88b}$$

on $\mathcal{F}(M_{X,T_1 \wedge T_2})$ and $\mathcal{F}(M_{Y,T_1 \wedge T_2})$, respectively.

Proof. See Appendix D.3. □

5.8.5   Commutative Diagrams

General principle. The formulas in Lemma 5.8 are of a strikingly similar shape, reminiscent to the notion of similarity between matrices. (Recall that two matrices $A, A' \in \mathbb{R}^{m \times m}$ are called similar if $A' = PAP^{-1}$ for some invertible matrix P.) This is not a coincidence.

Suppose one is given two spaces V and V' linked by a bijective map $P \colon V \to V'$, and an operator $A \colon V \to V$. Then the operator $A' \colon V' \to V'$ defined as $A' := PAP^{-1}$ has the property that $\forall v \in V \colon P^{-1}(A'(Pv)) = Av$, i.e., one can obtain the image of any $v \in V$ under A by considering its counterpart $v' := Pv$ in V' instead, computing its image $A'v'$ under the map A', and then finding the counterpart $P^{-1}(A'v')$ of the obtained solution in V. This is summarized in the following commutative diagram:

$$\begin{array}{ccc} V & \xrightarrow{\quad A \quad} & V \\ \downarrow{\scriptstyle P} & & \downarrow{\scriptstyle P} \\ V' & \xrightarrow{\ A':=PAP^{-1}\ } & V' \end{array} \tag{5.89}$$

Illustration of our conversion formulas. In this sense, our formula (5.87) should be understood as illustrated by the following commutative diagram:

$$\begin{array}{ccc} \mathcal{F}(M_{Y,T}) & \xrightarrow{\ \partial_t + D^{Y,T}\ } & \mathcal{F}(M_{Y,T}) \\ \downarrow{\scriptstyle \bar{U}} & & \downarrow{\scriptstyle \bar{U}} \\ \mathcal{F}(M_{X,T}) & \xrightarrow{\ \partial_t + D^{X,T}\ } & \mathcal{F}(M_{X,T}), \end{array} \tag{5.90}$$

i.e., we can compute the action of the Kolmogorov operator interchangeably in either parametrization: To compute the action of $\partial_t + D^{Y,T}$ on some (sufficiently well-behaved) function $f_{Y,T} \in \mathcal{F}(M_{Y,T})$, we may compute the action of $\partial_t + D^{X,T}$ on the function $f_{X,T} := \bar{U} f_{Y,T} \in \mathcal{F}(M_{X,T})$ instead and then map the result back via $\bar{U}^{-1}$.

Similarly, the formula (5.88a) is illustrated by the diagram

$$
\begin{array}{ccc}
\mathcal{F}(M_{X,T_1 \wedge T_2}) & \xrightarrow{\partial_t + D^{X,T_2}} & \mathcal{F}(M_{X,T_1 \wedge T_2}) \\
\downarrow{\scriptstyle B^X_{T_1,T_2}} & & \downarrow{\scriptstyle B^X_{T_1,T_2}} \\
\mathcal{F}(M_{X,T_1 \wedge T_2}) & \xrightarrow{\partial_t + D^{X,T_1}} & \mathcal{F}(M_{X,T_1 \wedge T_2}),
\end{array}
\tag{5.91}
$$

where the functions in the spaces in the top row represent account values measured in units of maturity-T_2 bonds, and the functions in the spaces in the bottom row represent account values measured in units of maturity-T_1 bonds. To compute the action of $\partial_t + D^{X,T_2}$ on some function $f_{X,T_2} \in \mathcal{F}(M_{X,T_2})$ representing the value of some account in units of maturity-T_2 bonds, we may therefore compute the action of $\partial_t + D^{X,T_1}$ on the function $f_{X,T_1} := B^X_{T_1,T_2} \cdot f_{X,T_2} \in \mathcal{F}(M_{X,T_1 \wedge T_2})$ instead, which represents the value of the same account in units of maturity-T_1 bonds, and then change the numeraire of the resulting function back to maturity-T_2 bonds by multiplying by $B^X_{T_2,T_1}$ (i.e., dividing by $B^X_{T_1,T_2}$). The formula (5.88b) is illustrated analogously.

Note, however, in the latter example that if $T_1 < T_2$ then this alternative way of computing $(\partial_t + D^{X,T_2}) f_{X,T_2}$ will make us lose the function values for $\forall t \in (T_1, T_2]$. In such cases where the operator P is not invertible, the diagram (5.89) still makes sense; it then more generally illustrates that the two ways of going from the top left of the diagram to the bottom right are equivalent, i.e., we have $A' \circ P = P \circ A$.

We encourage the reader to test their understanding by drawing the commutative diagram corresponding to the conversion rules (5.85)–(5.86).

Diagrams like these also explain why in our big conversion chart in Figure 5.1 it sufficed to annotate each arrow with only one map from which the associated conversion rules for both account value functions (or sections) and the operators acting on them can be read off: By their definitions, all the operators that we chose to list in that conversion chart convert in a way that is consistent with the commutative diagrams introduced here; therefore, the maps annotating the arrows in Figure 5.1, which correspond to the map P in the example diagram (5.89), imply the conversion rule

$$
A \mapsto P \circ A \circ P^{-1}
\tag{5.92}
$$

for each of these operators. For example, when following any of the three horizontal arrows in the middle row of Figure 5.1, the Kolmogorov operator converts according to one of the rules we proved in Lemma 5.8, which is consistent with (5.92).

5.8.6 Deterministic and Stochastic Part

Motivation. This freedom to switch between the Kolmogorov operators of the various representations of our process so effortlessly, by using the intuitive commutative diagrams above, is a consequence of their definition (5.80) and should not to be taken for granted: For example, the conversion of the infinitesimal generators $D^{X,T}$ and $D^{Y,T}$ themselves (without ∂_t) can generally *not* be carried out by following the arrows of an analogous diagram.

The only exception within our framework is the change of numeraire in the forward parametrization: Since $B^Y_{T_1,T_2}$ is independent of t, we have

$$B^Y_{T_2,T_1} \circ \partial_t \circ B^Y_{T_1,T_2} = B^Y_{T_2,T_1} \circ B^Y_{T_1,T_2} \circ \partial_t = \partial_t$$

and thus by (5.88b)

$$D^{Y,T_2} = B^Y_{T_2,T_1} \circ D^{Y,T_1} \circ B^Y_{T_1,T_2}, \tag{5.93}$$

and so the conversion of the generators $D^{Y,T}$ for various $T \in \mathbb{R}$ *is* consistent with a commutative diagram analogous to (5.91). In contrast, since the maps $B^X_{T_1,T_2}$ (for changing the numeraire in the spot parametrization) and $\bar{U}$ (for switching between the two parametrizations) are not time independent, these two conversions of the generator are *not* consistent with their corresponding diagrams.

This indicates that in our framework, which intends to provide maximal flexibility to switch between parametrizations and numeraires, with the eventual goal to derive a numeraire-free description of the process dynamics, the standard representation of the Kolmogorov operator $\partial_t + D^{X,T}$ as the sum of ∂_t and the generator $D^{X,T}$ is not ideal, since these two individual components do not convert in a way that is consistent with the corresponding commutative diagrams.

We will therefore now introduce a decomposition that splits each of our Kolmogorov operators into a *deterministic part* (denoted by ∇^X_t or ∇^Y_t) and a *stochastic part* (denoted by $\tilde{D}^{X,T}$ or $\tilde{D}^{Y,T}$), where

$$\begin{aligned}
\nabla^X_t &: \mathcal{F}(M_X) \to \mathcal{F}(M_X), & \tilde{D}^{X,T} &: \mathcal{F}(M_{X,T}) \to \mathcal{F}(M_{X,T}), \\
\nabla^Y_t &: \mathcal{F}(M_Y) \to \mathcal{F}(M_Y), & \tilde{D}^{Y,T} &: \mathcal{F}(M_{Y,T}) \to \mathcal{F}(M_{Y,T}),
\end{aligned}$$

and of course

$$\partial_t + D^{X,T} = \nabla^X_t + \tilde{D}^{X,T}, \tag{5.94a}$$
$$\partial_t + D^{Y,T} = \nabla^Y_t + \tilde{D}^{Y,T}, \tag{5.94b}$$

such that both of these individual components convert in line with the corresponding diagrams.

Our notation that we use to distinguish these two new operators for the various parametrization–numeraire combinations, along with some useful formulas derived later on that relate these operators to ∂_t and to the regular infinitesimal generators, respectively, can be found in the respective third and fourth rows of the large boxes in Figure 5.1.

Definition. The two parts are defined as

$$\nabla_t^X := \partial_t + D_{\sigma=0}^{X,T}, \qquad \tilde{D}^{X,T} := D^{X,T} - D_{\sigma=0}^{X,T}, \qquad \text{(5.95a)}$$

$$\nabla_t^Y := \partial_t + D_{\sigma=0}^{Y,T}, \qquad \tilde{D}^{Y,T} := D^{Y,T} - D_{\sigma=0}^{Y,T}; \qquad \text{(5.95b)}$$

in other words, the deterministic parts are the Kolmogorov operators in the zero-noise case $\sigma = 0$, and the stochastic parts are the remainders such that (5.94a)–(5.94b) hold. (Note that since the zero-noise case is a simple ODE independent of T, $D_{\sigma=0}^{X,T}$ and $D_{\sigma=0}^{Y,T}$ do not actually depend on T, which explains why we can suppress the index T in our notation ∇_t^X and ∇_t^Y for the deterministic parts, and why these two operators are defined for all times t.)

Conversion properties. The fact that under a change of variables and/or measure both of these components each transform consistently with the corresponding diagrams, just as the Kolmogorov operator does, is easy to see: The deterministic part transforms that way because it *is* a Kolmogorov operator (the one for the zero-noise case); the stochastic part does so because by (5.94a)–(5.94b) it is the difference of two Kolmogorov operators,

$$\tilde{D}^{X,T} = (\partial_t + D^{X,T}) - \nabla_t^X \qquad \text{and} \qquad \tilde{D}^{Y,T} = (\partial_t + D^{Y,T}) - \nabla_t^Y,$$

and because all the operators involved are linear. In particular, the analogues

$$\tilde{D}^{Y,T} = \bar{U}^{-1} \circ \tilde{D}^{X,T} \circ \bar{U} \qquad \text{(5.96a)}$$

$$\nabla_t^Y = \bar{U}^{-1} \circ \nabla_t^X \circ \bar{U} \qquad \text{(5.96b)}$$

$$\tilde{D}^{X,T_2} = B_{T_2,T_1}^X \circ \tilde{D}^{X,T_1} \circ B_{T_1,T_2}^X \qquad \text{(5.97a)}$$

$$\tilde{D}^{Y,T_2} = B_{T_2,T_1}^Y \circ \tilde{D}^{Y,T_1} \circ B_{T_1,T_2}^Y \qquad \text{(5.97b)}$$

of the relations in Lemma 5.8 hold, which is in accordance with Figure 5.1.

Explicit formulas. Computing the two components for the forward parametrization explicitly is easy: In Section 5.3.8 we found that the zero-noise process $(Y_t^{\sigma=0})$ in the forward parametrization is constant, and as a result we have $D_{\sigma=0}^{Y,T} \equiv 0$ for $\forall T \in \mathbb{R}$ by (5.79), and thus (5.95b) becomes

$$\nabla_t^Y = \partial_t \qquad \text{(5.98a)}$$

$$\tilde{D}^{Y,T} = D^{Y,T}. \qquad \text{(5.98b)}$$

In the spot parametrization, however, if for some given pair $(\mathbf{x}, t) \in M_X$ we denote by $(X_s^{\sigma=0})_{s \geq t}$ the deterministic process starting from the state $\mathbf{x} \in M$ at time t, we have by (5.80) and (5.8)

$$((\partial_t + D_{\sigma=0}^{X,T})f)(\mathbf{x}, t) = \lim_{\Delta t \searrow 0} \tfrac{1}{\Delta t}[f(X_{t+\Delta t}^{\sigma=0}, t + \Delta t) - f(\mathbf{x}, t)]$$

$$= \tfrac{d}{ds} f(X_s^{\sigma=0}, s)\big|_{s=t} = \big((\partial_t + \langle \mu, \nabla \rangle)f\big)(\mathbf{x}, t)$$

for $\forall f \in \mathcal{F}(M_X)$, and thus by (5.95a)

$$\nabla_t^X = \partial_t + \langle \mu, \nabla \rangle \qquad\qquad (5.99a)$$

$$\tilde{D}^{X,T} = D^{X,T} - \langle \mu, \nabla \rangle. \qquad\qquad (5.99b)$$

(Alternatively, we could have derived (5.99a) from the relation $\nabla_t^X = \bar{U}^{-1} \circ \nabla_t^Y \circ \bar{U} = \bar{U}^{-1} \circ \partial_t \circ \bar{U}$, which follows from the diagram (5.90) in the zero-noise case.)

The reader will find the formulas (5.98) and (5.99) summarized in the large boxes in the middle row of Figure 5.1.

5.9   The Numeraire-Free Kolmogorov Operator

Expressing our model dynamics in the classical way, i.e., in some fixed T-measure, has two decisive disadvantages:

- Since the numeraire (the maturity-T bond) ceases to exist at the process time $t = T$, the dynamics of the processes (X_t) and (Y_t) under the T-measure are no longer defined after that time.

- This representation is inconvenient to describe time homogeneity: Even if the market stochastically behaves in the same way at each process time t, the processes under any fixed T-measure will not, since the process time $t = T$ is special for the price evolution of the maturity-T bond (the bond price will approach the deterministic value 1). Thus, designing a truly time-homogeneous model of the market does *not* amount to defining the stochastic evolution of process (X_t) under some fixed T-measure in a time-independent way.

To remedy these issues, we will instead describe the model dynamics within our numeraire-free framework, by defining a *numeraire-free* infinitesimal generator that acts on sections instead of functions, and from which all classical generators under the various T-measures can be derived. Unlike these classical generators, this numeraire-free generator will define the dynamics for *all* process times $t \in \mathbb{R}$. It will come in two flavors—one for the spot and one for the forward parametrization—that are related to each other in a way analogous to (5.87), except with $\bar{U}$ replaced by $\hat{U}$.

Using this numeraire-free generator, we will then be able to derive a *numeraire-free Kolmogorov backward equation*, and thus to efficiently express our consistency requirement in our numeraire-free section-based notation.

5.9.1   Forward Parametrization

Stochastic part (numeraire-free generator). To define this operator in the forward parametrization, let us start by rewriting our conversion formula

(5.93) as

$$\tilde{D}^{Y,T_2} = \frac{\mathfrak{B}^Y_{T_1}}{\mathfrak{B}^Y_{T_2}} \circ \tilde{D}^{Y,T_1} \circ \frac{\mathfrak{B}^Y_{T_2}}{\mathfrak{B}^Y_{T_1}} \tag{5.100}$$

(recall also (5.98b)). Multiplying this equation from the left by the operator $\mathfrak{B}^Y_{T_2}$ and from the right by $\frac{1}{\mathfrak{B}^Y_{T_2}}$, we therefore find that

$$\mathfrak{B}^Y_{T_2} \circ \tilde{D}^{Y,T_2} \circ \frac{1}{\mathfrak{B}^Y_{T_2}} = \mathfrak{B}^Y_{T_1} \circ \tilde{D}^{Y,T_1} \circ \frac{1}{\mathfrak{B}^Y_{T_1}} \tag{5.101}$$

for $\forall T_1, T_2 \in \mathbb{R}$ wherever both operators are defined. In other words, the operator $\mathfrak{B}^Y_T \circ \tilde{D}^{Y,T} \circ \frac{1}{\mathfrak{B}^Y_T}$ does not depend on T, except that it acts on (and returns) sections $\psi \in \mathcal{S}(\mathcal{V}_{Y,T})$ whose domain depends on (and increases with) T. Since $\bigcup_{T \in \mathbb{R}} M_{Y,T} = M_Y$, this consistent family of operators therefore defines an operator $\tilde{D}^{Y,nf}$ that is defined for sections $\mathcal{S}(\mathcal{V}_Y)$ on our full space of state–time pairs, with

$$\tilde{D}^{Y,nf}\big|_{M_{Y,T}} = \mathfrak{B}^Y_T \circ \tilde{D}^{Y,T} \circ \frac{1}{\mathfrak{B}^Y_T} \tag{5.102}$$

for $\forall T \in \mathbb{R}$, or stated more precisely,

$$\left(\tilde{D}^{Y,nf} \psi \right)\big|_{M_{Y,T}} = \left(\mathfrak{B}^Y_T \circ \tilde{D}^{Y,T} \circ \frac{1}{\mathfrak{B}^Y_T} \right) \left(\psi\big|_{M_{Y,T}} \right)$$

for $\forall \psi \in \mathcal{S}(\mathcal{V}_Y)$ and $\forall T \in \mathbb{R}$.

The operator $\tilde{D}^{Y,nf}$, which we will call the *numeraire-free generator* in the forward parametrization, will be the *stochastic part* of our numeraire-free Kolmogorov operator. Its definition can be summarized with the diagram

$$\begin{array}{ccc} \mathcal{S}(\mathcal{V}_Y) & \xrightarrow{\tilde{D}^{Y,nf}} & \mathcal{S}(\mathcal{V}_Y) \\ \Big\downarrow{\scriptstyle 1/\mathfrak{B}^Y_T} & & \Big\downarrow{\scriptstyle 1/\mathfrak{B}^Y_T} \\ \mathcal{F}(M_{Y,T}) & \xrightarrow{\tilde{D}^{Y,T}} & \mathcal{F}(M_{Y,T}), \end{array} \tag{5.103}$$

where the sections in the top row represent numeraire-free account values, and where the functions in the bottom row represent the corresponding values in units of maturity-T bonds. (Note how this conversion between functions and sections, as well as the operator conversion rule (5.102), are in line with our usual interpretation of the bottom right vertical arrow in Figure 5.1.)

To be more precise, the operator $\tilde{D}^{Y,nf}$ is really only defined on the subset of those sections $\psi \in \mathcal{S}(\mathcal{V}_Y)$ for which the function $\frac{\psi}{\mathfrak{B}^Y_T}$ is sufficiently well-behaved for $\tilde{D}^{Y,T}\left(\frac{\psi}{\mathfrak{B}^Y_T} \right)$ to be defined. We do, however, apply our notational convention also to our numeraire-free operators defined here and in the following, and simply write $\tilde{D}^{Y,nf} : \mathcal{S}(\mathcal{V}_Y) \to \mathcal{S}(\mathcal{V}_Y)$.

Recovering the classical generators. Observe that from $\tilde{D}^{Y,nf}$ one can recover the classical generators $D^{Y,T}$ for $\forall T \in \mathbb{R}$: Solving (5.102) for $\tilde{D}^{Y,T}$, we obtain the relation

$$D^{Y,T} \overset{(5.98b)}{=} \tilde{D}^{Y,T} = \frac{1}{\mathfrak{B}^Y_T} \circ \tilde{D}^{Y,nf} \circ \mathfrak{B}^Y_T. \tag{5.104}$$

Our new section-based operator $\tilde{D}^{Y,nf}$ thus encodes all the information of our process dynamics, for all times $t \in \mathbb{R}$.

Deterministic part. Analogously we define the *deterministic part* of the numeraire-free Kolmogorov operator $\nabla_t^{Y,nf} : \mathcal{S}(\mathcal{V}_Y) \to \mathcal{S}(\mathcal{V}_Y)$ as

$$\nabla_t^{Y,nf} \quad := \quad \mathfrak{B}_T^Y \circ \nabla_t^Y \circ \frac{1}{\mathfrak{B}_T^Y} \tag{5.105}$$

$$\overset{(5.98a)}{=} \quad \mathfrak{B}_T^Y \circ \partial_t \circ \frac{1}{\mathfrak{B}_T^Y} = \partial_t \,, \tag{5.106}$$

where in the last step we used that $\mathfrak{B}_T^Y$ is independent of t.

Kolmogorov operator. The *numeraire-free Kolmogorov operator* in the forward parametrization, which we define to be the sum of the two, i.e., $\nabla_t^{Y,nf} + \tilde{D}^{Y,nf}$, has the following representations:

$$\nabla_t^{Y,nf} + \tilde{D}^{Y,nf} \quad \overset{(5.106)}{=} \quad \partial_t + \tilde{D}^{Y,nf} \tag{5.107}$$

$$\overset{(5.105),(5.102)}{=} \quad \mathfrak{B}_T^Y \circ (\nabla_t^Y + \tilde{D}^{Y,T}) \circ \frac{1}{\mathfrak{B}_T^Y} \tag{5.108}$$

$$\overset{(5.94b)}{=} \quad \mathfrak{B}_T^Y \circ (\partial_t + D^{Y,T}) \circ \frac{1}{\mathfrak{B}_T^Y} \,. \tag{5.109}$$

Note that the conversion rules (5.105) and (5.109) are as suggested by the bottom right vertical arrow in Figure 5.1, and that the formulas (5.106) and (5.107) can be found in the bottom-right large box of that figure.

5.9.2   Spot Parametrization

Stochastic part (numeraire-free generator). Following the commutative diagram

$$
\begin{array}{ccc}
\mathcal{S}(\mathcal{V}_Y) & \xrightarrow{\tilde{D}^{Y,nf}} & \mathcal{S}(\mathcal{V}_Y) \\
\downarrow{\hat{U}} & & \downarrow{\hat{U}} \\
\mathcal{S}(\mathcal{V}_X) & \xrightarrow{\tilde{D}^{X,nf}} & \mathcal{S}(\mathcal{V}_X)
\end{array}
$$

we now define the numeraire-free generator in the spot parametrization as

$$\tilde{D}^{X,nf} := \hat{U} \circ \tilde{D}^{Y,nf} \circ \hat{U}^{-1}. \tag{5.110}$$

It is then easy to show that this definition is consistent with the analogue of the diagram (5.103) for the spot parametrization, i.e., that the counterparts

$$\tilde{D}^{X,nf}\big|_{M_{X,T}} = \mathfrak{B}_T^X \circ \tilde{D}^{X,T} \circ \frac{1}{\mathfrak{B}_T^X}, \tag{5.111}$$

$$\tilde{D}^{X,T} = \frac{1}{\mathfrak{B}_T^X} \circ \tilde{D}^{X,nf} \circ \mathfrak{B}_T^X \tag{5.112}$$

of (5.102) and (5.104) hold. (The proof of (5.112) is given in Appendix D.4, (5.111) then follows by solving for $\tilde{D}^{X,nf}$.)

Recovering the classical generators. As one consequence, by combining (5.112) with (5.99b) we find that the operators $D^{X,T}$ can be recovered from $\tilde{D}^{X,nf}$ via the relation

$$D^{X,T} = \tfrac{1}{\mathfrak{B}_T^X} \circ \tilde{D}^{X,nf} \circ \mathfrak{B}_T^X + \langle \mu, \nabla \rangle. \tag{5.113}$$

Deterministic part. Analogously we define the numeraire-free deterministic part as

$$\nabla_t^{X,nf} := \hat{U} \circ \nabla_t^{Y,nf} \circ \hat{U}^{-1}, \tag{5.114}$$

which can be seen to have the representations

$$\nabla_t^{X,nf} = \mathfrak{B}_T^X \circ \nabla_t^X \circ \tfrac{1}{\mathfrak{B}_T^X} \tag{5.115}$$

$$= \partial_t - \nabla_{\hat{U}_t}, \tag{5.116}$$

$$\text{where} \quad \nabla_{\hat{U}_t} := \partial_s \hat{U}_s \big|_{s=0}. \tag{5.117}$$

Indeed, the representation (5.115) can be derived analogously to our proof of (5.111), the calculations leading to (5.116) are found in Appendix D.5.[21]

Kolmogorov operator. Finally, the numeraire-free Kolmogorov operator in the spot parametrization is defined as $\nabla_t^{X,nf} + \tilde{D}^{X,nf}$ and has the representations

$$\nabla_t^{X,nf} + \tilde{D}^{X,nf} \overset{(5.116)}{=} \partial_t - \nabla_{\hat{U}_t} + \tilde{D}^{X,nf} \tag{5.118}$$

$$\overset{(5.115),(5.111)}{=} \mathfrak{B}_T^X \circ (\nabla_t^X + \tilde{D}^{X,T}) \circ \tfrac{1}{\mathfrak{B}_T^X} \tag{5.119}$$

$$\overset{(5.94a)}{=} \mathfrak{B}_T^X \circ (\partial_t + D^{X,T}) \circ \tfrac{1}{\mathfrak{B}_T^X}. \tag{5.120}$$

Again note that all the conversion rules in this section are in agreement with Figure 5.1, and that the formulas (5.116) and (5.118) are shown in the bottom-left large box in that figure.

5.9.3 Numeraire-Free Kolmogorov Backward Equation

Our next goal is to translate also the classical Kolmogorov backward equation for functions in Lemma 5.6 into our numeraire-free section-based framework, using our newly defined numeraire-free Kolmogorov operator. In particular, this will allow us to express our consistency conditions (5.84a)–(5.84b) as a condition on our bond sections.

To do so, let the (sufficiently well-behaved) section $\psi_X \in \mathcal{S}(\mathcal{V}_X)$ represent the numeraire-free value of some account. Then for any fixed $T \in \mathbb{R}$ the function $\frac{\psi_X}{\mathfrak{B}_T^X} \in \mathcal{F}(M_{X,T})$ represents the value of that account measured in units

---

[21]Note that we could have used an analogous proof to rewrite (5.99a) as $\nabla_t^X = \partial_t - \nabla_{\bar{U}_t}$, where $\nabla_{\bar{U}_t} := \partial_s \bar{U}_s \big|_{s=0} = -\langle \mu, \nabla \rangle$.

of maturity-T bonds, and thus according to Section 5.8.2 the account is self-financing up until time T if and only if the Kolmogorov backward equation

$$(\partial_t + D^{X,T})\frac{\psi_X}{\mathfrak{B}_T^X} = 0 \qquad (5.121)$$

is satisfied on $M_{X,T}$, which in turn is equivalent to the section-based equation

$$0 = \mathfrak{B}_T^X(\partial_t + D^{X,T})\frac{\psi_X}{\mathfrak{B}_T^X} \overset{(5.120)}{=} \left(\nabla_t^{X,nf} + \tilde{D}^{X,nf}\right)\psi_X \qquad (5.122)$$

being fulfilled on $M_{X,T}$. This proves the following theorem:

Theorem 5.1 (Numeraire-free Kolmogorov backward equation). *Given a section $\psi_X \in \mathcal{S}(\mathcal{V}_X)$ representing the value of an account, the following three statements are equivalent:*

(i) The account is self-financing.

(ii) For $\forall T \in \mathbb{R}$ we have $(\partial_t + D^{X,T})\frac{\psi_X}{\mathfrak{B}_T^X} = 0$ on $M_{X,T}$.

(iii) The numeraire-free Kolmogorov backward equation

$$\left(\nabla_t^{X,nf} + \tilde{D}^{X,nf}\right)\psi_X = 0 \qquad (5.123)$$

holds on M_X.

If for some fixed $T_0 \in \mathbb{R}$ we are given a section $\psi_X \in \mathcal{S}(\mathcal{V}_{X,T_0})$ instead, then the statement holds with (ii) restricted to $T \leq T_0$ and with (iii) restricted to M_{X,T_0}.

The analogous statements for the forward parametrization hold as well, with X replaced by Y everywhere.

5.9.4 Consistency Revisited

In particular, this shows that our two alternative consistency requirements (5.84a)–(5.84b) can be expressed in our numeraire-free language as

$$\forall T \in \mathbb{R}: \quad \left(\nabla_t^{X,nf} + \tilde{D}^{X,nf}\right)\mathfrak{B}_T^X = 0 \qquad (5.124a)$$

$$\text{and} \quad \forall T \in \mathbb{R}: \quad \left(\nabla_t^{Y,nf} + \tilde{D}^{Y,nf}\right)\mathfrak{B}_T^Y = 0, \qquad (5.124b)$$

respectively. Now observe that since $\mathfrak{B}_T^Y$ is independent of t (the key property of the forward parametrization), we have[22]

$$\nabla_t^{Y,nf}\mathfrak{B}_T^Y \overset{(5.106)}{=} \partial_t\mathfrak{B}_T^Y = 0$$

and thus also

$$\nabla_t^{X,nf}\mathfrak{B}_T^X \overset{(5.114)}{=} (\hat{U} \circ \nabla_t^{Y,nf} \circ \hat{U}^{-1})\mathfrak{B}_T^X \overset{(5.52)}{=} (\hat{U} \circ \nabla_t^{Y,nf})\mathfrak{B}_T^Y = 0.$$

We therefore obtain the following corollary:

---

[22]Note that while $\mathbf{x}$-derivatives of sections $\psi(\mathbf{x},t)$ are not defined since we cannot compare vectors in different vector spaces $V_{\mathbf{x}}$ and $V_{\mathbf{x}'}$, t-derivatives *are* well defined since in Section 5.1.1 we chose to use the same vector spaces for different values of t (for exactly this reason).

Corollary 5.1 (Consistency). *The two alternative consistency requirements* (5.84a)–(5.84b) *are equivalent to*

$$\forall T \in \mathbb{R}: \quad \tilde{D}^{X,nf} \mathfrak{B}_T^X = 0 \qquad (5.125\text{a})$$

$$and \quad \forall T \in \mathbb{R}: \quad \tilde{D}^{Y,nf} \mathfrak{B}_T^Y = 0, \qquad (5.125\text{b})$$

respectively.

5.10   Cash Numeraire

The classical Kolmogorov backward equation (Lemma 5.6) and its numeraire-free equivalent for sections (Theorem 5.1 (iii)) tell us how the value of a self-financing account—expressed in units of some numeraire or as numeraire-free values, respectively—evolves over time. In practice, it is useful to have a corresponding PDE for the evolution of the *dollar* value of self-financing accounts as well. The derivation of this evolution equation within our framework is the content of the present section. The operator driving this equation (the *Kolmogorov operator in the cash numeraire*) will also give rise to the *risk-neutral measure* of our state process (X_t), which is used to write our model in the form of a short rate model.

5.10.1   Definition, Kolmogorov Backward Equation

Let $f_{X,\$} \in \mathcal{F}(M_X)$ be the value of an account, measured in units of dollars. Then by our discussion in Section 5.1.3 this function can be written as

$$f_{X,\$}(\mathbf{x}, t) = \frac{\psi_X(\mathbf{x}, t)}{\mathfrak{B}_t^X(\mathbf{x}, t)},$$

where $\psi_X \in \mathcal{S}(\mathcal{V}_X)$ represents the account's numeraire-free value. By Theorem 5.1 the account is self-financing if and only if the numeraire-free Kolmogorov backward equation

$$\left(\nabla_t^{X,nf} + \tilde{D}^{X,nf}\right)\psi_X = 0 \qquad (5.126)$$

is fulfilled. Defining the *Kolmogorov operator in the cash numeraire* as[23]

$$\partial_t + D^{X,\$} := \tfrac{1}{\mathfrak{B}_t^X} \circ \left(\nabla_t^{X,nf} + \tilde{D}^{X,nf}\right) \circ \mathfrak{B}_t^X, \qquad (5.127)$$

we find that (5.126) is equivalent to

$$\begin{aligned}
0 &= \tfrac{1}{\mathfrak{B}_t^X}\left(\nabla_t^{X,nf} + \tilde{D}^{X,nf}\right)\psi_X \\
&= \left(\tfrac{1}{\mathfrak{B}_t^X} \circ \left(\nabla_t^{X,nf} + \tilde{D}^{X,nf}\right) \circ \mathfrak{B}_t^X\right)\tfrac{\psi_X}{\mathfrak{B}_t^X} \\
&= \left(\partial_t + D^{X,\$}\right)f_{X,\$}.
\end{aligned} \qquad (5.128)$$

---

[23]We use the intuitive notation $\mathfrak{B}_t^X$ to refer to the more correctly named section $\mathfrak{B}_\$^X$ defined as $\mathfrak{B}_\$^X(\mathbf{x}, t) := \mathfrak{B}_t^X(\mathbf{x}, t)$.

This is called the *Kolmogorov backward equation in the cash numeraire*, which we have shown to hold if and only if $f_{X,\$}$ represents the dollar value of a self-financing account.

The corresponding equation in the forward parametrization, i.e.,

$$\left(\partial_t + D^{Y,\$}\right) f_{Y,\$} = 0, \tag{5.129}$$

$$\text{where} \qquad \partial_t + D^{Y,\$} := \tfrac{1}{\mathfrak{B}_t^Y} \circ \left(\nabla_t^{Y,nf} + \tilde{D}^{Y,nf}\right) \circ \mathfrak{B}_t^Y, \tag{5.130}$$

is derived analogously.

Note the important fact that by (5.40) we have

$$\mathfrak{B}_t^X(\mathbf{x}, t) = \mathfrak{B}_0^Y(\mathbf{x}) \tag{5.131}$$

for $\forall(\mathbf{x}, t) \in M_X$, and so the conversion factor $\mathfrak{B}_t^X$ in (5.127) for the spot parametrization does not actually depend on t.

5.10.2   Alternative Representations

In (5.127) and (5.130) we have separately defined the Kolmogorov operators in the cash numeraire for the forward and the spot parametrization, based on our numeraire-free Kolmogorov operator. Note how these definitions are in fact consistent with our big conversion chart in Figure 5.1, if we vertically connect the large boxes for the cash numeraire and for the numeraire-free representation with downward arrows that have the appropriate annotations $\mathfrak{B}_t^X$ and $\mathfrak{B}_t^Y$, respectively. It is therefore not surprising that our definitions also imply all the other relations that one would expect from Figure 5.1, in particular the relation

$$\partial_t + D^{Y,\$} = \bar{U}^{-1} \circ \left(\partial_t + D^{X,\$}\right) \circ \bar{U} \tag{5.132}$$

between the two, and the relations

$$\partial_t + D^{X,T} = B_{T,t}^X \circ \left(\partial_t + D^{X,\$}\right) \circ B_{t,T}^X$$
$$= \left(B_{\$,T}^X\right)^{-1} \circ \left(\partial_t + D^{X,\$}\right) \circ B_{\$,T}^X \tag{5.133a}$$
$$\partial_t + D^{Y,T} = B_{T,t}^Y \circ \left(\partial_t + D^{Y,\$}\right) \circ B_{t,T}^Y$$
$$= \left(B_{\$,T}^Y\right)^{-1} \circ \left(\partial_t + D^{Y,\$}\right) \circ B_{\$,T}^Y \tag{5.133b}$$

to the Kolmogorov operators under fixed T-measures. The reader who still demands formal proofs of these relations will find them in Appendix D.6.

Of course, these relations are also just what one would expect by drawing the corresponding commutative diagrams, since account value functions in the two parametrizations, using units of dollars and maturity-T bonds, are related via the formulas

$$f_{X,\$} = \bar{U} f_{Y,\$}, \qquad f_{X,\$} = B_{\$,T}^X \cdot f_{X,T} \qquad \text{and} \qquad f_{Y,\$} = B_{\$,T}^Y \cdot f_{Y,T}.$$

5.10.3 Computation of the Kolmogorov Operator

To compute the operators $D^{X,\$}$ and $D^{Y,\$}$ explicitly, let us split our Kolmogorov operators defined in (5.127) and (5.130) into their deterministic and their stochastic parts, by writing them as

$$\partial_t + D^{X,\$} = \nabla_t^{X,\$} + \tilde{D}^{X,\$}, \tag{5.134a}$$

$$\partial_t + D^{Y,\$} = \nabla_t^{Y,\$} + \tilde{D}^{Y,\$}, \tag{5.134b}$$

where we define

$$\nabla_t^{X,\$} := \tfrac{1}{\mathfrak{B}_t^X} \circ \nabla_t^{X,nf} \circ \mathfrak{B}_t^X, \qquad \tilde{D}^{X,\$} := \tfrac{1}{\mathfrak{B}_t^X} \circ \tilde{D}^{X,nf} \circ \mathfrak{B}_t^X, \tag{5.135a}$$

$$\nabla_t^{Y,\$} := \tfrac{1}{\mathfrak{B}_t^Y} \circ \nabla_t^{Y,nf} \circ \mathfrak{B}_t^Y, \qquad \tilde{D}^{Y,\$} := \tfrac{1}{\mathfrak{B}_t^Y} \circ \tilde{D}^{Y,nf} \circ \mathfrak{B}_t^Y. \tag{5.135b}$$

Explicitly, the deterministic parts $\nabla_t^{X,\$}$ and $\nabla_t^{Y,\$}$ are given by

$$
\begin{aligned}
\nabla_t^{Y,\$} \quad &= \quad \tfrac{1}{\mathfrak{B}_t^Y} \circ \nabla_t^{Y,nf} \circ \mathfrak{B}_t^Y \overset{(5.106)}{=} \tfrac{1}{\mathfrak{B}_t^Y} \circ \partial_t \circ \mathfrak{B}_t^Y = \partial_t + \left(\tfrac{1}{\mathfrak{B}_t^Y} \cdot \partial_t \mathfrak{B}_t^Y \right) \\
&= \quad \partial_t + \left(\tfrac{1}{\mathfrak{B}_t^Y} \cdot \partial_s \mathfrak{B}_{t+s}^Y \big|_{s=0} \right) = \partial_t + \left(\partial_s B_{t,t+s}^Y \big|_{s=0} \right) \\
&\overset{(5.38),(5.37)}{=} \partial_t - r^Y, \tag{5.136a}
\end{aligned}
$$

$$
\begin{aligned}
\nabla_t^{X,\$} \quad &\overset{(5.132)'}{=} \quad \bar{U} \circ \nabla_t^{Y,\$} \circ \bar{U}^{-1} \overset{(5.136a)}{=} \bar{U} \circ \partial_t \circ \bar{U}^{-1} - \bar{U} \circ r^Y \circ \bar{U}^{-1} \\
&= \quad \partial_t + \langle \mu, \nabla \rangle - r^X, \tag{5.136b}
\end{aligned}
$$

where the first step in our calculation for $\nabla_t^{X,\$}$ can be proven analogously to (5.132), and where the last step is shown in more detail in Appendix D.7. The stochastic parts can best be computed starting from relations analogous to (5.133):[24]

$$\tilde{D}^{X,\$} \overset{(5.133a)'}{=} B_{t,T}^X \circ \tilde{D}^{X,T} \circ B_{T,t}^X \overset{(5.97a)}{=} \tilde{D}^{X,t} \overset{(5.99b)}{=} D^{X,t} - \langle \mu, \nabla \rangle, \tag{5.137a}$$

$$\tilde{D}^{Y,\$} \overset{(5.133b)'}{=} B_{t,T}^Y \circ \tilde{D}^{Y,T} \circ B_{T,t}^Y \overset{(5.97b)}{=} \tilde{D}^{Y,t} \overset{(5.98b)}{=} D^{Y,t}. \tag{5.137b}$$

Note for future reference that by (5.135a) and (5.131) we also have the representation

$$\tilde{D}^{X,\$} = \tfrac{1}{\mathfrak{B}_0^Y} \circ \tilde{D}^{X,nf} \circ \mathfrak{B}_0^Y. \tag{5.138}$$

Now combining (5.134), (5.136), and (5.137), we find that the Kolmogorov operators in the cash numeraire are given by

$$\partial_t + D^{X,\$} = \partial_t + D^{X,t} - r^X, \tag{5.139a}$$

$$\partial_t + D^{Y,\$} = \partial_t + D^{Y,t} - r^Y, \tag{5.139b}$$

---

[24]In the respective second steps of (5.137) one needs to be careful when replacing the fixed tenor T_2 in (5.97) by the variable t. This is only possible because by (5.99b), (5.98b), and the very definition of the classical generators, the values $(\tilde{D}^{X,T} f)(\mathbf{x}, t)$ and $(\tilde{D}^{Y,T} f)(\mathbf{y}, t)$ for any fixed t only depend on $f(\cdot, t)$, and not on the function values of f for any other value $t' \neq t$.

and so the Kolmogorov backward equations (5.128)–(5.129) in the cash numeraire become

$$\left(\partial_t + D^{X,t} - r^X\right)f_{X,\$} = 0, \qquad (5.140a)$$

$$\left(\partial_t + D^{Y,t} - r^Y\right)f_{Y,\$} = 0. \qquad (5.140b)$$

The formulas (5.136), (5.137), and (5.139) are displayed in the two top-most large boxes in Figure 5.1.

5.10.4 Risk-Neutral Measure, Feynman–Kac

The operator $D^{X,t}$, obtained by replacing the parameter T in the formula for $D^{X,T}$ by the process time t, is the generator of a process (X_t) as well; that process's measure is called the *risk-neutral measure*. However, since the numeraire $\mathfrak{B}_t^X$ used for constructing it—the dollar—is not a self-financing account (recall our discussion in Section 5.8.2), the properties and use of $D^{X,t}$ differ from the ones of the generators $D^{X,T}$ corresponding to the T-measures decisively:

As we have shown in the preceding section, the PDE (5.140a) fulfilled by functions $f_{X,\$}$ representing the dollar values of self-financing accounts includes an additional rate term r^X. By Lemma 5.6 this means that the associated processes $(f_{X,\$}(X_t, t))$ cannot be local martingales, and therefore we can *not* expect that $\mathbb{E}^{X,\$}[f_{X,\$}(X_T, T)|X_t = \mathbf{x}] = f_{X,\$}(\mathbf{x}, t)$, where $\mathbb{E}^{X,\$}$ denotes expectation under the risk-neutral measure.

The well-known Feynman–Kac theorem provides the necessary correction term: The uniqueness statement of this theorem tells us that under certain technical conditions imposed on $(X_t)_{t \geq 0}$ and r^X, any solution $f_{X,\$}$ of the PDE (5.140a) whose growth as $\mathbf{x} \to \infty$ can be sufficiently controlled must satisfy the relation

$$f_{X,\$}(\mathbf{x}, t) = \mathbb{E}^{X,\$}\left[f_{X,\$}(X_T, T) \exp\left(-\int_t^T r^X(X_s)\,\mathrm{d}s\right) \Big| X_t = \mathbf{x}\right] \qquad (5.141)$$

for $\forall T \in \mathbb{R}$ and $\forall(\mathbf{x}, t) \in M_{X,T}$. Naturally, all of the above remains true in the forward parametrization.[25]

The Feynman–Kac theorem comes in various flavors, with weaker growth constraints on $f_{X,\$}$ requiring stronger assumptions on $(X_t)_{t \geq 0}$ and a more difficult proof; see for example [32, Theorem 8.2.1] for a version that requires uniform boundedness of $f_{X,\$}$, or [33, Theorem 5.7.6] for one that allows for polynomial growth. See Lemma 2.8 for a simple version that was tailored to our specific needs.

5.10.5 Short Rate Model

In particular, we can apply (5.141) to the functions $f_{X,\$} = B_{\$,T}^X$, $T \in \mathbb{R}$, that represent the dollar values of the maturity-T bonds. Since $B_{\$,T}^X(\mathbf{x}, T) = 1$ for

---

[25]Note, however, that r^Y depends on t, and so the integrand in the counterpart of (5.141) in the forward parametrization is the function $r^Y(Y_s, s)$.

$\forall \mathbf{x} \in M_0$, this formula then reduces to

$$B_{\$,T}^X(\mathbf{x}, t) = \mathbb{E}^{X,\$} \big[\exp \big(- \textstyle\int_t^T r^X(X_s) \, \mathrm{d}s \big) \, \big| \, X_t = \mathbf{x} \big]. \tag{5.142}$$

Using the technical conditions of the Feynman–Kac theorem presented in Lemma 2.8, this shows the following:

Lemma 5.9. *(i) As long as the functions $B_{\$,T}^X(\mathbf{x}, t)$ constructed in our framework are continuous and bounded on $M_0 \times [0, T]$, they satisfy the relation (5.142).*

(ii) Therefore, the measure encoded via (5.81)–(5.82) by the generator $D^{X,t}$, the risk-neutral measure, is the one providing the correct dynamics for the process (X_t) if one wants to write the ACE model in the form of a short rate model as described in Section 1.3.

5.11   Time Homogeneity

The time homogeneity property listed in Section 1.1 requires some further clarification, which we are now in the position to provide. Intuitively, what we want to ask is that in its basic form our model should have some sort of symmetry that causes it to behave "in the same way" when started from "the same state" at different times t. This, however, leaves us with two questions.

Choice of parametrization.   First, we have to decide for the parametrization in which we want our model to be time homogeneous, i.e., what it means to be in "the same state." In fact, in Section 5.2.2 we have already chosen to make the zero-noise process time homogeneous in the *spot* parametrization, and we shall now follow up with a justification of this choice.

To see why the choice of parametrization matters, consider the experiment of dropping a ball twice at two different times t. When performing this experiment in a stationary laboratory, it will lead to the same outcome both times; however, when carrying out the same experiment while riding on a roller coaster, the ball will fly off into a different direction each time. This shows that the time independence of model dynamics generally does not survive *time-dependent* changes of the reference frame, i.e., state space reparametrizations like our map U.[26]

In our case, we will want the model dynamics to be time independent in the *spot* parametrization, since by its very definition in Section 5.2.1 the state $\mathbf{x}$ is a parametrization of the forward rate curve $\tau \mapsto F_{\mathrm{rel}}(\mathbf{x}; \tau)$ *relative to the time t at which the state is observed*, and thus it encodes the traders' view on the market in their respective future. If one assumes that no information other than the current yield curve affects their trading decisions, then starting the model from the same state $\mathbf{x}$ at two different times t should lead to the same stochastic model behavior.

---

[26]However, depending on the specific system at hand, exceptions *can* exist: In the example above, time homogeneity does survive if the roller coaster is a closed car moving on a straight track at constant speed.

Choice of numeraire. Second, since we are dealing with a stochastic system, we need to further specify which numeraire (i.e., which measure of the process (X_t)) we want to consider, as this will decide what it means to stochastically behave "in the same way."

Since $B^X_{T_1, T_2}$ depends on t, one can see from (5.88a) that time-independent dynamics of (X_t) under one specific T-measure (which is equivalent to the generator $D^{X,T}$ being independent of the process time t) does *not* imply the same for all other T-measures. However, there is no reason why any specific value of T should deserve to be singled out, and so clearly we do not mean to ask that the dynamics of (X_t) be time independent under any specific T-measure. What we would rather have to ask is that delaying the process (X_t) under any T-measure by some $\Delta t \in \mathbb{R}$, i.e., considering the process $(X_t^{\Delta t})$ defined as $X_t^{\Delta t} := X_{t - \Delta t}$, should result in the $(T + \Delta t)$-measure of the original process (X_t).

As a simpler alternative, one may find it more natural to ask that any description of the process dynamics of (X_t) that does not utilize a symmetry-breaking numeraire should be independent of t, such as the Kolmogorov operator in the cash numeraire $\partial_t + D^{X,\$}$ (as we asked in Section 1.1) or our newly defined numeraire-free Kolmogorov operator $\nabla_t^{X,nf} + \tilde{D}^{X,nf}$. As the following lemma shows, all these three ideas are equivalent.

Lemma 5.10 (Time homogeneity). *For any $\Delta t \in \mathbb{R}$ let us denote by $S_{\Delta t}$ the operator that shifts a given function $f(\mathbf{x}, t)$ or section $\psi(\mathbf{x}, t)$ to the left by Δt, i.e., we set $(S_{\Delta t} f)(\mathbf{x}, t) := f(\mathbf{x}, t + \Delta t)$ and $(S_{\Delta t} \psi)(\mathbf{x}, t) := \psi(\mathbf{x}, t + \Delta t)$. Then the following properties are equivalent:*

(i) For $\forall T, \Delta t \in \mathbb{R}$, the measure obtained by delaying the process (X_t) under the T-measure by the time Δt coincides with the $(T + \Delta t)$-measure of (X_t).

(ii) $\forall T, \Delta t \in \mathbb{R}$: $S_{-\Delta t} \circ D^{X,T} \circ S_{\Delta t} = D^{X, T + \Delta t}$

(iii) $\forall \Delta t \in \mathbb{R}$: $S_{-\Delta t} \circ D^{X,\$} \circ S_{\Delta t} = D^{X,\$}$

(iv) $\forall \Delta t \in \mathbb{R}$: $S_{-\Delta t} \circ \tilde{D}^{X,nf} \circ S_{\Delta t} = \tilde{D}^{X,nf}$

Each of the properties (ii)–(iv) are equivalent to the corresponding statements with the generators replaced by the full Kolmogorov operators.

Proof. See Appendix D.8. $\square$

The condition (iv) says that in a time-homogeneous model the operator $\tilde{D}^{X,nf} : \mathcal{S}(\mathcal{V}_X) \to \mathcal{S}(\mathcal{V}_X)$ must act on sections $\psi(\mathbf{x}, t)$ in a way that treats every process time t equally.

In particular, this means that one can design a time-homogeneous model by defining $\tilde{D}^{X,nf}$ as an operator $\tilde{D}^{X,nf} : \mathcal{S}(\mathcal{V}_0) \to \mathcal{S}(\mathcal{V}_0)$ that acts on time-independent sections $\psi(\mathbf{x})$, and whose action can then naturally be extended to time-dependent sections $\psi(\mathbf{x}, t)$ by having it act on each time slice $\psi(\cdot, t)$ individually.

5.12   Summary: Generators

Introducing noise to our model based on the family of T-measures of our process would have had the decisive disadvantages that (i) each T-measure is only defined for the process times $t \leq T$, and that (ii) this approach is inconvenient for expressing our time homogeneity requirement (see Lemma 5.10 (i)–(ii)).

In Sections 5.8–5.11 we have therefore developed a way of resolving both of these issues within our numeraire-free framework: Starting from the family of classical infinitesimal generators that each encode the process dynamics under a different T-measure, we have defined a new section-based "numeraire-free generator" from which all the classical generators under the T-measures can be recovered via the formulas (5.104) and (5.113), and that therefore encodes the process dynamics for *all* process times $t \in \mathbb{R}$. Using this numeraire-free generator, we could then express both our consistency requirement (Corollary 5.1) and our time homogeneity requirement (Lemma 5.10 (iv)) within our numeraire-free framework. While the proof of the former in Section 5.9.4 relies on the fact that the bond sections $\mathfrak{B}_T^Y$ in the forward parametrization do not depend on t, the latter is by definition based on the spot parametrization; this explains why we have put so much effort into having our framework support both parametrizations.

The construction of the numeraire-free generator $\tilde{D}^{Y,nf}$ in the forward parametrization was based on the observation that when changing between different T-measures, the classical generator transforms as (5.93), in a way that is consistent with the corresponding commutative diagram and therefore reminiscent of the notion of similarity between matrices in linear algebra. Using our bond sections to write this conversion rule as (5.100)[27] has then led us to the beautifully symmetric section-based equation (5.101), which ultimately motivated our definition (5.102) of $\tilde{D}^{Y,nf}$.

Unfortunately, the construction of the analogous operator $\tilde{D}^{X,nf}$ in the spot parametrization is not that easy: Since B_{T_1,T_2}^X depends on t, one cannot immediately translate our derivation of (5.93) and thus ultimately our definition of $\tilde{D}^{Y,nf}$ to the spot parametrization. First we had to find a way to generalize the decomposition of the Kolmogorov operator $\partial_t + D$ into ∂_t and the generator D in such a way that its two components would have the much-desired conversion property analogous to (5.93) also in other parametrizations. We solved this problem with our definition of the deterministic and the stochastic part of the Kolmogorov operator.

5.13   Confining the Process to Its State Space

There is still some final issue to address: In the deterministic case we have already guaranteed in Section 5.2.2 that the state process $(X_t)_{t \geq 0}$ is defined for all process times $t \geq 0$, and that it cannot leave the set M_0, which means that

---

[27]For the purpose of this summary, the reader should replace the operators $\tilde{D}^{Y,T}$ by $D^{Y,T}$ in (5.100)–(5.102), justified by (5.98b).

M_0 is indeed the state space of the process, and that it suffices—as we did—to define our bond sections $\mathfrak{B}_T^X(\mathbf{x}, t)$ only for $\mathbf{x} \in M_0$. We now need to ensure that this remains true even if we add noise to the system, i.e., that the noise encoded in $\tilde{D}^{X,nf}$ cannot drive the process to ∞ in finite time or out of its intended state space M_0, either.

Noise constraint. When aiming for maximum generality, this can lead to different types of constraints depending on the nature of the noise. For example, to ensure that the process cannot escape through the boundary ∂M_0 of our state space, for Markov jump processes one could ask that at states $\mathbf{x}$ near ∂M_0 all possible jumps can only point away from the boundary and towards the interior of the state space, without necessarily requiring that their jump sizes need to vanish as $\mathbf{x} \to \partial M_0$. For diffusion processes, it would suffice to ask that at the boundary the component of the noise perpendicular to the boundary vanishes, while all components parallel to it may remain finite.

However, we will sacrifice some generality in this respect in favor of a simplified description of our framework, and simply *require that the size of the noise vanishes sufficiently fast near ∂M_0, and that it does not grow too fast as* $\mathbf{x} \to \infty$ (if M_0 is unbounded). To not get hung up in technicalities, we will not start elaborating at this point what "sufficiently fast" and "too fast" mean exactly; instead, we will defer this to the moment when we actually hold our final evolution equations in our hands.

Section-based formulation. To see how this requirement could fit into our section-based framework, note that having zero noise at any state $\mathbf{x}$ means that our stochastic evolution under any T-measure is locally reduced to the deterministic system (5.8), and thus by (5.82) (with $\tilde{\sigma} \equiv 0$) that locally we have $D^{X,T} = \langle \mu, \nabla \rangle$. We therefore see from (5.113) that our intended condition could be expressed in terms of the operator $\tilde{D}^{X,nf}$ alone, by asking that $\tilde{D}^{X,nf}$ *itself should vanish sufficiently fast near ∂M_0 and not grow too fast as* $\mathbf{x} \to \infty$.

While this intuitive section-based formulation will give us some very useful guidance during our construction, one should point out, however, that it is rather ill-defined: The problem with section-based statements like these is that since one cannot compare vectors in different vector spaces $V_\mathbf{x}$ and $V_{\mathbf{x}'}$, there is no notion of a limit $\lim \psi(\mathbf{x})$ as $\mathbf{x} \to \partial M_0$ or $\mathbf{x} \to \infty$, and thus no notion of a limit $\lim (\tilde{D}^{X,nf} \psi)(\mathbf{x}, t)$, either. Such limits only make sense if one chooses some fixed $\psi_0 \in \mathcal{S}_\varnothing(V)$ (a "numeraire") to compare the section of interest to, and then considers the classical limit of the ratio—in our case the function $\frac{1}{\psi_0} \tilde{D}^{X,nf} \psi$—instead. Unfortunately, now choosing $\psi_0 := \mathfrak{B}_T^X$ and taking another look at (5.113), we see that our search for a clean interpretation of our section-based condition has led us right back to a statement about the classical generators $D^{X,T}$.

Since we have decided not to make our condition more rigorous at this stage anyways, we will not worry about this right now.

5.14 Summary: The Mathematical Framework

Let us summarize our construction so far. Our general section-based interest rate modeling framework is based on the following mathematical objects:

- a signed line bundle $\mathcal{V} := (V_\mathbf{x})_{\mathbf{x} \in M}$ on a state space $M \subseteq \mathbb{R}^n$,

- a C^1-group $(U_s)_{s \in \mathbb{R}}$ of operators $U_s \colon M \to M$, and an extension $(\hat{U}_s)_{s \in \mathbb{R}}$, i.e., operators $\hat{U}_s \colon \mathcal{S}(\mathcal{V}) \to \mathcal{S}(\mathcal{V})$ with the properties (5.67)–(5.71),

- a subset $M_0 \subset M$ such that $\forall s \geq 0 \colon U_s(M_0) \subseteq M_0$,

- a positive section $\mathfrak{B}_0^Y \in \mathcal{S}_+(\mathcal{V}_0)$, where $\mathcal{V}_0 := (V_\mathbf{x})_{\mathbf{x} \in M_0}$, and

- an operator $\tilde{D}^{X,nf} \colon \mathcal{S}(\mathcal{V}_X) \to \mathcal{S}(\mathcal{V}_X)$ acting on sections on the line bundle $\mathcal{V}_X := (V_{\mathbf{x},t})_{(\mathbf{x},t) \in M_X}$, where $M_X := M_0 \times \mathbb{R}$ and $V_{\mathbf{x},t} := V_\mathbf{x}$ for $\forall (\mathbf{x},t) \in M_X$, which is such that the operators

$$\frac{1}{\mathfrak{B}_T^X} \circ \tilde{D}^{X,nf} \circ \mathfrak{B}_T^X, \qquad \text{where} \qquad \mathfrak{B}_T^X(\mathbf{x},t) := (\hat{U}_{t-T}\mathfrak{B}_0^Y)(\mathbf{x}),$$

are classical generators of Markov processes. In particular, this requires satisfying the consistency condition

$$\forall T \in \mathbb{R} \colon \quad \tilde{D}^{X,nf}\mathfrak{B}_T^X = 0.$$

To ensure that the process $(X_t)_{t \geq 0}$ is confined to M_0 and that it does not explode, (the noise encoded in) $\tilde{D}^{X,nf}$ should vanish sufficiently fast near ∂M_0 and not grow too fast as $\mathbf{x} \to \infty$.

To enforce positive short rates and/or time homogeneity, the following additional conditions need to be fulfilled:

- To constrain the model to simulate only positive short rates (and as a result also forward rates), one needs to ensure that the short rate function r^X defined in the next box is positive on all of M_0.

- To make the model time homogeneous, the operator $\tilde{D}^{X,nf}$ must treat all process times t equally (in the sense of Lemma 5.10 (iv)).

Once these objects have been defined, the classical model equations can be derived as follows:

- The bond exchange rates B_{T_1,T_2}^X, i.e., the value of the maturity-T_2 bond in units of maturity-T_1 bonds, can then be computed via the formula

$$B_{T_1,T_2}^X(\mathbf{x},t) = \frac{\mathfrak{B}_{T_2}^X(\mathbf{x},t)}{\mathfrak{B}_{T_1}^X(\mathbf{x},t)},$$

the dollar value $B_{\$,T}$ of the maturity-T bond is given by

$$B_{\$,T}^X(\mathbf{x},t) = B_{i,T}^X(\mathbf{x},t) = \frac{\mathfrak{B}_T^X(\mathbf{x},t)}{\mathfrak{B}_t^X(\mathbf{x},t)}.$$

- The short rate is given by

$$r^X(\mathbf{x}) = -\partial_\tau B_{0,\tau}^X(\mathbf{x},0)\big|_{\tau=0},$$

and the full relative forward rate curve $\tau \mapsto F_{\mathrm{rel}}(\mathbf{x};\tau)$, $\tau \geq 0$, is given by

$$F_{\mathrm{rel}}(\mathbf{x};\tau) = r^X(U_\tau(\mathbf{x})).$$

(Both formulas do not depend on the time t at which $\mathbf{x} \in M_0$ is observed.)

- The classical generator $D^{X,T}$ of the process (X_t) under the T-measure is given by

$$D^{X,T} = \tfrac{1}{\mathfrak{B}_T^X} \circ \tilde{D}^{X,nf} \circ \mathfrak{B}_T^X + \langle \mu, \nabla \rangle,$$

  where $\mu(\mathbf{x}) = \partial_s U_s(\mathbf{x})\big|_{s=0}.$

The classical generator of the process (X_t) under the risk-neutral measure (which is used when writing the model as a short rate model) is $D^{X,t}$, i.e., it is obtained by replacing the maturity parameter T in the formula for $D^{X,T}$ by the process time t.

- Finally, the actual process dynamics of (X_t) under some measure of interest can be read off from the formula of the corresponding generator. In particular, a generator of the form

$$D = \sum_{i=1}^n \tilde{\mu}_i(\mathbf{x},t)\, \partial_{\mathbf{x}_i} + \frac{1}{2}\sum_{i,j=1}^n a_{ij}(\mathbf{x},t)\, \partial_{\mathbf{x}_i}\partial_{\mathbf{x}_j}$$

corresponds to the process satisfying the n-dimensional SDE

$$\mathrm{d}X_t = \tilde{\mu}(X_t,t)\,\mathrm{d}t + \tilde{\sigma}(X_t,t)\,\mathrm{d}W_t,$$

where $\tilde{\sigma}$ is any matrix-valued function such that $\tilde{\sigma}\tilde{\sigma}^T = a$.

Observe that we have chosen to base our classical model description in the last box on the *spot* parametrization, due to its simpler, time-independent state space M_0. For all practical purposes, the larger set $M \supset M_0$ is therefore no longer relevant for us at this point: It was merely introduced to support the forward parametrization that we used during our derivation so far, and although it still appears as the domain of our evolution operators, a careful look at the formulas above reveals that we never actually request their values on $M \setminus M_0$.

Chapter 6

Solution Strategy

In this chapter we will devise a strategy for how to solve our problem of defining all the mathematical objects described in our summary in Section 5.14, with all their desired properties. From this point forward we will only consider diffusion processes with nondegenerate noise, i.e., we will only attempt to define an interest rate model in which the dynamics of the process (X_t) (and thus also of (Y_t)) is given in the form of an SDE, so that its infinitesimal generator is of the form (5.82), with the noise matrix function $\tilde{\sigma}\tilde{\sigma}^T$ taking only positive-definite matrices as values. In addition, we will make some purely technical assumptions that will allow us to get a handle on the problem (in particular the "scaling assumption" in Section 6.3), but which may restrict the generality of our solution.

To avoid notational overhead, we will assume from now on that the operator $\tilde{D}^{X,nf}$ that we are trying to construct extends to all of $M \times \mathbb{R}$ (instead of just $M_0 \times \mathbb{R}$), i.e., to an operator $\tilde{D}^{X,nf}: \mathcal{S}(\mathcal{V}_{M \times \mathbb{R}}) \to \mathcal{S}(\mathcal{V}_{M \times \mathbb{R}})$, which in turn implies the same for the operator $\tilde{D}^{Y,nf}$ defined via (5.110). Since our conditions on $\tilde{D}^{X,nf}$ ensure that our state process $(X_t)_{t \geq 0}$, when started from a point $\mathbf{x}_0 \in M_0$, does not leave M_0 almost surely, the specific choice of this extension has no influence on the actual model behavior, and it will not be necessary to extend also the bond section $\mathfrak{B}_0^Y$ beyond M_0.

We begin our construction by focussing on a fully time-homogeneous model, and so as described at the end of Section 5.11 we will assume that $\tilde{D}^{X,nf}$ is really an operator $\tilde{D}^{X,nf}: \mathcal{S}(\mathcal{V}) \to \mathcal{S}(\mathcal{V})$ acting on time-independent sections $\psi(\mathbf{x})$, and whose action on a given time-dependent section $\psi \in \mathcal{S}(\mathcal{V}_{M \times \mathbb{R}})$ is naturally defined, by having it act on each time slice $\psi(\cdot, t)$ individually. In a last step, we will then introduce our functional noise parameter α to the model by considering the operator $\tilde{D}_\alpha^{X,nf} := \alpha \cdot \tilde{D}^{X,nf}$ instead, for some positive function $\alpha(\mathbf{x}, t)$ or $\alpha(\mathbf{x})$ depending on whether we want to allow for time inhomogeneity in our model or not.

6.1   Generators of Diffusion Processes

First let us understand what we need to ask of the operator $\tilde{D}^{X,nf}$ in order to ensure that all the operators $D^{X,T}\colon \mathcal{F}(M_{X,T}) \to \mathcal{F}(M_{X,T})$, $T \in \mathbb{R}$, given by (5.113) are indeed of the form (5.82), with a nondegenerate noise coefficient matrix $\tilde{\sigma}\tilde{\sigma}^T$. We begin by defining the general notion of a differential operator mapping one space of sections to another.

Definition 6.1. *An operator $D\colon \mathcal{S}(\mathcal{V}^1) \to \mathcal{S}(\mathcal{V}^2)$, for two line bundles $\mathcal{V}^1$ and $\mathcal{V}^2$ on M, is called a kth-order differential operator ($k \in \mathbb{N}$) if the operator $\frac{1}{\psi_2} \circ D \circ \psi_1\colon \mathcal{F}(M) \to \mathcal{F}(M)$ is a kth-order differential operator in the classical sense for some $\psi_1 \in \mathcal{S}_\varnothing(\mathcal{V}^1)$ and $\psi_2 \in \mathcal{S}_\varnothing(\mathcal{V}^2)$.*

Lemma 6.1. *$\tilde{D}^{X,nf}$ is a kth-order differential operator on M_0 if and only if the operators $D^{X,T}$ given by (5.113) are (classical) k^{th}-order differential operators on M_0 for $\forall T \in \mathbb{R}$.[1]*

Proof. See Appendix E.1.   □

This shows that in the language of Definition 6.1 we have to construct a second-order differential operator $\tilde{D}^{X,nf}$. The following lemma tells us how to ensure that the multiplicative (i.e., zeroth-order) terms of the operators $D^{X,T}$ vanish.

Lemma 6.2. *The multiplicative term in $D^{X,T}$ vanishes for $\forall T \in \mathbb{R}$ if and only if $\tilde{D}^{X,nf}$ fulfills the consistency requirement (5.125a).*

Proof. Denoting by 1 the constant function that takes the value 1 on all of M_0, the multiplicative term of $D^{X,T}$ is given by

$$D^{X,T}1 \overset{(5.99b)}{=} \tilde{D}^{X,T}1 \overset{(5.112)}{=} \left(\tfrac{1}{\mathfrak{B}_T^X} \circ \tilde{D}^{X,nf} \circ \mathfrak{B}_T^X\right)1 = \tfrac{1}{\mathfrak{B}_T^X} \cdot (\tilde{D}^{X,nf}\mathfrak{B}_T^X),$$

which vanishes if and only if $\tilde{D}^{X,nf}\mathfrak{B}_T^X = 0$.   □

Therefore, as long as our model fulfills the consistency requirement, we do not need to worry about the multiplicative terms.

Finally, we would like to rephrase also our second desired property, namely that the second-order coefficient matrices of the operators $D^{X,T}$ should be positive definite everywhere, in terms of $\tilde{D}^{X,nf}$. We therefore make the following definition.

Definition 6.2. *Given any second-order differential operator $D\colon \mathcal{S}(\mathcal{V}) \to \mathcal{S}(\mathcal{V})$, we define the matrix-valued function $A_D\colon M \to \mathbb{R}^{n \times n}$ as the symmetrized matrix containing the second-order coefficients of $\frac{1}{\psi} \circ D \circ \psi$, for any $\psi \in \mathcal{S}_\varnothing(\mathcal{V})$ such that this is a classical second-order differential operator.*

We call the operator D positive definite if A_D takes values in the subset $\mathbb{R}_+^{n \times n}$ of all positive-definite $(n \times n)$-matrices.

---

[1]To prove the direction "⇒", we have to assume that $\frac{\psi_1}{\mathfrak{B}_T^X}, \frac{\psi_2}{\mathfrak{B}_T^X} \in C^k(M_0, \mathbb{R})$ for $\forall T \in \mathbb{R}$, where $\psi_1, \psi_2 \in \mathcal{S}(\mathcal{V}_0)$ are sections associated to $\tilde{D}^{X,nf}$ via Definition 6.1. As usual, we will not obsess over such smoothness conditions here.

It is not hard to see that A_D indeed does not depend on the specific choice of ψ: For $\forall \psi, \psi' \in \mathcal{S}_\varnothing(\mathcal{V})$ and $\forall f \in C^2(M, \mathbb{R})$ we have

$$\left(\tfrac{1}{\psi'} \circ D \circ \psi'\right) f = \tfrac{\psi}{\psi'} \cdot \left(\tfrac{1}{\psi} \circ D \circ \psi\right)\left(\tfrac{\psi'}{\psi} \cdot f\right),$$

and after applying the product rule on the right, the only second-derivative terms of f that we obtain are those where no derivative of the operator $\left(\tfrac{1}{\psi} \circ D \circ \psi\right)$ acted on $\tfrac{\psi'}{\psi}$. As a result, the prefactor $\tfrac{\psi'}{\psi}$ of these second-derivative terms cancels with the factor $\tfrac{\psi}{\psi'}$, and so the second-derivative terms of $\left(\tfrac{1}{\psi'} \circ D \circ \psi'\right) f$ indeed coincide with those of $\left(\tfrac{1}{\psi} \circ D \circ \psi\right) f$.

Note that for any $f \in \mathcal{F}(M)$ we have

$$A_{f \cdot D} = f \cdot A_D. \tag{6.1}$$

Denoting the function associated to $\tilde{D}^{X,nf}$ by

$$A^X := A_{\tilde{D}^{X,nf}}, \tag{6.2}$$

because of (5.113) $A^X\big|_{M_0}$ is just the matrix of second-order coefficients of the classical generators $D^{X,T}$, which (as a well-known consequence of Girsanov's theorem) is independent of T.[2]

Note that we can extend this definition to time-dependent second-order differential operators $D: \mathcal{S}(\mathcal{V}_{M \times \mathbb{R}}) \to \mathcal{S}(\mathcal{V}_{M \times \mathbb{R}})$, which leads us to analogously define

$$A^Y := A_{\tilde{D}^{Y,nf}}. \tag{6.3}$$

In summary: We need to construct a second-order derivative operator $\tilde{D}^{X,nf}$ whose associated coefficient function A^X given by (6.2) is positive definite everywhere, and we need to ensure that the bond sections $\mathfrak{B}_T^X$ fulfill the consistency requirement (5.125a). This will ensure that all the operators $D^{X,T}$ given by (5.113) are the infinitesimal generators of a diffusion process with nondegenerate noise.

6.2 Some Technical Preparations

Before tackling the question of how to construct $\tilde{D}^{X,nf}$, $(\hat{U}_s)_{s \in \mathbb{R}}$, and $\mathfrak{B}_T^X$ such that the consistency requirement (5.125a) is fulfilled, we need to learn some necessary techniques that not every reader will be familiar with.

First, we will discuss the definition and key properties of tensor products of two or more vector spaces, since they will play a vital role in our construction of $\tilde{D}^{X,nf}$ in Section 6.4. Then we will review the action of the various flavors ("representations") $\hat{U}_s$, $\bar{U}_s$, and U_s of our evolution operators, and we will extend

---

[2]Indeed, by writing Girsanov's formula in a form analogous to (5.88a), this general invariance of the noise under a change of measure can be shown with arguments analogous to the preceding paragraph.

this list by adding further representations that act on operators, on bilinear forms, on sections on the dual line bundle, and on sections on tensor product bundles. Finally, we will introduce the reader to the concept of a conformal structure, and we will show how it can be interpreted as a line bundle as well.

6.2.1   Tensor Products

Definition and properties.   Given any two real vector spaces V_1 and V_2, the tensor product $\otimes\colon (v_1, v_2) \mapsto v_1 \otimes v_2$ between any two vectors $v_1 \in V_1$ and $v_2 \in V_2$ is a bilinear operation in the sense that

$$\alpha(v_1 \otimes v_2) = (\alpha v_1) \otimes v_2 = v_1 \otimes (\alpha v_2),$$
$$(v_1 \otimes v_2) + (v_1' \otimes v_2) = (v_1 + v_1') \otimes v_2,$$
$$(v_1 \otimes v_2) + (v_1 \otimes v_2') = v_1 \otimes (v_2 + v_2')$$

for $\forall \alpha \in \mathbb{R}$, $\forall v_1, v_1' \in V_1$, and $\forall v_2, v_2' \in V_2$. Sums $(v_1 \otimes v_2) + (v_1' \otimes v_2')$ between any two tensor products are then defined in an abstract way (using "abstract sums") that respects the commutative and the distributive laws (with scalar multiplication); however, such a sum does not necessarily need to be a tensor product of two vectors again. The space of all the objects obtained in this way—called two-tensors—is a vector space, denoted by $V_1 \otimes V_2$. This is the tensor product of the vector spaces V_1 and V_2.

For two chosen bases $e_1^1, \ldots, e_1^p$ and $e_2^1, \ldots, e_2^q$ of V_1 and V_2, respectively, the tensor products $e_1^i \otimes e_2^j$ for $i = 1, \ldots, p$ and $j = 1, \ldots, q$ form a basis of $V_1 \otimes V_2$, and so we have

$$\dim(V_1 \otimes V_2) = \dim(V_1) \cdot \dim(V_2). \tag{6.4}$$

The extension of the mapping $v_1 \otimes v_2 \mapsto r_1^T r_2$ to $V_1 \otimes V_2$, where r_1 and r_2 are the respective coefficient row vectors of v_1 and v_2 in these bases, respectively, is then a linear bijection from $V_1 \otimes V_2$ to $\mathbb{R}^{p \times q}$.

Tensor products $V_1 \otimes \cdots \otimes V_k$ of more than two vector spaces are defined analogously; objects in such spaces are called k-tensors. If $V_1 = \cdots = V_k = V$ then we write

$$V^{\otimes k} := \underbrace{V \otimes \cdots \otimes V}_{k \text{ times}}.$$

Dual spaces, one-dimensional vector spaces.   Of particular importance for us will be the tensor products involving only copies of some one-dimensional vector space V and its dual space

$$V^* := \{\lambda \colon V \to \mathbb{R} \mid \lambda \text{ linear}\},$$

i.e., tensor products of the form

$$(V^*)^{\otimes k_1} \otimes V^{\otimes k_2} = \underbrace{(V^* \otimes \cdots \otimes V^*)}_{k_1 \text{ times}} \otimes \underbrace{(V \otimes \cdots \otimes V)}_{k_2 \text{ times}}.$$

Therefore *let us assume for the rest of this section that* $\dim(V) = 1$.

Since by (6.4) tensor products $V_1 \otimes V_2$ of two one-dimensional vector spaces are one-dimensional themselves, all their elements are of the form $\alpha(e_1^1 \otimes e_2^1) = (\alpha e_1^1) \otimes e_2^1$, i.e., they are all tensor products. (Recall that in higher dimensions, elements of $V_1 \otimes V_2$ are generally *sums* of tensor products.)

Given any one-dimensional vector space V, its dual space V^* and by (6.4) also the space $V \otimes V^*$ are one-dimensional as well. The *tensor contraction* mapping

$$V^* \otimes V \ni \lambda \otimes v \mapsto \lambda(v) \in \mathbb{R} \tag{6.5}$$

is therefore a linear bijection from $V^* \otimes V$ to $\mathbb{R}$, i.e., we have

$$V^* \otimes V \cong \mathbb{R}$$

if $\dim(V) = 1$. Denoting

$$V^{\otimes(-k)} := \underbrace{V^* \otimes \cdots \otimes V^*}_{k \text{ times}} \quad \text{for } \forall k \in \mathbb{N}$$

$$\text{and} \quad V^{\otimes 0} := \mathbb{R}, \tag{6.6}$$

repeated application of this rule then implies that

$$(V^*)^{\otimes k_1} \otimes V^{\otimes k_2} \cong V^{\otimes k_2 - k_1}, \tag{6.7}$$

where we identify

$$(\lambda_1 \otimes \cdots \otimes \lambda_{k_1}) \otimes (x_1 \otimes \cdots \otimes x_{k_2})$$
$$\cong \begin{cases} \lambda_1(x_1) \cdots \lambda_{k_1}(x_{k_1}) \cdot (x_{k_1+1} \otimes \cdots \otimes x_{k_2}) & \text{if } k_1 < k_2, \\ \lambda_1(x_1) \cdots \lambda_{k_1}(x_{k_1}) & \text{if } k_1 = k_2, \\ \lambda_1(x_1) \cdots \lambda_{k_2}(x_{k_2}) \cdot (\lambda_{k_2+1} \otimes \cdots \otimes \lambda_{k_1}) & \text{if } k_1 > k_2. \end{cases} \tag{6.8}$$

(Note that because of the linearity of all the expressions involved and since $\dim(V) = 1$, we can reshuffle the order of the λ_i's and the x_i's at will, and so it does not matter which λ_i's we pair up with which x_i's, and which leftover factors we keep in the tensor product on the right, as long as each factor appears exactly once.)

In particular, we have $(V^*)^{\otimes k} \otimes V^{\otimes k} = V^{\otimes 0} = \mathbb{R}$, which implies that

$$(V^{\otimes k})^* = (V^*)^{\otimes k}, \tag{6.9}$$

where for any $\lambda = \lambda_1 \otimes \cdots \otimes \lambda_k \in (V^*)^{\otimes k}$ and any $x = x_1 \otimes \cdots \otimes x_k \in V^{\otimes k}$ one defines

$$\lambda(x) := \lambda_1(x_1) \cdots \lambda_k(x_k)$$
$$\overset{(6.8)}{=} (\lambda_1 \otimes \cdots \otimes \lambda_k) \otimes (x_1 \otimes \cdots \otimes x_k) = \lambda \otimes x. \tag{6.10}$$

We want to stress that none of this holds in higher dimensions.

Multiplication and division. Given a vector $v \in V \setminus \{0\}$, where we still assume that $\dim(V) = 1$, we denote by $\frac{1}{v}$ the unique functional $\lambda \in V^*$ such that $\lambda(v) = 1$, and we use the notation $\frac{v'}{v} := \frac{1}{v}(v')$ for $\forall v' \in V$. Note that clearly we have $\frac{1}{cv} = \frac{1}{c}\frac{1}{v}$ for $\forall c \in \mathbb{R} \setminus \{0\}$.

Given any other vector space W (with arbitrary dimension), we can then define the division operator $\frac{1}{v}\otimes$ and the multiplication operator $v\otimes$ as

$$\frac{1}{v}\otimes\colon\; W \qquad \to V^* \otimes W, \qquad\qquad w \mapsto \frac{1}{v} \otimes w,$$
$$v\otimes\colon\; V^* \otimes W \to V \otimes V^* \otimes W \cong W, \qquad \lambda \otimes w \mapsto v \otimes \lambda \otimes w \cong \lambda(v) \cdot w.$$

Then we have $\frac{1}{cv}\otimes = \frac{1}{c} \cdot (\frac{1}{v}\otimes)$ and $(cv)\otimes = c \cdot (v\otimes)$ for $\forall c \in \mathbb{R} \setminus \{0\}$ (as one would expect from multiplication and division operators), and since $\frac{1}{v}(v) = 1$, the concatenation of these two operators is the identity.

Note the essential fact that none of these definitions required the choice of a specific basis, which makes them well suited for our section-based framework.

Sections. All of the above definitions naturally extend to line bundles: Given any line bundle such as our wealth line bundle $\mathcal{V} = (V_\mathbf{x})_{\mathbf{x} \in M}$, the *dual line bundle* is the family $\mathcal{V}^* := (V_\mathbf{x}^*)_{\mathbf{x} \in M}$ of the dual vector spaces, and more generally, one can consider the line bundles $\mathcal{V}^{\otimes k} := (V_\mathbf{x}^{\otimes k})_{\mathbf{x} \in M}$ of the kth powers for $k \in \mathbb{Z}$. The case $k = 0$ is special: By our convention (6.6) we have $V_\mathbf{x}^{\otimes 0} = \mathbb{R}$ for $\forall \mathbf{x} \in M$, and so the sections $\psi \in \mathcal{S}(\mathcal{V}^{\otimes 0})$ are simply regular real-valued functions, i.e., we have

$$\mathcal{S}(\mathcal{V}^{\otimes 0}) = \mathcal{F}(M).$$

More generally, the tensor product between any two sections $\psi_1 \in \mathcal{S}(\mathcal{V}^1)$ and $\psi_2 \in \mathcal{S}(\mathcal{V}^2)$ on two different line bundles $\mathcal{V}^1 = (V_\mathbf{x}^1)_{\mathbf{x} \in M}$ and $\mathcal{V}^2 = (V_\mathbf{x}^2)_{\mathbf{x} \in M}$, respectively, is defined in a pointwise way, i.e.,

$$(\psi_1 \otimes \psi_2)(\mathbf{x}) := \psi_1(\mathbf{x}) \otimes \psi_2(\mathbf{x}),$$

and so $\psi_1 \otimes \psi_2$ is a section on the product line bundle, i.e., $\psi_1 \otimes \psi_2 \in \mathcal{S}(\mathcal{V}^1 \otimes \mathcal{V}^2)$, where $\mathcal{V}^1 \otimes \mathcal{V}^2 := (V_\mathbf{x}^1 \otimes V_\mathbf{x}^2)_{\mathbf{x} \in M}$.

Note that in the language introduced here, the action of the division operator $1/\mathfrak{B}_T^Y\colon \mathcal{S}(\mathcal{V}_T) \to \mathcal{F}(\mathcal{V}_T)$ defined in (5.77b) is really the tensor product multiplication by the section $1/\mathfrak{B}_T^Y \in \mathcal{S}(\mathcal{V}_T^*)$.

Signed vector spaces. When computing the tensor product between two or more *signed* one-dimensional vector spaces $V^1, \dots, V^k$, the natural way of choosing the positive half of the resulting product space is to set

$$(V^1 \otimes \cdots \otimes V^k)_+ := \left\{ v_1 \otimes \cdots \otimes v_k \,\middle|\, v_1 \in V_+^1, \dots, v_k \in V_+^k \right\}. \tag{6.11}$$

In particular, this implies that the positive half of the kth power ($k \in \mathbb{N}$) of a signed one-dimensional vector space V is given by

$$(V^{\otimes k})_+ = \{\, \underbrace{v \otimes \cdots \otimes v}_{k \text{ times}} \mid v \in V_+\}. \tag{6.12}$$

This can be generalized also to negative powers k by setting

$$(V^*)_+ := \left\{ \tfrac{1}{x} \,\middle|\, x \in V_+ \right\}. \tag{6.13}$$

Note, however, that the representations in (6.11)–(6.12) are not the only ways in which positive vectors in product spaces can be encountered, as we may for example multiply any two factors by -1 (resulting in non-positive vectors) without altering their product.

For sections on signed line bundles the representations (6.11)–(6.12) imply the following:

Lemma 6.3. *(i) The product of two or more positive sections is positive.*

(ii) Given any signed line bundle such as $\mathcal{V}$, we have

$$\forall k \in \mathbb{N}: \quad \mathcal{S}_+(\mathcal{V}^{\otimes k}) = \{ \underbrace{\psi \otimes \cdots \otimes \psi}_{k \text{ times}} \,|\, \psi \in \mathcal{S}_+(\mathcal{V}) \}, \tag{6.14}$$

$$\forall k = 2, 4, 6, \ldots : \quad \mathcal{S}_+(\mathcal{V}^{\otimes k}) = \{ \underbrace{\psi \otimes \cdots \otimes \psi}_{k \text{ times}} \,|\, \psi \in \mathcal{S}_\varnothing(\mathcal{V}) \}. \tag{6.15}$$

Proof. Part (i) and (6.14) are direct consequences of (6.11)–(6.12). To see (6.15), observe that if for any given $\psi \in \mathcal{S}_\varnothing(\mathcal{V})$ we define the function $f(\mathbf{x}) := \text{sign}(\psi(\mathbf{x}))$, we have $f\psi \in \mathcal{S}_+(\mathcal{V})$ and therefore $\psi \otimes \cdots \otimes \psi = f^k \cdot (\psi \otimes \cdots \otimes \psi) = (f\psi) \otimes \cdots \otimes (f\psi) \in \mathcal{S}_+(\mathcal{V}^{\otimes k})$ by (6.14). □

6.2.2 Representations of the Evolution Operators

Recap. In Chapter 5 we have introduced a group $(\hat{U}_s)_{s \in \mathbb{R}}$ of operators $\hat{U}_s$ acting on sections $\psi \in \mathcal{S}(\mathcal{V})$, which encoded the information for operators acting on other mathematical objects as well, namely the group $(\bar{U}_s)_{s \in \mathbb{R}}$ of operators acting on functions $f \in \mathcal{F}(M)$, and the group $(U_s)_{s \in \mathbb{R}}$ of operators U_s acting on states $\mathbf{x} \in M$. These three operations then gave further rise to operations $\hat{U}$, $\bar{U}$, and U acting on the corresponding time-dependent objects $\psi \in \mathcal{S}(\mathcal{V}_{M \times \mathbb{R}})$, $f \in \mathcal{F}(M \times \mathbb{R})$, and $(\mathbf{y}, t) \in M \times \mathbb{R}$; they are generally used to convert objects from the forward to the spot parametrization. This conversion is consistent in the sense that one can carry out any calculation in either parametrization and arrive at corresponding results; for example, writing $(\mathbf{x}, t) = U(\mathbf{y}, t)$ and $f_X = \bar{U} f_Y$, we have $f_X(\mathbf{x}, t) = f_Y(\mathbf{y}, t)$ (see (5.33a) and (5.41)–(5.42)).

Representations of U_s. Another way of viewing this setup is to think of the operators U_s and $\bar{U}_s$ as the actions of $\hat{U}_s$ on other mathematical objects (namely, functions and states). Although these two operators contain less information than $\hat{U}_s$ itself, they are simply called different *representations* of $\hat{U}_s$. Going one step further, one may in fact use the same symbol U_s for all of these mappings, as well as the symbol U for their time-dependent counterparts. This does not lead to any ambiguities, since one can always see from the context (i.e., from the object that the operator acts on) which flavor of the operator one intends to use. In this simplified notation, the rules (5.58) and (5.71) can be written as

$$U_s(f \cdot g) = (U_s f) \cdot (U_s g) \qquad\qquad (6.16)$$

$$\text{and} \qquad U_s(f \cdot \psi) = (U_s f) \cdot (U_s \psi) \qquad\qquad (6.17)$$

for $\forall f, g \in \mathcal{F}(M)$ and $\forall \psi \in \mathcal{S}(\mathcal{V})$, i.e., the action of U_s (or rather, of the family of representations of U_s) is consistent w.r.t. the multiplication between functions, and w.r.t. the multiplication of sections with functions:

$$
\begin{array}{ccc}
f, g \xrightarrow{\;\;\cdot\;\;} f \cdot g & \qquad & f, \psi \xrightarrow{\;\;\cdot\;\;} f \cdot \psi \\
\Big\downarrow{\scriptstyle U_s} \qquad \Big\downarrow{\scriptstyle U_s} & & \Big\downarrow{\scriptstyle U_s} \qquad \Big\downarrow{\scriptstyle U_s} \\
U_s f, U_s g \xrightarrow{\;\;\;\;} (6.16) & & U_s f, U_s \psi \xrightarrow{\;\;\;\;} (6.17)
\end{array}
$$

Furthermore, defining the representation of U_s on real numbers as the identity, i.e.,

$$\forall a \in \mathbb{R}: \quad U_s(a) := a, \qquad\qquad (6.18)$$

(5.44a) can be written as

$$U_s(f(\mathbf{x})) = (U_s f)(U_s(\mathbf{x})) \qquad\qquad (6.19)$$

for $\forall f \in \mathcal{F}(M)$ and $\mathbf{x} \in M$, i.e., U_s is consistent w.r.t. the operation of evaluating a function at some argument:

$$
\begin{array}{ccc}
f, \mathbf{x} \xrightarrow{\;\;\text{evaluation}\;\;} f(\mathbf{x}) \\
\Big\downarrow{\scriptstyle U_s} \qquad\qquad \Big\downarrow{\scriptstyle U_s} \\
U_s f, U_s \mathbf{x} \xrightarrow{\;\;\text{evaluation}\;\;} (6.19)
\end{array}
$$

We will now start defining the action of U_s (and thus of U) also on other mathematical objects, always driven by the goal of making these definitions consistent w.r.t. as many operations as possible, as this will allow us to carry out each of these operations in either parametrization and arrive at corresponding results. In the process, we may at times still use our old notation $\hat{U}_s$ and $\bar{U}_s$ for clarity, but we will then slowly adjust to our new unified notation, denoting all representations simply by U_s.

Action on sections on dual line bundles. We will begin by deriving from the action of $\hat{U}_s$ on sections $\psi \in \mathcal{S}(\mathcal{V})$ an action on sections $\psi^* \in \mathcal{S}(\mathcal{V}^*)$ on the dual line bundle.

Given any line bundle such as $\mathcal{V}$, sections $\psi^* \in \mathcal{S}(\mathcal{V}^*)$ on the dual line bundle (i.e., mappings that assign to each $\mathbf{x} \in M$ an element $\psi^*(\mathbf{x}) \in V_{\mathbf{x}}^*$) can equivalently be thought of as mappings

$$\tilde{\psi}^* : \mathcal{S}(\mathcal{V}) \to \mathcal{F}(M), \qquad (\tilde{\psi}^*(\psi))(\mathbf{x}) := (\psi^*(\mathbf{x}))(\psi(\mathbf{x}))$$

that have to fulfill

$$\forall \psi \in \mathcal{S}(\mathcal{V}) \ \forall f \in \mathcal{F}(M): \quad \tilde{\psi}^*(f \cdot \psi) = f \cdot \tilde{\psi}^*(\psi).$$

(The latter condition is due to the linearity of the elements in the dual spaces $V_{\mathbf{x}}^*$.) Simply writing ψ^* instead $\tilde{\psi}^*$ from now on, we want our representation of U acting on such sections to be consistent w.r.t. this equivalence, and so it should fulfill the relation

$$\bar{U}_s(\psi^*(\psi)) = (U_s\psi^*)(\hat{U}_s\psi), \tag{6.20}$$

which is illustrated by the following diagram:

$$
\begin{array}{ccc}
\psi^*, \psi & \xrightarrow{\ \text{evaluation}\ } & \psi^*(\psi) \\
\ \downarrow{\scriptstyle U_s} & & \ \downarrow{\scriptstyle U_s} \\
U_s\psi^*, \, U_s\psi & \xrightarrow{\ \text{evaluation}\ } & (6.20)
\end{array}
$$

Substituting $\psi' := \hat{U}_s\psi$, we therefore define

$$(U_s\psi^*)(\psi') := \bar{U}_s(\psi^*(\hat{U}_{-s}\psi')) \tag{6.21}$$

for $\forall \psi' \in \mathcal{S}(\mathcal{V})$.

Observe that (just like $\hat{U}_s$) this new representation $U_s \colon \mathcal{S}(\mathcal{V}^*) \to \mathcal{S}(\mathcal{V}^*)$ defined in (6.21) is linear and fulfills the group property

$$\forall s_1, s_2 \in \mathbb{R} \ \forall \psi^* \in \mathcal{S}(\mathcal{V}^*): \quad U_{s_1}(U_{s_2}\psi^*) = U_{s_1+s_2}\psi^*,$$

as well as

$$U_s(f \cdot \psi^*) = (\bar{U}_s f) \cdot (U_s\psi^*). \tag{6.22}$$

Furthermore, note that if $\psi^* = \frac{1}{\psi''}$ for some $\psi'' \in \mathcal{S}_{\varnothing}(\mathcal{V})$, then we have $\psi^*(\psi) = \frac{\psi}{\psi''}$ and thus $(U_s\psi^*)(\psi') = \bar{U}_s\big(\frac{\hat{U}_{-s}\psi'}{\psi''}\big) = \frac{\psi'}{\hat{U}_s\psi''}$ by (6.21) and (5.72), i.e.,

$$U_s\left(\frac{1}{\psi''}\right) = \frac{1}{\hat{U}_s\psi''}. \tag{6.23}$$

This makes our definition of U_s consistent w.r.t. the operation of taking the inverse of a non-zero section:

$$
\begin{array}{ccc}
\psi & \xrightarrow{\ \text{inversion}\ } & \frac{1}{\psi} \\
\ \downarrow{\scriptstyle U_s} & & \ \downarrow{\scriptstyle U_s} \\
U_s\psi & \xrightarrow{\ \text{inversion}\ } & (6.23)
\end{array}
$$

Finally, we can use this group of operators $U_s \colon \mathcal{S}(\mathcal{V}^*) \to \mathcal{S}(\mathcal{V}^*)$ to define a representation $U \colon \mathcal{S}(\mathcal{V}_{M\times\mathbb{R}}^*) \to \mathcal{S}(\mathcal{V}_{M\times\mathbb{R}}^*)$ as

$$(U\psi^*)(\mathbf{x}, t) := (U_t\psi^*(\cdot, t))(\mathbf{x}),$$

which fulfills the time-dependent analogues of (6.22) and (6.23).

Action on tensor products. Suppose we are given two line bundles $\mathcal{V}^1$ and $\mathcal{V}^2$ on M (these line bundles could be our wealth line bundle $\mathcal{V}$ or its dual, but we will soon apply this technique to other line bundles as well), and that there is a linear action of U_s defined on the associated section spaces $\mathcal{S}(\mathcal{V}^1)$ and $\mathcal{S}(\mathcal{V}^2)$. Then we can define an action on sections in $\mathcal{S}(\mathcal{V}^1 \otimes \mathcal{V}^2)$ as

$$U_s(\psi_1 \otimes \psi_2) := (U_s\psi_1) \otimes (U_s\psi_2) \tag{6.24}$$

for $\forall \psi_1 \in \mathcal{S}(\mathcal{V}^1)$ and $\forall \psi_2 \in \mathcal{S}(\mathcal{V}^2)$, where the linearity of U_s ensures that this representation is well defined and linear itself. In this way, the action of U_s is consistent w.r.t. the operation of taking tensor products:

$$
\begin{array}{ccc}
\psi_1, \psi_2 & \xrightarrow{\ \otimes\ } & \psi_1 \otimes \psi_2 \\
\downarrow{\scriptstyle U_s} & & \downarrow{\scriptstyle U_s} \\
U_s\psi_1, U_s\psi_2 & \xrightarrow{\ \otimes\ } & (6.24)
\end{array}
$$

This definition can be extended to tensor products with more than just two factors, which in particular leads us to representations of U_s acting on the sections in any space $\mathcal{S}(\mathcal{V}^{\otimes k})$, $k \in \mathbb{Z}$. Note in this context that if $\mathcal{V}^1 := \mathcal{V}^*$ and $\mathcal{V}^2 := \mathcal{V}$ then for any $\psi^* \in \mathcal{S}(\mathcal{V}^*)$ and any $\psi \in \mathcal{S}(\mathcal{V})$ we have

$$U_s(\psi^* \otimes \psi) = (U_s\psi^*) \otimes (U_s\psi) \cong (U_s\psi^*)(U_s\psi),$$

and so our definition is consistent with the contraction (6.5) of tensor products:

$$
\begin{array}{ccc}
\psi^* \otimes \psi & \xrightarrow{\ \cong\ } & \psi^*(\psi) \\
\downarrow{\scriptstyle U_s} & & \downarrow{\scriptstyle U_s} \\
U_s(\psi^* \otimes \psi) & \xrightarrow{\ \cong\ } & (6.24) + (6.20)
\end{array}
$$

Again, this definition naturally gives rise to a time-dependent representation $U \colon \mathcal{S}(\mathcal{V}^1_{M \times \mathbb{R}} \otimes \mathcal{V}^2_{M \times \mathbb{R}}) \to \mathcal{S}(\mathcal{V}^1_{M \times \mathbb{R}} \otimes \mathcal{V}^2_{M \times \mathbb{R}})$ fulfilling the time-dependent counterpart of (6.24), and to similar representations on spaces consisting of more than just two factors.

Action on operators. Similarly to the conversion of states, functions, and sections, we have also derived rules to consistently convert *operators* from the forward to the spot parametrization: Equations (5.96a)–(5.96b) describe how to convert operators acting on functions (such as the two components ∇^Y_t and $\tilde{D}^{Y,T}$ of our classical Kolmogorov operators), while (5.110) and (5.114) describe how to convert operators acting on sections (such as the two components $\nabla^{Y,nf}_t$ and $\tilde{D}^{Y,nf}$ of our numeraire-free Kolmogorov operators).

To express this conversion using our evolution operators, let us define the representations U_s acting on operators $D_{\text{fct}} \colon \mathcal{F}(M) \to \mathcal{F}(M)$ and

$D_{\text{sec}} \colon \mathcal{S}(\mathcal{V}) \to \mathcal{S}(\mathcal{V})$ as

$$U_s D_{\text{fct}} := \bar{U}_s \circ D_{\text{fct}} \circ \bar{U}_{-s} \tag{6.25a}$$

$$\text{and} \quad U_s D_{\text{sec}} := \hat{U}_s \circ D_{\text{sec}} \circ \hat{U}_{-s}, \tag{6.25b}$$

respectively, which ensures that

$$U_s(D_{\text{fct}} f) = (U_s D_{\text{fct}})(U_s f) \tag{6.26a}$$

$$\text{and} \quad U_s(D_{\text{sec}} \psi) = (U_s D_{\text{sec}})(U_s \psi) \tag{6.26b}$$

for $\forall f \in \mathcal{F}(M)$ and $\forall \psi \in \mathcal{S}(\mathcal{V})$, i.e., that this representation is consistent w.r.t. the operation of applying an operator to a function or section:

$$
\begin{array}{ccc}
D_{\text{fct}}, f \xrightarrow{\text{apply operator}} D_{\text{fct}} f & \qquad & D_{\text{sec}}, \psi \xrightarrow{\text{apply operator}} D_{\text{sec}} \psi \\
\downarrow{\scriptstyle U_s} \qquad \qquad \downarrow{\scriptstyle U_s} & & \downarrow{\scriptstyle U_s} \qquad \qquad \downarrow{\scriptstyle U_s} \\
U_s D_{\text{fct}}, U_s f \xrightarrow{\text{apply operator}} (6.26a) & & U_s D_{\text{sec}}, U_s \psi \xrightarrow{\text{apply operator}} (6.26b)
\end{array}
$$

Lemma 6.4. *The two representations of U_s defined in (6.25) are linear and again form groups, and they map differential operators to differential operators of the same degree.*

Proof. The linearity and the group property follow right from their definitions in (6.25). The last statement is also easy to see from (6.25a) for the action on function-based operators D_{fct}; the proof for the action on section-based operators D_{sec} can be found in Appendix E.2. $\qquad \square$

Finally, defining the time-dependent counterparts U acting on operators $D_{\text{fct}} \colon \mathcal{F}(M \times \mathbb{R}) \to \mathcal{F}(M \times \mathbb{R})$ and $D_{\text{sec}} \colon \mathcal{S}(\mathcal{V}_{M \times \mathbb{R}}) \to \mathcal{S}(\mathcal{V}_{M \times \mathbb{R}})$ as

$$U D_{\text{fct}} := \bar{U} \circ D_{\text{fct}} \circ \bar{U}^{-1} \tag{6.27a}$$

$$\text{and} \quad U D_{\text{sec}} := \hat{U} \circ D_{\text{sec}} \circ \hat{U}^{-1}, \tag{6.27b}$$

the time-dependent counterparts of (6.26) hold as well, and so (5.96a) and (5.110) can then be written as

$$\tilde{D}^{X,T} = U \tilde{D}^{Y,T}$$

$$\text{and} \quad \tilde{D}^{X,nf} = U \tilde{D}^{Y,nf}. \tag{6.28}$$

Action on bilinear forms. A bilinear form m on M is a family of bilinear maps $\langle \cdot, \cdot \rangle_{\mathbf{x}}^m$ on $\mathbb{R}^n$ (one for each $\mathbf{x} \in M$), or more precisely, on the tangent spaces $T_{\mathbf{x}} M$. If such a map m is positive definite then it induces a metric on M, i.e., it can be used to measure the length of curves $\gamma \subset M$ via the formula

$$|\gamma|_m := \int_\gamma \sqrt{\langle \mathrm{d}\mathbf{x}, \mathrm{d}\mathbf{x} \rangle_{\mathbf{x}}^m}. \tag{6.29}$$

Note, however, that not all metrics are derived from positive-definite bilinear forms.

Since the image under U_s of an infinitesimal curve segment $[\mathbf{x}, \mathbf{x} + d\mathbf{x}]$ in M is $[U_s(\mathbf{x}), U_s(\mathbf{x}) + \nabla U_s(\mathbf{x})\, d\mathbf{x}]$, we would like to define a new bilinear form $U_s m$ as the unique bilinear form that fulfills

$$\left\langle \nabla U_s(\mathbf{x})\, d\mathbf{x}_1, \nabla U_s(\mathbf{x})\, d\mathbf{x}_2 \right\rangle_{U_s(\mathbf{x})}^{U_s m} = \left\langle d\mathbf{x}_1, d\mathbf{x}_2 \right\rangle_{\mathbf{x}}^{m}, \tag{6.30}$$

since this would make this representation of U_s consistent with respect to the operation of evaluating a bilinear form at infinitesimal curve segments:

$$
\begin{array}{ccc}
m,\, \mathbf{x},\, d\mathbf{x}_{1,2} & \xrightarrow{\quad \text{evaluate form} \quad} & \langle d\mathbf{x}_1, d\mathbf{x}_2 \rangle_{\mathbf{x}}^{m} \\[2pt]
\Big\downarrow{\scriptstyle U_s} & & \Big\downarrow{\scriptstyle U_s} \\[2pt]
U_s m,\, U_s(\mathbf{x}),\, \nabla U_s(\mathbf{x})\, d\mathbf{x}_{1,2} & \xrightarrow{\quad \text{evaluate form} \quad} & (6.30) + (6.18)
\end{array}
$$

By replacing the curve segments $[\mathbf{x}, \mathbf{x} + d\mathbf{x}_i]$ by their images under U_{-s}, we find that (6.30) requires us to define $U_s m$ as

$$\langle d\mathbf{x}_1, d\mathbf{x}_2 \rangle_{\mathbf{x}}^{U_s m} := \left\langle \nabla U_{-s}(\mathbf{x})\, d\mathbf{x}_1, \nabla U_{-s}(\mathbf{x})\, d\mathbf{x}_2 \right\rangle_{U_{-s}(\mathbf{x})}^{m}. \tag{6.31}$$

Note that the operators U_s defined in (6.31) form a group, and that they are linear in the sense that

$$
\begin{aligned}
U_s(m_1 + m_2) &= U_s m_1 + U_s m_2\,, \\
U_s(f \cdot m) &= U_s f \cdot U_s m
\end{aligned}
\tag{6.32}
$$

for any two bilinear forms m_1, m_2, and for any function $f \in \mathcal{F}(M)$.[3] Furthermore, since the matrices $\nabla U_{-s}(\mathbf{x})$ in (6.31) are invertible, with inverse

$$\nabla U_{-s}(\mathbf{x})^{-1} = \nabla U_s(U_{-s}(\mathbf{x})) \tag{6.33}$$

(this is shown by differentiating the relation $U_s(U_{-s}(\mathbf{x})) = \mathbf{x}$ with respect to $\mathbf{x}$), $U_s m$ is positive definite if and only if m is, and so

the operators U_s map the subset of all *positive-definite* bilinear forms onto itself. $\tag{6.34}$

Finally, as usual, this representation U_s extends to a map U acting on time-dependent bilinear forms.

Action on coefficient functions. In Section 6.1, Definition 6.2 had associated to any second-order differential operator $D \colon \mathcal{S}(\mathcal{V}) \to \mathcal{S}(\mathcal{V})$ a symmetric matrix-valued function $A_D \colon M \to \mathbb{R}^{n \times n}$, and in (6.25b) we have defined a representation of U_s acting on such operators D. We would now like to define a

---

[3] The sum of two bilinear forms m_1 and m_2 is defined as $\langle \cdot, \cdot \rangle_{\mathbf{x}}^{m_1 + m_2} := \langle \cdot, \cdot \rangle_{\mathbf{x}}^{m_1} + \langle \cdot, \cdot \rangle_{\mathbf{x}}^{m_2}$; the product of a function $f \in \mathcal{F}_+(M)$ and a bilinear form m is defined as $\langle \cdot, \cdot \rangle_{\mathbf{x}}^{f \cdot m} := f(\mathbf{x}) \cdot \langle \cdot, \cdot \rangle_{\mathbf{x}}^{m}$.

representation of U_s acting on matrix-valued functions $A \colon M \to \mathbb{R}^{n \times n}$ that is consistent with the operation of extracting the coefficient function from a given second-order differential operator, i.e.,

$$U_s A_D = A_{U_s D}. \tag{6.35}$$

$$
\begin{array}{ccc}
D & \xrightarrow{\;\text{extract coefficients}\;} & A_D \\
\downarrow{\scriptstyle U_s} & & \downarrow{\scriptstyle U_s} \\
U_s D & \xrightarrow{\;\text{extract coefficients}\;} & (6.35)
\end{array}
$$

Lemma 6.5. *Assuming that the operators U_s are C^2, to achieve the relation (6.35) we need to define*

$$(U_s A)(\mathbf{x}) := \nabla U_s(\mathbf{y}) \, A(\mathbf{y}) \, \nabla U_s(\mathbf{y})^T, \quad \text{where } \mathbf{y} := U_{-s}(\mathbf{x}), \tag{6.36}$$

for $\forall s \in \mathbb{R}$ and $\forall \mathbf{x} \in M$.

Proof. See Appendix E.3. $\qquad \square$

Note that the representation of U_s defined in (6.36) is linear, forms a group (this easily follows from (6.35) and the group property of the representation acting on operators D), and fulfills

$$U_s(f \cdot A) = (U_s f) \cdot (U_s A) \tag{6.37}$$

for any $f \in \mathcal{F}(M)$, which means that it is consistent w.r.t. the multiplication of A by a function.

The corresponding operator U acting on time-dependent matrix-valued functions $A(\mathbf{x}, t)$ fulfills the counterpart $U A_D = A_{UD}$ of (6.35) for time-dependent operators D, and applying this formula to $D = \tilde{D}^{Y,nf}$ tells us that by (6.2)–(6.3) and (6.28) we have

$$A^X = U A^Y. \tag{6.38}$$

This action of U_s on matrix-valued functions is also consistent with its action on bilinear forms in the following sense. First note that any function $A \colon M \to \mathbb{R}^{n \times n}_+$ with values in the set of positive-definite matrices defines a positive-definite bilinear form m_A on M via

$$\langle d\mathbf{x}_1, d\mathbf{x}_2 \rangle^{m_A}_{\mathbf{x}} := \langle d\mathbf{x}_1, A(\mathbf{x})^{-1} d\mathbf{x}_2 \rangle_{\mathbb{R}^n}, \tag{6.39}$$

where $\langle \cdot, \cdot \rangle_{\mathbb{R}^n}$ on the right-hand side denotes the regular Euclidean inner product; in particular, this associates to every positive-definite second-order differential operator $D \colon \mathcal{S}(\mathcal{V}) \to \mathcal{S}(\mathcal{V})$ a positive-definite bilinear form m_{A_D}.

Lemma 6.6. *For any function $A\colon M \to \mathbb{R}^{n \times n}_+$ the consistency relation*

$$U_s m_A = m_{U_s A} \qquad (6.40)$$

holds, which is illustrated by the following diagram:

$$
\begin{array}{ccc}
A & \xrightarrow{\ \text{define bilinear form}\ } & m_A \\
\downarrow{\scriptstyle U_s} & & \downarrow{\scriptstyle U_s} \\
U_s A & \xrightarrow{\ \text{define bilinear form}\ } & (6.40)
\end{array}
$$

Proof. See Appendix E.4. $\qquad\qquad\qquad\qquad\qquad\qquad\qquad$ $\square$

6.2.3   Conformal Structures

Definition and purpose. A *conformal structure* c on M is an equivalence class of metrics ν on M that are positive multiples of one another, i.e.,[4]

$$\nu_1 \sim \nu_2 \qquad \Leftrightarrow \qquad \nu_1 = f \cdot \nu_2 \quad \text{for some } f \in \mathcal{F}_+(M).$$

However, since in our construction we will only encounter metrics that are derived from positive-definite bilinear forms as in (6.29), we will instead think of a conformal structure c as an equivalence class of *positive-definite bilinear forms* that are positive multiples of one another,[5] i.e.,

$$m_1 \sim m_2 \qquad \Leftrightarrow \qquad m_1 = f \cdot m_2 \quad \text{for some } f \in \mathcal{F}_+(M).$$

While the purpose of a metric is to define curve lengths, a conformal class uniquely specifies *angles* between two curves: Indeed, given any two infinitesimal curve segments $[\mathbf{x}, \mathbf{x} + d\mathbf{x}_1]$ and $[\mathbf{x}, \mathbf{x} + d\mathbf{x}_2]$, the angle between the two under a positive-definite bilinear form m,

$$\angle(d\mathbf{x}_1, d\mathbf{x}_2)^m_{\mathbf{x}} := \arccos\left(\frac{\langle d\mathbf{x}_1, d\mathbf{x}_2 \rangle^m_{\mathbf{x}}}{|d\mathbf{x}_1|^m_{\mathbf{x}}|d\mathbf{x}_2|^m_{\mathbf{x}}}\right), \quad \text{where } |d\mathbf{x}|^m_{\mathbf{x}} := \sqrt{\langle d\mathbf{x}, d\mathbf{x} \rangle^m_{\mathbf{x}}},$$

does not change if m is replaced by $f \cdot m$, and so it is the same for all members m of a given conformal structure c.

Representation of U_s acting on conformal structures. The representation of U_s acting on bilinear forms m on M gives further rise to a representation acting on conformal structures, defined as

$$U_s c := \{U_s m \mid m \in c\}.$$

This image is indeed a conformal structure because (i) by (6.34) $U_s m$ is positive definite whenever m is, and (ii) we have $U_s m_1 \sim U_s m_2 \Leftrightarrow U_s m_1 = f \cdot U_s m_2$ for some $f \in \mathcal{F}_+(M) \Leftrightarrow m_1 = (U_{-s}f) \cdot m_2$ for some f by (6.32) $\Leftrightarrow m_1 \sim m_2$.

---

[4] For simplicity, we again decide to not get specific about any smoothness requirements for f.

[5] The use of bilinear forms will be more natural in our construction since the scaling factor f between our bilinear forms of interest will turn out to be just the one we need, while the factor $\sqrt{f}$ between the associated metrics will not work for us.

Conformal maps. A map such as $U_s \colon M \to M$ is called *conformal* w.r.t. a conformal structure c if its induced representation acting on conformal structures fulfills

$$U_s c = c,$$

i.e., if for $\forall m \in c$ we have $U_s m = f \cdot m$ for some $f \in \mathcal{F}_+(M)$. Using (6.31) to write this more explicitly as

$$f(\mathbf{x}) \cdot \langle \mathrm{d}\mathbf{x}_1, \mathrm{d}\mathbf{x}_2 \rangle^m_{\mathbf{x}} = \langle \mathrm{d}\mathbf{x}_1, \mathrm{d}\mathbf{x}_2 \rangle^{f \cdot m}_{\mathbf{x}} = \langle \mathrm{d}\mathbf{x}_1, \mathrm{d}\mathbf{x}_2 \rangle^{U_s m}_{\mathbf{x}}$$
$$= \big\langle \nabla U_{-s}(\mathbf{x}) \, \mathrm{d}\mathbf{x}_1, \nabla U_{-s}(\mathbf{x}) \, \mathrm{d}\mathbf{x}_2 \big\rangle^m_{U_{-s}(\mathbf{x})}$$

shows that this is the case if and only if for $\forall m \in c$ we have

$$\angle (\mathrm{d}\mathbf{x}_1, \mathrm{d}\mathbf{x}_2)^m_{\mathbf{x}} = \angle \big(\nabla U_{-s}(\mathbf{x}) \, \mathrm{d}\mathbf{x}_1, \nabla U_{-s}(\mathbf{x}) \, \mathrm{d}\mathbf{x}_2 \big)^m_{U_{-s}(\mathbf{x})},$$

i.e., if the angle under c between any two curves $\gamma_1, \gamma_2 \subset M$ is the same as the angle between the images $U_{-s}\gamma_1$ and $U_{-s}\gamma_2$ of these curves under U_{-s}. Now replacing the curves γ_i by $U_s\gamma_i$ for $i = 1, 2$, we arrive at the following statement:

U_s *is conformal if and only if it preserves the angles under c between any two curves $\gamma_1, \gamma_2 \subset M$.*

Conformal structures as line bundles. There is an interesting alternative viewpoint on conformal structures: Let c be some fixed conformal structure on M, let $m_0 \in c$, and for any $\mathbf{x} \in M$ let $W_{\mathbf{x}}$ be the one-dimensional vector space of bilinear maps on $T_{\mathbf{x}}M = \mathbb{R}^n$ given by

$$W_{\mathbf{x}} := \big\{ a \cdot \langle \cdot, \cdot \rangle^{m_0}_{\mathbf{x}} \,\big|\, a \in \mathbb{R} \big\}.$$

Since the spaces $W_{\mathbf{x}}$ do not depend on our choice of $m_0 \in c$, this associates to each conformal structure c a line bundle $\mathcal{W}_c := (W_{\mathbf{x}})_{\mathbf{x} \in M}$.

Note that we allow for non-positive values of a as well, so that only half of the space $W_{\mathbf{x}}$ consists of bilinear maps that are positive definite. This provides us with a natural way of making $\mathcal{W}_c$ a *signed* line bundle, by defining

$$W_{\mathbf{x}}^+ := \big\{ a \cdot \langle \cdot, \cdot \rangle^{m_0}_{\mathbf{x}} \,\big|\, a > 0 \big\}$$
$$= \big\{ \langle \cdot, \cdot \rangle \in W_{\mathbf{x}} \,\big|\, \langle \cdot, \cdot \rangle \text{ is positive definite} \big\}.$$

Sections on $\mathcal{W}_c$ assign to each $\mathbf{x} \in M$ a bilinear map on $T_{\mathbf{x}}M = \mathbb{R}^n$, and so $\mathcal{S}(\mathcal{W}_c)$ is a space of bilinear forms on M, given by

$$\mathcal{S}(\mathcal{W}_c) = \{ f \cdot m_0 \,|\, f \in \mathcal{F}(M) \}. \tag{6.41}$$

The subset $\mathcal{S}_+(\mathcal{W}_c)$ of all positive sections on $\mathcal{W}_c$ consists of all those bilinear forms that assign to each $\mathbf{x} \in M$ a *positive-definite* bilinear map, i.e., it consists of all the positive-definite bilinear forms $m \in \mathcal{S}(\mathcal{W}_c)$; in other words we have

$$\mathcal{S}_+(\mathcal{W}_c) = \{ f \cdot m_0 \,|\, f \in \mathcal{F}_+(M) \}$$
$$= \{ m \in \mathcal{S}(\mathcal{W}_c) \,|\, m \text{ is positive definite} \}$$
$$= c.$$

In summary: Any conformal structure c on M defines a signed line bundle $\mathcal{W}_c$ on M. The sections on $\mathcal{W}_c$ are bilinear forms on M, and we have $\mathcal{S}_+(\mathcal{W}_c) = c$.

6.3   The Scaling Assumption

Let us now get back to our actual task of devising a strategy for constructing $\tilde{D}^{X,nf}$, $(\hat{U}_s)_{s\in\mathbb{R}}$, and $\mathfrak{B}_T^X$ such that all the requirements on our framework (in particular the consistency requirement) are met.

6.3.1   Motivation and Definition

Using (5.75), the consistency requirement (5.125a) can be rewritten as

$$0 = \tilde{D}^{X,nf}\mathfrak{B}_T^X = \tilde{D}^{X,nf}\hat{U}_{t-T}\mathfrak{B}_0^Y$$

for $\forall T \in \mathbb{R}$ and $\forall t \leq T$, i.e., as

$$\forall\tau \leq 0: \ \left(\tilde{D}^{X,nf} \circ \hat{U}_\tau\right)\mathfrak{B}_0^Y = 0. \tag{6.42}$$

Satisfying it is a task that is hard to fulfill, since these are infinitely many operator equations that the section $\mathfrak{B}_0^Y$ has to fulfill. If, however, we make the additional purely technical *scaling assumption*

$$\forall\tau \in \mathbb{R}: \quad \tilde{D}^{X,nf} \circ \hat{U}_\tau = d \cdot (\hat{U}_\tau \circ \tilde{D}^{X,nf}) \tag{6.43}$$

for some scalar-valued function $d \in \mathcal{F}_+(M \times \mathbb{R})$,[6] then (6.42) holds if and only if

$$\tilde{D}^{X,nf}\mathfrak{B}_0^Y = 0. \tag{6.44}$$

While this additional assumption may restrict the generality of our model, it is our way of getting a handle on the problem.

This splits our construction into two tasks that are more manageable: First construct $\tilde{D}^{X,nf}$ and $(\hat{U}_s)_{s\in\mathbb{R}}$ such that (6.43) holds for some function d, and then find a section $\mathfrak{B}_0^Y$ that solves (6.44).

6.3.2   Consequences

Let us now look for hints on how to solve the first of these two problems, i.e., to ensure that the scaling assumption (6.43) holds.

To get a first hint, let us multiply (6.43) from the right by $\hat{U}_{-\tau}$, substitute $\tau := -t$, and use (5.110) to rewrite this assumption as

$$\tilde{D}^{X,nf} = d_- \cdot \tilde{D}^{Y,nf}, \tag{6.45}$$

---

[6]In fact, we only need to require this relation for $\forall(\mathbf{x},\tau) \in M_0 \times (-\infty, 0]$, but our construction will fulfill it everywhere.

where $d_-(\mathbf{x}, t) := d(\mathbf{x}, -t)$. In particular, this implies that for $\forall \psi \in \mathcal{S}_\varnothing(\mathcal{V}_{M \times \mathbb{R}})$ we have

$$\frac{1}{\psi} \circ \tilde{D}^{X, nf} \circ \psi = d_- \cdot \frac{1}{\psi} \circ \tilde{D}^{Y, nf} \circ \psi,$$

and comparing the coefficients of the second-order derivatives on both sides, we find that

$$A^X = d_- \cdot A^Y. \tag{6.46}$$

Now abbreviating $A^{Y,t} := A^Y(\cdot, t)$ and $d_t := d(\cdot, t)$, and using (6.38), which says that for $\forall t \in \mathbb{R}$ we have $A^X = U_t A^{Y,t}$, this implies that

$$U_t A^X \stackrel{(6.38)}{=} A^{Y,-t} \stackrel{(6.46)}{=} \frac{1}{d_t} \cdot A^X, \tag{6.47}$$

and thus that

$$U_t m_{A^X} \stackrel{(6.40)}{=} m_{U_t A^X} \stackrel{(6.47)}{=} m_{(1/d_t) \cdot A^X} \stackrel{(6.39)}{=} d_t \cdot m_{A^X}. \tag{6.48}$$

We therefore find that for every bilinear form m of the form $m = f \cdot m_{A^X}$ for some $f \in \mathcal{F}(M)$ we have

$$U_t m = U_t(f \cdot m_{A^X}) \stackrel{(6.32)}{=} (U_t f) \cdot U_t m_{A^X} \stackrel{(6.48)}{=} (U_t f) \cdot d_t \cdot m_{A^X}.$$

Denoting by c the conformal structure that includes the positive-definite bilinear form m_{A^X} associated to $\tilde{D}^{X, nf}$, this defines a representation of U_t acting on the section space $\mathcal{S}(\mathcal{W}_c)$ (which we have seen in Section 6.2.3 to be the space of just these bilinear forms m). Furthermore, considering only the positive functions $f \in \mathcal{F}_+(M)$, we see that this representation maps positive multiples of m_{A^X} to positive multiples of m_{A^X} again (i.e., it maps all elements of $c = \mathcal{S}_+(\mathcal{W}_c)$ into c), and in fact that $U_t c = c$, i.e., c is conformal. In summary:

Lemma 6.7. *The scaling assumption (6.43) implies that the maps U_t, $t \in \mathbb{R}$, are conformal w.r.t. the conformal structure associated to $\tilde{D}^{X, nf}$.*

This teaches us two things:

(i) In order to have our construction fulfill the scaling assumption, we need to start by defining our group $(\hat{U}_s)_{s \in \mathbb{R}}$ in such a way that the maps U_s acting on states $\mathbf{x} \in M$ are conformal w.r.t. the conformal structure c associated to $\tilde{D}^{X, nf}$. Note, however, that this is not a sufficient condition.

(ii) The conformal property of the operators U_s says that their representations acting on bilinear forms map the space $c = \mathcal{S}_+(\mathcal{W}_c)$ and thus also the set $\mathcal{S}(\mathcal{W}_c)$ of all functional multiples of m_{A^X} (see (6.41)) onto itself, respectively, i.e., we have a positivity-preserving representation of these operators on $\mathcal{S}(\mathcal{W}_c)$.

6.3.3   The Diamond Operator – Definition & Invariance

Using (6.28), we can rewrite (6.45) as $\tilde{D}^{X,nf} = d_- \cdot U^{-1}\tilde{D}^{X,nf}$; then dividing by d_-, restricting the equation to the time slice at $-t$, and substituting $-t$ by t, we therefore find that under the scaling assumption we have both

$$U_t\tilde{D}^{X,nf} = \tfrac{1}{d_t} \cdot \tilde{D}^{X,nf} \qquad \text{and} \qquad U_t\, m_{A^X} = d_t \cdot m_{A^X} \qquad (6.49)$$

(the second equation is just (6.48)). These two relations imply that the *diamond operator* $\diamond\colon \mathcal{S}(\mathcal{V}) \to \mathcal{S}(\mathcal{W}_c \otimes \mathcal{V})$ defined as

$$\diamond := m_{A^X} \otimes \tilde{D}^{X,nf} \qquad (6.50)$$

(i.e., $\diamond\psi := m_{A^X} \otimes (\tilde{D}^{X,nf}\psi)$) fulfills

$$
\begin{aligned}
U_t(\diamond\psi) \;&=\; U_t\big(m_{A^X} \otimes (\tilde{D}^{X,nf}\psi)\big) \overset{(6.24)}{=} (U_t m_{A^X}) \otimes \big(U_t(\tilde{D}^{X,nf}\psi)\big) \\
&\overset{(6.26b)}{=} (U_t m_{A^X}) \otimes \big((U_t\tilde{D}^{X,nf})(U_t\psi)\big) \overset{(6.49)}{=} m_{A^X} \otimes \big(\tilde{D}^{X,nf}(U_t\psi)\big) \\
&=\; \diamond(U_t\psi)
\end{aligned}
$$

for $\forall\psi \in \mathcal{S}(\mathcal{V})$. In short: Under the scaling assumption we have

$$U_t \circ \diamond = \diamond \circ U_t\,, \qquad (6.51)$$

where the two operators U_t are the representations acting on $\mathcal{S}(\mathcal{W}_c \otimes \mathcal{V})$ and on $\mathcal{S}(\mathcal{V})$, respectively.

To phrase this observation differently, we can generalize our definition (6.25b) to define a representation of U_t acting on operators $D\colon \mathcal{S}(\mathcal{V}) \to \mathcal{S}(\mathcal{W}_c \otimes \mathcal{V})$ as $U_t D := U_t \circ D \circ U_{-t}$, where again the operators U_t and U_{-t} on the right are the representations acting on $\mathcal{S}(\mathcal{W}_c \otimes \mathcal{V})$ and on $\mathcal{S}(\mathcal{V})$, respectively. This allows us to rewrite (6.51) as

$$U_t \diamond = \diamond\,,$$

i.e., the operator $\diamond$ defined in (6.50) is invariant under U_t.

To summarize: *Under the scaling assumption, the diamond operator $\diamond$ commutes with (i.e., is invariant under) the operators U_t.*

6.4   Construction of $\tilde{D}^{X,nf}$

Solving the definition (6.50) of $\diamond$ for $\tilde{D}^{X,nf}$ shows that

$$\tilde{D}^{X,nf} = \frac{1}{m_{A^X}} \otimes \diamond\,, \qquad (6.52)$$

where $\frac{1}{m_{A^X}}$ is a section in $\mathcal{S}(\mathcal{W}_c^*)$. The simple commutativity property (6.51) of $\diamond$ (in comparison with the scaling assumption (6.43) on $\tilde{D}^{X,nf}$) motivates us to base our construction of $\tilde{D}^{X,nf}$ on the construction of $\diamond$, as described in more detail later in Section 6.4.2.

6.4.1 Differential Operators Mapping Between Two Line Bundles

To prepare, let us start by learning a bit more about differential operators $D\colon \mathcal{S}(\mathcal{V}_1) \to \mathcal{S}(\mathcal{V}_2)$ (such as $\diamond\colon \mathcal{S}(\mathcal{V}) \to \mathcal{S}(\mathcal{W}_c \otimes \mathcal{V})$) that—in contrast to $\tilde{D}^{X,nf}$— map from one line bundle to another.

Reduction to the case of one line bundle. First note that given any such operator D and any $\psi \in \mathcal{S}(\mathcal{V}_1 \otimes \mathcal{V}_2^*)$, the operator $\psi \otimes D\colon \mathcal{S}(\mathcal{V}_1) \to \mathcal{S}(\mathcal{V}_1)$ maps $\mathcal{S}(\mathcal{V}_1)$ to itself again, which allows us to apply our techniques from the preceding sections to it. In the case of the diamond operator, we have

$$\mathcal{V}_1 \otimes \mathcal{V}_2^* = \mathcal{V} \otimes (\mathcal{W}_c \otimes \mathcal{V})^* = \mathcal{V} \otimes \mathcal{V}^* \otimes \mathcal{W}_c^* \cong \mathcal{W}_c^*,$$

and so we can take $\psi = \frac{1}{m}$ for any bilinear form $m \in c$; see (6.52) for an example.

Lemma 6.8. *Let $D\colon \mathcal{S}(\mathcal{V}_1) \to \mathcal{S}(\mathcal{V}_2)$ and $\psi \in \mathcal{S}_\varnothing(\mathcal{V}_1 \otimes \mathcal{V}_2^*)$. Then D is a kth-order differential operator according to Definition 6.1 if and only if $\psi \otimes D$ is.*

Proof. See Appendix E.5. □

Applied to $D := \diamond$ and $\psi := \frac{1}{m_A X}$, this lemma shows that $\tilde{D}^{X,nf}$ is a second-order differential operator if and only if $\diamond$ is (in the sense of Definition 6.1).

Coefficient functions. Definition 6.2 had associated to any second-order differential operator $D\colon \mathcal{S}(\mathcal{V}) \to \mathcal{S}(\mathcal{V})$ the matrix-valued function $A_D\colon M \to \mathbb{R}^{n \times n}$ given by the second-order coefficients of the classical differential operator $\frac{1}{\psi} \circ D \circ \psi$ for any $\psi \in \mathcal{S}_\varnothing(\mathcal{V})$, where A_D was seen to not depend on the specific choice of ψ. However, this definition does not directly extend to general second-order differential operators $D\colon \mathcal{S}(\mathcal{V}_1) \to \mathcal{S}(\mathcal{V}_2)$ such as $\diamond$, since here the sections ψ_1 and ψ_2 in Definition 6.1 come from different spaces, which makes the expression $\frac{1}{\psi} \circ D \circ \psi$ ill-defined. Instead, for such operators we need to define a *family* A_D of matrix-valued functions as follows:

Definition 6.3. *For any two non-coinciding line bundles $\mathcal{V}_1$ and $\mathcal{V}_2$ on M and any second-order differential operator $D\colon \mathcal{S}(\mathcal{V}_1) \to \mathcal{S}(\mathcal{V}_2)$ we define*

$$A_D := \big\{ A_{\psi \otimes D} \,\big|\, \psi \in \mathcal{S}(\mathcal{V}_1 \otimes \mathcal{V}_2^*) \big\}, \tag{6.53}$$

where $A_{\psi \otimes D}$ are the matrix-valued functions given in Definition 6.2.[7]

---

[7] Equivalently, we can define A_D as the family of symmetrized coefficient functions of the classical second-order differential operators $\psi_2^* \circ D \circ \psi_1$, where $\psi_1 \in \mathcal{S}_\varnothing(\mathcal{V}_1)$ and $\psi_2^* \in \mathcal{S}(\mathcal{V}_2^*)$. To see this, note that $\psi_2^* \circ D \circ \psi_1 = \frac{1}{\psi_1} \circ (\psi \circ D) \circ \psi_1$ for $\psi := \psi_1 \otimes \psi_2^* \in \mathcal{S}(\mathcal{V}_1 \otimes \mathcal{V}_2^*)$, and that the latter relation can be inverted, by $\psi_2^* = \frac{1}{\psi_1} \otimes \psi$.

Fixing any $\psi_0 \in \mathcal{S}_\varnothing(\mathcal{V}_1 \otimes \mathcal{V}_2^*)$ and using (6.1), we can rewrite this as

$$A_D := \left\{ \alpha \cdot A_{\psi_0 \otimes D} \,\middle|\, \alpha \in \mathcal{F}(M) \right\}.$$

For our operator $D = \diamond$, choosing $\psi_0 := \frac{1}{m_{A^X}}$ and using (6.52) and (6.2) therefore shows that

$$A_\diamond = \left\{ \alpha \cdot A^X \,\middle|\, \alpha \in \mathcal{F}(M) \right\}.$$

Positive definiteness, conformal structure. If for some (and thus all) $\psi_0 \in \mathcal{S}_+(\mathcal{V}_1 \otimes \mathcal{V}_2^*)$ the coefficient function $A_{\psi_0 \otimes D}$ actually maps into the subset $\mathbb{R}_+^{n \times n}$ of positive-definite matrices then we call the operator D itself *positive definite*, and the family A_D of coefficient functions further specifies a conformal structure, given by

$$c_D := \left\{ \alpha \cdot m_{A_{\psi_0 \otimes D}} \,\middle|\, \alpha \in \mathcal{F}_+(M) \right\}.$$

For our operator $D = \diamond$, again choosing $\psi_0 := \frac{1}{m_{A^X}}$ as above shows that our operator $\diamond$ is positive definite in this sense, and that

$$c_\diamond = \left\{ \alpha \cdot m_{A^X} \,\middle|\, \alpha \in \mathcal{F}_+(M) \right\} = c.$$

Summary. While to second-order differential operators $D \colon \mathcal{S}(\mathcal{V}) \to \mathcal{S}(\mathcal{V})$ we could associate a matrix-valued function A_D, which, if it takes positive-definite values, further defines a positive-definite bilinear form m_{A_D}, a second-order differential operator $D \colon \mathcal{S}(\mathcal{V}_1) \to \mathcal{S}(\mathcal{V}_2)$ only specifies a *family* A_D of matrix-valued functions that are functional multiples of one another, and that, if it contains a function with positive-definite values, specifies a *conformal structure*.

6.4.2   Representation Theorem

We are now ready to obtain our key representation theorem, which will tell us how to construct our operator $\tilde{D}^{X,nf}$.

Theorem 6.1. *Let $(\hat{U}_s)_{s \in \mathbb{R}}$ be given. Then an operator $D \colon \mathcal{S}(\mathcal{V}) \to \mathcal{S}(\mathcal{V})$ is a positive-definite second-order differential operator and fulfills the scaling assumption*

$$\forall s \in \mathbb{R}: \quad D \circ \hat{U}_s = d \cdot (\hat{U}_s \circ D)$$

for some function $d \in \mathcal{F}_+(M \times \mathbb{R})$ if and only if there exists a conformal structure c on M such that

(i) the representations U_s, $s \in \mathbb{R}$, acting on M are conformal w.r.t. c, and

(ii) there exist a bilinear form $m \in c$ and a positive-definite second-order differential operator $\diamond \colon \mathcal{S}(\mathcal{V}) \to \mathcal{S}(\mathcal{W}_c \otimes \mathcal{V})$ that commutes with the operators U_s, such that

$$D = \tfrac{1}{m} \otimes \diamond. \tag{6.54}$$

Proof. We already showed the first direction during our treatment of $D = \tilde{D}^{X,nf}$: If D is positive definite then we can associate with it a positive-definite bilinear form m_{A_D} and thus the conformal structure c that m_{A_D} belongs to. If in addition it fulfills the scaling assumption, then we have shown that the operators U_s are conformal with respect to c (Lemma 6.7), and that the operator $\diamond := m_{A_D} \otimes D$ is a positive-definite second-order differential operator (Section 6.4.1) that commutes with the operators U_s (Section 6.3.3). Solving the definition of $\diamond$ for D, D is then of the form (6.54), with $m = m_{A_D}$.

For the reverse direction, suppose that D is such that (i) and (ii) hold for some conformal structure c on M. Then since $\diamond$ is assumed to be a positive-definite second-order differential operator, so is D (by Lemma 6.8 and the definition of positive definiteness). Furthermore, since $\diamond$ is assumed to commute with the operators U_s, we have

$$U_s(D\psi) = U_s(\tfrac{1}{m} \otimes \diamond\psi) = (U_s\tfrac{1}{m}) \otimes (U_s\diamond\psi)$$
$$= \tfrac{1}{U_s m} \otimes (\diamond U_s \psi) = \tfrac{m}{U_s m} \cdot (\tfrac{1}{m} \otimes \diamond(U_s\psi))$$
$$= \tfrac{m}{U_s m} \cdot D(U_s\psi)$$

for $\forall s \in \mathbb{R}$ (where in the third step we used (6.23)), and thus

$$D \circ U_s = d \cdot (U_s \circ D), \tag{6.55}$$

where the function $d := \frac{U_s m}{m}$ is positive by (6.34). (The two occurrences of U_s in (6.55) are the representations $\hat{U}_s$ acting on $\mathcal{S}(\mathcal{V})$.) $\qquad\square$

One should point out that while on the first glimpse it appears that property (i) is not required for the second direction of the proof, this is not actually the case: Without it, we would not have a representation of the operators U_s acting on $\mathcal{S}(\mathcal{W}_c)$, which is necessary for property (ii) to even make sense, since the commutativity property (6.51) of $\diamond$ requires a representation of U_s acting on $\mathcal{S}(\mathcal{W}_c \otimes \mathcal{V})$.

6.4.3 Construction

This representation theorem suggests the following procedure for constructing the group $(\hat{U}_s)_{s\in\mathbb{R}}$ and our operator $\tilde{D}^{X,nf}$:

- Choose a conformal structure c on M.

- Define a group of operators U_s acting on $\mathcal{S}(\mathcal{V})$ (ideally depending on some scalar model parameters) such that their representations acting on M are conformal w.r.t. c.

- Define a positive-definite second-order differential operator $\diamond : \mathcal{S}(\mathcal{V}) \to \mathcal{S}(\mathcal{W}_c \otimes \mathcal{V})$ that commutes with the operators U_s.

- Set $\tilde{D}^{X,nf} := \frac{1}{m} \otimes \diamond$ for some bilinear form $m \in c = \mathcal{S}_+(\mathcal{W}_c)$.

The operator $\tilde{D}^{X,nf}$ thus obtained is then guaranteed to be a positive-definite second-order differential operator that fulfills the scaling assumption.

The bilinear form $m \in c$ is a model parameter. To turn it into a more familiar *functional model parameter* α, we can choose a fixed $m_0 \in c$ and write any other $m \in c$ as $\frac{m_0}{\alpha}$ for some $\alpha \in \mathcal{F}_+(M)$, which leads us to the definition

$$\tilde{D}^{X,nf}_\alpha := \alpha \cdot \left(\tfrac{1}{m_0} \otimes \diamond \right). \tag{6.56}$$

As already described in our outline at the beginning of this chapter, we can then allow the user to infuse time inhomogeneity into our model by allowing for functions $\alpha \in \mathcal{F}_+(M \times \mathbb{R})$ that also depend on the process time t. Clearly, this generalization will not break the consistency relation (5.125a).

Observe that according to our discussion in Section 5.13, to confine our state process $(X_t)_{t \geq 0}$ to M_0 even in the presence of noise, α would need to be chosen such that it vanishes sufficiently fast near ∂M_0, and that it does not grow too fast as $\mathbf{x} \to \infty$ (in a way that we will make rigorous only at the end of our construction).

6.4.4 Consistency, Unspanned Volatility, Multiplicative Terms

By plugging our definition (6.56) of $\tilde{D}^{X,nf}_\alpha$ into our consistency condition (6.44) under the scaling assumption, and then multiplying it by $\frac{m_0}{\alpha}$, we find that in our construction the consistency condition is equivalent to

$$\diamond \mathfrak{B}^Y_0 = 0, \tag{6.57}$$

independently of our choice of α or m_0. This shows that we can choose the section $\mathfrak{B}^Y_0$ to be independent of α as well, and as a result, by (5.74)–(5.75) in fact the entire two families $(\mathfrak{B}^X_T)_{T \in \mathbb{R}}$ and $(\mathfrak{B}^Y_T)_{T \in \mathbb{R}}$ of bond sections will be independent of α. This shows that our construction will make our model unspanned w.r.t. our functional model parameter α, which will control the volatility of our state process $(X_t)_{t \geq 0}$ (see (8.61) and (8.65)).

Furthermore, recalling Lemma 6.2 and the equivalence of (5.125a) with (6.44) under the scaling assumption, fulfilling (6.57) will also guarantee that the operators $D^{X,T}_\alpha$ obtained from $\tilde{D}^{X,nf}_\alpha$ via (5.113) have vanishing multiplicative terms, so that they are indeed the infinitesimal generators of diffusion processes.

6.4.5 Choosing the Conformal Structure

Obtaining the desired conformal structure. Note that it is not guaranteed that the conformal structure $c_\diamond$ associated to $\diamond$ (and thus also to $\tilde{D}^{X,nf}$) actually coincides with the conformal structure c that we chose at the beginning of our construction. While in fact we would not care about this (see the paragraph "Gauge fixing" below for a detailed explanation), this would be an important observation to make:

Since our constructed operator $\tilde{D}^{X,nf}$ fulfills the scaling assumption, by Lemma 6.7 the operators U_s are conformal w.r.t. its associated conformal structure $c_\diamond$ as well; therefore, if we have $c \neq c_\diamond$ then our operators U_s are actually conformal w.r.t. two different conformal structures. This is fairly unlikely to happen by mere luck during our construction, since finding maps that are conformal with respect to *any* conformal structure is already hard enough. If it does happen, it would indicate that the obtained model is not yet as general as it could be, and that we might be able extend our family of groups $(U_s)_{s \in \mathbb{R}}$ (i.e., to add a model parameter) until the conformal property only holds for a unique conformal structure c on M. If at that point we still succeed in constructing an operator $\diamond$ that commutes also with the additional operators U_s, then the uniqueness of c would guarantee that the conformal structure associated to $\tilde{D}^{X,nf}$ must indeed be c itself.

Gauge fixing. This leads us to the question of which conformal structure $c_\diamond$ we should design our operator $\tilde{D}^{X,nf}$ to give rise to. A first thought may be to try to make $c_\diamond$ a model parameter, i.e., to define a model for any given conformal structure $c_\diamond$.

Observe, however, that the noise coefficients A^X by themselves—and thus $c_\diamond$—have no physical meaning unless we specify the meaning of our state variable $\mathbf{x}$. Indeed, given any model that fits into our framework in Chapter 5, we can use any diffeomorphism $f \colon M \to \tilde{M}$ to define a new process $\tilde{X}_t := f(X_t)$ on the state space $\tilde{M}_0 := f(M_0)$, and instead consider the bond sections $\tilde{\mathfrak{B}}_T^X(\tilde{\mathbf{x}}, t) := \mathfrak{B}_T^X(f^{-1}(\tilde{\mathbf{x}}), t)$ over the line bundle $\tilde{\mathcal{V}}_0 := (\tilde{V}_{\tilde{\mathbf{x}}})_{\tilde{\mathbf{x}} \in \tilde{M}_0}$ consisting of the vector spaces $\tilde{V}_{\tilde{\mathbf{x}}} := V_{f^{-1}(\tilde{\mathbf{x}})}$. We then obtain a model that is equivalent to the original one in the sense that the exchange rate processes $(B_{T_1,T_2}^X(X_t))_t$ and $(\tilde{B}_{T_1,T_2}^X(\tilde{X}_t))_t$ coincide. However, the function $A^{\tilde{X}}$ associated to the process $(\tilde{X}_t)$ generally need not coincide with A^X; instead, it is modified according to Itô's Lemma.

In fact, given any point $\mathbf{x}$, under some smoothness condition on A^X we can always achieve that $A^{\tilde{X}}$ will be a multiple of the identity matrix in some neighborhood of $\mathbf{x}$, by choosing f appropriately. Now if in fact this were always possible *globally*, then building a model with $A^X(\mathbf{x}) = a(\mathbf{x})I$ for $\forall \mathbf{x} \in M$ and for some function $a \in \mathcal{F}_+(M)$ would not make us lose generality, since any other model would be equivalent to a model of this type.

In our construction we will design our model such that A^X is of the form $A^X(\mathbf{x}) = a(\mathbf{x})I$ for $\forall \mathbf{x} \in M$ (i.e., that $c_\diamond$ is the Euclidean conformal structure), and so we will restrict our class of models to only those that are globally *equivalent to a model of this type.* We believe that this modeling decision does not lead to any substantial constraints on the model behavior.

Note that constraining the conformal structure c in some way is also necessary for achieving a stable calibration, since not specifying it a priori and instead introducing it as a model parameter would lead to redundancies in our parameter space. This is called gauge fixing.

We chose the Euclidean conformal structure on M as our gauge since this will make our equations the simplest: It will mean that the second-order part of the generators $D^{X,T}$ is a functional multiple of the Laplacian $\Delta = \partial_1^2 + \cdots + \partial_n^2$.

6.5 State Space Manifold

6.5.1 Representation of Conformal Maps

So what is the class of operators that are conformal w.r.t. the Euclidean conformal structure? In two dimensions, this class of maps is simply the class of holomorphic functions. In higher dimensions, however, this class is drastically reduced, namely to the compositions of shifts $\mathbf{x} \mapsto \mathbf{x} + \mathbf{a}$, turns $\mathbf{x} \mapsto O\mathbf{x}$ (with O orthogonal, i.e., $O^T O = I$), scalar multiplications $\mathbf{x} \mapsto a\mathbf{x}$ (with $a > 0$), and (less intuitively) inversions $\mathbf{x} \mapsto \frac{\mathbf{x}}{\|\mathbf{x}\|^2}$.

Since we want our model to be as general as possible, we would like to allow for *all* such mappings (and the compositions thereof).[8] This will introduce further model parameters that enter our group $(U_s)_{s \in \mathbb{R}}$; however, the model will not be unspanned w.r.t. these parameters since our bond sections $\mathfrak{B}_T^X(\mathbf{x}, t) := \hat{U}_{T-t} \mathfrak{B}_0^Y(\mathbf{x})$ will depend on these parameters as well.

Representing these seemingly very different types of maps in a more unified way would certainly make it not only easier to parametrize this family of maps, but more importantly, it would make our task of defining an operator $\diamond$ that commutes with all these maps more manageable. Fortunately, this representation problem has been well studied, and it has been found that it is best to rather think of this problem on a certain n-dimensional manifold $\mathscr{M}$ that is equipped with a conformal structure $\mathfrak{c}$: As we will see in Chapter 7 (where we will properly define $\mathscr{M}$ and $\mathfrak{c}$), the class of all maps $\mathcal{U}: \mathscr{M} \to \mathscr{M}$ that are conformal w.r.t. $\mathfrak{c}$ can simply be represented by the set of "pseudo-orthogonal" matrices in the dimension $n + 2$.

If one then defines a bijective map

$$h: \mathscr{M} \to M$$

that associates to each point $\mathbf{p} \in \mathscr{M}$ one of our states $\mathbf{x} \in M \subset \mathbb{R}^n$, in such a way that the induced conformal structure $c = h(\mathfrak{c})$ on M is the Euclidean one, then the set of maps

$$U := h \circ \mathcal{U} \circ h^{-1}: M \to M$$

is the complete set of maps on M that are conformal with respect to the Euclidean conformal structure. See the following commutative diagram for an illustration:

$$
\begin{array}{ccc}
(\mathscr{M}, \mathfrak{c}) & \xrightarrow{\ \mathcal{U}\ (\text{conformal w.r.t. } \mathfrak{c})\ } & (\mathscr{M}, \mathfrak{c}) \\
\downarrow{\scriptstyle h} & & \downarrow{\scriptstyle h} \\
(M, c) & \xrightarrow{\ U\ (\text{conformal w.r.t. } c)\ } & (M, c)
\end{array}
$$

---

[8]However, the continuity of the group $(U_s)_{s \in \mathbb{R}}$ will exclude the inversions from this group.

6.5.2 The ACE Model on $\mathcal{M}$

Due to the simplicity of the set of conformal maps in $\mathcal{M}$, we will therefore carry out our entire construction on $\mathcal{M}$ and then map our solution back to M at the very end. By defining the conversion of the various mathematical objects between the two spaces in a consistent way, all operations involving these objects can be carried out in either space, with corresponding results, and so all their properties or conditions will translate into the corresponding equivalent statements in $\mathcal{M}$. Operators acting on functions or sections on $\mathcal{M}$ will be called "second-order differential operators" and "positive definite" if and only if their corresponding counterpart in M is (this is independent of the choice of h). Table 6.1 lists the notation for all of our key objects in the new state space $\mathcal{M}$, along with their conversion rules.

In addition to this, from now on the notation for the various representations of our evolution operators ($\hat{U}_s$, $\bar{U}_s$, U_s, etc.) will be simplified more rigorously: In $\mathcal{M}$, we will refer to all of these operators as $\mathcal{U}_s$. As mentioned before, this will not lead to any ambiguities since one can always see from the object they act on which representation of the operator one intends use.

description	notation in M	notation in $\mathcal{M}$	conversion
state space	M	$\mathcal{M}$	$M = h(\mathcal{M})$
state variable	$\mathbf{x}$	$\mathbf{p}$	$\mathbf{x} = h(\mathbf{p})$
bilinear forms	m	$\mathfrak{m}$	$\langle d\mathbf{x}_1, d\mathbf{x}_2 \rangle_{\mathbf{x}}^m = \langle d\mathbf{p}_1, d\mathbf{p}_2 \rangle_{\mathbf{p}}^{\mathfrak{m}}$
conformal structure	c	$\mathfrak{c}$	$c = \{ h(\mathfrak{m}) \mid \mathfrak{m} \in \mathfrak{c} \}$
1D vector spaces	$V_{\mathbf{x}}$ $W_{\mathbf{x}}$	$V'_{\mathbf{p}}$ $W'_{\mathbf{p}}$	$V_{\mathbf{x}} = V'_{\mathbf{p}}$ $W_{\mathbf{x}} = W'_{\mathbf{p}}$
line bundles	$\mathcal{V}$ $\mathcal{W}_c$	$\mathscr{V}$ $\mathscr{W}_{\mathfrak{c}}$	$\mathcal{V} = (V_{\mathbf{x}})_{\mathbf{x} \in M},\ \mathscr{V} = (V'_{\mathbf{p}})_{\mathbf{p} \in \mathcal{M}}$ $\mathcal{W}_c = (W_{\mathbf{x}})_{\mathbf{x} \in M},\ \mathscr{W}_{\mathfrak{c}} = (W'_{\mathbf{p}})_{\mathbf{p} \in \mathcal{M}}$
functions	f_1	f_2	$f_1(\mathbf{x}, t) = f_2(\mathbf{p}, t)$
sections	ψ_1 $\mathfrak{B}_T^X$ $\mathfrak{B}_T^Y$	ψ_2 $\mathcal{B}_T^X$ $\mathcal{B}_T^Y$	$\psi_1(\mathbf{x}, t) = \psi_2(\mathbf{p}, t)$ $\mathfrak{B}_T^X(\mathbf{x}, t) = \mathcal{B}_T^X(\mathbf{p}; t)$ $\mathfrak{B}_T^Y(\mathbf{x}) = \mathcal{B}_T^Y(\mathbf{p})$
evolution operators	U_s $\bar{U}_s$ $\hat{U}_s$	$\mathcal{U}_s$ $\mathcal{U}_s$ $\mathcal{U}_s$	$U_s(\mathbf{x}) = h(\mathcal{U}_s(\mathbf{p}))$ $(\bar{U}_s f_1)(\mathbf{x}, t) = (\mathcal{U}_s f_2)(\mathbf{p}; t)$ $(\hat{U}_s \psi_1)(\mathbf{x}, t) = (\mathcal{U}_s \psi_2)(\mathbf{p}; t)$
generators	$\tilde{D}^{X,nf}$ $\tilde{D}^{X,T}$ $\tilde{D}^{X,\$}$	$\tilde{\mathcal{D}}^{X,nf}$ $\tilde{\mathcal{D}}^{X,T}$ $\tilde{\mathcal{D}}^{X,\$}$	$(\tilde{D}^{X,nf} \psi_1)(\mathbf{x}, t) = (\tilde{\mathcal{D}}^{X,nf} \psi_2)(\mathbf{p}, t)$ $(\tilde{D}^{X,T} f_1)(\mathbf{x}, t) = (\tilde{\mathcal{D}}^{X,T} f_2)(\mathbf{p}, t)$ $(\tilde{D}^{X,\$} f_1)(\mathbf{x}, t) = (\tilde{\mathcal{D}}^{X,\$} f_2)(\mathbf{p}, t)$
diamond operator	$\Diamond$	$\Diamond$	$(\Diamond \psi_1)(\mathbf{x}, t) = (\Diamond \psi_2)(\mathbf{p}, t)$

Table 6.1: The conversion between the various objects of our framework on the state space $M \subset \mathbb{R}^n$ and on the manifold $\mathcal{M}$.

6.6   Summary

Our general interest rate framework requires the construction of the following mathematical objects:

- a state space manifold $\mathcal{M}$ with a conformal structure $\mathfrak{c}$ defined on it (which we think of as a set of positive-definite bilinear forms on $\mathcal{M}$),

- a map $h\colon \mathcal{M} \to \mathbb{R}^n$ that maps $\mathfrak{c}$ to the Euclidean conformal structure on $M := h(\mathcal{M}) \subset \mathbb{R}^n$,

- a signed line bundle $\mathscr{V} := (V'_{\mathbf{p}})_{\mathbf{p} \in \mathcal{M}}$,

- a group $(\mathcal{U}_s)_{s \in \mathbb{R}}$ of operators $\mathcal{U}_s$ with consistent representations acting on $\mathcal{M}$, $\mathcal{F}(\mathcal{M})$, and $\mathcal{S}(\mathscr{V})$, such that their representations acting on $\mathcal{M}$ are conformal w.r.t. $\mathfrak{c}$,

- a second-order differential operator $\Diamond\colon \mathcal{S}(\mathscr{V}) \to \mathcal{S}(\mathscr{W}_{\mathfrak{c}} \otimes \mathscr{V})$—where $\mathscr{W}_{\mathfrak{c}}$ is the line bundle given by $\mathfrak{c}$—that commutes with the operators $\mathcal{U}_s$,

- a subset $\mathcal{M}_0 \subset \mathcal{M}$ such that $\forall s \geq 0\colon \mathcal{U}_s(\mathcal{M}_0) \subseteq \mathcal{M}_0$, and

- a section $\mathcal{B}_0^Y \in \mathcal{S}_+(\mathscr{V}_0)$ fulfilling $\Diamond \mathcal{B}_0^Y = 0$, where $\mathscr{V}_0 := (V'_{\mathbf{p}})_{\mathbf{p} \in \mathcal{M}_0}$.

Given those objects, we then define a family of second-order differential operators operators $\tilde{\mathcal{D}}_\alpha^{X,nf}\colon \mathcal{S}(\mathscr{V}) \to \mathcal{S}(\mathscr{V})$ as

$$\tilde{\mathcal{D}}_\alpha^{X,nf} := \alpha \cdot \left(\tfrac{1}{\mathfrak{m}_0} \otimes \Diamond \right)$$

for some fixed choice $\mathfrak{m}_0 \in \mathfrak{c}$, where $\alpha \in \mathcal{F}_+(\mathcal{M})$ is a functional model parameter. If the user does not require the model to be time homogeneous then one can more generally allow for time-dependent functions $\alpha \in \mathcal{F}_+(\mathcal{M} \times \mathbb{R})$. The function α needs to be chosen such that it vanishes sufficiently fast near ∂M_0, and that it does not grow too fast as $\mathbf{x} \to \infty$.

To obtain our classical interest rate model on $M_0 \subset \mathbb{R}^n$, we then need to first compute all relevant functions and operators on $\mathcal{M}_0$, analogously to our instructions in the last box in Section 5.14, and then translate them back to $M_0 \subset \mathbb{R}^n$ via the map h, as described in Table 6.1. Note that in practice it therefore suffices to define h only on $\mathcal{M}_0$.

As discussed in Section 6.4.5, we should then check whether the second-order part of the operators $D^{X,T}$ is indeed a positive function times the Laplacian, as this would mean that A^X is a positive function times the identity matrix, so that $\tilde{D}^{X,nf}$ is positive definite and that the conformal structure associated to it is the Euclidean one as well, just like $h(\mathfrak{c})$.[9]

---

[9]It would then further imply that $\tilde{\mathcal{D}}^{X,nf}$ and thus also $\Diamond$ are positive definite as well, as we had asked in Theorem 6.1.

Chapter 7

Solution

7.1 State Space

We begin our construction of the objects listed in Section 6.6 with the state space manifold $\mathscr{M}$.

7.1.1 Minkowski Space

The *Minkowski space* in $m + 1$ dimensions is the vector space $\mathbb{T} := \mathbb{R}^{m+1}$, equipped with the *"Minkowski inner product"*

$$\langle x, y \rangle_{\mathbb{T}} := -x^0 y^0 + x^1 y^1 + \cdots + x^m y^m \tag{7.1}$$

$$:= -x^0 y^0 + \langle \vec{x}, \vec{y} \rangle_{\mathbb{R}^m}$$

$$:= \sum_{i,j=0}^{m} Q_{ij} x^i y^j$$

$$= \langle x, (Q_{ij}) y \rangle_{\mathbb{R}^{m+1}}$$

for $\forall x, y \in \mathbb{T}$, where we use the notation $x =: (x^0, x^1, \ldots, x^m) =: (x^0, \vec{x})$ for $\vec{x} := (x^1, \ldots, x^m) \in \mathbb{R}^m$,[1] where $\langle \cdot, \cdot \rangle_{\mathbb{R}^m}$ and $\langle \cdot, \cdot \rangle_{\mathbb{R}^{m+1}}$ denote the standard dot product in the Euclidean space, and where the matrix[2] $(Q_{ij}) \in \mathbb{R}^{(m+1) \times (m+1)}$ is defined as

$$(Q_{ij}) := \begin{pmatrix} -1 & 0 & 0 & \cdots & 0 \\ 0 & 1 & 0 & & \vdots \\ 0 & 0 & 1 & \ddots & \vdots \\ \vdots & & \ddots & \ddots & 0 \\ 0 & \cdots & \cdots & 0 & 1 \end{pmatrix}. \tag{7.2}$$

---

[1] The use of *upper* indices for the coefficients of state variables and similar vectors is common in physics and will be explained in more detail in Section 7.3.

[2] We use the notation (Q_{ij}) to denote the matrix with the elements Q_{ij}, $1 \leq i, j \leq m + 1$.

Referring to the signs of the eigenvalues of the matrix (Q_{ij}), we say that the Minkowski space has the signature $(-, +, \ldots, +)$, or simply $(1, m)$.

We will usually denote the elements of $\mathbb{T}$ by the symbols x or y, i.e., in non-fat characters; this is to distinguish them from our states $\mathbf{x}, \mathbf{y} \in M \subset \mathbb{R}^n$ used in our general mathematical framework in Section 5. The Minkowski space itself will *not* serve as our state space; it is only the starting point for our definition of the state space manifold $\mathscr{M}$ later in Sections 7.1.4–7.1.5, whose elements we will denote by the fat character $\mathbf{p}$.

Observe that while a proper inner product $\langle \cdot, \cdot \rangle$ defined on some vector space V over $\mathbb{R}$ is required to fulfill the three properties

1. **symmetry:** $\forall x, y \in V \colon \langle x, y \rangle = \langle y, x \rangle$,

2. **linearity**[3]**:** $\forall x, x', y \in V \ \forall a, a' \in \mathbb{R} \colon \langle ax + a'x', y \rangle = a\langle x, y \rangle + a'\langle x', y \rangle$,

3. **positive definiteness:** $\forall x \in V \colon \langle x, x \rangle \geq 0$, with equality only for $x = 0$,

the Minkowski inner product violates the positive-definiteness requirement, i.e., there are vectors $x \in \mathbb{T} \setminus \{0\}$ with $\langle x, x \rangle_\mathbb{T} = 0$, and in fact we even have $\langle x, x \rangle_\mathbb{T} < 0$ for some vectors $x \in \mathbb{T}$. Instead, it only has the weaker property

3'. **nondegeneracy:** $\forall x \in V \colon \big((\forall y \in V \colon \langle x, y \rangle = 0) \ \Rightarrow \ x = 0 \big)$.

Therefore, although the function defined in (7.1) is commonly referred to in the literature as the "Minkowski inner product," technically speaking it is only a nondegenerate symmetric bilinear form.

For ease of notation, from this point forward we will simply write the Minkowski inner product of two vectors $x, y \in \mathbb{T}$ as $\langle x, y \rangle$, i.e., we will infer from the space that x and y reside in which inner product is to be used. Only at places with an increased risk of confusion or with a specific significance, we may be more explicit and decide to write $\langle x, y \rangle_\mathbb{T}$ or $\langle \vec{u}, \vec{v} \rangle_{\mathbb{R}^n}$ for $x, y \in \mathbb{T}$ and $\vec{u}, \vec{v} \in \mathbb{R}^n$.[4]

7.1.2   Special Relativity, Light Cones, Isotropic Space

The four-dimensional Minkowski space (i.e., $m = 3$) is used by physicists in the context of Einstein's Special Theory of Relativity to describe spacetime, where x_0 denotes time and where $\vec{x} = (x^1, x^2, x^3)$ denotes the three-dimensional space vector. A photon (i.e., light particle) emitted at time $x^0 = 0$ at the spacial origin $\vec{x} = (0, 0, 0)$—an event occurring at the spacetime $x = (x^0, \vec{x}) = (0, 0, 0, 0) \in \mathbb{T}$—will travel at the constant speed of light (which we can set equal to 1 by deciding

---

[3]Note that linearity in the *second* argument then follows from the symmetry property.

[4]When still in doubt, recall that in our notation any bold-faced variables (such as $\mathbf{x}$ and $\mathbf{y}$, with the exception of $\mathbf{p}$) and variables with an arrow on the top (such as $\vec{x}$) will always belong to a Euclidean space and will therefore require the use of the Euclidean inner product. Furthermore, since the Minkowski inner product is not positive definite and therefore does not induce a norm on $\mathbb{T}$, all norms $\| \cdot \|$ in this book will be the standard Euclidean ones.

to measure the temporal dimension in appropriate units), and thus at each fixed time $x^0 > 0$ it can only be located at any of those spacial locations $\vec{x}$ with

$$\frac{\|\vec{x}\|}{x^0} = 1 \quad \Leftrightarrow \quad \|\vec{x}\|^2 = (x^0)^2 \quad \Leftrightarrow \quad \langle x, x \rangle = 0.$$

We therefore call the set

$$\begin{aligned}
\mathbb{T}_0^\uparrow &:= \left\{ x \in \mathbb{T} \mid \langle x, x \rangle = 0, \; x^0 > 0 \right\} \\
&= \left\{ (x^0, \vec{x}) \in \mathbb{T} \setminus \{0\} \mid x^0 = \|\vec{x}\| \right\},
\end{aligned}$$

the *forward light cone* (although to be precise, this is only the *surface* of a cone). Similarly, we call the set

$$\begin{aligned}
\mathbb{T}_0^\downarrow &:= \left\{ x \in \mathbb{T} \mid \langle x, x \rangle = 0, \; x^0 < 0 \right\} \\
&= \left\{ (x^0, \vec{x}) \in \mathbb{T} \setminus \{0\} \mid x^0 = -\|\vec{x}\| \right\},
\end{aligned}$$

which is the set of all spacetimes that a photon passing through the spacial origin at time $x^0 = 0$ may have come from, the *backward light cone*.

We say that a vector $x \in \mathbb{T}$ is *isotropic* if $\langle x, x \rangle = 0$, and we call the set

$$\begin{aligned}
\mathbb{T}_0 &:= \left\{ x \in \mathbb{T} \setminus \{0\} \mid \langle x, x \rangle = 0 \right\} \\
&= \left\{ (x^0, \vec{x}) \in \mathbb{T} \setminus \{0\} \mid |x^0| = \|\vec{x}\| \right\} \\
&= \mathbb{T}_0^\uparrow \cup \mathbb{T}_0^\downarrow
\end{aligned} \tag{7.3}$$

of all non-zero isotropic vectors the *isotropic space*. The set $\mathbb{T}_0$ is the union of the surfaces $\mathbb{T}_0^\uparrow$ and $\mathbb{T}_0^\downarrow$ of two cones whose tips meet at the origin; we excluded the origin to make $\mathbb{T}_0$ a proper manifold. Note that since $\mathbb{T}_0 \cup \{0\}$ is not closed under the addition operation on $\mathbb{T}$, it is not a linear space. However, we have $\forall x \in \mathbb{T}_0 \cup \{0\} \; \forall a \in \mathbb{R} \colon ax \in \mathbb{T}_0 \cup \{0\}$, and so $\mathbb{T}_0 \cup \{0\}$ is the union of certain straight lines through the origin.

Finally, we denote the cone-shaped regions enclosed by $\mathbb{T}_0^\uparrow$ and $\mathbb{T}_0^\downarrow$, respectively, i.e., the interiors of the two cones, by

$$\begin{aligned}
\mathbb{T}^\uparrow &:= \left\{ x \in \mathbb{T} \mid \langle x, x \rangle < 0, \; x^0 > 0 \right\} \tag{7.4a} \\
&= \left\{ (x^0, \vec{x}) \in \mathbb{T} \mid x^0 > \|\vec{x}\| \right\}
\end{aligned}$$

$$\text{and} \quad \begin{aligned}
\mathbb{T}^\downarrow &:= \left\{ x \in \mathbb{T} \mid \langle x, x \rangle < 0, \; x^0 < 0 \right\} \tag{7.4b} \\
&= \left\{ (x^0, \vec{x}) \in \mathbb{T} \mid x^0 < -\|\vec{x}\| \right\}.
\end{aligned}$$

(In physics the vectors in these two sets are called *time-like*.)

We want to emphasize that our decision to work in Minkowski space is motivated solely by its geometric properties. Although we will use some notation and expressions that are common in the literature on the Special Theory of Relativity, our actual modeling problem does not really have anything in common with the problem addressed by that theory. In particular, we will *not* interpret the x^0-component of a point $x \in \mathbb{T}$ as our process time t.

7.1.3   The Sign of the Minkowski Inner Product

Let us understand the behavior of the sign of the Minkowski inner product a bit better; we will need the results shown here later in Sections 7.7.1 and 7.8.3.

Lemma 7.1. *(i) Let $x, y \in \mathbb{T}_0$. Then referring to the cones $\mathbb{T}_0^\uparrow$ and $\mathbb{T}_0^\downarrow$, we have*

$$\langle x, y \rangle = 0 \quad \text{if} \quad x \parallel y, \tag{7.5a}$$

$$\langle x, y \rangle > 0 \quad \text{if} \quad x \nparallel y \text{ and if } x \text{ and } y \text{ lie in opposite cones,} \tag{7.5b}$$

$$\langle x, y \rangle < 0 \quad \text{if} \quad x \nparallel y \text{ and if } x \text{ and } y \text{ lie in the same cone.} \tag{7.5c}$$

(ii) Let $x \in \mathbb{T}_0^\uparrow$ and $y \in \mathbb{T}^\downarrow \cup \mathbb{T}^\uparrow$. Then we have

$$\langle x, y \rangle > 0 \quad \text{if} \quad y \in \mathbb{T}^\downarrow, \tag{7.6a}$$

$$\langle x, y \rangle < 0 \quad \text{if} \quad y \in \mathbb{T}^\uparrow. \tag{7.6b}$$

(iii) Let $x \in \mathbb{T}_0^\downarrow$ and $y \in \mathbb{T}^\downarrow \cup \mathbb{T}^\uparrow$. Then we have

$$\langle x, y \rangle > 0 \quad \text{if} \quad y \in \mathbb{T}^\uparrow, \tag{7.7a}$$

$$\langle x, y \rangle < 0 \quad \text{if} \quad y \in \mathbb{T}^\downarrow. \tag{7.7b}$$

Proof. See Appendix F.1. $\square$

The lemma does not make any statement about those $y \in \mathbb{T}$ with $\langle y, y \rangle > 0$, since for them the sets of all $x \in \mathbb{T}_0$ with positive, negative, and vanishing Minkowski inner product $\langle x, y \rangle$ are non-trivial.

7.1.4   Projective Space

Let $\mathbb{P}$ be the *projective space* of $\mathbb{T}$, that is the space of all the one-dimensional linear subspaces (i.e., straight lines through the origin) of $\mathbb{T}$. Let

$$\pi \colon \mathbb{T} \setminus \{0\} \to \mathbb{P}$$

be the natural projection that maps any non-zero point $x \in \mathbb{T} \setminus \{0\}$ to the unique line that it is contained in. Finally, let us define the *isotropic projective space* as

$$\mathbb{P}_0 := \pi(\mathbb{T}_0), \tag{7.8}$$

i.e., as the set of all straight lines on the double light cone $\mathbb{T}_0 \cup \{0\}$.

Recall that although $\mathbb{T}$ is a linear space, $\mathbb{T}_0$ is not (even when including the origin), since it is not closed under addition. Furthermore, since there is no unique natural way of defining an addition operation on $\mathbb{P}$, the projective space $\mathbb{P}$ as well as its subset $\mathbb{P}_0$ are not linear spaces, either. Instead, we should merely view the sets $\mathbb{T}_0$, $\mathbb{P}$, and $\mathbb{P}_0$ as manifolds, with

$$\mathbb{T}_0 \subset \mathbb{T}, \qquad \dim(\mathbb{T}_0) = m, \qquad \dim(\mathbb{T}) = m + 1, \tag{7.9a}$$

$$\mathbb{P}_0 \subset \mathbb{P}, \qquad \dim(\mathbb{P}_0) = m - 1, \qquad \dim(\mathbb{P}) = m. \tag{7.9b}$$

For future reference, note that

$$\forall x \in \mathbb{T} \setminus \{0\} \; \forall a \in \mathbb{R} \setminus \{0\}: \quad \pi(ax) = \pi(x). \tag{7.10}$$

7.1.5   State Space Manifold, Topology

We have now essentially completed the construction of our state space manifold $\mathcal{M}$: It will be an open subset of the isotropic projective space $\mathbb{P}_0$, i.e.,

$$\mathcal{M} \subsetneq \mathbb{P}_0 .$$

Given any desired dimension

$$n = \dim(\mathcal{M})$$

for our state space, by (7.9b) we need to choose

$$m = \dim(\mathbb{P}_0) + 1 = \dim(\mathcal{M}) + 1 = n + 1. \tag{7.11}$$

As the topology of $\mathbb{P}_0$ and thus of our state space manifold $\mathcal{M}$ we choose the one that it naturally inherits from the Euclidean topology on $\mathbb{T}_0$ by way of the projection π; in other words, we say that a set

$$B \subset \mathbb{P}_0 \text{ is open} \quad \Leftrightarrow \quad \pi^{-1}(B) \subset \mathbb{T}_0 \text{ is open in } \mathbb{T}_0,$$

where $\pi^{-1}(B)$ denotes the preimage of B under π. (This is the finest topology under which $\pi|_{\mathbb{T}_0}$ is continuous.)

7.1.6   Line Bundle

This construction of $\mathcal{M}$ as a projective space now gives rise to a natural choice of a line bundle $\mathscr{P}$ on $\mathbb{P}_0$, namely the one defined as

$$\mathscr{P} := (\mathbf{p})_{\mathbf{p} \in \mathbb{P}_0}; \tag{7.12}$$

in other words, for every state $\mathbf{p} \in \mathbb{P}_0$ we choose $\mathbf{p}$ itself as our one-dimensional linear vector space. We make $\mathscr{P}$ a *signed* line bundle by specifying the positive half of each space $\mathbf{p} \in \mathbb{P}_0$ as

$$\mathbf{p}_+ := \mathbf{p} \cap \mathbb{T}_0^\uparrow = \left\{ (x^0, \dots, x^m) \in \mathbf{p} \,\middle|\, x^0 > 0 \right\}. \tag{7.13}$$

However, this is not yet our wealth line bundle $\mathscr{V}$ that we will use to model account values. Instead, we will use carefully chosen powers $\mathscr{P}^{\otimes k}$, both as our wealth line bundle $\mathscr{V}$ and as a representation of the line bundle $\mathscr{W}_{\mathfrak{c}}$ associated to our conformal structure.

7.2   Conformal Structure

Now that we have constructed the manifold $\mathbb{P}_0$ that will form the basis for our state space $\mathcal{M}$, let us construct a conformal structure $\mathfrak{c}$ on it. As we will see, there is a unique conformal structure on $\mathbb{P}_0$ that is naturally derived from the Minkowski inner product on the space $\mathbb{T}$.

7.2.1   Tangent Spaces

To begin, we need to gain some understanding of the tangent spaces $T_{\mathbf{p}}\mathbb{P}_0$ of $\mathbb{P}_0$, and to do so, we will start with those of the manifold $\mathbb{T}_0$ that we used to define $\mathbb{P}_0$ in (7.8).

The tangent spaces of $\mathbb{T}_0$.   Let $x \in \mathbb{T}_0$, and let $\tilde{\gamma}$ be a C^1-path in $\mathbb{T}_0$ with $\tilde{\gamma}(0) = x$. Then by definition of $\mathbb{T}_0$ we have $\langle \tilde{\gamma}(s), \tilde{\gamma}(s) \rangle = 0$ for every s, and thus $0 = \partial_s \frac{1}{2} \langle \tilde{\gamma}(s), \tilde{\gamma}(s) \rangle|_{s=0} = \langle \tilde{\gamma}'(0), x \rangle$, i.e., $\tilde{\gamma}'(0) \in \pi(x)^{\perp}$.[5] Since $\dim(\pi(x)^{\perp}) = \dim(\mathbb{T}) - 1$ coincides with the expected dimension of the manifold $\mathbb{T}_0$, this shows that

$$\forall x \in \mathbb{T}_0: \quad T_x(\mathbb{T}_0) = \pi(x)^{\perp}. \tag{7.14}$$

Quotient spaces.   As a further preparation, let us quickly review the notion of a quotient space. Given any vector space V and any linear subspace $N \subseteq V$, the quotient space V/N is defined as follows: Consider the equivalence relation on V defined as $v_1 \sim v_2$ iff $v_1 - v_2 \in N$; then V/N is defined as the vector space of all equivalence classes $[v]$, $v \in V$, with the vector addition and scalar multiplication naturally defined as $[v_1] + [v_2] := [v_1 + v_2]$ and $\alpha[v] := [\alpha v]$. One can also think of V/N as the vector space of all the sets of the form $v + N$, $v \in V$. It it not hard to see that

$$\dim(V/N) = \dim(V) - \dim(N). \tag{7.15}$$

The tangent spaces of $\mathbb{P}_0$.   Let $\mathbf{p} \in \mathbb{P}_0$, and let γ be a C^1-path in $\mathbb{P}_0$ with $\gamma(0) = \mathbf{p}$. One can show that such paths must be of the form $\gamma := \pi \circ \tilde{\gamma}$ for some C^1-path $\tilde{\gamma}$ in $\mathbb{T}_0$, and so we have $\gamma'(0) = \pi'(\tilde{\gamma}(0))\tilde{\gamma}'(0)$. The element $\gamma'(0) \in T_{\mathbf{p}}\mathbb{P}_0$ is therefore uniquely determined by the pair $(\tilde{\gamma}(0), \tilde{\gamma}'(0))$.

However, the choice of $\tilde{\gamma}$ is not unique, and so if we want to abstractly represent $\gamma'(0)$ by such pairs then we should rather write

$$\gamma'(0) \cong \big\{ (\tilde{\gamma}(0), \tilde{\gamma}'(0)) \,\big|\, \tilde{\gamma} \text{ such that } \gamma = \pi \circ \tilde{\gamma} \big\}.$$

In particular, for any fixed such path $\tilde{\gamma}_0$, also the paths $\tilde{\gamma}_{a,b}(s) := a(1 + bs)\tilde{\gamma}_0(s)$ for $\forall a, b \in \mathbb{R}$, $a \neq 0$, all lead to the same path $\gamma = \pi \circ \tilde{\gamma}_{a,b}$ (for s small enough that $1 + bs \neq 0$). In fact, this subfamily of paths $\tilde{\gamma}$ already gives rise to our full set of pairs $(\tilde{\gamma}(0), \tilde{\gamma}'(0))$, since the parameter a lets $\tilde{\gamma}_{a,b}(0)$ explore all non-zero points on the given line $\mathbf{p}$ (recall that we need $\pi(\tilde{\gamma}(0)) = \gamma(0) = \mathbf{p}$), and the parameter b lets the nearby point $\tilde{\gamma}(s)$ for small s move independently in the direction of its own line through the origin, thus exploring also our freedom in $\tilde{\gamma}'(0)$. This shows that

$$\gamma'(0) \cong \big\{ (\tilde{\gamma}_{a,b}(0), \tilde{\gamma}'_{a,b}(0)) \,\big|\, a, b \in \mathbb{R}, \ a \neq 0 \big\}$$

---

[5]We will use the symbol $\perp$ to indicate *pseudo*-orthogonality, i.e., orthogonality w.r.t. to the Minkowski inner product.

$$\cong \big\{ \big(a\tilde{\gamma}_0(0), a(\tilde{\gamma}_0'(0) + b\tilde{\gamma}_0(0)) \big) \,\big|\, a, b \in \mathbb{R}, \ a \neq 0 \big\}$$
$$= \big\{ \big(ax, a(\tilde{\gamma}_0'(0) + bx) \big) \,\big|\, a, b \in \mathbb{R}, \ a \neq 0 \big\},$$

where we abbreviate $x := \tilde{\gamma}_0(0)$. Since

$$\tilde{\gamma}_0'(0) \in T_x(\mathbb{T}_0) = \pi(x)^\perp = \pi(\tilde{\gamma}_0(0))^\perp = \mathbf{p}^\perp$$

and since bx traverses all of $\mathbf{p} \subset \mathbf{p}^\perp$,[6] the class of all vectors $\tilde{\gamma}_0'(0) + bx$, $b \in \mathbb{R}$, is in fact an element in the quotient space $\mathbf{p}^\perp/\mathbf{p}$, which shows that

$$\gamma'(0) \cong \big\{ \big(ax, a[\tilde{\gamma}_0'(0)] \big) \,\big|\, a \in \mathbb{R} \setminus \{0\} \big\},$$

where $[\tilde{\gamma}_0'(0)]$ is the equivalence class in $\mathbf{p}^\perp/\mathbf{p}$ containing $\tilde{\gamma}_0'(0)$. Finally, considering the map $f \colon (\mathbf{p} \setminus \{0\}) \times \mathbf{p}^\perp/\mathbf{p} \to \mathbf{p}^* \otimes \mathbf{p}^\perp/\mathbf{p}$ given by $f(v_1, v_2) := \frac{1}{v_1} \otimes v_2$, this set can be identified as the preimage $f^{-1}\big(\big\{ \frac{1}{x} \otimes [\tilde{\gamma}_0'(0)] \big\}\big)$, and so in this sense we have

$$\gamma'(0) \cong \frac{1}{x} \otimes [\tilde{\gamma}_0'(0)] = \frac{1}{\tilde{\gamma}_0(0)} \otimes [\tilde{\gamma}_0'(0)] \in \mathbf{p}^* \otimes \mathbf{p}^\perp/\mathbf{p}. \tag{7.16}$$

Since

$$\dim(\mathbf{p}^* \otimes \mathbf{p}^\perp/\mathbf{p}) \stackrel{(6.4)}{=} \dim(\mathbf{p}^*) \cdot \dim(\mathbf{p}^\perp/\mathbf{p}) \stackrel{(7.15)}{=} 1 \cdot \big(\dim(\mathbf{p}^\perp) - \dim(\mathbf{p}) \big)$$
$$= (\dim(\mathbb{T}) - 1) - 1 \stackrel{(7.9a)}{=} m - 1$$

coincides with the expected dimension of the manifold $\mathbb{P}_0$ (recall (7.9b)), we have thus shown that the tangent space $T_{\mathbf{p}}\mathbb{P}_0$ has the representation

$$T_{\mathbf{p}}\mathbb{P}_0 \cong \mathbf{p}^* \otimes \mathbf{p}^\perp/\mathbf{p}. \tag{7.17}$$

7.2.2 Bilinear Forms on $\mathbb{P}_0$, Conformal Structure

A bilinear form $\mathfrak{m}$ on $\mathbb{P}_0$ is therefore a collection of inner products $\langle \cdot, \cdot \rangle_{\mathbf{p}}$ on $\mathbf{p}^* \otimes \mathbf{p}^\perp/\mathbf{p}$, one for each $\mathbf{p} \in \mathbb{P}_0$. To construct one, let us begin by first constructing a proper (i.e., positive-definite) inner product on $\mathbf{p}^\perp/\mathbf{p}$ for some fixed $\mathbf{p} \in \mathbb{P}_0$, naturally derived from the Minkowski inner product on $\mathbb{T}$.

Inner product on $\mathbf{p}^\perp/\mathbf{p}$. Let $\mathbf{p} \in \mathbb{P}_0$ be fixed. Then for $\forall x_1, x_2 \in \mathbf{p}^\perp$ and $\forall y_1, y_2 \in \mathbf{p}$ we have

$$\langle x_1 + y_1, x_2 + y_2 \rangle_{\mathbb{T}} = \langle x_1, x_2 \rangle_{\mathbb{T}} + \langle x_1, y_2 \rangle_{\mathbb{T}} + \langle y_1, x_2 \rangle_{\mathbb{T}} + \langle y_1, y_2 \rangle_{\mathbb{T}}$$
$$= \langle x_1, x_2 \rangle_{\mathbb{T}} + 0 + 0 + 0,$$

since $x_1 \perp \mathbf{p} \ni y_2$ and $x_2 \perp \mathbf{p} \ni y_1$, and by (7.5a). Since the right-hand side does not depend on y_1 and y_2, this shows that we can define a symmetric bilinear form on $\mathbf{p}^\perp/\mathbf{p}$ as

$$\langle w_1, w_2 \rangle_{\mathbf{p}^\perp/\mathbf{p}} := \langle x_1, x_2 \rangle_{\mathbb{T}} \quad \text{for } \forall w_1, w_2 \in \mathbf{p}^\perp/\mathbf{p},$$

---

[6]This inclusion holds because $\mathbf{p} \in \mathbb{P}_0$ consists of isotropic vectors, i.e., of points $x \in \mathbb{T}$ with $\langle x, x \rangle = 0$, which implies that $\langle x, y \rangle = 0$ for $\forall y \in \mathbf{p}$, i.e., $x \in \mathbf{p}^\perp$.

where $x_1, x_2 \in \mathbf{p}^\perp$ are arbitrary members of the given two equivalence classes w_1 and w_2, respectively; in other words, we define

$$\langle [x_1], [x_2] \rangle_{\mathbf{p}^\perp/\mathbf{p}} := \langle x_1, x_2 \rangle_{\mathbb{T}} . \qquad (7.18)$$

To interpret this bilinear form, observe that the straight line $\mathbf{p}$ cannot be a subset of the $(x^0 = 0)$-hyperplane since by (7.3) the only isotropic point in that hyperplane is the origin. We can therefore pick those unique representations $x_1 = (x_1^0, \vec{x}_1)$ and $x_2 = (x_2^0, \vec{x}_2)$ with $x_1^0 = x_2^0 = 0$; for these representations we have

$$\langle w_1, w_2 \rangle_{\mathbf{p}^\perp/\mathbf{p}} = \langle \vec{x}_1, \vec{x}_2 \rangle_{\mathbb{R}^m} .$$

To see that this bilinear form is positive definite and thus indeed a proper inner product, let $w \in \mathbf{p}^\perp/\mathbf{p}$, and let $x \in \mathbf{p}^\perp$ be the representation of w with $x^0 = 0$. Then we have $\langle w, w \rangle_{\mathbf{p}^\perp/\mathbf{p}} = \langle \vec{x}, \vec{x} \rangle_{\mathbb{R}^m} = \|\vec{x}\|^2 \geq 0$, with equality if and only if $\vec{x} = 0$, i.e., if $x = 0$ and thus $w = 0$.

Bilinear forms on $\mathbb{P}_0$. Since by (7.14) the tangent spaces of $\mathbb{T}_0$ are simply subspaces of $\mathbb{T}$, and since the Minkowski inner product is positive semidefinite[7] on these subspaces (this is another consequence of our calculations in the preceding paragraph), the Minkowski inner product naturally induces a pseudometric[8] on $\mathbb{T}_0$, defined as

$$|\mathrm{d}x|_x := \langle \mathrm{d}x, \mathrm{d}x \rangle_{\mathbb{T}}^{1/2} \quad \text{for} \quad \mathrm{d}x \in T_x(\mathbb{T}_0) = \pi(x)^\perp. \qquad (7.19)$$

(Consider $\mathrm{d}x \parallel x \in \pi(x)^\perp$ to see that we can have $|\mathrm{d}x|_x = 0$ although $\mathrm{d}x \neq 0$.)

On $\mathbb{P}_0$, however, there is no *unique* way of making such a construction. (On the positive side, these different ways will all lead to proper metrics, as opposed to just pseudometrics.) Indeed, following the train of thought from Section 7.2.1, to assign a length to a given infinitesimal curve segment $\mathrm{d}\gamma \subset \mathbb{P}_0$ with some starting point $\mathbf{p} \in \mathbb{P}_0$, the natural idea would be to find a curve segment $\mathrm{d}\tilde{\gamma} \subset \mathbb{T}_0$ with $\mathrm{d}\gamma = \pi \circ \mathrm{d}\tilde{\gamma}$, and then to define its length via (7.19) as $|\mathrm{d}\gamma|_{\mathbf{p}} := |\mathrm{d}\tilde{\gamma}|_x$, where $x \in \mathbf{p} \setminus \{0\}$ is the starting point of $\mathrm{d}\tilde{\gamma}$. However, as we saw in Section 7.2.1, there are various ways of choosing $\mathrm{d}\tilde{\gamma}$, and in particular one has the freedom to scale the entire segment (and thus also its length) by any factor $a > 0$. Only once its starting point $x \in \mathbf{p} \setminus \{0\}$ is decided upon, the length of the segment $\mathrm{d}\tilde{\gamma}$ will be fixed (since it is unaffected by the choice of b).

This argument manifests itself in the representation (7.17) of an infinitesimal curve segment $\mathrm{d}\gamma \in T_{\mathbf{p}}\mathbb{P}_0$ with starting point $\mathbf{p}$ by an element $\mathrm{d}\mathbf{p} \in \mathbf{p}^* \otimes \mathbf{p}^\perp/\mathbf{p}$: Once a point $x \in \mathbf{p} \setminus \{0\}$ is fixed, there is a unique way of writing any given element $\mathrm{d}\mathbf{p}$ of that product space in the form $\mathrm{d}\mathbf{p} = \frac{1}{x} \otimes \mathrm{d}v$ for some $\mathrm{d}v \in \mathbf{p}^\perp/\mathbf{p}$,

---

[7]A bilinear form $\langle \cdot, \cdot \rangle \colon V \times V \to \mathbb{R}$ on some given vector space V is *positive semidefinite* if $\forall x \in V \colon \langle x, x \rangle \geq 0$. It is *positive definite* if the only vector with $\langle x, x \rangle = 0$ is $x = 0$.

[8]A pseudometric is a generalized metric that allows the distance between two distinct points to be zero. One can therefore construct a pseudometric by defining a seminorm (as opposed to a proper norm) on each tangent space, and thus by defining a positive-semidefinite (as opposed to positive-definite) bilinear form.

and we can recover this unique dv simply by computing $x \otimes d\mathbf{p} = x \otimes \frac{1}{x} \otimes dv \cong \left(\frac{x}{x}\right) \cdot dv = dv$. Since we have a proper inner product on $\mathbf{p}^\perp/\mathbf{p}$, we can thus define a proper norm on $T_\mathbf{p}\mathbb{P}_0$ as $|d\mathbf{p}|_\mathbf{p} := |x \otimes d\mathbf{p}| = |dv|$, where the norm on the right-hand side is the one induced by our inner product on $\mathbf{p}^\perp/\mathbf{p}$.

More generally, for any fixed $\mathbf{p} \in \mathbb{P}_0$, any choice of $x_1, x_2 \in \mathbf{p}$ defines a bilinear form

$$\langle d\mathbf{p}_1, d\mathbf{p}_2 \rangle_\mathbf{p}^{x_1 \otimes x_2} := \langle x_1 \otimes d\mathbf{p}_1, x_2 \otimes d\mathbf{p}_2 \rangle_{\mathbf{p}^\perp/\mathbf{p}}$$

for $\forall d\mathbf{p}_1, d\mathbf{p}_2 \in T_\mathbf{p}\mathbb{P}_0 = \mathbf{p}^* \otimes \mathbf{p}^\perp/\mathbf{p}$, where the inner product on the right is the one on $\mathbf{p}^\perp/\mathbf{p}$. Note that this bilinear form indeed only depends on $x_1 \otimes x_2$ (as opposed to the pair (x_1, x_2)) since the right-hand side remains unchanged under the transition $(x_1, x_2) \to (ax_1, \frac{1}{a}x_2)$ for any $a \neq 0$.

To specify a bilinear form $\mathfrak{m}$ on $\mathbb{P}_0$, we therefore need to choose an element $x_1 \otimes x_2 \in \mathbf{p}^{\otimes 2}$ for every point $\mathbf{p} \in \mathbb{P}_0$, i.e., we need to specify a section $\psi_\mathfrak{m} = \mathfrak{x}_1 \otimes \mathfrak{x}_2 \in \mathcal{S}(\mathscr{P}^{\otimes 2})$ (where $\mathfrak{x}_1, \mathfrak{x}_2 \in \mathcal{S}(\mathscr{P})$). This will define a bilinear form as

$$\langle d\mathbf{p}_1, d\mathbf{p}_2 \rangle_\mathbf{p}^\mathfrak{m} := \left\langle \mathfrak{x}_1(\mathbf{p}) \otimes d\mathbf{p}_1, \mathfrak{x}_2(\mathbf{p}) \otimes d\mathbf{p}_2 \right\rangle_{\mathbf{p}^\perp/\mathbf{p}}. \tag{7.20}$$

The mapping $\psi_\mathfrak{m} \mapsto \mathfrak{m}$ is linear, which makes the representation of a bilinear form $\mathfrak{m}$ on $\mathbb{P}_0$ by a section $\psi_\mathfrak{m} \in \mathcal{S}(\mathscr{P}^{\otimes 2})$ the most natural one.

The bilinear form $\mathfrak{m}$ given by some section $\psi_\mathfrak{m} \in \mathcal{S}(\mathscr{P}^{\otimes 2})$ is positive definite (and thus defines a proper metric) if and only if $\psi_\mathfrak{m}$ is of the form $\psi_\mathfrak{m} = \mathfrak{x} \otimes \mathfrak{x}$ for some $\mathfrak{x} \in \mathcal{S}_\varnothing(\mathscr{P})$, i.e., if $\psi_\mathfrak{m} \in \mathcal{S}_+(\mathscr{P}^{\otimes 2})$ (recall (6.15)).

Conformal structure. Our definition (7.20) implies that

$$\forall \psi_{\mathfrak{m}_1}, \psi_{\mathfrak{m}_2} \in \mathcal{S}_+(\mathscr{P}^{\otimes 2}): \quad \mathfrak{m}_1 = \frac{\psi_{\mathfrak{m}_1}}{\psi_{\mathfrak{m}_2}} \cdot \mathfrak{m}_2,$$

i.e., that the various positive-definite bilinear forms given by the different sections in $\mathcal{S}_+(\mathscr{P}^{\otimes 2})$—and thus also their associated metrics—are all positive multiples of one another; in other words, they all belong to the same conformal structure. We have therefore shown that the Minkowski inner product induces a unique *conformal structure* $\mathfrak{c}$ on $\mathbb{P}_0$, and that every section $\psi_\mathfrak{m} \in \mathcal{S}_+(\mathscr{P}^{\otimes 2})$ corresponds to a specific positive-definite bilinear form in $\mathfrak{c}$.

It is this conformal structure $\mathfrak{c}$ that we will use for the construction of our interest rate model, and so by our discussion in this section the one-to-one correspondence $\psi_\mathfrak{m} \leftrightarrow \mathfrak{m}$ allows us to identify

$$\mathcal{S}(\mathscr{P}^{\otimes 2}) \cong \mathcal{S}(\mathscr{W}_\mathfrak{c}). \tag{7.21}$$

7.3 Covariant and Contravariant Vectors

The next step in our construction will be the definition of our group of operators $(\mathcal{U}_s)_{s \in \mathbb{R}}$. Before we begin, however, we need to learn some basic tools and notation that have been proven useful in the physics literature, in particular when performing calculations in Minkowski space in the context of the Special Theory of Relativity.

7.3.1 Dual Space, Representation Theorem

As in any vector space equipped with a bilinear form, every element $x \in \mathbb{T}$ naturally gives rise to an element $x^* \in \mathbb{T}^*$ in the dual space

$$\mathbb{T}^* := \{\lambda \colon \mathbb{T} \to \mathbb{R} \mid \lambda \text{ linear}\},$$

defined as

$$x^*(y) := \langle x, y \rangle \qquad (7.22)$$

for $\forall y \in \mathbb{T}$. In fact, *every* $\lambda \in \mathbb{T}^*$ can be obtained in this way: Since $\mathbb{T}$ is finite dimensional and since the Minkowski inner product is nondegenerate, there exists a pseudo-orthogonal basis $(e_i)_i$ of $\mathbb{T}$ with $\langle e_i, e_i \rangle \neq 0$ for $\forall i$, and thus every given $\lambda \in \mathbb{T}^*$ can be written as $\lambda = x^*$ for

$$x := \sum_i \frac{\lambda(e_i)}{\langle e_i, e_i \rangle} e_i. \qquad (7.23)$$

The mapping $x \mapsto x^*$ therefore defines a linear bijection $Q \colon \mathbb{T} \to \mathbb{T}^*$ such that

$$(Qx)(y) = \langle x, y \rangle \qquad (7.24)$$

for $\forall x, y \in \mathbb{T}$.[9]

7.3.2 Einstein Notation

Definition. Now let $(e_i)_i$ be *any* basis of $\mathbb{T}$, let $(e^i)_i$ be the associated dual basis, i.e., the unique basis of $\mathbb{T}^*$ composed of those elements $e^i \in \mathbb{T}^*$ with

$$e^j(e_i) = \delta_i^j$$

for $\forall i, j$, where

$$\delta_i^j := \begin{cases} 1 & \text{if } i = j, \\ 0 & \text{if } i \neq j \end{cases}$$

is the Kronecker symbol. Furthermore, let us denote the coefficients of any $x \in \mathbb{T}$ and of its associated vector $x^* := Qx \in \mathbb{T}^*$ in these two bases by x^i and x_i, respectively, i.e.,

$$x = x^i e_i := \sum_i x^i e_i \qquad \text{and} \qquad x^* = x_i e^i := \sum_i x_i e^i, \qquad (7.25)$$

where *from now on it is understood that indices that appear twice within a product are summed over* (this is called the *summation convention*). The coefficient vector of x is called the *contravariant* vector, while the one of x^* is called the *covariant* vector.

The use of upper and lower indices to denote contravariant and covariant vectors, respectively, in combination with the use of the summation convention, is called *Einstein notation*.

---

[9] In this context the reader may also recall the related Riesz Representation Theorem, which says that such a mapping $Q \colon H \to H^*$ exists for general Hilbert spaces H with a proper inner product.

Lowering and raising indices. With this notational convention we have the relation

$$
\begin{aligned}
x_i &= x_j \delta_i^j = x_j e^j(e_i) = x^*(e_i) \\
&\stackrel{(7.22)}{=} \langle x, e_i \rangle = \langle x^j e_j, e_i \rangle = x^j \langle e_j, e_i \rangle = \langle e_i, e_j \rangle x^j \\
&= Q_{ij} x^j,
\end{aligned} \tag{7.26}
$$

where the coefficients Q_{ij} (note: with lower indices) are defined as

$$
Q_{ij} := \langle e_i, e_j \rangle. \tag{7.27}
$$

In short, (7.26) says that the *covariant* vector is obtained from the associated *contravariant* one by multiplying the latter by the matrix (Q_{ij}).

In $\mathbb{R}^n$, equipped with the standard Euclidean inner product and with the standard basis $(e_i)_i$, the matrix (Q_{ij}) would be the identity matrix, and so the two coefficient vectors would coincide, which explains why one typically does not distinguish between $\mathbb{R}^n$ and $(\mathbb{R}^n)^*$. In the Minkowski space $\mathbb{T}$, equipped with the same standard basis $(e_i)_i$, however, (Q_{ij}) is given by (7.2), and so the two coefficient vectors differ by the sign of their zeroth component. When changing to a different basis $(e_i)_i$, (Q_{ij}) will change further, and so the two coefficient vectors can differ in less obvious ways.

The Einstein notation introduced here (and common in the physics literature) is therefore a useful way to make the transition from a vector $x \in \mathbb{T}$ to its associated element $x^* \in \mathbb{T}^*$ of the dual space as notationally convenient as possible. One says that the coefficients Q_{ij} can be used to "lower indices" by way of (7.26).

Furthermore, if we define the coefficients Q^{ij} (note: with upper indices) such that the matrix (Q^{ij}) is the inverse of the matrix (Q_{ij}),[10] i.e., such that

$$
Q_{ij} Q^{jk} = \delta_i^k \qquad \text{and} \qquad Q^{ij} Q_{jk} = \delta_k^i, \tag{7.28}
$$

then (7.26) implies that

$$
Q^{ij} x_j \stackrel{(7.26)}{=} Q^{ij} Q_{jk} x^k \stackrel{(7.28)}{=} \delta_k^i x^k = x^i, \tag{7.29}
$$

and so one says that the coefficients Q^{ij} can be used to "raise indices."

7.3.3 Inner Product, Symmetry

By (7.25)–(7.27), the Minkowski inner product can conveniently be written as

$$
\langle x, y \rangle = \langle x^i e_i, y^j e_j \rangle = x^i y^j \langle e_i, e_j \rangle = x^i y^j Q_{ij} = x^i y_i
$$

for $\forall x, y \in \mathbb{T}$, which in particular implies that

$$
x^i y_i = \langle x, y \rangle = \langle y, x \rangle = x_i y^i. \tag{7.30}
$$

---

[10]The invertibility of (Q_{ij}) follows from the fact that the mapping $x \leftrightarrow x^*$ is bijective.

This shows that we can switch the upper and the lower position of any pair of equal indices within a product at will.

Note also that by our definition of Q_{ij} and the symmetry of the Minkowski inner product, and by the definition of (Q^{ji}) we have

$$Q_{ij} = Q_{ji} \qquad \text{and} \qquad Q^{ij} = Q^{ji}.$$

7.3.4   Operators

By way of our linear bijection $Q\colon \mathbb{T} \to \mathbb{T}^*$, we can identify any operator $A\colon \mathbb{T} \to \mathbb{T}$ with their alternative representations

$$
\begin{aligned}
A' &: \mathbb{T}^* \to \mathbb{T}, & A' &:= A \circ Q^{-1}, \\
A'' &: \mathbb{T} \to \mathbb{T}^*, & A'' &:= Q \circ A, & \text{(7.31)} \\
A''' &: \mathbb{T}^* \to \mathbb{T}^*, & A''' &:= Q \circ A \circ Q^{-1},
\end{aligned}
$$

corresponding to the different ways to map from the left side to the right side of the following commutative diagram:

$$
\begin{array}{ccc}
\mathbb{T} & \xrightarrow{\ A\ } & \mathbb{T} \\
\downarrow{\scriptstyle Q} & & \downarrow{\scriptstyle Q} \\
\mathbb{T}^* & \longrightarrow & \mathbb{T}^*
\end{array}
$$

We will denote the matrix representations of A, A', A'', and A''', w.r.t. our bases of $\mathbb{T}$ and $\mathbb{T}^*$, by $A^i{}_j$, A^{ij}, A_{ij}, and $A_i{}^j$, respectively, i.e., we have

$$(Ax)^i = A^i{}_j x^j, \qquad (A'x^*)^i = A^{ij} x_j, \qquad (A''x)_i = A_{ij} x^j, \qquad (A'''x^*)_i = A_i{}^j x_j$$

for $\forall x \in \mathbb{T}$ and $\forall x^* \in \mathbb{T}^*$. In other words, the positions of the left and the right index of A indicate the target and the source space, respectively, with an upper left and a lower right index corresponding to $\mathbb{T}$, and with a lower left and an upper right index corresponding to $\mathbb{T}^*$.

Since rewriting the definitions (7.31) of A', A'', and A''' in Einstein notation yields

$$A^{ij} = A^i{}_k Q^{kj}, \qquad A_{ij} = Q_{ik} A^k{}_j, \qquad A_i{}^j = Q_{ik} A^k{}_l Q^{lj},$$

this notation is consistent with our notion of raising and lowering indices via the matrices Q^{ij} and Q_{ij}. Note though that in general we do *not* have $(A^{ij}) = (A_{ij})^{-1}$.

7.3.5   Adjoint

The adjoint of an operator $A\colon \mathbb{T} \to \mathbb{T}$ is the unique operator $A^*\colon \mathbb{T} \to \mathbb{T}$ such that $\langle A^*x, y \rangle = \langle x, Ay \rangle$ for $\forall x, y \in \mathbb{T}$, or in index notation,

$$(A^*)^i{}_j x^j y_i = (A^*x)^i y_i = \langle A^*x, y \rangle = \langle x, Ay \rangle = x^j (Ay)_j = x^j A_j{}^i y_i \,.$$

This shows that the coefficients of A^* are given by

$$(A^*)^i{}_j := A_j{}^i \,. \tag{7.32}$$

7.4 Pseudo-Orthogonal Operators & Evolution Group

7.4.1 Outline

In this section we will describe the construction of our operators $\mathcal{U}_s$ with all of its representations. The key to this construction will be the group $\hat{O}(\mathbb{T})$ of certain pseudo-orthogonal operators $A\colon \mathbb{T} \to \mathbb{T}$ and their multiples.

After defining the group $\hat{O}(\mathbb{T})$ in Section 7.4.2, in Sections 7.4.3–7.4.4 we will describe how every operator $A \in \hat{O}(\mathbb{T})$ has natural representations acting on states $\mathbf{p} \in \mathbb{P}_0$, on functions, on sections, and on other mathematical objects. Since these representations preserve the group property of $\hat{O}(\mathbb{T})$,[11] this will allow us to specify our group $(\mathcal{U}_s)_{s\in\mathbb{R}}$ (with all of its various representations) by way of a group $(A_s)_{s\in\mathbb{R}}$ of operators $A_s \in \hat{O}(\mathbb{T})$.

In Section 7.4.5 we will then show that the representation of any operator $A \in \hat{O}(\mathbb{T})$ (and thus of our operators $\mathcal{U}_s$) acting on states $\mathbf{p} \in \mathbb{P}_0$ preserves the conformal structure of $\mathbb{P}_0$, as desired. The reason for this is simple: Since by definition pseudo-orthogonal operators preserve the Minkowski inner product, their multiples preserve it up to a factor; their representation acting on $\mathbb{P}_0$ thus preserves the conformal structure on $\mathbb{P}_0$ that was naturally derived from the Minkowski inner product.

Finally, in Section 7.4.6, Lemma 7.7 (ii), we will then derive the conditions on an operator $\bar{\varepsilon}\colon \mathbb{T} \to \mathbb{T}$ under which the operators $\mathrm{e}^{s\bar{\varepsilon}}$ are in $\hat{O}(\mathbb{T})$ for $\forall s \in \mathbb{R}$. We can then define

$$\mathcal{U}_s := \mathrm{e}^{s\bar{\varepsilon}} \quad \text{for } \forall s \in \mathbb{R},$$

where the generator $\bar{\varepsilon}$ is a model parameter.

7.4.2 Definition: Pseudo-Orthogonal Operators

Motivation. In the regular Euclidean space $\mathbb{R}^n$ with the standard basis, linear maps $A\colon \mathbb{R}^n \to \mathbb{R}^n$ that preserve the Euclidean inner product are represented by orthogonal matrices, i.e., those matrices with $A^T A = I$. Their counterparts in our Minkowski space $\mathbb{T}$, namely linear maps $A\colon \mathbb{T} \to \mathbb{T}$ that preserve the *Minkowski* inner product, are those maps fulfilling $A^* A = I$; they are called *pseudo*-orthogonal.

Pseudo-orthogonal operators. To make our definition more explicit, we call a linear operator $A\colon \mathbb{T} \to \mathbb{T}$ *pseudo-orthogonal* if it is orthogonal w.r.t. the Minkowski inner product, i.e., if

$$\forall x \in \mathbb{T}\colon \quad \langle Ax, Ax \rangle = \langle x, x \rangle. \tag{7.33}$$

---

[11]E.g., denoting by $A_{\mathbb{P}_0}\colon \mathbb{P}_0 \to \mathbb{P}_0$ the representation of some $A \in \hat{O}(\mathbb{T})$ acting on $\mathbb{P}_0$, we have $(A^1 \cdot A^2)_{\mathbb{P}_0} = A^1_{\mathbb{P}_0} \circ A^2_{\mathbb{P}_0}$ and $I_{\mathbb{P}_0} = \mathrm{id}$. As a result, the set $\hat{O}(\mathbb{P}_0) := \{A_{\mathbb{P}_0} \mid A \in \hat{O}(\mathbb{T})\}$ forms a group, and every subgroup of $\hat{O}(\mathbb{T})$ gives rise to a subgroup of $\hat{O}(\mathbb{P}_0)$.

We denote the set of pseudo-orthogonal operators by $O(\mathbb{T})$. It is easy to see that these operators form a group, called the *Lorentz group*.

Note that in fact (7.33) implies that

$$\forall A \in O(\mathbb{T}) \; \forall x, y \in \mathbb{T}: \quad \langle Ax, Ay \rangle = \langle x, y \rangle \tag{7.34}$$

(this can be shown by replacing x in (7.33) by $x + y$). Rewriting this as $\langle A^*Ax, y \rangle = \langle x, y \rangle$ and then using the nondegeneracy property of the Minkowski inner product, this shows that a linear operator $A \colon \mathbb{T} \to \mathbb{T}$ is pseudo-orthogonal if and only if $A^*A = I$, i.e., if it is invertible with inverse A^*:

$$A \in O(\mathbb{T}) \quad \Leftrightarrow \quad A^{-1} = A^* . \tag{7.35}$$

Since the right inverse is equal to the left inverse, we have $A^*A = I$ if and only if $I = AA^* = (A^*)^*(A^*)$, so that

$$A \in O(\mathbb{T}) \quad \Leftrightarrow \quad A^* \in O(\mathbb{T}) . \tag{7.36}$$

Subgroups. Observe that by (7.33) pseudo-orthogonal operators map the set $\mathbb{T}_0 \cup \{0\}$ and thus because of their invertibility also the set $\mathbb{T}_0$ onto itself. Since the latter is the disjoint union of the two connected components $\mathbb{T}_0^\uparrow$ and $\mathbb{T}_0^\downarrow$, we can therefore write

$$O(\mathbb{T}) = O^\uparrow(\mathbb{T}) \cup O^\downarrow(\mathbb{T}), \tag{7.37}$$

where the subgroups $O^\uparrow(\mathbb{T})$ and $O^\downarrow(\mathbb{T})$ are defined as

$$O^\uparrow(\mathbb{T}) := \{A \in O(\mathbb{T}) \,|\, A(\mathbb{T}_0^\uparrow) = \mathbb{T}_0^\uparrow\}, \tag{7.38a}$$

$$O^\downarrow(\mathbb{T}) := \{A \in O(\mathbb{T}) \,|\, A(\mathbb{T}_0^\uparrow) = \mathbb{T}_0^\downarrow\}. \tag{7.38b}$$

Lemma 7.2. *The operators $A \in O^\uparrow(\mathbb{T})$ map the sets $\mathbb{T}_0$, $\mathbb{T}_0^\uparrow$, $\mathbb{T}^\uparrow$, $\mathbb{T}_0^\downarrow$, and $\mathbb{T}^\downarrow$ onto themselves, respectively.*

Proof. Operators $A \in O^\uparrow(\mathbb{T})$ map $\mathbb{T}_0^\uparrow$ onto itself by the very definition of $O^\uparrow(\mathbb{T})$ in (7.38a), and by considering convex combinations of points in $\mathbb{T}_0^\uparrow$ this implies that they also map $\mathbb{T}^\uparrow$ onto itself. By multiplying the equations $A(\mathbb{T}_0^\uparrow) = \mathbb{T}_0^\uparrow$ and $A(\mathbb{T}^\uparrow) = \mathbb{T}^\uparrow$ by -1 on both sides we then find that they map also the sets $\mathbb{T}_0^\downarrow$ and $\mathbb{T}^\downarrow$ onto themselves, respectively, which finally allows us to conclude that also $\mathbb{T}_0 = \mathbb{T}_0^\uparrow \cup \mathbb{T}_0^\downarrow$ is mapped onto itself. $\square$

Scalar multiples. We extend the subgroup $O^\uparrow(\mathbb{T})$ to the group of all positive multiples of its elements, i.e.,

$$\hat{O}(\mathbb{T}) := \{\nu A_0 \,|\, A_0 \in O^\uparrow(\mathbb{T}), \; \nu > 0\} .$$

Lemma 7.3. *For each operator $A \in \hat{O}(\mathbb{T})$ there is only exactly one way of writing it in the form νA_0 for some $\nu > 0$ and some $A_0 \in O^\uparrow(\mathbb{T})$.*

Given any $A \in \hat{O}(\mathbb{T})$, this allows us to use the notation $|A| > 0$ to denote its unique associated scaling factor ν.

Proof. For any given operator of this form we have $\langle Ax, Ax \rangle = \nu^2 \langle A_0 x, A_0 x \rangle = \nu^2 \langle x, x \rangle$ by (7.34), and so fixing any non-isotropic vector $x \in \mathbb{T}$, we see that ν and A_0 must be given by $\nu = \sqrt{\langle Ax, Ax \rangle / \langle x, x \rangle}$ and $A_0 = \frac{1}{\nu} A$. □

As a result of the scaling, the operators in this extended class no longer fulfill (7.34)–(7.35), but instead

$$\forall A \in \hat{O}(\mathbb{T}) \; \forall x, y \in \mathbb{T}: \quad \langle Ax, Ay \rangle = |A|^2 \cdot \langle x, y \rangle, \tag{7.39}$$

$$\forall A \in \hat{O}(\mathbb{T}): \qquad A^{-1} = |A|^{-2} \cdot A^*, \tag{7.40}$$

as well as the counterparts of (7.36), i.e.,

$$A \in \hat{O}(\mathbb{T}) \quad \Leftrightarrow \quad A^* \in \hat{O}(\mathbb{T}), \tag{7.41}$$

and of Lemma 7.2:

Lemma 7.4. *The operators $A \in \hat{O}(\mathbb{T})$ map the sets $\mathbb{T}_0$, $\mathbb{T}_0^\uparrow$, $\mathbb{T}^\uparrow$, $\mathbb{T}_0^\downarrow$, and $\mathbb{T}^\downarrow$ onto themselves, respectively.*

7.4.3   Actions on States, Functions, and Sections

Action on states. Any map $A \in \hat{O}(\mathbb{T})$ gives rise to a representation $A \colon \mathbb{P}_0 \to \mathbb{P}_0$ acting on $\mathbb{P}_0$, defined as

$$\forall \mathbf{p} \in \mathbb{P}_0: \quad A\mathbf{p} := \pi(Ax) \quad \text{for any arbitrary } x \in \mathbf{p} \setminus \{0\}. \tag{7.42}$$

This map is well defined since by (7.10) the expression $\pi(Ax)$ does not depend on the specific choice of $x \in \mathbf{p} \setminus \{0\}$, and it indeed maps into $\mathbb{P}_0$ since by Lemma 7.4 A maps $\mathbb{T}_0$ onto itself.

Note that in order to obtain a representation acting on our state space $\mathcal{M} \subsetneq \mathbb{P}_0$, later in (7.124) we will have to ensure that for the specific maps A that we will be using, this representation on $\mathbb{P}_0$ defined in (7.42) actually maps $\mathcal{M}$ into itself.

Also observe that because of (7.10) the action on $\mathbb{P}_0$ does not depend on the multiplying factor $\nu > 0$; this representation (as well as all other representations derived from it in the following) therefore only uses part of the information encoded in $A \in \hat{O}(\mathbb{T})$. The factor ν will, however, enter the representation acting on sections defined later on, and so—as expected—that representation will indeed carry more information.

Action on functions. Given any operator $A \in \hat{O}(\mathbb{T})$, its representation acting on functions $f \in \mathcal{F}(\mathbb{P}_0)$ can be derived directly from the one acting on states $\mathbf{p} \in \mathbb{P}_0$, as these two representations contain the same information. In line with (5.54), it is defined as

$$Af := f \circ A^{-1}, \tag{7.43}$$

where the operator A^{-1} on the right is the inverse of the representation of A acting on $\mathbb{P}_0$. In this way, if as in (6.18) we also define the action of A on real numbers as

$$\forall a \in \mathbb{R}: \quad Aa := a, \tag{7.44}$$

we have

$$(Af)(A\mathbf{p}) = A(f(\mathbf{p})) \tag{7.45}$$

for $\forall f \in \mathcal{F}(\mathbb{P}_0)$ and $\forall \mathbf{p} \in \mathbb{P}_0$, and so we obtain the following diagram:

$$
\begin{array}{ccc}
f, \mathbf{p} & \xrightarrow{\text{evaluation}} & f(\mathbf{p}) \\
\downarrow{\scriptstyle A} & & \downarrow{\scriptstyle A} \\
Af, A\mathbf{p} & \xrightarrow{\text{evaluation}} & (7.45) + (7.44)
\end{array}
$$

Observe that—just like the action on $\mathbb{P}_0$—this action on functions also does not depend on the multiplier ν, and that its definition (7.43) implies that positive (non-zero) functions are mapped to positive (non-zero) functions again.

Action on sections in $\mathcal{S}(\mathscr{P})$. To define a representation acting on sections $\psi \in \mathcal{S}(\mathscr{P})$, which contains more information, we also need to make use of the remaining information contained in the given $A \in \hat{O}(\mathbb{T})$ that got lost in our definition (7.42). In analogy to (7.45), we want to define our representation so that

$$(A\psi)(A\mathbf{p}) = A(\psi(\mathbf{p})) \tag{7.46}$$

for $\forall \psi \in \mathcal{S}(\mathscr{P})$ and $\forall \mathbf{p} \in \mathbb{P}_0$, as illustrated in the following diagram:

$$
\begin{array}{ccc}
\psi, \mathbf{p} & \xrightarrow{\text{evaluation}} & \psi(\mathbf{p}) \\
\downarrow{\scriptstyle A} & & \downarrow{\scriptstyle A} \\
A\psi, A\mathbf{p} & \xrightarrow{\text{evaluation}} & (7.46)
\end{array}
$$

Given any $\psi \in \mathcal{S}(\mathscr{P})$, we therefore set

$$A\psi := A \circ \psi \circ A^{-1}, \tag{7.47}$$

where again the operator A^{-1} on the right is the inverse of the representation of A acting on $\mathbb{P}_0$, and where the first occurrence of the operator A on the right is the original map acting on $\mathbb{T}$ (or more precisely, on $\mathbb{T}_0$).[12] Note that since by our definition of the line bundle $\mathscr{P}$ in (7.12) we have $\psi(A^{-1}\mathbf{p}) \in A^{-1}\mathbf{p}$ and thus $(A \circ \psi \circ A^{-1})(\mathbf{p}) = A(\psi(A^{-1}\mathbf{p})) \in \mathbf{p}$, the image $A\psi$ defined in (7.47) is indeed a section in $\mathcal{S}(\mathscr{P})$.

---

[12]The definition (7.47) is in fact analogous to (7.43); this becomes more evident if we rewrite the latter as $Af := A \circ f \circ A^{-1}$, where the first occurrence of A on the right is the identity operator on $\mathbb{R}$ defined in (7.44).

Observe that because of the linearity of the given map $A \in \hat{O}(\mathbb{T})$, the representations acting on functions $f \in \mathcal{F}(\mathbb{P}_0)$ and on sections $\psi \in \mathcal{S}(\mathscr{P})$ as defined in (7.43) and (7.47) fulfill the relation that we asked for in (5.71), namely

$$A(f \cdot \psi) = (Af) \cdot (A\psi).$$

Further observe that together with Lemma 7.4 the definition (7.47) implies that positive (non-zero) sections are mapped to positive (non-zero) sections again.

Action on sections in $\mathcal{S}(\mathscr{P}^{\otimes k})$. Since (as mentioned in Section 7.1.6) we will use a specific power $\mathscr{P}^{\otimes k}$ as our wealth line bundle $\mathscr{V}$, and since by Section 7.2.2 the bilinear forms in $\mathfrak{c}$ are represented by sections $\mathfrak{m} \in \mathcal{S}_+(\mathscr{P}^{\otimes 2})$, we will also need to define representations of A acting on sections $\mathcal{S}(\mathscr{P}^{\otimes k})$. To do so, we will follow the same logic as in Section 6.2.2:

First we define the action of A on sections $\psi^* \in \mathcal{S}(\mathscr{P}^*)$ on the dual line bundle as in (6.21) as $(A\psi^*)(\psi) := A(\psi^*(A^{-1}\psi))$ for $\forall \psi^* \in \mathcal{S}(\mathscr{P}^*)$ and $\forall \psi \in \mathcal{S}(\mathscr{P})$, i.e., as

$$A\psi^* := A \circ \psi^* \circ A^{-1}. \tag{7.48}$$

As in (6.23) this in particular implies that

$$A\left(\frac{1}{\psi}\right) = \frac{1}{A\psi}$$

for $\forall \psi \in \mathcal{S}_\varnothing(\mathscr{P})$, and by (6.13) therefore also that positive (non-zero) sections in $\mathcal{S}(\mathscr{P}^*)$ are mapped to positive (non-zero) sections again.

Then we define the action of A on sections in $\mathcal{S}(\mathscr{P}^{\otimes k})$, $k \in \mathbb{Z} \setminus \{0\}$, as in (6.24) by applying A to each factor separately, i.e., for $k > 0$ we set

$$A(\psi_1 \otimes \cdots \otimes \psi_k) := (A\psi_1) \otimes \cdots \otimes (A\psi_k) \tag{7.49a}$$

for any $\psi_1, \ldots, \psi_k \in \mathcal{S}(\mathscr{P})$, and for $k < 0$ we set

$$A(\psi_1^* \otimes \cdots \otimes \psi_{|k|}^*) := (A\psi_1^*) \otimes \cdots \otimes (A\psi_{|k|}^*) \tag{7.49b}$$

for any $\psi_1^*, \ldots, \psi_{|k|}^* \in \mathcal{S}(\mathscr{P}^*)$. Again, by (6.14) this implies that positive (non-zero) sections in $\mathcal{S}(\mathscr{P}^{\otimes k})$ are mapped to positive (non-zero) sections again.

7.4.4 Action on Bilinear Forms on $\mathbb{P}_0$

The desired key property of our maps $\mathcal{U}_s$ is that their representations acting on $\mathbb{P}_0$ preserve the conformal structure $\mathfrak{c}$ on $\mathbb{P}_0$ defined in Section 7.2.2, or more precisely, that their representations acting on bilinear forms map the set $\mathfrak{c}$ onto itself. To prove this property for our operators $A \in \hat{O}(\mathbb{T})$ from which we will construct our group $(\mathcal{U}_s)_{s\in\mathbb{R}}$, let us now begin by carefully defining this representation acting on bilinear forms. This is slightly more tricky on our manifold $\mathbb{P}_0$ than in regular Euclidean space, where the image of an infinitesimal curve segment under a given map was more trivial to compute (see the lines above (6.30)). To begin, let us fix any $A \in \hat{O}(\mathbb{T})$ and any $\mathbf{p} \in \mathbb{P}_0$.

Action on $\mathbf{p}^*$. First, we define an action $A\colon \mathbf{p}^* \to (A\mathbf{p})^*$ as

$$Ax^* := x^* \circ A^{-1}$$

for $\forall x^* \in \mathbf{p}^*$, which in particular implies that

$$\forall x \in \mathbf{p} \setminus \{0\}\colon \quad A\tfrac{1}{x} = \tfrac{1}{Ax}\,. \tag{7.50}$$

Action on $\mathbf{p}^\perp/\mathbf{p}$. Second, note that as a direct consequence of (7.39), the original action of A maps the set $\mathbf{p}^\perp \subset \mathbb{T}$ onto $(A\mathbf{p})^\perp \subset \mathbb{T}$. Since A is linear and maps $\mathbf{p} \subset \mathbf{p}^\perp$ to $A\mathbf{p} \subset (A\mathbf{p})^\perp$, it therefore induces a mapping $A\colon \mathbf{p}^\perp/\mathbf{p} \to (A\mathbf{p})^\perp/(A\mathbf{p})$, naturally defined as

$$A[x] := [Ax] \tag{7.51}$$

for $\forall x \in \mathbf{p}^\perp$, where as in Section 7.2.1 $[x]$ denotes the equivalence class in $\mathbf{p}^\perp/\mathbf{p}$ containing x. Observe that for $\forall x_1, x_2 \in \mathbf{p}^\perp$ we have

$$
\begin{aligned}
\big\langle A[x_1], A[x_2] \big\rangle_{A\mathbf{p}^\perp/A\mathbf{p}} &\overset{(7.51)}{=} \big\langle [Ax_1], [Ax_2] \big\rangle_{A\mathbf{p}^\perp/A\mathbf{p}} \\
&\overset{(7.18)}{=} \big\langle Ax_1, Ax_2 \big\rangle_{\mathbb{T}} \\
&\overset{(7.39)}{=} |A|^2 \cdot \big\langle x_1, x_2 \big\rangle_{\mathbb{T}} \\
&\overset{(7.18)}{=} |A|^2 \cdot \big\langle [x_1], [x_2] \big\rangle_{\mathbf{p}^\perp/\mathbf{p}}\,.
\end{aligned}
\tag{7.52}
$$

Action on $T_\mathbf{p}\mathbb{P}_0$. Combining these two actions of A, we thus obtain an action $A\colon \mathbf{p}^* \otimes \mathbf{p}^\perp/\mathbf{p} \to (A\mathbf{p})^* \otimes (A\mathbf{p})^\perp/(A\mathbf{p})$ from the tangent space $T_\mathbf{p}\mathbb{P}_0$ to the tangent space $T_{A\mathbf{p}}\mathbb{P}_0$, by applying the appropriate action of A to each factor individually:

$$\forall x^* \otimes w \in \mathbf{p}^* \otimes \mathbf{p}^\perp/\mathbf{p}\colon \quad A(x^* \otimes w) := Ax^* \otimes Aw. \tag{7.53}$$

Action on infinitesimal path segments at $\mathbf{p}$. Now let $\gamma \subset \mathbb{P}_0$ be a C^1-path in $\mathbb{P}_0$ with $\gamma(0) = \mathbf{p}$, and let $\mathrm{d}\mathbf{p} \in T_\mathbf{p}\mathbb{P}_0$ be its infinitesimal path segment at $\mathbf{p}$. Denoting the infinitesimal path segments of the image path $A\gamma$ at $A\mathbf{p}$ by $\mathrm{d}\mathbf{p}' \in T_{A\mathbf{p}}\mathbb{P}_0$, we claim that

$$\mathrm{d}\mathbf{p}' = A\,\mathrm{d}\mathbf{p}, \tag{7.54}$$

where the action A on the right is the one defined in the preceding paragraph.

To see this, let $\tilde{\gamma}$ be any C^1-path in $\mathbb{T}_0$ with $\gamma = \pi \circ \tilde{\gamma}$. Then by our definition (7.42) of the representation of A acting on $\mathbb{P}_0$, the path $A\gamma$ can be written as $A\gamma = \pi \circ (A\tilde{\gamma})$, and so in the notation of (7.16) we have

$$(A\gamma)'(0) \overset{(7.16)}{\cong} \frac{1}{(A\tilde{\gamma})(0)} \otimes [(A\tilde{\gamma})'(0)] \overset{\substack{(7.50),\\ A \text{ linear}}}{=} A\big(\tfrac{1}{\tilde{\gamma}(0)}\big) \otimes [A(\tilde{\gamma}'(0))]$$

$$\overset{(7.51)}{=} A\left(\tfrac{1}{\tilde{\gamma}(0)}\right) \otimes A[\tilde{\gamma}'(0)] \overset{(7.53)}{=} A\left(\tfrac{1}{\tilde{\gamma}(0)} \otimes [\tilde{\gamma}'(0)]\right)$$

$$\overset{(7.16)}{\cong} A(\gamma'(0)).$$

Now multiplying both sides by ds, where s denotes the argument of the path $\gamma(s)$, we thus obtain (7.54).

Action on bilinear forms. In analogy to our definition (6.30) in Euclidean space, we therefore define the image under A of a bilinear form $\mathfrak{m}$ on $\mathbb{P}_0$ as the unique bilinear form $A\mathfrak{m}$ fulfilling

$$\langle A\,d\mathbf{p}_1, A\,d\mathbf{p}_2\rangle_{A\mathbf{p}}^{A\mathfrak{m}} = \langle d\mathbf{p}_1, d\mathbf{p}_2\rangle_{\mathbf{p}}^{\mathfrak{m}} \tag{7.55}$$

for $\forall \mathbf{p} \in \mathbb{P}_0$ and $\forall d\mathbf{p}_1, d\mathbf{p}_2 \in T_{\mathbf{p}}\mathbb{P}_0 = \mathbf{p}^* \otimes \mathbf{p}^\perp/\mathbf{p}$.

7.4.5 Conformal Property

We are now ready to prove the conformal property of the maps $A \in \hat{O}(\mathbb{T})$.

Lemma 7.5 (Conformal property). *The maps $A \in \hat{O}(\mathbb{T})$ (or more precisely, their representations acting on bilinear forms on $\mathbb{P}_0$ defined in (7.55)) preserve the conformal structure $\mathfrak{c}$, i.e., we have*

$$\forall \mathfrak{m} \in \mathfrak{c}: \quad A\mathfrak{m} \in \mathfrak{c}. \tag{7.56}$$

Furthermore, the sections $\psi_{\mathfrak{m}}, \psi_{A\mathfrak{m}} \in \mathcal{S}_+(\mathscr{P}^{\otimes 2})$ associated to a given bilinear form $\mathfrak{m} \in \mathfrak{c}$ and to its image $A\mathfrak{m} \in \mathfrak{c}$ as described in Section 7.2.2 satisfy the relation

$$\psi_{A\mathfrak{m}} = |A|^{-2} \cdot A\psi_{\mathfrak{m}}. \tag{7.57}$$

Proof. Let $\mathfrak{m} \in \mathfrak{c}$, and let $\psi_{\mathfrak{m}} = \mathfrak{r} \otimes \mathfrak{r} \in \mathcal{S}_+(\mathscr{P}^{\otimes 2})$ be its associated section, for some section $\mathfrak{r} \in \mathcal{S}_+(\mathscr{P})$ (recall (6.14)). Then we have

$$\langle A\,d\mathbf{p}_1, A\,d\mathbf{p}_2\rangle_{A\mathbf{p}}^{A\mathfrak{m}} \overset{(7.55)}{=} \langle d\mathbf{p}_1, d\mathbf{p}_2\rangle_{\mathbf{p}}^{\mathfrak{m}}$$

$$\overset{(7.20)}{=} \langle \mathfrak{r}(\mathbf{p}) \otimes d\mathbf{p}_1, \mathfrak{r}(\mathbf{p}) \otimes d\mathbf{p}_2\rangle_{\mathbf{p}^\perp/\mathbf{p}}$$

$$\overset{(7.52)}{=} |A|^{-2} \cdot \langle A(\mathfrak{r}(\mathbf{p}) \otimes d\mathbf{p}_1), A(\mathfrak{r}(\mathbf{p}) \otimes d\mathbf{p}_2)\rangle_{A\mathbf{p}^\perp/A\mathbf{p}}$$

$$= |A|^{-2} \cdot \langle A(\mathfrak{r}(\mathbf{p})) \otimes A\,d\mathbf{p}_1, A(\mathfrak{r}(\mathbf{p})) \otimes A\,d\mathbf{p}_2\rangle_{A\mathbf{p}^\perp/A\mathbf{p}}$$

$$\overset{(7.46)}{=} |A|^{-2} \cdot \langle (A\mathfrak{r})(A\mathbf{p}) \otimes A\,d\mathbf{p}_1, (A\mathfrak{r})(A\mathbf{p}) \otimes A\,d\mathbf{p}_2\rangle_{A\mathbf{p}^\perp/A\mathbf{p}},$$

where the fourth step is a consequence of (7.53) and (7.50). Comparing this with (7.20), this shows that $A\mathfrak{m}$ is in fact the bilinear form specified by the section

$$\psi_{A\mathfrak{m}} = |A|^{-2} \cdot (A\mathfrak{r} \otimes A\mathfrak{r}) = |A|^{-2} \cdot A(\mathfrak{r} \otimes \mathfrak{r}) = |A|^{-2} \cdot A\psi_{\mathfrak{m}},$$

and since this section is positive (as the image of a positive section), this in turn means that $A\mathfrak{m} \in \mathfrak{c}$. $\qquad \square$

7.4.6   Evolution Group and Generator

As outlined in Section 7.4.1, we would now like to define our evolution group operators as $\mathcal{U}_s := \mathrm{e}^{\bar{\varepsilon}s}$ for $\forall s \in \mathbb{R}$, and for some operator $\bar{\varepsilon} \colon \mathbb{T} \to \mathbb{T}$ that serves as the infinitesimal generator of our group. We therefore now need to find the conditions under which the operators $\mathrm{e}^{\bar{\varepsilon}s}$ are in fact in $\hat{O}(\mathbb{T})$, as this would imply that the operators $\mathcal{U}_s$ thus defined have representations acting on states, functions, sections, etc. (as described in Sections 7.4.3–7.4.4), and that the conformal structure $\mathfrak{c}$ on $\mathbb{P}_0$ is invariant under the maps $\mathcal{U}_s$ (as proven in Section 7.4.5).

To prepare, we need to introduce the concept of antisymmetric operators.

Definition 7.1. *A linear operator* $\varepsilon \colon \mathbb{T} \to \mathbb{T}$ *is called* antisymmetric *if*

$$\varepsilon^* = -\varepsilon. \tag{7.58}$$

This property has the following two alternative characterizations:

Lemma 7.6. *Given a linear operator* $\varepsilon \colon \mathbb{T} \to \mathbb{T}$, *the following three statements are equivalent:*

  (i) The operator ε *is antisymmetric in the sense of Definition 7.1.*

  (ii) The matrix (ε^{ij}) *is antisymmetric, i.e.,* $\forall i, j = 0, \ldots, m\colon \varepsilon^{ij} = -\varepsilon^{ji}$.

  (iii) $\forall x \in \mathbb{T}\colon \langle x, \varepsilon x \rangle = 0$

Proof. (i) $\Leftrightarrow$ (ii): By (7.32), the condition (7.58) can be rewritten as $\varepsilon^i{}_j = (-\varepsilon^*)^i{}_j = -\varepsilon_j{}^i$, and raising the index j on both sides, we therefore find that (7.58) is equivalent to (ii).

(i) $\Leftrightarrow$ (iii): Since $\langle x, \varepsilon x \rangle = \langle \varepsilon x, x \rangle = \langle x, \varepsilon^* x \rangle$, we can write $\langle x, \varepsilon x \rangle = \frac{1}{2} \langle x, (\varepsilon + \varepsilon^*) x \rangle$, which shows that (7.58) implies (iii). On the other hand, if (iii) holds, i.e., if $\langle x, (\varepsilon + \varepsilon^*) x \rangle = 0$ for $\forall x \in \mathbb{T}$, then in fact we have $\langle x, (\varepsilon + \varepsilon^*) y \rangle = 0$ for $\forall x, y \in \mathbb{T}$ (this can be shown by replacing x by $x + y$); the nondegeneracy of the Minkowski inner product then implies that $(\varepsilon + \varepsilon^*) y = 0$ for $\forall y \in \mathbb{T}$, i.e., we have $\varepsilon + \varepsilon^* = 0$ and thus (7.58). $\square$

The following lemma provides our desired criteria.

Lemma 7.7. *Given any linear operators* $\varepsilon, \bar{\varepsilon} \colon \mathbb{T} \to \mathbb{T}$, *we have*

  (i) $\forall s \in \mathbb{R}\colon \mathrm{e}^{s\varepsilon} \in O(\mathbb{T})$ *if and only if* ε *is antisymmetric, and*

  (ii) $\forall s \in \mathbb{R}\colon \mathrm{e}^{s\bar{\varepsilon}} \in \hat{O}(\mathbb{T})$ *if and only if* $\bar{\varepsilon} = \varepsilon + \rho$ *for some antisymmetric operator* ε *and some* $\rho \in \mathbb{R}$, *where the occurrence of* ρ *in* $\bar{\varepsilon}$ *should be interpreted as the multiplication operator* $x \mapsto \rho x$.

Proof. Here we will only show that the conditions on ε and $\bar{\varepsilon}$ are sufficient. The proofs that they are also necessary can be found in Appendix F.2.

(i) For any antisymmetric operator ε and $\forall s \in \mathbb{R}$ we have

$$(\mathrm{e}^{s\varepsilon})^* = \mathrm{e}^{s\varepsilon^*} = \mathrm{e}^{-s\varepsilon} = (\mathrm{e}^{s\varepsilon})^{-1},$$

and thus $\mathrm{e}^{s\varepsilon} \in O(\mathbb{T})$ by (7.35).

(ii) In fact, since $\mathrm{e}^{0 \cdot \varepsilon} = I \in O^{\uparrow}(\mathbb{T})$, by a continuity argument we have

$$\mathrm{e}^{s\varepsilon} \in O^{\uparrow}(\mathbb{T}) \quad \text{for } \forall s \in \mathbb{R}.$$

Indeed, otherwise by (7.37) there would be an $\hat{s} \in \mathbb{R}$ such that $\mathrm{e}^{\hat{s}\varepsilon} \in O^{\downarrow}(\mathbb{T})$, i.e., $\exists x \in \mathbb{T}_0^{\uparrow}$: $\mathrm{e}^{\hat{s}\varepsilon} x \in \mathbb{T}_0^{\downarrow}$; the path $s \mapsto \mathrm{e}^{s\varepsilon} x$, which by (7.37) lies entirely in $\mathbb{T}_0^{\uparrow} \cup \mathbb{T}_0^{\downarrow}$, would therefore connect the points $\mathrm{e}^{0 \cdot \varepsilon} x = x \in \mathbb{T}_0^{\uparrow}$ and $\mathrm{e}^{\hat{s}\varepsilon} x \in \mathbb{T}_0^{\downarrow}$. But this is not possible since $\mathbb{T}_0^{\uparrow}$ and $\mathbb{T}_0^{\downarrow}$ are not connected.

We therefore find that for $\forall \rho \in \mathbb{R}$ we have

$$\mathrm{e}^{s(\varepsilon+\rho)} = \mathrm{e}^{s\rho} \cdot \mathrm{e}^{s\varepsilon} \in \hat{O}(\mathbb{T}). \qquad \square$$

Definition 7.2. *We define our group* $(\mathcal{U}_s)_{s \in \mathbb{R}}$ *of operators* $\mathcal{U}_s \in \hat{O}(\mathbb{T})$ *as*

$$\mathcal{U}_s := \mathrm{e}^{s(\varepsilon+\rho)} \tag{7.59}$$

for some antisymmetric operator $\varepsilon \colon \mathbb{T} \to \mathbb{T}$ *and some* $\rho \in \mathbb{R}$*, which are model parameters. Their representations acting on the various types of mathematical objects—in particular, on states* $\mathbf{p} \in \mathbb{P}_0$*, functions* $f \in \mathcal{F}(\mathbb{P}_0)$*, and sections* $\psi \in \mathcal{S}(\mathscr{P}^{\otimes k})$*—are then derived as laid out in Sections 7.4.3–7.4.4.*

7.5 Sections as Homogeneous Functions

To prepare for our next step, the construction of the diamond operator $\diamondsuit$, we first need to address two issues:

(i) It is not immediately obvious how to get a handle on the task of defining such a "second-derivative-like" operator that acts not on regular functions, but on sections. To address this problem, we will construct a map that for each fixed $k \in \mathbb{Z}$ identifies the space of sections $\mathcal{S}(\mathscr{P}^{\otimes k})$ with the class of functions on $f \colon \mathbb{T}_0^{\uparrow} \to \mathbb{R}$ that are homogeneous of degree $-k$. We are thus left with the more intuitive task of defining a second-derivative-like operator on this function space.

(ii) As we will see later, the construction of $\diamondsuit$ can only succeed for the specific value $k = 1 - \frac{n}{2}$ of our exponent, which for odd dimensions n is non-integer. Luckily, after making the transition to our function spaces, such fractional values of k will no longer be a problem.

7.5.1   Elements of $\mathbf{p}^{\otimes k}$ as Homogeneous Functions

To begin our construction, let for $\forall k \in \mathbb{R}$ and $\forall \mathbf{p} \in \mathbb{P}_0$

$$H^k(\mathbf{p}_+) := \left\{ f \in \mathcal{F}(\mathbf{p}_+) \,\middle|\, \forall x \in \mathbf{p}_+ \; \forall a > 0 \colon \; f(ax) = a^k f(x) \right\} \tag{7.60}$$

denote the one-dimensional vector space of all k-homogeneous functions on $\mathbf{p}^+$. Note that

$$\forall f_1 \in H^{k_1}(\mathbf{p}_+) \; \forall f_2 \in H^{k_2}(\mathbf{p}_+) \colon \quad f_1 \cdot f_2 \in H^{k_1 + k_2}(\mathbf{p}_+).$$

Now fixing any $\mathbf{p} \in \mathbb{P}_0$, we begin by claiming that we can construct an invertible mapping

$$g_{\mathbf{p}} \colon \; \bigcup_{k \in \mathbb{Z}} \mathbf{p}^{\otimes k} \to \bigcup_{k \in \mathbb{Z}} H^{-k}(\mathbf{p}_+)$$

that maps every space $\mathbf{p}^{\otimes k}$ to the function space $H^{-k}(\mathbf{p}_+)$, and that is such that for $\forall k, k_1, k_2 \in \mathbb{Z}$ we have

$$\forall c \in \mathbb{R} \; \forall v \in \mathbf{p}^{\otimes k} \colon \qquad g_{\mathbf{p}}(cv) = c \cdot g_{\mathbf{p}}(v), \tag{7.61a}$$

$$\forall v_1 \in \mathbf{p}^{\otimes k_1} \; \forall v_2 \in \mathbf{p}^{\otimes k_2} \colon \; g_{\mathbf{p}}(v_1 \otimes v_2) = g_{\mathbf{p}}(v_1) \cdot g_{\mathbf{p}}(v_2). \tag{7.61b}$$

Note that $v_1 \otimes v_2 \in \mathbf{p}^{\otimes (k_1 + k_2)}$ and $g_{\mathbf{p}}(v_1) \cdot g_{\mathbf{p}}(v_2) \in H^{k_1 + k_2}(\mathbf{p}_+)$, so the right-hand side of (7.61b) is in the right space.

As we will see in the following construction, this idea is rather natural for integer values of k if in (7.60) one replaces $\mathbf{p}_+$ by $\mathbf{p}$ and allows for $\forall a \in \mathbb{R}$ instead of only $a > 0$; we had to make these two adjustments in order for $H^k(\mathbf{p}_+)$ to be well defined also for non-integer values of k, which will be relevant later on.

Powers of $\mathbf{p}^*$. For any fixed $\mathbf{p} \in \mathbb{P}_0$, an element $\lambda \in \mathbf{p}^*$ is by definition a linear function $\lambda \colon \mathbf{p} \to \mathbb{R}$. Since $\mathbf{p}$ is one-dimensional, linearity is equivalent to asking that

$$\forall x \in \mathbf{p} \; \forall a \in \mathbb{R} \colon \quad \lambda(ax) = a\,\lambda(x), \tag{7.62}$$

so that $\lambda|_{\mathbf{p}_+} \in H^1(\mathbf{p}_+)$. It is therefore natural to set

$$\forall \lambda \in \mathbf{p}^* \colon \quad g_{\mathbf{p}}(\lambda) := \lambda|_{\mathbf{p}_+}, \tag{7.63}$$

and repeated application of (7.61b) thus tells us that we need to set

$$\forall \lambda_1, \ldots, \lambda_k \in \mathbf{p}^* \colon \; g_{\mathbf{p}}(\lambda_1 \otimes \cdots \otimes \lambda_k) := (\lambda_1 \cdots \lambda_k)|_{\mathbf{p}_+} \in H^k(\mathbf{p}_+). \tag{7.64}$$

Note that the map $g_{\mathbf{p}}$ so far is well defined since

$$g_{\mathbf{p}}((c\lambda_1) \otimes \cdots \otimes \lambda_k) = \cdots = g_{\mathbf{p}}(\lambda_1 \otimes \cdots \otimes (c\lambda_k))$$

for $\forall c \in \mathbb{R}$. For every negative integer k we can therefore identify $\mathbf{p}^{\otimes k} = (\mathbf{p}^*)^{\otimes (-k)}$ with $H^{-k}(\mathbf{p}_+)$.

Constants. For functions $f \in H^0(\mathbf{p}_+)$ we have $f(ax) = a^0 f(x) = f(x)$ for $\forall x \in \mathbf{p}_+$ and $\forall a > 0$, and so $H^0(\mathbf{p}_+)$ is the space of all constant functions on $\mathbf{p}_+$. This motivates us to extend our definition of g to $\mathbf{p}^{\otimes 0} = \mathbb{R}$ by defining

$$\forall c \in \mathbb{R} = \mathbf{p}^{\otimes 0}: \quad g_{\mathbf{p}}(c) := c, \tag{7.65}$$

where the right-hand side denotes the function on $\mathbf{p}_+$ that takes the constant value c. Note that if one identifies $c \otimes v \cong c \cdot v$ for $\forall c \in \mathbf{p}^{\otimes 0}$ and $\forall v \in \mathbf{p}^{\otimes k}$, this is consistent with (7.61).

Powers of p. Now given $x \in \mathbf{p}$, for our definition of $g_{\mathbf{p}}(x)$ to be consistent with (6.5) and as a result also with (6.7), we need to have

$$\lambda(x) \overset{(7.65)}{=} g_{\mathbf{p}}(\lambda(x)) \overset{(6.5)}{=} g_{\mathbf{p}}(\lambda \otimes x) \overset{(7.61b)}{=} g_{\mathbf{p}}(\lambda) \cdot g_{\mathbf{p}}(x) \overset{(7.63)}{=} \lambda|_{\mathbf{p}_+} \cdot g_{\mathbf{p}}(x) \tag{7.66}$$

for $\forall \lambda \in \mathbf{p}^*$. Solving for $g_{\mathbf{p}}(x)$, this shows that we need to define

$$g_{\mathbf{p}}(x) := \lambda_x, \tag{7.67}$$

where the function $\lambda_x \in H^{-1}(\mathbf{p}_+)$ is given by

$$\lambda_x(y) := \frac{\lambda(x)}{\lambda(y)} = \frac{\lambda(\frac{x}{y} \cdot y)}{\lambda(y)} = \frac{x}{y} \qquad \text{for } \forall y \in \mathbf{p}_+. \tag{7.68}$$

Finally, repeated application of (7.61b) tells us that we need to set

$$\forall x_1, \ldots, x_k \in \mathbf{p}: \quad g_{\mathbf{p}}(x_1 \otimes \cdots \otimes x_k) := \lambda_{x_1} \cdots \lambda_{x_k}, \tag{7.69}$$

which maps $\mathbf{p}^{\otimes k}$ to $H^{-k}(\mathbf{p}_+)$ for any $k \in \mathbb{N}$. Note that the property $\lambda_{cx} = c\lambda_x$ for $\forall c \in \mathbb{R}$ implies that the mapping (7.69) is well defined, i.e.,

$$g_{\mathbf{p}}((cx_1) \otimes \cdots \otimes x_k) = \cdots = g_{\mathbf{p}}(x_1 \otimes \cdots \otimes (cx_k))$$

for $\forall c \in \mathbb{R}$.

Homomorphism. We have now fully defined $g_{\mathbf{p}}(v)$ for $\forall v \in \mathbf{p}^{\otimes k}$, $k \in \mathbb{Z}$, and by way of the equivalence (6.7)–(6.8) also $g_{\mathbf{p}}(v_1 \otimes v_2)$ for $\forall v_1 \in \mathbf{p}^{\otimes k_1}$ and $\forall v_2 \in \mathbf{p}^{\otimes k_2}$, $k_1, k_2 \in \mathbb{Z}$. In our definitions we have made sure that (7.61b) holds whenever k_1 and k_2 have the same sign, or when $k_1 = -1$ and $k_2 = 1$ (see (7.66)). The remaining cases can then be proven as in the following example for the case $k_1 < 0 < k_2$ with $|k_1| < k_2$:

$$g_{\mathbf{p}}\big((\lambda_1 \otimes \cdots \otimes \lambda_{|k_1|}) \otimes (x_1 \otimes \cdots \otimes x_{k_2})\big)$$

$$\overset{(6.8)}{=} g_{\mathbf{p}}\big(\lambda_1(x_1) \cdots \lambda_{|k_1|}(x_{|k_1|}) \cdot (x_{|k_1|+1} \otimes \cdots \otimes x_{k_2})\big)$$

$$\overset{(7.61a)}{=} \lambda_1(x_1) \cdots \lambda_{|k_1|}(x_{|k_1|}) \cdot g_{\mathbf{p}}(x_{|k_1|+1} \otimes \cdots \otimes x_{k_2})$$

$$\overset{(7.66)}{=} g_{\mathbf{p}}(\lambda_1) \cdot g_{\mathbf{p}}(x_1) \cdots g_{\mathbf{p}}(\lambda_{|k_1|}) \cdot g_{\mathbf{p}}(x_{|k_1|}) \cdot g_{\mathbf{p}}(x_{|k_1|+1} \otimes \cdots \otimes x_{k_2})$$

$$\overset{(7.61b)}{=} g_{\mathbf{p}}(\lambda_1 \otimes \cdots \otimes \lambda_{|k_1|}) \cdot g_{\mathbf{p}}(x_1 \otimes \cdots \otimes x_{k_2}).$$

7.5.2   Evaluating Elements of the Dual Space

For $\forall k \in \mathbb{N}$, the value of a linear function $\lambda = \lambda_1 \otimes \cdots \otimes \lambda_k \in (\mathbf{p}^*)^{\otimes k} = (\mathbf{p}^{\otimes k})^*$ at a point $x = x_1 \otimes \cdots \otimes x_k \in \mathbf{p}^{\otimes k}$ is given by

$$\lambda(x) \overset{(7.65)}{=} g_{\mathbf{p}}(\lambda(x)) \overset{(6.10)}{=} g_{\mathbf{p}}(\lambda \otimes x) \overset{(7.61b)}{=} g_{\mathbf{p}}(\lambda) \cdot g_{\mathbf{p}}(x),$$

i.e., it can be obtained by simply multiplying the associated functions $g_{\mathbf{p}}(\lambda) \in H^k(\mathbf{p}_+)$ and $g_{\mathbf{p}}(x) \in H^{-k}(\mathbf{p}_+)$. More precisely, multiplying these two functions results in a function $g_{\mathbf{p}}(\lambda) \cdot g_{\mathbf{p}}(x) \in H^{k+(-k)}(\mathbf{p}_+) = H^0(\mathbf{p}_+)$ that is constant on $\mathbf{p}_+$, with value $\lambda(x)$.

In particular, this implies that given any $x \in \mathbf{p}^{\otimes k} \setminus \{0\}$, the "inverse" $\frac{1}{x}$ (i.e., the unique $\lambda \in (\mathbf{p}^{\otimes k})^*$ with $\lambda(x) = 1$) is associated to the function

$$g_{\mathbf{p}}\left(\tfrac{1}{x}\right) = \frac{1}{g_{\mathbf{p}}(x)} \, . \tag{7.70}$$

7.5.3   Equivalence: $\mathcal{S}(\mathscr{P}^{\otimes k}) \cong H^{-k}(\mathbb{T}_0^\uparrow)$

Finally, fixing $k \in \mathbb{Z}$, recall that a section $\psi \in \mathcal{S}(\mathscr{P}^{\otimes k})$ on the line bundle $\mathscr{P}^{\otimes k} = (\mathbf{p}^{\otimes k})_{\mathbf{p} \in \mathbb{P}_0}$ is a map that assigns to each state $\mathbf{p} \in \mathbb{P}_0$ an element $\psi(\mathbf{p}) \in \mathbf{p}^{\otimes k}$, and thus a function $g_{\mathbf{p}}(\psi(\mathbf{p})) \in H^{-k}(\mathbf{p}^+)$ on $\mathbf{p}_+$. It therefore defines a function $\hat{\psi} \in \mathcal{F}(\mathbb{T}_0^\uparrow)$ that is $(-k)$-homogeneous, i.e., $\hat{\psi} \in H^{-k}(\mathbb{T}_0^\uparrow)$, where we use the notation

$$H^\kappa(\mathbb{T}_0^\uparrow) := \left\{ f \in \mathcal{F}(\mathbb{T}_0^\uparrow) \,\middle|\, \forall x \in \mathbb{T}_0^\uparrow \; \forall a > 0 \colon \; f(ax) = a^\kappa f(x) \right\} \tag{7.71}$$

for $\forall \kappa \in \mathbb{R}$. Explicitly, this function is given by

$$\hat{\psi}(x) := \big(g_{\pi(x)}\big(\psi(\pi(x)) \big) \big)(x). \tag{7.72}$$

Conversely, every function $\hat{\psi} \in H^{-k}(\mathbb{T}_0^\uparrow)$ defines for each $\mathbf{p} \in \mathbb{P}_0$ a function in $H^{-k}(\mathbf{p}_+)$ and thus (by way of $g_{\mathbf{p}}^{-1}$) an element in $\mathbf{p}^{\otimes k}$. It therefore uniquely specifies a section $\psi \in \mathcal{S}(\mathscr{P}^{\otimes k})$, given by

$$\psi(\mathbf{p}) := g_{\mathbf{p}}^{-1}\big(\hat{\psi}|_{\mathbf{p}_+} \big).$$

We can therefore identify the space of sections $\psi \in \mathcal{S}(\mathscr{P}^{\otimes k})$ with the space of $(-k)$-homogeneous functions $\hat{\psi} \in H^{-k}(\mathbb{T}_0^\uparrow)$ for $\forall k \in \mathbb{R}$:

$$\psi \in \mathcal{S}(\mathscr{P}^{\otimes k}) \qquad \cong \qquad \hat{\psi} \in H^{-k}(\mathbb{T}_0^\uparrow).$$

In particular, the case $k = 0$ reduces to the intuitive statement that we can identify functions $f \in \mathcal{F}(\mathbb{P}_0)$ with functions $\hat{f} \in H^0(\mathbb{T}_0^\uparrow)$, i.e., with functions on $\mathbb{T}_0^\uparrow$ that for each $\mathbf{p} \in \mathbb{P}_0$ are constant along the ray $\mathbf{p}_+$:

$$\forall f \in \mathcal{F}(\mathbb{P}_0) \; \forall x \in \mathbb{T}_0^\uparrow \colon \quad \hat{f}(x) = f(\pi(x)). \tag{7.73}$$

7.5.4 Working with Homogeneous Functions

Using this equivalence, we can construct our framework with $(-k)$-homogeneous functions instead of sections in $\mathcal{S}(\mathscr{P}^{\otimes k})$. Let us quickly summarize how the various objects and operations in the original section-based framework translate into their counterparts in the world of $(-k)$-homogeneous functions.

Positive and non-zero sections. First, for any fixed $\mathbf{p} \in \mathbb{P}_0$ and any $x \in \mathbf{p}_+$ the function λ_x defined in (7.68) fulfills $\lambda_x(y) = \frac{x}{y} > 0$ for $\forall y \in \mathbf{p}_+$. By (6.12), (7.67) and (7.69) this shows that elements of $(\mathbf{p}^{\otimes k})_+$ are associated with *positive* functions in $H^{-k}(\mathbf{p}_+)$ for $\forall k \in \mathbb{N}$, and by (6.13) and (7.70) this also holds for negative $k \in \mathbb{Z}$. This shows that for any $\psi \in \mathcal{S}(\mathscr{P}^{\otimes k})$ we have

$$\psi \in \mathcal{S}_+(\mathscr{P}^{\otimes k}) \qquad \Leftrightarrow \qquad \hat{\psi} > 0. \tag{7.74}$$

(For $k = 0$ this immediately follows from (7.73).) Similarly, one can see that

$$\psi \in \mathcal{S}_\varnothing(\mathscr{P}^{\otimes k}) \qquad \Leftrightarrow \qquad \hat{\psi} \neq 0. \tag{7.75}$$

Products. By (7.61b) and our definition (7.72) we have

$$\widehat{\psi_1 \otimes \psi_2} = \hat{\psi}_1 \cdot \hat{\psi}_2 \tag{7.76}$$

for $\forall \psi_1 \in \mathcal{S}(\mathscr{P}^{\otimes k_1})$ and $\forall \psi_2 \in \mathcal{S}(\mathscr{P}^{\otimes k_2})$.

Inverses. In particular, given a section $\psi \in \mathcal{S}_\varnothing(\mathscr{P}^{\otimes k})$, by (6.10) the inverse section $\frac{1}{\psi} \in \mathcal{S}(\mathscr{P}^{\otimes(-k)})$ is the unique section fulfilling $1 \equiv (\frac{1}{\psi}(\mathbf{p}))(\psi(\mathbf{p})) = \frac{1}{\psi}(\mathbf{p}) \otimes \psi(\mathbf{p}) = (\frac{1}{\psi} \otimes \psi)(\mathbf{p})$ for $\forall \mathbf{p} \in \mathbb{P}_0$. Using (7.76), this shows that

$$\widehat{\left(\frac{1}{\psi}\right)} = \frac{1}{\hat{\psi}}.$$

Operators. Finally, the action of an operator $A \in \hat{O}(\mathbb{T})$ on sections $\psi \in \mathcal{S}(\mathscr{P}^{\otimes k})$ as described in Section 7.4.3 translates into an action on the corresponding functions $\hat{\psi}$, given by

$$\widehat{A\psi} = A\hat{\psi}. \tag{7.77}$$

where the right-hand side is as usual defined as $A\hat{\psi} := \hat{\psi} \circ A^{-1}$. The proof can be found in Appendix F.3.

7.5.5   Fractional Powers

By carrying out our entire construction based on functions $H^{-k}(\mathbb{T}_0^{\uparrow})$ instead of sections $\mathcal{S}(\mathscr{P}^{\otimes k})$, we can now allow for powers $k \in \mathbb{R}$ as well that may not be integers (since the definition (7.71) of $H^{\kappa}(\mathbb{T}_0^{\uparrow})$ makes sense for $\forall \kappa \in \mathbb{R}$), with all the above rules still intact.

From this point on we will no longer distinguish between these two spaces as strictly: We may use the notation $\psi \in H^{-k}(\mathbb{T}_0^{\uparrow})$ and still refer to ψ as a "section," and we may switch between the notations $\psi(x)$ (for $x \in \mathbb{T}_0^{\uparrow}$) and $\psi(\mathbf{p})$ (for $\mathbf{p} \in \mathbb{P}_0$) at our convenience.

7.6   The Diamond Operator

7.6.1   The Diamond Operator Acting on k-Homogeneous Functions

In our model we will define the wealth line bundle as

$$\mathscr{V} := \mathscr{P}^{\otimes(-k)}$$

for some specific value $k \in \mathbb{R}$ to be determined later in (7.86).[13] Since by (7.21) we can identify $\mathscr{W}_{\mathfrak{c}} \cong \mathscr{P}^{\otimes 2}$ via the map $\mathfrak{m} \leftrightarrow \psi_{\mathfrak{m}}$, we can then think of our operator $\Diamond \colon \mathcal{S}(\mathscr{V}) \to \mathcal{S}(\mathscr{W}_{\mathfrak{c}} \otimes \mathscr{V})$ equivalently as an operator

$$\Diamond \colon \mathcal{S}(\mathscr{P}^{\otimes(-k)}) \to \mathcal{S}(\mathscr{P}^{\otimes(2-k)}), \tag{7.78}$$

and thus by our equivalence introduced in Section 7.5 as an operator

$$\Diamond \colon H^{k}(\mathbb{T}_0^{\uparrow}) \to H^{k-2}(\mathbb{T}_0^{\uparrow}). \tag{7.79}$$

The requirement of having the original operator $\Diamond \colon \mathcal{S}(\mathscr{V}) \to \mathcal{S}(\mathscr{W}_{\mathfrak{c}} \otimes \mathscr{V})$ commute with the appropriate representations of the operators $\mathcal{U}_s \in \hat{O}(\mathbb{T})$ then turns into corresponding requirements for the operators (7.79). However, there is one subtle difference:

Lemma 7.8. *The original operator $\Diamond \colon \mathcal{S}(\mathscr{V}) \to \mathcal{S}(\mathscr{W}_{\mathfrak{c}} \otimes \mathscr{V})$ commutes with an operator $A \in \hat{O}(\mathbb{T})$ if and only if its representation in (7.79) satisfies*

$$\forall f \in H^{k}(\mathbb{T}_0^{\uparrow}): \qquad \Diamond(Af) = |A|^{-2} \cdot A(\Diamond f). \tag{7.80}$$

In other words, while for pseudo-orthogonal operators $A \in O^{\uparrow}(\mathbb{T})$ (where $|A| = 1$) the commutativity property is preserved exactly, for *multiples* of such operators an additional factor appears in the equation.

---

[13]The permitted value of k will turn out to be negative except in the dimensions $n = 1$ and $n = 2$; we introduce the minus sign here for notational convenience later on once we switch over to working with functions in $H^{k}(\mathbb{T}_0^{\uparrow})$.

Proof of Lemma 7.8. For the purpose of this proof let us denote the original operator as $\Diamond\colon \mathcal{S}(\mathcal{V}) \to \mathcal{S}(\mathcal{W}_{\mathfrak{c}} \otimes \mathcal{V})$, and let us denote its representations in (7.78) and (7.79) by $\Diamond'$ and $\Diamond''$, respectively.

If for some given $\psi \in \mathcal{S}(\mathcal{V})$ we write $\Diamond\psi = \mathfrak{m} \otimes \psi'$ for some $\mathfrak{m} \in \mathcal{S}_+(\mathcal{W}_{\mathfrak{c}})$ and some $\psi' \in \mathcal{S}(\mathcal{V})$, the original commutativity requirement asks that $\Diamond(A\psi) = A(\Diamond\psi) = A\mathfrak{m} \otimes A\psi'$, and so the corresponding image of $\Diamond'$ should be

$$
\begin{aligned}
\Diamond'(A\psi) &= \psi_{A\mathfrak{m}} \otimes A\psi' \\
&\overset{(7.57)}{=} (|A|^{-2} \cdot A\psi_{\mathfrak{m}}) \otimes A\psi' = |A|^{-2} \cdot A(\psi_{\mathfrak{m}} \otimes \psi') \\
&= |A|^{-2} \cdot A(\Diamond'\psi).
\end{aligned}
$$

Using the notation introduced in Section 7.5.3, the operator in (7.79) should therefore fulfill $\Diamond''(\widehat{A\psi}) = |A|^{-2} \cdot \widehat{A(\Diamond'\psi)}$, and by (7.77) and the definition of $\Diamond''$ thus

$$
\Diamond''(A\hat{\psi}) = |A|^{-2} \cdot A(\widehat{\Diamond'\psi}) = |A|^{-2} \cdot A(\Diamond''\hat{\psi}),
$$

which is (7.80) for $f := \hat{\psi}$. $\qquad\square$

7.6.2 The Commutativity Property of the d'Alembert Operator

The requirement of having our operator $\Diamond$ in (7.79) properly commute with all pseudo-orthogonal operators immediately leads us to the idea of basing its construction on the d'Alembert operator $\Box\colon C^2(\mathbb{T}, \mathbb{R}) \to C(\mathbb{T}, \mathbb{R})$, which is defined as

$$
\Box := \partial^i\partial_i = -\partial_0^2 + \partial_1^2 + \cdots + \partial_m^2\,, \tag{7.81}
$$

where we abbreviate $\partial_i := \partial_{x^i}$ and $\partial^i := Q^{ij}\partial_j$. The d'Alembert operator can be seen as the Minkowski space analogue of the Laplace operator $\Delta\colon C^2(\mathbb{R}^n, \mathbb{R}) \to C(\mathbb{R}^n, \mathbb{R})$ in the regular Euclidean space $\mathbb{R}^n$, which is defined as $\Delta := \partial_1^2 + \cdots + \partial_n^2$.

Just like the Laplace operator commutes with regular orthogonal operators (i.e., rotations and reflections), the d'Alembert operator commutes with *pseudo*-orthogonal operators, and in fact we have the following:

Lemma 7.9. *For $\forall A \in \hat{O}(\mathbb{T})$ and $\forall f \in C^2(\mathbb{T}, \mathbb{R})$ we have*

$$
\Box(Af) = |A|^{-2} \cdot A(\Box f), \tag{7.82}
$$

where the action of A on functions is the one defined by (7.43).

Proof. See Appendix F.4. $\qquad\square$

To define $\Diamond$, we now need to "project" this operator from $\mathbb{T}$ to $\mathbb{T}_0^{\uparrow}$ to have it act on our functions in $H^k(\mathbb{T}_0^{\uparrow})$.

7.6.3 Definition of the Diamond Operator

For each neighborhood $G \subset \mathbb{T} \setminus \{0\}$ of $\mathbb{T}_0^\uparrow$ with the property that

$$\forall x \in G \ \forall a > 0: \quad ax \in G, \tag{7.83}$$

let $H^k(G)$ denote the set of all k-homogeneous functions on G, let $H_\infty^k(\mathbb{T}_0^\uparrow)$ and $H_\infty^k(G)$ denote the subsets consisting of only those functions in $H^k(\mathbb{T}_0^\uparrow)$ and $H^k(G)$, respectively, that are analytic (i.e., that can locally be written as a power series), and let

$$H_{\infty,0}^k(G) := \{ f \in H_\infty^k(G) \,|\, \forall x \in \mathbb{T}_0^\uparrow: \ f(x) = 0 \}.$$

By utilizing its power series representation, it is not hard to see that every function $f \in H_\infty^k(\mathbb{T}_0^\uparrow)$ can be extended to a function $\tilde{f} \in H_\infty^k(G)$ for some neighborhood G with the property (7.83). We would like to define

$$\diamond f := (\Box \tilde{f})\big|_{\mathbb{T}_0^\uparrow}. \tag{7.84}$$

Not only would this be a second-derivative operator that we will easily see to commute with pseudo-orthogonal operators (just like $\Box$ does), it would also map into the desired target space $H^{k-2}(\mathbb{T}_0^\uparrow)$ found in (7.79), since

$$(\Box \tilde{f})(ax) = \tfrac{1}{a^2}\Box\big[\tilde{f}(ax)\big] = \tfrac{1}{a^2}\Box\big[a^k \tilde{f}(x)\big] = a^{k-2}(\Box \tilde{f})(x).$$

However, the extension $\tilde{f}$ is not unique, and so to show that the right-hand side of (7.84) is independent of the specific extension, we need to show that

$$\Box \text{ maps } H_{\infty,0}^k(G) \text{ into } H_{\infty,0}^{k-2}(G). \tag{7.85}$$

Indeed, once this is established, since any two extensions $\tilde{f}_1, \tilde{f}_2 \in H_\infty^k(G)$ of a given $f \in H_\infty^k(\mathbb{T}_0^\uparrow)$ fulfill $\tilde{f}_1 - \tilde{f}_2 \in H_{\infty,0}^k(G)$, this would allow us to conclude that $\Box(\tilde{f}_1 - \tilde{f}_2) \in H_{\infty,0}^{k-2}(G)$, and thus $\Box\tilde{f}_1 = \Box\tilde{f}_2$ on $\mathbb{T}_0^\uparrow$.

To show (7.85), let $f \in H_{\infty,0}^k(G)$. Then since the function $x \mapsto \langle x, x \rangle$ for $x \in \mathbb{T}$ vanishes on $\mathbb{T}_0^\uparrow$ as well (with a gradient $x \mapsto 2(Q_{ij})x$ that does not vanish on $\mathbb{T}_0^\uparrow$), we can write $f(x) = \hat{f}(x) \cdot \langle x, x \rangle$ for some function $\hat{f} \in H_\infty^{k-2}(G)$. Since

$$\partial_i \langle x, x \rangle = \partial_i x^j x_j = Q_{jl}\partial_i x^j x^l = Q_{jl}(\delta_i^j x^l + x^j \delta_i^l)$$
$$= Q_{il}x^l + Q_{ji}x^j = Q_{il}x^l + Q_{ij}x^j = 2x_i$$
$$\partial^i \langle x, x \rangle = 2x^i$$
$$\partial^i \partial_i \langle x, x \rangle = 2\partial_i x^i = 2(m+1) = 2(n+2)$$
$$(\partial_i \hat{f})(x)\partial^i \langle x, x \rangle = 2x^i \partial_i \hat{f}(x) = 2\partial_a \hat{f}(ax)\big|_{a=1} = 2\partial_a\big(a^{k-2}\hat{f}(x)\big)\big|_{a=1}$$
$$= 2\big[(k-2)a^{k-3}\hat{f}(x)\big]\big|_{a=1} = 2(k-2)\hat{f}(x)$$

(where in the last calculation we used that $\hat{f} \in H^{k-2}(G)$), we then find that

$$\begin{aligned}
\Box f(x) &= \partial^i \partial_i \big(\hat{f}(x) \cdot \langle x, x \rangle\big) \\
&= \langle x, x \rangle \, \partial^i \partial_i \hat{f}(x) + 2(\partial_i \hat{f})(x) \partial^i \langle x, x \rangle + \hat{f}(x) \partial^i \partial_i \langle x, x \rangle \\
&= \langle x, x \rangle \, \Box \hat{f}(x) + 4(k-2)\hat{f}(x) + 2(n+2)\hat{f}(x) \\
&= \langle x, x \rangle \, \Box \hat{f}(x) + 4\big[(k-2) + \tfrac{n+2}{2}\big]\hat{f}(x) \\
&= \langle x, x \rangle \, \Box \hat{f}(x) + 4\big[k - (1 - \tfrac{n}{2})\big]\hat{f}(x)
\end{aligned}$$

for $\forall x \in G$. In particular, this shows that (7.85) is fulfilled and thus our construction (7.84) of our diamond operator $\Diamond$ is well defined if and only if

$$k = 1 - \frac{n}{2}. \tag{7.86}$$

Given any desired dimension n for our model, we will therefore choose this particular value of k and model account values as functions ("sections") $\psi \in H^k(\mathbb{T}_0^\uparrow)$.

7.6.4 The Commutativity Property of the Diamond Operator

It now only remains to show explicitly that our newly defined operator $\Diamond$ indeed inherits the desired commutativity property from the d'Alembert operator.

Lemma 7.10. *The operator* $\Diamond \colon H^k(\mathbb{T}_0^\uparrow) \to H^{k-2}(\mathbb{T}_0^\uparrow)$ *defined in* (7.84) *satisfies the commutativity property stated in Lemma 7.8.*

Proof. Let $A \in \hat{O}(\mathbb{T})$ and $f \in H_\infty^k(\mathbb{T}_0^\uparrow)$, and let $\tilde{f} \in H_\infty^k(G)$ be a k-homogeneous analytic extension of f to some neighborhood G of $\mathbb{T}_0^\uparrow$ with the property (7.83). Then $A\tilde{f} \in H_\infty^k(AG)$ is a k-homogeneous analytic extension of f to the neighborhood AG of $\mathbb{T}_0^\uparrow$, and so we have

$$\Diamond(Af) \overset{(7.84)}{=} (\Box A\tilde{f})\big|_{\mathbb{T}_0^\uparrow} \overset{(7.82)}{=} \big(|A|^{-2} \cdot A\Box\tilde{f}\big)\big|_{\mathbb{T}_0^\uparrow} \overset{\text{L.}7.2}{=} |A|^{-2} \cdot A(\Box\tilde{f})\big|_{\mathbb{T}_0^\uparrow}$$

$$\overset{(7.84)}{=} |A|^{-2} \cdot A\Diamond f. \qquad \Box$$

7.7 Bond Sections and Bond Prices

7.7.1 Harmonic Sections

Let us now find a section $\mathcal{B}_0^Y \in \mathcal{S}_+(\mathscr{P}^{\otimes(-k)})$, i.e., a positive function in $H^k(\mathbb{T}_0^\uparrow)$, fulfilling

$$\Diamond \mathcal{B}_0^Y \equiv 0.$$

Sections satisfying this equation are called *harmonic*. To be more precise, recall that we only require the section $\mathcal{B}_0^Y$ to be defined on the state space

$\mathscr{M}_0 \subset \mathscr{M} \subsetneq \mathbb{P}_0$ of our process in the spot parametrization (we will decide how to define our subsets $\mathscr{M}_0$ and $\mathscr{M}$ later in Section 7.8.2). Naturally, by the rotational symmetry around the x_0-axis of our construction of both $\mathbb{P}_0$ and $\Diamond$ so far, we expect in fact to find an entire family of elementary solutions to this equation.

To find a family of such harmonic sections, let $b \in \mathbb{T}_0^\downarrow$ be fixed, and consider the linear functional $Qb \in \mathbb{T}^*$ defined in (7.24). By (7.5b) this function is strictly positive on all of $\mathbb{T}_0^\uparrow$, except on the ray $\pi(b)$ parallel to b, on which it vanishes by (7.5a). By its linearity it is therefore positive on an open neighborhood G of $\mathbb{T}_0^\uparrow \setminus \pi(b)$ with the property (7.83), and so the function $[b] \in \mathcal{F}_+(G)$ given by[14]

$$[b](x) := (Qb)^k(x) = \langle b, x \rangle^k = \langle b, x \rangle^{1-n/2} \tag{7.87}$$

is well defined and positive on G for $\forall n \in \mathbb{N}$. It is clearly k-homogeneous and analytic, and so we have $[b] \in H_\infty^k(G)$. Furthermore, we have

$$\begin{aligned}
\Box[b](x) &= \partial^i \partial_i (b_j x^j)^k = \partial^i k (b_j x^j)^{k-1} b_i \\
&= Q^{il} \partial_l k (b_j x^j)^{k-1} b_i = Q^{il} k(k-1)(b_j x^j)^{k-2} b_l b_i \\
&= k(k-1)(b_j x^j)^{k-2} b^i b_i = k(k-1)\langle b, x \rangle^{k-2} \langle b, b \rangle \\
&= 0
\end{aligned}$$

for $\forall x \in G$, since by our choice of b we have $\langle b, b \rangle = 0$.

In other words, the function $[b] \in \mathcal{F}_+(\mathbb{T}_0^\uparrow \setminus \pi(b))$ has an analytic continuation to G that vanishes under the application of the $\Box$-operator, and so by our definition (7.84) we have

$$\Diamond[b] = 0.$$

For every fixed $b \in \mathbb{T}_0^\downarrow$, *the function* $[b]$ *is therefore positive and harmonic on* $\mathbb{T}_0^\uparrow \setminus \pi(b)$.

7.7.2 Bond Sections

Now that we have found a family $\{[b] \mid b \in \mathbb{T}_0^\downarrow\}$ of elementary positive harmonic sections, we can model our basis bond section as

$$\mathcal{B}_0^Y(x) := [b_0](x) = \langle b_0, x \rangle^{1-n/2} \tag{7.88}$$

for any fixed choice of an isotropic vector

$$b_0 \in \mathbb{T}_0^\downarrow \tag{7.89}$$

in the backward light cone. Since this function is only defined for $x \in \mathbb{T}_0^\uparrow \setminus \pi(b_0)$, the associated section $\mathcal{B}_0^Y(\mathbf{p})$ is only defined for $\mathbf{p} \in \mathbb{P}_0 \setminus \{\pi(b_0)\}$, and so it is important that later in Lemma 7.13 we will ensure that $\pi(b_0) \notin \mathscr{M}_0$.

---

[14]To prevent confusion, we want to point out that here and in the calculation to come, all superscripts containing k and n take on the meaning of an exponent, while all occurrences of i, j, and l serve as indices of vectors.

By the symmetry of our construction so far, the specific choice of b_0 has no impact on our model. However, note that we could in fact also choose any positive linear combination of these elementary functions, but for simplicity of our model we decide not to at this point. (See also our conclusions in Section 9.2.)

With this choice of $\mathcal{B}_0^Y$ we then find that our bond sections for all other maturities $T \in \mathbb{R}$ are given by

$$\mathcal{B}_T^Y(x) \overset{(5.74)}{=} (\mathcal{U}_{-T}\mathcal{B}_0^Y)(x) = (\mathcal{U}_{-T}[b_0])(x) \overset{(7.77)}{=} [b_0](\mathcal{U}_T x)$$
$$\overset{(7.87)}{=} \langle b_0, \mathcal{U}_T x\rangle^{1-n/2} = \langle \mathcal{U}_T^* b_0, x\rangle^{1-n/2} \overset{(7.87)}{=} [\mathcal{U}_T^* b_0](x).$$

In short, we have for $\forall T \in \mathbb{R}$

$$\mathcal{B}_T^Y = [b_T], \tag{7.90}$$

$$\text{where} \quad b_T := \mathcal{U}_T^* b_0 = \mathrm{e}^{T(-\varepsilon+\rho)} b_0 \tag{7.91}$$

by (7.59) and (7.58), and thus by (5.40) the bond sections in the spot parametrization are given by

$$\mathcal{B}_T^X(\cdot, t) = \mathcal{B}_{T-t}^Y = [b_{T-t}].$$

We will call the vectors b_T the *bond vectors*. Since $\forall t \in \mathbb{R}: \mathcal{U}_t^* \in \hat{O}(\mathbb{T})$ by (7.41), we have

$$\forall t \in \mathbb{R}: \quad b_t \in \mathbb{T}_0^\downarrow \tag{7.92}$$

by (7.89) and Lemma 7.4, and so these sections are well defined, positive, and harmonic on $\mathbb{P}_0 \setminus \{\pi(b_T)\}$ and $\mathbb{P}_0 \setminus \{\pi(b_{T-t})\}$, respectively.

7.7.3 Bond Prices, Short and Forward Rates

The bond exchange rates in the two parametrizations are then given by

$$B_{T_1,T_2}^Y = \frac{\mathcal{B}_{T_2}^Y}{\mathcal{B}_{T_1}^Y} = \frac{[b_{T_2}]}{[b_{T_1}]} \quad \text{and} \quad B_{T_1,T_2}^X(\cdot, t) = \frac{\mathcal{B}_{T_2}^X(\cdot, t)}{\mathcal{B}_{T_1}^X(\cdot, t)} = \frac{[b_{T_2-t}]}{[b_{T_1-t}]} \tag{7.93}$$

(defined on $\mathbb{P}_0 \setminus \{\pi(b_{T_1}), \pi(b_{T_2})\}$ and $\mathbb{P}_0 \setminus \{\pi(b_{T_1-t}), \pi(b_{T_2-t})\}$, respectively), and so the bond dollar prices in the spot parametrization are given by

$$B_{\$,t+\tau}^X(\cdot, t) = B_{t,t+\tau}^X(\cdot, t) = \frac{[b_\tau]}{[b_0]}, \tag{7.94}$$

defined on $\mathbb{P}_0 \setminus \{\pi(b_0), \pi(b_\tau)\}$.

Note how the k-homogeneity of the functions $[b_T]$ implies that $B_{T_1,T_2}^Y(ax) = B_{T_1,T_2}^Y(x)$ for $\forall x$ and $\forall a > 0$, i.e., that B_{T_1,T_2}^Y is 0-homogeneous, or in other words constant along each ray $\mathbf{p}_+$. In the sense of (7.73), the exchange rates B_{T_1,T_2}^Y (and similarly $B_{T_1,T_2}^X(\cdot, t)$ and $B_{\$,t+\tau}^X(\cdot, t)$) are therefore indeed just standard functions of our state variable $\mathbf{p}$.

Finally, the short rate function $r^X \in \mathcal{F}\big(\mathbb{P}_0 \setminus \{\pi(b_0)\}\big)$, again written as a function in $H^0\big(\mathbb{T}_0^{\uparrow} \setminus \pi(b_0)\big)$, is given by

$$r^X(x) \overset{(5.62)}{=} -\partial_\tau B^Y_{0,\tau}(x)\big|_{\tau=0} \overset{(7.93)}{=} -\partial_\tau \frac{[b_\tau](x)}{[b_0](x)}\bigg|_{\tau=0} \overset{(7.87)}{=} -\partial_\tau \frac{\langle b_\tau, x\rangle^{1-n/2}}{\langle b_0, x\rangle^{1-n/2}}\bigg|_{\tau=0}$$

$$= \left(\tfrac{n}{2}-1\right)\frac{\langle \dot b_0, x\rangle}{\langle b_0, x\rangle} \tag{7.95}$$

$$\overset{(7.91)}{=} \left(\tfrac{n}{2}-1\right)\frac{\langle(-\varepsilon+\rho)b_0, x\rangle}{\langle b_0, x\rangle} = \left(\tfrac{n}{2}-1\right)\left(\rho - \frac{\langle \varepsilon b_0, x\rangle}{\langle b_0, x\rangle}\right), \tag{7.96}$$

where we use the notation $\dot b_t := \partial_t b_t$. Observe that this function, too, remains invariant under the substitution $x \mapsto ax$, i.e., r^X is indeed a function of $\mathbf{p}$.

For future reference, note that if we also introduce the notation $\ddot b_t := \partial_t^2 b_t$, we have the representations

$$\dot b_t = (-\varepsilon+\rho)\; e^{t(-\varepsilon+\rho)}b_0 = (-\varepsilon+\rho)\; b_t = \mathcal{U}_t^* \dot b_0\,, \tag{7.97a}$$

$$\ddot b_t = (-\varepsilon+\rho)^2 e^{t(-\varepsilon+\rho)}b_0 = (-\varepsilon+\rho)^2 b_t = \mathcal{U}_t^* \ddot b_0\,. \tag{7.97b}$$

The following lemma summarizes our main results, namely the formulas (7.94) and (7.96), in a more accurate form, without our recently adopted ambiguity between sections and homogeneous functions.

Lemma 7.11. *The bond dollar price and the short rate function in the spot parametrization are given by*

$$B^X_{\$,t+\tau}(\mathbf{p}, t) = \frac{[b_\tau](x)}{[b_0](x)} = \left(\frac{\langle b_\tau, x\rangle}{\langle b_0, x\rangle}\right)^{1-n/2} \tag{7.98}$$

$$r^X(\mathbf{p}) = \left(\tfrac{n}{2}-1\right)\frac{\langle \dot b_0, x\rangle}{\langle b_0, x\rangle} = \left(\tfrac{n}{2}-1\right)\left(\rho - \frac{\langle \varepsilon b_0, x\rangle}{\langle b_0, x\rangle}\right) \tag{7.99}$$

for any arbitrary $x \in \mathbf{p}_+$.

While these expressions are formally defined for all states $\mathbf{p}$ in the sets $\mathbb{P}_0 \setminus \{\pi(b_0), \pi(b_\tau)\}$ and $\mathbb{P}_0 \setminus \{\pi(b_0)\}$, respectively, in our model we will only end up using their restrictions to the subset $\mathcal{M}_0$ that will be defined later in Section 7.8.2.

7.7.4 Bond Vector Constraint

We conclude this section with an observation about the quantity $\langle \dot b_0, \dot b_0\rangle$ that will lead to an important parameter constraint for our model.

Lemma 7.12. *We have $\langle \dot b_0, \dot b_0\rangle_{\mathbb{T}} \geq 0$. Equality holds if and only if b_0 is an eigenvector of ε, and in that case our model degenerates to the deterministic case in which r^X does not actually depend on $\mathbf{p}$, so that the process (X_t) becomes irrelevant to our model.*

Proof. First, by Lemma 7.1 (i) we have for $\forall x, y \in \mathbb{T}_0^\downarrow$

$$\langle x - y, x - y \rangle = \langle x, x \rangle - 2\langle x, y \rangle + \langle y, y \rangle = -2\langle x, y \rangle \geq 0,$$

and using (7.92) we can therefore conclude that

$$\langle \dot{b}_0, \dot{b}_0 \rangle = \lim_{t \to 0} \tfrac{1}{t^2} \langle b_t - b_0, b_t - b_0 \rangle \geq 0.$$

Now if b_0 is an eigenvector of ε with some eigenvalue $\eta \in \mathbb{R}$ then we have

$$\langle \dot{b}_0, \dot{b}_0 \rangle = \langle (-\varepsilon + \rho)b_0, (-\varepsilon + \rho)b_0 \rangle = (-\eta + \rho)^2 \langle b_0, b_0 \rangle = 0.$$

To prove the other direction, suppose that $\langle \dot{b}_0, \dot{b}_0 \rangle = 0$, i.e., $\dot{b}_0 \in \mathbb{T}_0 \cup \{0\}$. Since by (7.92) we have

$$\langle \dot{b}_0, b_0 \rangle = \tfrac{1}{2} \partial_t \langle b_t, b_t \rangle |_{t=0} = \tfrac{1}{2} \partial_t 0 |_{t=0} = 0, \tag{7.100}$$

(7.89) and Lemma 7.1 (i) then imply that

$$\dot{b}_0 = \eta' b_0 \qquad \text{for some} \qquad \eta' \in \mathbb{R}; \tag{7.101}$$

writing this as $(-\varepsilon + \rho)b_0 = \eta' b_0$, we therefore find that b_0 is then indeed an eigenvector of ε, with eigenvalue $\rho - \eta'$. Furthermore, plugging (7.101) into (7.99) shows that in this case we have $r^X(\mathbf{p}) \equiv (\tfrac{n}{2} - 1)\eta'$ for every state $\mathbf{p}$. $\square$

To exclude this degenerate deterministic case, we will therefore from now on assume that $\langle \dot{b}_0, \dot{b}_0 \rangle > 0$. In fact, since we can multiply b_0 by any positive scalar without affecting the model (as such a factor would cancel out in all of our formulas in Lemma 7.11), by choosing this scalar as $(\zeta / \langle \dot{b}_0, \dot{b}_0 \rangle)^{1/2}$ for some arbitrary fixed constant $\zeta > 0$ (which we will choose later to simplify our equations), *we can from now on assume that*

$$\langle \dot{b}_0, \dot{b}_0 \rangle_\mathbb{T} = \zeta. \tag{7.102}$$

Note that by Lemma 7.12 this assumption implies that b_0 is not an eigenvector of ε. Furthermore, observe that by writing $\dot{b}_0 = (-\varepsilon + \rho)b_0$ (recall (7.97a)) and using that $\langle b_0, \varepsilon b_0 \rangle = 0$ (by the antisymmetry of ε) and $\langle b_0, b_0 \rangle = 0$ (by (7.92)), we can rewrite (7.102) as

$$\langle \varepsilon b_0, \varepsilon b_0 \rangle_\mathbb{T} = \zeta. \tag{7.103}$$

7.8 State Space

7.8.1 Dimension

We now see from (7.99) that in the dimension $n = 2$ our construction reduces to the trivial model where $r^X(\mathbf{p}) = 0$ for every state $\mathbf{p}$. This dimension will therefore no longer be of any interest to us.[15]

---

[15]Observe also that for this value of n we have $k = 0$, and so $\mathcal{S}(\mathcal{V}) = \mathcal{S}(\mathscr{P}^{\otimes(-k)})$ degenerates to the regular function space $\mathcal{F}(M)$.

The dimension $n = 1$, which is not rich enough to support a model with sufficient dynamic flexibility anyways, would be tedious for us to keep track of from this point forward since it is the only dimension in which the factor $\frac{n}{2} - 1$ in our short rate formula (7.99) is negative. *From now on we will therefore assume that*

$$n \geq 3. \tag{7.104}$$

We will revisit the case $n = 1$ again briefly at the end of Section 7.8.3.

7.8.2 Short Rate Constraint, State Space Family

Next we will decide on our state space $\mathscr{M}_0 \subset \mathscr{M} \subset \mathbb{P}_0$ of the spot parametrization. Guided by our plan to restrict our process to only those states at which the short rate function is positive, we define

$$\mathscr{M}_0 := \left\{ \mathbf{p} \in \mathbb{P}_0 \setminus \{\pi(b_0)\} \,\middle|\, r^X(\mathbf{p}) > 0 \right\}. \tag{7.105}$$

(As one can see from (7.99), at the point $\pi(b_0)$ the short rate function r^X is undefined.) Using (7.99), (7.104), (7.5b), (7.89), and (7.100), this can be rewritten as

$$\mathscr{M}_0 = \pi\left(\left\{ x \in \mathbb{T}_0^\uparrow \,\middle|\, \langle b_0, x \rangle > 0 \right\}\right). \tag{7.106}$$

Recall that by (5.14) and (5.9), our definition (7.105) implies that in fact the entire forward rate curve $\tau \mapsto F_{\mathrm{rel}}(\mathbf{p}; \tau)$ is guaranteed to be positive at each state $\mathbf{p} \in \mathscr{M}_0$.

Lemma 7.13. *The time-dependent state space $\mathscr{M}_t$, $t \in \mathbb{R}$, of the forward parametrization is given by*

$$\mathscr{M}_t := \mathcal{U}_{-t}(\mathscr{M}_0) = \pi\left(\left\{ x \in \mathbb{T}_0^\uparrow \,\middle|\, \langle \dot{b}_t, x \rangle > 0 \right\}\right), \tag{7.107}$$

and so in particular we have

$$\partial \mathscr{M}_t = \pi\left(\left\{ x \in \mathbb{T}_0^\uparrow \,\middle|\, \langle \dot{b}_t, x \rangle = 0 \right\}\right) \tag{7.108}$$

and thus $\pi(b_t) \in \partial \mathscr{M}_t$ for $\forall t \in \mathbb{R}$.

Proof. See Appendix F.5.   □

Finally, as in (5.32) we define

$$\mathscr{M} := \bigcup_{t \geq 0} \mathscr{M}_t. \tag{7.109}$$

7.8.3 Closure under the Zero-Noise Flow

It is now time to find out how to choose ε and ρ in such a way that (5.9) holds, i.e., that $\forall t \geq 0$: $\mathcal{U}_t(\mathcal{M}_0) \subseteq \mathcal{M}_0$. Our criterion is based on the vector

$$\ddot{b}_0^\perp := \varepsilon(\varepsilon - \rho)b_0, \tag{7.110}$$

which is the component of $\ddot{b}_0$ perpendicular to $\dot{b}_0$, as it fulfills

$$\ddot{b}_0^\perp = \ddot{b}_0 - \rho\dot{b}_0 \tag{7.111}$$

$$\langle \ddot{b}_0^\perp, \dot{b}_0 \rangle_\mathbb{T} = -\langle \varepsilon(\varepsilon - \rho)b_0, (\varepsilon - \rho)b_0 \rangle_\mathbb{T} = 0 \tag{7.112}$$

by (7.97) and Lemma 7.6 (iii).

Lemma 7.14. *Under the condition*

$$\langle \ddot{b}_0^\perp, \ddot{b}_0^\perp \rangle_\mathbb{T} \leq 0 \tag{7.113}$$

we have the desired property

$$\forall t \geq 0: \quad \mathcal{U}_t(\mathcal{M}_0) \subseteq \mathcal{M}_0, \tag{7.114}$$

and so the state space family $(\mathcal{M}_t)_{t\in\mathbb{R}}$ is increasing, i.e.,

$$\mathcal{M}_{t_1} \subseteq \mathcal{M}_{t_2} \qquad for \qquad t_1 \leq t_2. \tag{7.115}$$

Furthermore, for all $t_1 \neq t_2$ we have

$$\partial\mathcal{M}_{t_1} \cap \partial\mathcal{M}_{t_2} = \begin{cases} \{\pi(\ddot{b}_0^\perp)\} & if \ \langle \ddot{b}_0^\perp, \ddot{b}_0^\perp \rangle_\mathbb{T} = 0 \ and \ if \ \ddot{b}_0^\perp \ is \ an \ eigenvector \ of \ \varepsilon, \\ \varnothing & else, \end{cases} \tag{7.116}$$

i.e., the sets $\mathcal{M}_t$ are in fact strictly increasing everywhere on their boundaries, except (in some degenerate case) at a single critical point.

Proof. First note that the condition (7.113) is equivalent to

$$\ddot{b}_0^\perp \in \mathbb{T}^\downarrow \cup \mathbb{T}_0^\downarrow. \tag{7.117}$$

Indeed, (7.117) trivially implies (7.113) by our definitions of $\mathbb{T}^\downarrow$ and $\mathbb{T}_0^\downarrow$. For the reverse direction, let us assume that (7.113) holds. By (7.110), the antisymmetry of ε, Lemma 7.6 (iii), (7.97a), (7.102), and (7.89) we then have

$$\langle \ddot{b}_0^\perp, b_0 \rangle = \langle \varepsilon^2 b_0, b_0 \rangle - \rho\langle \varepsilon b_0, b_0 \rangle = -\langle \varepsilon b_0, \varepsilon b_0 \rangle - 0$$

$$= -\langle (-\varepsilon + \rho)b_0, (-\varepsilon + \rho)b_0 \rangle - 2\rho\langle \varepsilon b_0, b_0 \rangle + \rho^2\langle b_0, b_0 \rangle$$

$$= -\langle \dot{b}_0, \dot{b}_0 \rangle - 0 + 0 = -\zeta < 0. \tag{7.118}$$

As a first consequence, this shows that $\ddot{b}_0^\perp \neq 0$, and so (7.113) implies that $\ddot{b}_0^\perp \in \mathbb{T}^\uparrow \cup \mathbb{T}^\downarrow \cup \mathbb{T}_0$; since by Lemma 7.1 (i) and (iii) (with $x := b_0 \in \mathbb{T}_0^\downarrow$) this estimate also implies that $\ddot{b}_0^\perp \notin \mathbb{T}^\uparrow \cup \mathbb{T}_0^\uparrow$, we have thus shown that (7.113) implies (7.117).

Let us therefore assume that (7.117) is fulfilled. Our proof presented here will only treat the case $\ddot{b}_0^\perp \in \mathbb{T}^\downarrow$; the remaining case $\ddot{b}_0^\perp \in \mathbb{T}_0^\downarrow$ is more technical and is postponed to Appendix F.6.

If $\ddot{b}_0^\perp \in \mathbb{T}^\downarrow$ then by (7.97a), (7.111), (7.108), and Lemma 7.1 (ii) we have for $\forall \mathbf{p} \in \partial \mathscr{M}_0$ and $\forall x \in \mathbf{p} \cap \mathbb{T}_0^\uparrow$

$$\partial_t \langle \dot{b}_0, \mathcal{U}_t x \rangle \big|_{t=0} = \partial_t \langle \mathcal{U}_t^* \dot{b}_0, x \rangle \big|_{t=0} = \partial_t \langle \dot{b}_t, x \rangle \big|_{t=0} = \langle \ddot{b}_0, x \rangle$$
$$= \langle \ddot{b}_0^\perp, x \rangle + \rho \langle \dot{b}_0, x \rangle = \langle \ddot{b}_0^\perp, x \rangle > 0, \qquad (7.119)$$

and so by (7.106) and our definition (7.42) of $\mathcal{U}_t(\mathbf{p})$ there $\exists \eta > 0$ such that

$$\forall t \in (-\eta, 0]: \quad \mathcal{U}_t(\mathbf{p}) \notin \mathscr{M}_0 \qquad (7.120\text{a})$$
$$\text{and} \qquad \forall t \in (0, \eta): \quad \mathcal{U}_t(\mathbf{p}) \in \mathscr{M}_0. \qquad (7.120\text{b})$$

Now if for any $\hat{\mathbf{p}} \in \mathscr{M}_0$ the path $\{\mathcal{U}_t(\hat{\mathbf{p}}) \,|\, t \geq 0\}$ would lead out of $\mathscr{M}_0$, then we could apply the above to that path's first exit point $\mathbf{p}$ from $\mathscr{M}_0$, and (7.120a) would contradict the fact that $\mathbf{p}$ is the *first* exit point. This proves (7.114) for this case, and the monotonicity (7.115) of the sets $\mathscr{M}_t$ follows just like back in Section 5.3.4:

$$\mathscr{M}_{t_1} \overset{(7.107)}{=} \mathcal{U}_{-t_1}(\mathscr{M}_0) = \mathcal{U}_{-t_2}(\mathcal{U}_{t_2-t_1}(\mathscr{M}_0)) \overset{(7.114)}{\subseteq} \mathcal{U}_{-t_2}(\mathscr{M}_0) \overset{(7.107)}{=} \mathscr{M}_{t_2}.$$

To show (7.116) (for which in our case the right-hand side is $\varnothing$), note that (7.120b) can be written as $\forall t \in (0, \eta): \mathbf{p} \in \mathcal{U}_{-t}(\mathscr{M}_0) = \mathscr{M}_t$, which due to the monotonicity (7.114) of the sets $\mathscr{M}_t$ implies that in fact we have $\forall t > 0: \mathbf{p} \in \mathscr{M}_t$. Since this holds for $\forall \mathbf{p} \in \partial \mathscr{M}_0$, this shows that $\forall t > 0: \partial \mathscr{M}_0 \subset \mathscr{M}_t$, and since the sets $\mathscr{M}_t$ are open, this implies that $\forall t > 0: \partial \mathscr{M}_0 \cap \partial \mathscr{M}_t = \varnothing$. Finally, for any given $t_1 < t_2$ choosing $t := t_2 - t_1 > 0$ and then applying $\mathcal{U}_{-t_1}$ to both sides, we obtain $\partial \mathscr{M}_{t_1} \cap \partial \mathscr{M}_{t_2} = \varnothing$. $\qquad \square$

Corollary 7.1. *Under the assumptions* (7.102) *and* (7.113) *we have*

$$\forall \tau_1, \tau_2 \in \mathbb{R}: \quad \tau_1 \neq \tau_2 \implies \pi(b_{\tau_1}) \neq \pi(b_{\tau_2}).$$

Proof. Suppose that for some $\tau_1 \neq \tau_2$ we had $\pi(b_{\tau_1}) = \pi(b_{\tau_2})$. Then by Lemma 7.13 we would have $\pi(b_{\tau_1}) \in \partial \mathscr{M}_{\tau_1} \cap \partial \mathscr{M}_{\tau_2}$, and so (7.116) would allow us to conclude that $\pi(b_{\tau_1}) = \pi(\ddot{b}_0^\perp)$, and that $\ddot{b}_0^\perp$ and thus also b_{τ_1} must be eigenvectors of ε. But denoting the eigenvalue by η, this would imply that $b_0 = \mathcal{U}_{-\tau_1}^* b_{\tau_1} = \mathrm{e}^{(-\tau_1)(-\varepsilon + \rho)} b_{\tau_1} = \mathrm{e}^{(-\tau_1)(-\eta + \rho)} b_{\tau_1}$, so that also b_0 would be an eigenvector of ε. However, as we stated below our assumption (7.102), this is not the case. $\qquad \square$

Remarks on the vector $\ddot{b}_0^\perp$. As becomes evident in the proof of Lemma 7.14, we could have based our condition (7.113) on any vector $\ddot{b}_0 + c \cdot \dot{b}_0$, $c \in \mathbb{R}$, instead of $\ddot{b}_0^\perp$. The specific value $c = -\rho$ used in the definition of $\ddot{b}_0^\perp$ was chosen to minimize $\langle \ddot{b}_0 + c \cdot \dot{b}_0, \ddot{b}_0 + c \cdot \dot{b}_0 \rangle$, and thus to make the condition (7.113) as weak as possible.

Another reason why our use of $\ddot{b}_0^\perp$ was the most natural choice is the formula

$$\partial_t r^X(\mathcal{U}_t(\mathbf{p}))\big|_{t=0} = \left(\tfrac{n}{2} - 1\right) \cdot \frac{\langle \ddot{b}_0^\perp, x \rangle}{\langle b_0, x \rangle} \qquad (7.121)$$

for $\forall \mathbf{p} \in \partial \mathcal{M}_0$ and $\forall x \in \mathbf{p}_+$. (The derivation can be found in Appendix F.7.) Since by Lemma 7.1 (i) the denominator on the right is always positive, this formula could be used as an alternative to our estimate (7.119); we chose not to use it, in order to avoid the problems at the point $\mathbf{p} = \pi(b_0)$ at which the denominator vanishes.

Reformulation of the condition (7.113). By substituting the definition (7.110) in (7.113) and then using that $\langle \varepsilon b_0, \varepsilon^2 b_0 \rangle = 0$ (by the antisymmetry of ε) and $\langle \varepsilon b_0, \varepsilon b_0 \rangle = \zeta$ (by our assumption (7.103)), we can rewrite the condition (7.113) as

$$\langle \varepsilon^2 b_0, \varepsilon^2 b_0 \rangle + \zeta \rho^2 \leq 0. \qquad (7.122)$$

From now on we will assume that this condition (7.122) — and thus also (7.113) — is fulfilled.

The one-dimensional case $n = 1$. In one dimension, the prefactor $\frac{n}{2} - 1$ in the short rate function r^X in (7.99) is negative, and so the $>$-sign in our state space formula (7.107) must be flipped. As a result, our technique that we use in the proof of Lemma 7.14 to ensure that the sets $\mathcal{M}_t$ are increasing no longer works: Indeed, looking at (7.119) (or alternatively, (7.121)), we would now need to ensure that

$$\langle \ddot{b}_0^\perp, x \rangle < 0 \qquad \text{for} \quad \forall x \in \mathbb{T}_0^\uparrow \quad \text{such that} \quad \pi(x) \in \partial \mathcal{M}_0 \qquad (7.123)$$

(instead of $>$), and so our technique based on Lemma 7.1 (ii) would now lead us to the condition $\ddot{b}_0^\perp \in \mathbb{T}^\uparrow \cup \mathbb{T}_0^\uparrow$ (instead of $\ddot{b}_0^\perp \in \mathbb{T}^\downarrow \cup \mathbb{T}_0^\downarrow$); however, this condition can never be fulfilled, since it implies (7.113), which our proof has shown to be equivalent to (7.117). Since $\ddot{b}_0^\perp \in \mathbb{T}^\downarrow \cup \mathbb{T}_0^\downarrow$ would lead to the wrong type of flow, and since we generally have $\ddot{b}_0^\perp \notin \mathbb{T}^\uparrow \cup \mathbb{T}_0^\uparrow$, we would therefore have to ask that $\langle \ddot{b}_0^\perp, \ddot{b}_0^\perp \rangle > 0$, and for such vectors there is no simple rule about the sign of $\langle \ddot{b}_0^\perp, x \rangle$ for arbitrary $x \in \mathbb{T}_0^\uparrow$ (which is why Lemma 7.1 does not mention this case).

Fortunately, in one dimension the boundary $\partial \mathcal{M}_0$ only consists of two isolated points, and so in this case we do not need to resort to this rather crude method; instead, we can just manually ensure that (7.123) happens to hold for these two points once we have derived our final equations. It is, however, not worth the effort to work out the details now in this $\mathcal{M}_0$-based description of the model; this is done much more easily once we have our $\mathbb{R}^n$-based model description for all dimensions $n \geq 3$ at our disposal. See Section 2.9.

Closure of $\mathscr{M}$ under the maps $\mathcal{U}_t$. Finally, observe that the monotonicity (7.115) of the state space family $(\mathscr{M}_t)_{t \in \mathbb{R}}$ allows us to rewrite (7.109) as

$$\mathscr{M} = \bigcup_{t \in \mathbb{R}} \mathscr{M}_t .$$

Recalling the definition (7.107) of the sets $\mathscr{M}_t$ and the group property of the operators $\mathcal{U}_t$, this shows that under the condition (7.113) we have

$$\forall t \in \mathbb{R}: \quad \mathcal{U}_t(\mathscr{M}) = \mathscr{M} . \tag{7.124}$$

Chapter 8

Parametrization in $\mathbb{R}^n$

Now that the construction of our model on the manifold $\mathscr{M} \subset \mathbb{P}_0$ is complete (we have defined $\mathscr{M}$ and $\mathscr{M}_0$, $\mathfrak{c}$, $\mathscr{V}$, $(\mathcal{U}_s)_{s \in \mathbb{R}}$, $\diamondsuit$, and $\mathcal{B}_0^Y$), the last task on our to-do list in Section 6.6 is to define a map $h \colon \mathscr{M} \to \mathbb{R}^n$ in such a way that the conformal structure $c := h(\mathfrak{c})$ is the Euclidean one, and then to use this map to transport our model to $\mathbb{R}^n$, by utilizing the rules listed in Table 6.1. Any such parametrization of our model in $\mathbb{R}^n$ is called a *flat parametrization*.

We will begin by defining a whole family of maps $h_{\mathfrak{I}} \colon \mathbb{P}_0 \to \bar{\mathbb{R}}^n$ (one for each "isotropic basis" $\mathfrak{I}$ of $\mathbb{T}$) that map the manifold $\mathbb{P}_0$ that contains our space $\mathscr{M}$ down to the compactified Euclidean space $\bar{\mathbb{R}}^n := \mathbb{R}^n \cup \{\infty\}$.[1] While for the right choice of $\mathfrak{I}$ we would indeed have $h_{\mathfrak{I}}(\mathscr{M}) = \mathbb{R}^n$,[2] the resulting $\mathbb{R}^n$-based model description would turn out to be unnecessarily complicated. Instead, since our final model is based only on M_0, we can allow $h_{\mathfrak{I}}(\mathscr{M})$ to contain the point ∞, as long as we ensure that $M_0 = h_{\mathfrak{I}}(\mathscr{M}_0) \subset \mathbb{R}^n$.

8.1 Isotropic Bases

To facilitate our construction of h, we will first express our model in a more convenient basis. The following definition describes a class of bases that is tailored to suit our situation.

Definition 8.1. *An* isotropic basis *of $\mathbb{T}$ is a collection of independent vectors*

$$\mathfrak{I} := \{e_\odot, e_\times, e_1, \ldots, e_n\} \subset \mathbb{T}$$

with $e_\odot \in \mathbb{T}_0^\downarrow$ and $e_\times \in \mathbb{T}_0^\uparrow$, such that

$$\langle e_\mu, e_\nu \rangle = \tilde{Q}_{\mu\nu} \qquad for \qquad \forall \mu, \nu \in \{\odot, \times, 1, \ldots, n\}, \tag{8.1}$$

---

[1] Note that it is not possible to map $\mathbb{P}_0$ onto $\mathbb{R}^n$, since $h(\mathbb{P}_0)$ is the continuous image of a compact set and therefore compact itself.

[2] All we would need to do is in Section 8.5 to not choose b_0 as our basis vector $e_\odot$, but instead the eigenvector x' in Lemma 8.10 (ii).

where

$$(\tilde{Q}_{\mu\nu}) := \begin{pmatrix} 0 & 1 & 0 & \cdots & 0 \\ 1 & 0 & 0 & \cdots & 0 \\ 0 & 0 & 1 & \ddots & \vdots \\ \vdots & \vdots & \ddots & \ddots & 0 \\ 0 & 0 & \cdots & 0 & 1 \end{pmatrix}. \tag{8.2}$$

Observe that by (7.11) and (7.9a) the number of basis vectors is $n + 2 = m + 1 = \dim(\mathbb{T})$, as required. To better understand this definition, consider the specific example

$$\mathfrak{I} = \{e_\odot, e_\times, e_1, \ldots, e_n\} = \left\{ \begin{pmatrix} -\frac{1}{\sqrt{2}} \\ 0 \\ 0 \\ \vdots \\ 0 \\ 0 \\ \frac{1}{\sqrt{2}} \end{pmatrix}, \begin{pmatrix} \frac{1}{\sqrt{2}} \\ 0 \\ 0 \\ \vdots \\ 0 \\ 0 \\ \frac{1}{\sqrt{2}} \end{pmatrix}, \begin{pmatrix} 0 \\ 1 \\ 0 \\ \vdots \\ 0 \\ 0 \\ 0 \end{pmatrix}, \ldots, \begin{pmatrix} 0 \\ 0 \\ 0 \\ \vdots \\ 0 \\ 1 \\ 0 \end{pmatrix} \right\}.$$

The following lemma explores this concept in more generality.

Lemma 8.1. *(i) Given any isotropic basis $\mathfrak{I} = \{e_\odot, e_\times, e_1, \ldots, e_n\}$ of $\mathbb{T}$ and any operator $A \in O^\uparrow(\mathbb{T})$, the set $\mathfrak{I}' := \{Ae_\odot, Ae_\times, Ae_1, \ldots, Ae_n\}$ is an isotropic basis of $\mathbb{T}$ as well.*

(ii) Given any two non-collinear vectors $e_\odot \in \mathbb{T}_0^\downarrow$ and $x \in \mathbb{T}_0^\uparrow$, there exist a scaling factor $c > 0$ and vectors $e_1, \ldots, e_n \in \mathbb{T}$ such that $\{e_\odot, c \cdot x, e_1, \ldots, e_n\}$ is an isotropic basis of $\mathbb{T}$.

While the value c is unique, the vectors $e_1, \ldots, e_n$ are not: For any orthogonal matrix $P \in \mathbb{R}^{n \times n}$ they can be replaced by the vectors $e_1', \ldots, e_n'$ computed via the matrix product $(e_1', \ldots, e_n') := (e_1, \ldots, e_n)P$. In fact, every other possible choice of vectors $e_1', \ldots, e_n'$ can be obtained in this way.

Proof. (i) Since by (7.35) pseudo-orthogonal operators are invertible, $\mathfrak{I}'$ is a basis of $\mathbb{T}$, and by Lemma 7.2 we have $Ae_\odot \in \mathbb{T}_0^\downarrow$ and $Ae_\times \in \mathbb{T}_0^\uparrow$. Finally, by (7.34) we have $\langle Ae_\mu, Ae_\nu \rangle = \langle e_\mu, e_\nu \rangle = \tilde{Q}_{\mu\nu}$ for $\forall \mu, \nu = \odot, \times, 1, \ldots, n$.

(ii) To prove the first statement, let two non-collinear vectors $e_\odot \in \mathbb{T}_0^\downarrow$ and $x \in \mathbb{T}_0^\uparrow$ be given. Since $e_\odot, x \in \mathbb{T}_0$, we have $\langle e_\odot, e_\odot \rangle = \langle x, x \rangle = 0$, and by (7.5b) we have $\langle e_\odot, x \rangle > 0$. This shows that for the unique choice $c := 1/\langle e_\odot, x \rangle > 0$, the vectors $e_\odot$ and $e_\times := c \cdot x$ have all the required inner products. It therefore remains to construct a set of mutually pseudo-orthogonal vectors $e_1, \ldots, e_n$ that are spanning the space $\{e_\odot, e_\times\}^\perp$ of vectors that are pseudo-orthogonal to $e_\odot$ and $e_\times$. For this it suffices to show that

$$\forall y \in \{e_\odot, e_\times\}^\perp \setminus \{0\}: \quad \langle y, y \rangle > 0, \tag{8.3}$$

since this would allow us to construct these vectors with the standard Gram–Schmidt procedure, modified to use the Minkowski inner product instead of the Euclidean one.

To see (8.3), note that no vector $y \in \{e_\odot, e_\times\}^\perp \setminus \{0\}$ can be in $\mathbb{T}_0$ since by (7.5a) this would imply that $y \parallel e_\odot$ and $y \parallel e_\times$, and thus $e_\odot \parallel e_\times$; furthermore, by Lemma 7.1 (ii) no vector in $\mathbb{T}^\downarrow \cup \mathbb{T}^\uparrow$ can be pseudo-orthogonal to $e_\times$. But these two observations just say that no vector $y \in \{e_\odot, e_\times\}^\perp \setminus \{0\}$ can fulfill $\langle y, y \rangle = 0$ or $\langle y, y \rangle < 0$, respectively, and so we have proven (8.3).

For the second statement of part (ii), let us write $P = (p^i{}_j)$ and use Einstein notation to sum over the indices $k, l = 1, \ldots, n$ only. We then find for $\forall i, j = 1, \ldots, n$ that

$$\langle e'_i, e'_j \rangle = \langle p^k{}_i e_k, p^l{}_j e_l \rangle = p^k{}_i p^l{}_j \langle e_k, e_l \rangle \overset{(8.1)}{=} p^k{}_i p^l{}_j \tilde{Q}_{kl} \overset{(8.2)}{=} p^k{}_i p^k{}_j$$
$$= (P^T P)_{ij} = I_{ij} = \delta_{ij} \, ,$$
$$\langle e'_i, e_\odot \rangle = \langle p^k{}_i e_k, e_\odot \rangle = p^k{}_i \underbrace{\langle e_k, e_\odot \rangle}_{=0} = 0 \, ,$$

and similarly $\langle e'_i, e_\times \rangle = 0$, i.e., the vectors $e'_1, \ldots, e'_n$ have all the correct inner products, and we have $e'_1, \ldots, e'_n \in \{e_\odot, e_\times\}^\perp$. Finally, since P is invertible, we have $\mathrm{rank}(e'_1, \ldots, e'_n) = \mathrm{rank}(e_1, \ldots, e_n) = n = \dim\{e_\odot, e_\times\}^\perp$ and thus indeed $\mathrm{span}\{e'_1, \ldots, e'_n\} = \{e_\odot, e_\times\}^\perp$ as desired.

To see the reverse direction, note that since $\{e_1, \ldots, e_n\}$ is a basis of $\{e_\odot, e_\times\}^\perp$, every other basis $\{e'_1, \ldots, e'_n\}$ of $\{e_\odot, e_\times\}^\perp$ can be written as $(e'_1, \ldots, e'_n) := (e_1, \ldots, e_n) P$ for some matrix $P \in \mathbb{R}^{n \times n}$. A calculation just like the one above then shows that $(P^T P)_{ij} = \cdots = \langle e'_i, e'_j \rangle = \delta_{ij}$, i.e., $P^T P = I$, so that P is an orthogonal matrix. $\square$

8.2 Isotropic Coefficients

Given any fixed isotropic basis $\mathfrak{J}$, we will denote the (contravariant) coefficients of a point $x \in \mathbb{T}$ with respect to $\mathfrak{J}$ by $(x^\odot, x^\times, x^1, \ldots, x^n)$, or in short, by $(x^\odot, x^\times, \vec{x})$, where $\vec{x} := (x^1, \ldots, x^n)$. Adopting the Einstein notation to such bases by summing over all indices $\mu = \odot, \times, 1, \ldots, n$ instead, this can in short be written as

$$x = x^\odot e_\odot + x^\times e_\times + x^1 e_1 + \cdots + x^n e_n = x^\mu e_\mu \, . \tag{8.4}$$

Of course, these coefficients depend on the specific choice of $\mathfrak{J}$ and should therefore rather be denoted as $x^\mu_{\mathfrak{J}}$, also to distinguish them from the original coefficients x^i and the vector $\vec{x} \in \mathbb{R}^m$ introduced in Section 7.1.1. However, since from now on we will no longer work in the standard basis of $\mathbb{T}$ (as we did throughout Chapter 7), the notational ambiguity arising from this simplified notation will not be a problem for us; in fact, *from this point forward we will allow ourselves to write "$x = (x^\odot, x^\times, \vec{x})$."* The isotropic basis $\mathfrak{J}$ that we work with will be kept general at first, and a specific such basis will be chosen later on.

The *covariant* coefficients of a point $x \in \mathbb{T}$ for a given isotropic basis $\mathfrak{J}$ are defined as

$$x_\mu := \tilde{Q}_{\mu\nu}x^\nu, \tag{8.5a}$$

i.e., $\quad (x_\odot, x_\times, x_1, \ldots, x_n) := (x^\times, x^\odot, x^1, \ldots, x^n). \tag{8.5b}$

In other words: Lowering (and thus also raising) an index exchanges the $\odot$- and the $\times$-coordinate and leaves the remaining coordinates unchanged.

With this notation, we see from (8.4) that the Minkowski inner product can be written as

$$\begin{aligned}
\langle x, y \rangle_\mathbb{T} = \langle x^\mu e_\mu, y^\nu e_\nu \rangle_\mathbb{T} &= x^\mu y^\nu \langle e_\mu, e_\nu \rangle_\mathbb{T} \\
&= \tilde{Q}_{\mu\nu} x^\mu y^\nu \\
&= x^\mu y_\mu = x_\nu y^\nu \tag{8.6} \\
&= x^\odot y_\odot + x^\times y_\times + \langle \vec{x}, \vec{y} \rangle_{\mathbb{R}^n} \\
&= x^\odot y^\times + x^\times y^\odot + \langle \vec{x}, \vec{y} \rangle_{\mathbb{R}^n} \tag{8.7}
\end{aligned}$$

for $\forall x, y \in \mathbb{T}$. In particular, we have

$$\langle x, x \rangle_\mathbb{T} = 2x^\odot x^\times + \|\vec{x}\|^2 \tag{8.8}$$

and therefore

$$\forall x \in \mathbb{T}_0: \quad 2x^\odot x^\times + \|\vec{x}\|^2 = 0. \tag{8.9}$$

Lemma 8.2. *(i) For any isotropic basis $\mathfrak{I}$, the forward and the backward light cone are given by*

$$\mathbb{T}_0^\uparrow = \{x \in \mathbb{T}_0 \mid x^\times > x^\odot\} \tag{8.10a}$$

$$\mathbb{T}_0^\downarrow = \{x \in \mathbb{T}_0 \mid x^\times < x^\odot\}. \tag{8.10b}$$

(ii) For $\forall x \in \mathbb{T}_0$ we have

$$\begin{aligned}
x^\times > 0 &\quad \Leftrightarrow \quad x \in \mathbb{T}_0^\uparrow \setminus \pi(e_\odot), \\
x^\times = 0 &\quad \Leftrightarrow \quad x \in \pi(e_\odot), \tag{8.11} \\
x^\times < 0 &\quad \Leftrightarrow \quad x \in \mathbb{T}_0^\downarrow \setminus \pi(e_\odot),
\end{aligned}$$

which in particular implies that $\forall x \in \mathbb{T}_0 \setminus \pi(e_\odot): \ x^\times \neq 0$.

Proof. See Appendix G.1. $\qquad\square$

8.3 Mapping the State Space Manifold to $\mathbb{R}^n$

We are now ready to define the function $h: \mathbb{P}_0 \to \bar{\mathbb{R}}^n$. We will denote the images of our states $\mathbf{p} \in \mathbb{P}_0$ by $\mathbf{x} := h(\mathbf{p}) \in \bar{\mathbb{R}}^n$.

Our map h will work by intersecting a given line $\mathbf{p} \in \mathbb{P}_0$ with the $(x^\times = 1)$-hyperplane and then returning the $(x^1, \ldots, x^n)$-components of the intersection point. Naturally, this construction will not work for any line $\mathbf{p}$ that is parallel to the $(x^\times = 1)$-hyperplane, i.e., only for $\mathbf{p} \in \mathbb{P}_0 \setminus \{\pi(e_\odot)\}$,[3] and so we will set $h(\pi(e_\odot)) := \infty$.

More explicitly, since for $\mathbf{p} \neq \pi(e_\odot)$ the desired intersection can be found by taking any point $x \in \mathbf{p} \setminus \{0\} \subset \mathbb{T}_0 \setminus \pi(e_\odot)$ and scaling it by $\frac{1}{x^\times}$, we define our map h as

$$h(\mathbf{p}) := \mathbf{x} := \begin{cases} \dfrac{\vec{x}}{x^\times} & \text{for any } x \in \mathbf{p} \setminus \{0\} & \text{if } \mathbf{p} \neq \pi(e_\odot), \\ \infty & & \text{if } \mathbf{p} = \pi(e_\odot). \end{cases} \qquad (8.12)$$

(Note that the first ratio is well defined by Lemma 8.2 (ii).) We denote the components of this vector by $\mathbf{x} =: (\mathbf{x}^1, \ldots, \mathbf{x}^n) = \left(\frac{x^1}{x^\times}, \ldots, \frac{x^n}{x^\times}\right)$, and as already for the vector $\vec{x}$ (see (8.5b)) we set $\mathbf{x}_i := \mathbf{x}^i$ for $i = 1, \ldots, n$.

The inverse mapping h^{-1} takes a point $\mathbf{x} \in \mathbb{R}^n$ and returns the unique ray $\mathbf{p} \in \mathbb{P}_0$ whose intersection point with the $(x^\times = 1)$-hyperplane is of the form $x = (x^\odot, 1, \mathbf{x})$. Since $\mathbf{p} \in \mathbb{P}_0$, the value $x^\odot$ must be chosen so that x is isotropic, which by (8.9) means that $x^\odot = -\frac{1}{2}\|\mathbf{x}\|^2$. This shows that

$$h^{-1}(\mathbf{x}) = \begin{cases} \pi\left((-\frac{1}{2}\|\mathbf{x}\|^2, 1, \mathbf{x})\right) & \text{if } \mathbf{x} \in \mathbb{R}^n, \\ \pi(e_\odot) & \text{if } \mathbf{x} = \infty. \end{cases} \qquad (8.13)$$

8.4 The Conformal Structure Induced on $\mathbb{R}^n$

Let us quickly confirm that the conformal structure $h(\mathfrak{c})$ that our mapping h induces on $\mathbb{R}^n$ is indeed the Euclidean one, as we had intended.

Lemma 8.3. *For* $\forall \mathfrak{m} \in \mathfrak{c}$ $\exists f_{\mathfrak{m}} \in \mathcal{F}_+(\mathbb{R}^n)$ *such that* $h(\mathfrak{m}) = f_{\mathfrak{m}} \cdot \langle \cdot, \cdot \rangle_{\mathbb{R}^n}$, *i.e.,*

$$\forall \mathbf{x}, d\mathbf{x}_1, d\mathbf{x}_2 \in \mathbb{R}^n: \quad \langle d\mathbf{x}_1, d\mathbf{x}_2 \rangle_{\mathbf{x}}^{h(\mathfrak{m})} = f_{\mathfrak{m}}(\mathbf{x}) \cdot \langle d\mathbf{x}_1, d\mathbf{x}_2 \rangle_{\mathbb{R}^n}.$$

In other words, $c := h(\mathfrak{c})$ *is the Euclidean conformal structure on* $\mathbb{R}^n$.

Proof. Let $\mathfrak{m} \in \mathfrak{c}$, defined as in (7.20) via some section $\psi_{\mathfrak{m}} \in \mathcal{S}_+(\mathscr{P}^{\otimes 2})$, which according to the remarks below (7.20) can be written as $\psi_{\mathfrak{m}} = \mathfrak{x} \otimes \mathfrak{x}$ for some section $\mathfrak{x} \in \mathcal{S}_\varnothing(\mathscr{P})$. Let $\mathbf{x}, d\mathbf{x}_1, d\mathbf{x}_2 \in \mathbb{R}^n$ be given, and let $\hat{\gamma}_i : [0, 1] \to \mathbb{R}^n$, $i = 1, 2$, be two paths such that $\hat{\gamma}_i(0) = \mathbf{x}$ and $\hat{\gamma}_i'(0)\,ds = d\mathbf{x}_i$.

Let us define the paths $\tilde{\gamma}_i : [0, 1] \to \mathbb{T}_0$ and $\gamma_i : [0, 1] \to \mathbb{P}_0$, $i = 1, 2$, as

$$\tilde{\gamma}_i(s) := \left(-\tfrac{1}{2}\|\hat{\gamma}_i(s)\|^2, 1, \hat{\gamma}_i(s)\right) \qquad \text{and} \qquad \gamma_i(s) := \pi(\tilde{\gamma}_i(s)).$$

Then we have $\tilde{\gamma}_i(0) = (-\frac{1}{2}\|\mathbf{x}\|^2, 1, \mathbf{x}) =: x$ and therefore

$$\begin{aligned} d\mathbf{p}_i := \gamma_i'(0)\,ds &\overset{(7.16)}{=} \tfrac{1}{\tilde{\gamma}_i(0)} \otimes [\tilde{\gamma}_i'(0)]\,ds \\ &= \tfrac{1}{x} \otimes \left[\left(-\langle \hat{\gamma}_i(0), \hat{\gamma}_i'(0)\rangle_{\mathbb{R}^n}, 0, \hat{\gamma}_i'(0)\right)\right]\,ds \\ &= \tfrac{1}{x} \otimes \left[\left(-\langle \mathbf{x}, d\mathbf{x}_i\rangle_{\mathbb{R}^n}, 0, d\mathbf{x}_i\right)\right] \end{aligned}$$

---

[3]Indeed, any line $\mathbf{p} \in \mathbb{P}_0$ that is parallel to the $(x^\times = 1)$-hyperplane consists of points of the form $x = (x^\odot, 0, \vec{x})$, and since by (8.9) these points must fulfill $\|\vec{x}\|^2 = 0$, they are of the form $x = (x^\odot, 0, \vec{0}) \in \pi(e_\odot)$.

for $i = 1, 2$, and so abbreviating $\mathbf{p} := h^{-1}(\mathbf{x}) = \pi(x)$, we find that

$$\langle \mathrm{d}\mathbf{x}_1, \mathrm{d}\mathbf{x}_2 \rangle_{\mathbf{x}}^{h(\mathfrak{m})}$$

$$\overset{\text{Tab. 6.1}}{=} \langle \mathrm{d}\mathbf{p}_1, \mathrm{d}\mathbf{p}_2 \rangle_{\mathbf{p}}^{\mathfrak{m}}$$

$$\overset{(7.20)}{=} \langle \mathfrak{r}(\mathbf{p}) \otimes \mathrm{d}\mathbf{p}_1, \mathfrak{r}(\mathbf{p}) \otimes \mathrm{d}\mathbf{p}_2 \rangle_{\mathbf{p}^\perp/\mathbf{p}}$$

$$= \frac{\mathfrak{r}(\mathbf{p})}{x} \cdot \frac{\mathfrak{r}(\mathbf{p})}{x} \cdot \left\langle \left[\left(-\langle \mathbf{x}, \mathrm{d}\mathbf{x}_1 \rangle_{\mathbb{R}^n}, 0, \mathrm{d}\mathbf{x}_1 \right) \right], \left[\left(-\langle \mathbf{x}, \mathrm{d}\mathbf{x}_2 \rangle_{\mathbb{R}^n}, 0, \mathrm{d}\mathbf{x}_2 \right) \right] \right\rangle_{\mathbf{p}^\perp/\mathbf{p}}$$

$$\overset{(7.18)}{=} \frac{\psi_{\mathfrak{m}}(\mathbf{p})}{x \otimes x} \cdot \left\langle \left(-\langle \mathbf{x}, \mathrm{d}\mathbf{x}_1 \rangle_{\mathbb{R}^n}, 0, \mathrm{d}\mathbf{x}_1 \right), \left(-\langle \mathbf{x}, \mathrm{d}\mathbf{x}_2 \rangle_{\mathbb{R}^n}, 0, \mathrm{d}\mathbf{x}_2 \right) \right\rangle_{\mathbb{T}}$$

$$\overset{(8.7)}{=} f_{\mathfrak{m}}(\mathbf{x}) \cdot \langle \mathrm{d}\mathbf{x}_1, \mathrm{d}\mathbf{x}_2 \rangle_{\mathbb{R}^n} ,$$

where $f_{\mathfrak{m}}(\mathbf{x}) := \frac{\psi_{\mathfrak{m}}(\mathbf{p})}{x \otimes x} = \left(\frac{\mathfrak{r}(\mathbf{p})}{x} \right)^2 > 0$ for $\forall \mathbf{x} \in \mathbb{R}^n$. $\qquad\qquad \square$

8.5   The Choice of the Basis Vector $e_\odot$

Since the definition of $h(\mathbf{p})$ utilizes the isotropic coefficients of the points $x \in \mathbf{p}$, the map $h = h_{\mathfrak{I}}$ depends on our specific choice of the isotropic basis $\mathfrak{I}$, which we haven't decided upon yet. We will now take the first step of fixing this basis (and thus h), by choosing the first basis vector $e_\odot$; the remaining vectors will be chosen later.

Since $h(\pi(e_\odot)) = \infty$, in order to translate our model in the spot parametrization into $\mathbb{R}^n$, we need to choose $e_\odot$ such that $\pi(e_\odot) \notin \mathcal{M}_0$. A natural choice is

$$e_\odot := b_0 . \tag{8.14}$$

This choice is a valid one since $b_0 \in \mathbb{T}_0^\downarrow$ (as is required of $e_\odot$ in Definition 8.1), and since by Lemma 7.13 we have $\pi(b_0) \in \partial \mathcal{M}_0$. Furthermore, it has the following added benefits:

(i)   All bond-related calculations will simplify in the isotropic coordinates.

(ii) Our flat parametrization h will map the point $\mathbf{p} = \pi(b_0)$, on which $r^X(\mathbf{p})$ is undefined, to ∞, and as a result our short rate function $r^X(\mathbf{x})$ in flat coordinates will be defined on all of $\mathbb{R}^n$.

(iii) Some parametric redundancy in our model is removed, by essentially fixing the choice of b_0 when looking at the model in isotropic coordinates.

(iv) Our assumptions (7.103) and (7.122) will turn into simple constraints on the counterpart of our matrix $(\varepsilon^\mu_{\ \nu})$ in isotropic coordinates (see Section 8.11 for details).

8.6 The Zero-Noise Process, Part 1

To compute the action $\mathbf{x} \mapsto U_t(\mathbf{x}) = (h \circ \mathcal{U}_t \circ h^{-1})(\mathbf{x})$ of our evolution group on any given state $\mathbf{x} \in \bar{\mathbb{R}}^n$, we first write (8.13) as $h^{-1}(\mathbf{x}) = \pi(x)$ for

$$x := \begin{cases} (-\frac{1}{2}\|\mathbf{x}\|^2, 1, \mathbf{x}) & \text{if } \mathbf{x} \in \mathbb{R}^n, \\ e_\odot & \text{if } \mathbf{x} = \infty. \end{cases} \tag{8.15}$$

Now successively combining this with (7.42) and (8.12), and using that

$$\pi(\mathcal{U}_t x) = \pi(e_\odot) \quad \Leftrightarrow \quad \mathcal{U}_t x \in \pi(e_\odot) \quad \Leftrightarrow \quad (\mathcal{U}_t x)^\times = 0$$

by (8.11), we find that

$$U_t(\mathbf{x}) = (h \circ \mathcal{U}_t \circ h^{-1})(\mathbf{x}) = (h \circ \mathcal{U}_t)(\pi(x)) = h(\pi(\mathcal{U}_t x))$$

$$= \begin{cases} \dfrac{((\mathcal{U}_t x)^1, \dots, (\mathcal{U}_t x)^n)}{(\mathcal{U}_t x)^\times} & \text{if } (\mathcal{U}_t x)^\times \neq 0, \\ \infty & \text{if } (\mathcal{U}_t x)^\times = 0. \end{cases} \tag{8.16}$$

Finally, rewriting this expression explicitly in terms of the given vector $\mathbf{x}$, we thus obtain the formula

$$(U_t(\mathbf{x}))^i = \frac{(\mathcal{U}_t x)^i}{(\mathcal{U}_t x)^\times} = \frac{(\mathcal{U}_t)^i{}_\mu x^\mu}{(\mathcal{U}_t)^\times{}_\mu x^\mu}$$

$$= \begin{cases} \dfrac{-(\mathcal{U}_t)^i{}_\odot \frac{1}{2}\|\mathbf{x}\|^2 + (\mathcal{U}_t)^i{}_\times + (\mathcal{U}_t)^i{}_j \mathbf{x}^j}{-(\mathcal{U}_t)^\times{}_\odot \frac{1}{2}\|\mathbf{x}\|^2 + (\mathcal{U}_t)^\times{}_\times + (\mathcal{U}_t)^\times{}_j \mathbf{x}^j} & \text{if } \mathbf{x} \in \mathbb{R}^n, \\ \dfrac{(\mathcal{U}_t)^i{}_\odot}{(\mathcal{U}_t)^\times{}_\odot} & \text{if } \mathbf{x} = \infty, \end{cases} \tag{8.17}$$

for $\forall i = 1, \dots, n$ whenever the respective denominators are non-zero, and $U_t(\mathbf{x}) = \infty$ otherwise.

We will make this formula more explicit later in Section 8.14, once we have obtained explicit expressions for the matrix entries $(\mathcal{U}_t)^\mu{}_\nu$ in terms of the individual components of the matrix representation of $\bar{\varepsilon}$, as introduced in the following.

8.7 Generator Components

Let us now get a first idea of how the maps U_t operate, and see how the various components of the generator $\bar{\varepsilon} = \varepsilon + \rho$ of $\mathcal{U}_t$ affect the action of these maps $U_t(\mathbf{x})$ computed above, and thus ultimately the action of the drift vector field $\mu(\mathbf{x})$ derived later in Section 8.9.

First, the value of $\rho \in \mathbb{R}$ does not affect the action of the operators $\mathcal{U}_t = e^{\rho t} e^{t\varepsilon}$ on states (as defined via (7.42) and ultimately given by (8.17)), but only the one on sections. We may therefore set $\rho = 0$ for the sake of this analysis, i.e., we will consider only $\mathcal{U}_t = \mathrm{e}^{t\varepsilon} \in O^{\uparrow}(\mathbb{T})$.

As we had established in Lemma 7.7 (i), ε must be antisymmetric, (i.e., $\varepsilon^* = -\varepsilon$), which in Lemma 7.6 was found to be equivalent to its coefficients fulfilling $\varepsilon^{ij} = -\varepsilon^{ji}$, for $\forall i, j = 0, \ldots, m$. Now while at the time we were still working in the standard basis of $\mathbb{T}$, this equivalence actually carries over to isotropic bases, i.e., we have

$$\varepsilon^* = -\varepsilon \qquad \Leftrightarrow \qquad \varepsilon^{\mu\nu} = -\varepsilon^{\nu\mu} \quad \text{for} \quad \forall \mu, \nu = \odot, \times, 1, \ldots, n.$$

Indeed, since by definition (8.5a) of the lower-index coefficients x_μ the index representation (7.30) of the Minkowski inner product still holds in our new basis (see (8.6)), so does the index formula (7.32) for adjoint operators on which the proof of the equivalence (i) $\Leftrightarrow$ (ii) in Lemma 7.6 is based.

We can therefore write the matrix $(\varepsilon^{\mu\nu})$ in the form

$$(\varepsilon(\lambda, E, \mathbf{u}, \bar{\mathbf{v}})^{\mu\nu}) = \begin{pmatrix} 0 & -\lambda & - & -\mathbf{u} & - \\ \lambda & 0 & - & -\bar{\mathbf{v}} & - \\ | & | & & & \\ \mathbf{u} & \bar{\mathbf{v}} & & \Large{E} & \\ | & | & & & \end{pmatrix}, \tag{8.18}$$

depending on the four parameters $\lambda \in \mathbb{R}$, $E \in \mathbb{R}^{n \times n}$, and $\mathbf{u}, \bar{\mathbf{v}} \in \mathbb{R}^n$,[4] where the matrix E is assumed to be antisymmetric, i.e., it fulfills

$$E^T = -E. \tag{8.19}$$

Note, however, that this matrix cannot be used *as is* to compute the image of a point x (given in isotropic coordinates) under the operator ε; for this we need to use the matrix with coordinates $\varepsilon^\mu{}_\nu = \varepsilon^{\mu\alpha} \tilde{Q}_{\alpha\nu}$. To obtain it, we multiply the matrix in (8.18) from the right by the one in (8.2), resulting in

$$(\varepsilon(\lambda, E, \mathbf{u}, \bar{\mathbf{v}})^\mu{}_\nu) = \begin{pmatrix} -\lambda & 0 & - & -\mathbf{u} & - \\ 0 & \lambda & - & -\bar{\mathbf{v}} & - \\ | & | & & & \\ \bar{\mathbf{v}} & \mathbf{u} & & \Large{E} & \\ | & | & & & \end{pmatrix}. \tag{8.20}$$

The following lemma gives meaning to the four ingredients of ε.

Lemma 8.4. *The isolated effects of the four parameters* λ, E, $\mathbf{u}$, *and* $\bar{\mathbf{v}}$ *on the function* $U_t^{\lambda, E, \mathbf{u}, \bar{\mathbf{v}}}(\mathbf{x})$ *(i.e., on the action of the operators* $\mathcal{U}_t^{\lambda, E, \mathbf{u}, \bar{\mathbf{v}}} := \exp(t\,\varepsilon(\lambda, E, \mathbf{u}, \bar{\mathbf{v}}))$ *on states* $\mathbf{x}$*), are given by*

$$U_t^{\lambda, 0, \vec{0}, \vec{0}}(\mathbf{x}) = \mathrm{e}^{-\lambda t} \mathbf{x}, \tag{8.21a}$$

$$U_t^{0, E, \vec{0}, \vec{0}}(\mathbf{x}) = \mathrm{e}^{tE} \mathbf{x}, \tag{8.21b}$$

---

[4]The reason for denoting the second vector as $\bar{\mathbf{v}}$ is that in Section 8.8 we will start working with a rescaled version of this vector, which we will call $\mathbf{v}$.

$$U_t^{0,0,\mathbf{u},\vec{0}}(\mathbf{x}) = \mathbf{x} + t\mathbf{u}, \tag{8.21c}$$

$$U_t^{0,0,\vec{0},\bar{\mathbf{v}}}(\mathbf{x}) = \frac{\frac{\mathbf{x}}{\|\mathbf{x}\|^2} - \frac{1}{2}t\bar{\mathbf{v}}}{\left\|\frac{\mathbf{x}}{\|\mathbf{x}\|^2} - \frac{1}{2}t\bar{\mathbf{v}}\right\|^2}, \tag{8.21d}$$

where from here on forward it is understood that $\frac{\vec{0}}{\|\vec{0}\|^2} = \infty$ *and* $\frac{\infty}{\|\infty\|^2} = \vec{0}$.

Proof. The formulas are derived by explicitly evaluating the power series defining $\mathcal{U}_t^{\lambda,E,\mathbf{u},\bar{\mathbf{v}}}$ for these isolated cases and then plugging the result into (8.17). See Appendix G.2 for the calculations. $\qquad\square$

In other words: The parameters λ, E, and $\mathbf{u}$ lead to a scaling by $\mathrm{e}^{-\lambda t}$, a rotation by the orthogonal matrix e^{tE}, and a shift by $t\mathbf{u}$, respectively, while the parameter $\bar{\mathbf{v}}$ leads to the composition of (i) an inversion $\mathbf{x} \mapsto \frac{\mathbf{x}}{\|\mathbf{x}\|^2}$, (ii) a shift by $-\frac{1}{2}t\bar{\mathbf{v}}$, and (iii) another inversion.

8.8 Short Rate, State Space

We are now ready to derive formulas for the short rate function r^X and the state space M_0 in the spot parametrization, both written in terms of our state variable $\mathbf{x} \in \mathbb{R}^n$ and of course our model parameters, namely ρ and the ingredients of ε.

Lemma 8.5. *The short rate function in the spot parametrization is given by*

$$r^X(\mathbf{x}) = r_\infty + \langle \mathbf{v}, \mathbf{x} \rangle \tag{8.22}$$

for $\forall \mathbf{x} \in \mathbb{R}^n$, *where we abbreviate*

$$r_\infty := \left(\tfrac{n}{2} - 1\right)(\rho + \lambda) = \tfrac{n-2}{2}(\rho + \lambda), \tag{8.23}$$

$$\mathbf{v} := -\left(\tfrac{n}{2} - 1\right)\bar{\mathbf{v}} \quad = -\tfrac{n-2}{2}\bar{\mathbf{v}}. \tag{8.24}$$

The relative forward rates $F_{\mathrm{rel}}(\mathbf{x}; \tau)$ *can then be computed by combining this formula with* (5.14) *and* (8.17).

Proof. Temporarily denoting the short rate function on $\mathbb{R}^n$ by $\tilde{r}^X(\mathbf{x})$, this function is related to the function $r^X(\mathbf{p})$, which is defined on $\mathbb{P}_0 \setminus \{\pi(e_\odot)\}$ and given in (7.99), by

$$\tilde{r}^X(\mathbf{x}) = r^X(h^{-1}(\mathbf{x})) = r^X(\pi(x))$$

for $\forall \mathbf{x} \in \mathbb{R}^n$, where $x := (-\frac{1}{2}\|\mathbf{x}\|^2, 1, \mathbf{x})$ (recall (8.13)). Using (7.99), (8.14), (8.20), and finally (8.7), we therefore find that

$$\tilde{r}^X(\mathbf{x}) = \left(\tfrac{n}{2} - 1\right)\left(\rho - \frac{\langle \varepsilon e_\odot, x \rangle_{\mathbb{T}}}{\langle e_\odot, x \rangle_{\mathbb{T}}}\right) = \left(\tfrac{n}{2} - 1\right)\left(\rho - \frac{\langle (-\lambda, 0, \bar{\mathbf{v}}), (-\frac{1}{2}\|\mathbf{x}\|^2, 1, \mathbf{x}) \rangle_{\mathbb{T}}}{\langle (1, 0, \vec{0}), (-\frac{1}{2}\|\mathbf{x}\|^2, 1, \mathbf{x}) \rangle_{\mathbb{T}}}\right)$$

$$= \left(\tfrac{n}{2} - 1\right)\left(\rho - \frac{-\lambda + \langle \bar{\mathbf{v}}, \mathbf{x} \rangle}{1}\right) = \left(\tfrac{n}{2} - 1\right)\left(\rho + \lambda - \langle \bar{\mathbf{v}}, \mathbf{x} \rangle\right)$$

$$= r_\infty + \langle \mathbf{v}, \mathbf{x} \rangle.$$

Now dropping the tilde again for simplicity of notation, this is (8.22). $\qquad\square$

We have introduced r_∞ and $\mathbf{v}$ in order to simplify the formula for our short rate function r^X as much as possible. To keep the number of parameters in our model the same, we will continue to use r_∞ and $\mathbf{v}$ in all of our formulas derived in the following wherever possible, thus eliminating the need for the parameters ρ and $\bar{\mathbf{v}}$.

Lemma 8.6. *The state space in the spot parametrization is given by*

$$M_0 = \{\mathbf{x} \in \mathbb{R}^n \mid \langle \mathbf{v}, \mathbf{x} \rangle > -r_\infty\}. \tag{8.25}$$

Proof. This now follows by combining (8.22) with our definition (7.105): Again denoting the short rate function on $\mathbb{R}^n$ by $\tilde{r}^X(\mathbf{x})$ to distinguish it from its analogue $r^X(\mathbf{p})$ defined on $\mathbb{P}_0 \setminus \{\pi(b_0)\}$, we have

$$\begin{aligned}
M_0 &= h(\mathscr{M}_0) \\
&= h\big(\{\mathbf{p} \in \mathbb{P}_0 \setminus \{\pi(b_0)\} \mid r^X(\mathbf{p}) > 0\}\big) \\
&= \big\{\mathbf{x} \in \mathbb{R}^n \mid r^X\big(h^{-1}(\mathbf{x})\big) > 0\big\} \\
&= \big\{\mathbf{x} \in \mathbb{R}^n \mid \tilde{r}^X(\mathbf{x}) > 0\big\} \\
&= \big\{\mathbf{x} \in \mathbb{R}^n \mid r_\infty + \langle \mathbf{v}, \mathbf{x} \rangle > 0\big\} \\
&= \big\{\mathbf{x} \in \mathbb{R}^n \mid \langle \mathbf{v}, \mathbf{x} \rangle > -r_\infty\big\}. \qquad \square
\end{aligned}$$

8.9 The Drift Vector Field

We proceed by deriving also an explicit formula for the drift $\mu(\mathbf{x})$ of the deterministic system, as defined in (5.12) and used in our formulas for the deterministic parts of our generators (e.g., (5.99a) and (5.136b)). Observe how the terms in (8.26) can be seen to directly correspond to the four maps given in Lemma 8.4.

Lemma 8.7. *The drift* $\mu\colon \mathbb{R}^n \to \mathbb{R}^n$ *of the deterministic system is given by*

$$\mu(\mathbf{x}) = \mathbf{u} + (-\lambda I + E)\mathbf{x} + \tfrac{1}{n-2}\|\mathbf{x}\|^2 \mathbf{v} - \tfrac{2}{n-2}\langle \mathbf{v}, \mathbf{x} \rangle \mathbf{x}. \tag{8.26}$$

Proof. The formula is derived directly from (5.12) and (8.17). To differentiate (8.17), we apply the quotient rule, using that the operators $\mathcal{U}_t = e^{\varepsilon t}$ fulfill $\partial_t(\mathcal{U}_t)^\mu{}_\nu|_{t=0} = \varepsilon^\mu{}_\nu$ and $(\mathcal{U}_t)^\mu{}_\nu|_{t=0} = \delta^\mu_\nu$, and we then plug in the values of $\varepsilon^\mu{}_\nu$ found in (8.20) and make the substitution (8.24):

$$\begin{aligned}
\mu(\mathbf{x})^i = \partial_t(U_t(\mathbf{x}))^i\big|_{t=0} &= \partial_t\left(\frac{-(\mathcal{U}_t)^i{}_\odot \tfrac{1}{2}\|\mathbf{x}\|^2 + (\mathcal{U}_t)^i{}_\times + (\mathcal{U}_t)^i{}_j \mathbf{x}^j}{-(\mathcal{U}_t)^\times{}_\odot \tfrac{1}{2}\|\mathbf{x}\|^2 + (\mathcal{U}_t)^\times{}_\times + (\mathcal{U}_t)^\times{}_j \mathbf{x}^j}\right)\bigg|_{t=0} \\
&= \frac{\big(-\varepsilon^i{}_\odot \tfrac{1}{2}\|\mathbf{x}\|^2 + \varepsilon^i{}_\times + \varepsilon^i{}_j \mathbf{x}^j\big)\cdot 1 - \mathbf{x}^i \cdot \big(-\varepsilon^\times{}_\odot \tfrac{1}{2}\|\mathbf{x}\|^2 + \varepsilon^\times{}_\times + \varepsilon^\times{}_j \mathbf{x}^j\big)}{1^2} \\
&= \big(\bar{\mathbf{v}}^i(-\tfrac{1}{2}\|\mathbf{x}\|^2) + \mathbf{u}^i + E^i{}_j \mathbf{x}^j\big) - \mathbf{x}^i\big(0 + \lambda + (-\bar{\mathbf{v}}_j)\mathbf{x}^j\big) \\
&= \big(\mathbf{u} - \lambda\mathbf{x} + E\mathbf{x} - \tfrac{1}{2}\|\mathbf{x}\|^2\bar{\mathbf{v}} + \langle \bar{\mathbf{v}}, \mathbf{x}\rangle\mathbf{x}\big)^i \\
&= \big(\mathbf{u} - \lambda\mathbf{x} + E\mathbf{x} + \tfrac{1}{n-2}\|\mathbf{x}\|^2\mathbf{v} - \tfrac{2}{n-2}\langle \mathbf{v}, \mathbf{x}\rangle\mathbf{x}\big)^i \qquad \square
\end{aligned}$$

8.10 Bond Prices, Part 1

Next, we will take the first step towards translating also our bond price formula into $\mathbb{R}^n$. We will complete our derivation later in Section 8.13, by replacing also all occurrences of the bond vector b_τ in (8.28) with explicit expressions based on our model parameters.

To prepare for our calculations, we begin with a short technical lemma. Note how (8.27) is consistent with Lemma 7.1 (i).

Lemma 8.8. *For $\forall x, y \in \mathbb{T}_0 \setminus \pi(e_\odot)$ we have*

$$\langle x, y \rangle = -\tfrac{1}{2} x^\times y^\times \|\mathbf{x} - \mathbf{y}\|^2 . \tag{8.27}$$

Proof. See Appendix G.3. $\qquad\qquad\qquad\square$

Lemma 8.9. *The bond dollar prices for any tenor $\tau \geq 0$, any process time $t \in \mathbb{R}$, and any state $\mathbf{x} \in M_0$ in the spot parametrization are (independently of t) given by*

$$B^X_{\$,t+\tau}(\mathbf{x}, t) = \begin{cases} 1 & \text{for } \tau = 0, \\ \left(-\tfrac{1}{2} b^\times_\tau \|\mathbf{x} - \vec{b}_\tau / b^\times_\tau \|^2 \right)^{1-n/2} & \text{for } \tau > 0. \end{cases} \tag{8.28}$$

Proof. Temporarily denoting the bond price functions on $\mathbb{R}^n$ by $\tilde{B}^X_{\$,t+\tau}(\mathbf{x}, t)$, these functions are related to the functions $B^X_{\$,t+\tau}(\mathbf{p}, t)$ defined on $\mathscr{M}_0$ by

$$\tilde{B}^X_{\$,t+\tau}(\mathbf{x}, t) = B^X_{\$,t+\tau}(h^{-1}(\mathbf{x}), t) = B^X_{\$,t+\tau}(\pi(x), t);$$

where $x := (-\tfrac{1}{2} \|\mathbf{x}\|^2, 1, \mathbf{x})$ (again recall (8.13), and note that $\infty \notin M_0$). Now using (7.98) and that $\langle b_0, x \rangle = \langle e_\odot, x \rangle = x^\times = 1$ (which follows from (8.14) and (8.7)), we therefore find that

$$\tilde{B}^X_{\$,t+\tau}(\mathbf{x}, t) = \left(\frac{\langle b_\tau, x \rangle}{\langle b_0, x \rangle} \right)^{1-n/2} = \langle b_\tau, x \rangle^{1-n/2} .$$

For $\tau = 0$ the expressions on the right are 1, which proves the first case of (8.28). For $\tau > 0$ we have $\pi(b_\tau) \neq \pi(b_0)$ by Corollary 7.1, i.e., $b_\tau \notin \pi(b_0) = \pi(e_\odot)$; since we also have $x \notin \pi(e_\odot)$, we can therefore apply (8.27) to obtain the expression for the second case of (8.28). $\qquad\square$

The exchange rates between bonds of two different tenors can then be obtained from (8.28) via the formula

$$B^X_{t+\tau_1, t+\tau_2}(\mathbf{x}, t) = \frac{B^X_{\$,t+\tau_2}(\mathbf{x}, t)}{B^X_{\$,t+\tau_1}(\mathbf{x}, t)} . \tag{8.29}$$

8.11 Parameter Constraints, Choosing the Basis

Let us now translate our previously derived assumptions (7.103) and (7.113) into constraints on the ingredients of the matrix $(\bar{\varepsilon}^\mu{}_\nu)$, i.e., on our parameters ρ, λ, $\mathbf{u}$, $\bar{\mathbf{v}}$, and E (or equivalently, on r_∞, λ, $\mathbf{u}$, $\mathbf{v}$, an E), and let us remove all the remaining redundancy in our parametrization by imposing additional constraints on the choice of our isotropic basis $\mathfrak{I}$.

Nondegeneracy, symmetry under bond vector rescaling. First consider our assumption (7.102), which was introduced both to exclude the degenerate deterministic case and to break the symmetry of our model under rescaling of the bond vector. It translates into a simple constraint on the $\bar{\mathbf{v}}$-component of the generator ε: Indeed, by (8.14) and (8.20) we have

$$\varepsilon b_0 = \varepsilon e_\odot = (-\lambda, 0, \bar{\mathbf{v}}),$$

and so by (8.8) we have $\langle \varepsilon b_0, \varepsilon b_0 \rangle = 2 \cdot (-\lambda) \cdot 0 + \|\bar{\mathbf{v}}\|^2 = \|\bar{\mathbf{v}}\|^2$; our assumption (7.103) can therefore be written as

$$\|\bar{\mathbf{v}}\| = \sqrt{\zeta}. \tag{8.30}$$

Now choosing

$$\zeta := \frac{4}{(n-2)^2}, \tag{8.31}$$

this translates into the simple constraint

$$\|\mathbf{v}\| = 1 \tag{8.32}$$

for our rescaled parameter $\mathbf{v}$ defined in (8.24).

Choice of $e_\times$. Next, let us address some further redundancy in our parametrization by specifying our choice of the second isotropic basis vector, namely $e_\times$. Just as we did for the first basis vector, we will guide our choice by our desire to make our future calculations simpler. The following lemma gives us a key insight.

Lemma 8.10. *(i) In general, if an antisymmetric operator such as $\varepsilon \colon \mathbb{T} \to \mathbb{T}$ has two non-pseudo-orthogonal eigenvectors, then their two eigenvalues fulfill the relation $\lambda_1 = -\lambda_2$. In particular, this implies the following:*

(i.1)   Any eigenvector with non-zero eigenvalue must be isotropic.

(i.2)   The eigenvalues of any two non-collinear eigenvectors in $\mathbb{T}_0$ must fulfill the relation $\lambda_1 = -\lambda_2$.

(i.3)   There cannot be more than two distinct lines of eigenvectors in $\mathbb{T}_0$ with non-zero eigenvalues.

(ii) Under our specific parameter constraints, there exist two eigenvectors $x, x' \in \mathbb{T}_0^\uparrow$ of ε with $\pi(x) \in \overline{\mathscr{M}_0}$ and $\pi(x') \in \mathbb{P}_0 \setminus \mathscr{M}_0$ and with eigenvalues $\hat{\lambda}$ and $-\hat{\lambda}$, respectively, for some $\hat{\lambda} \geq |\rho|$. (We may have $x \parallel x'$ and $\hat{\lambda} = 0$.)

Proof. (i) Denoting these two eigenvectors by x_1 and x_2, we have

$$\lambda_2 \langle x_1, x_2 \rangle = \langle x_1, \varepsilon x_2 \rangle = \langle \varepsilon^* x_1, x_2 \rangle = -\langle \varepsilon x_1, x_2 \rangle = -\lambda_1 \langle x_1, x_2 \rangle,$$

and so $\langle x_1, x_2 \rangle \neq 0$ implies that $\lambda_1 = -\lambda_2$.

(i.1) Given any eigenvector $x \in \mathbb{T}$ with non-zero eigenvalue, we can now choose $x_1 := x_2 := x$, and since in this case we have $\lambda_1 = \lambda_2 \neq -\lambda_2$, our first statement implies that $0 = \langle x_1, x_2 \rangle = \langle x, x \rangle$, i.e., that x must be isotropic.

(i.2) This follows from our first statement since any two non-collinear vectors $x_1, x_2 \in \mathbb{T}_6$ satisfy $\langle x_1, x_2 \rangle \neq 0$ by Lemma 7.1 (i).

(i.3) By part (i.2) the eigenvalues of any two distinct lines of eigenvectors in $\mathbb{T}_0$ must fulfill $\lambda_1 = -\lambda_2$; however, in any set of more than two non-zero numbers there must be a pair with coinciding signs.

(ii) See Appendix G.4. □

Let us now fix any eigenvector x with the properties described in Lemma 8.10 (ii) (in fact, Corollary 8.2 below will show that this eigenvector is unique even if $\hat{\lambda} = 0$). Then since by our remark following (7.102) b_0 is *not* an eigenvector of ε and thus not parallel to x, we can apply Lemma 8.1 (ii) in order to build an isotropic basis with $e_\odot = b_0$ and $e_\times = c \cdot x$ for some $c > 0$. In other words, we can choose $e_\times$ to be an eigenvector of ε with eigenvalue $\hat{\lambda}$, i.e., we can have it fulfill $\varepsilon e_\times = \hat{\lambda} e_\times$, or written in the isotropic coordinates for such a basis, $(0, \lambda, \mathbf{u}) = (0, \hat{\lambda}, \vec{0})$ (recall (8.20)). This shows that

$$\mathbf{u} = \vec{0} \tag{8.33}$$

and that $\lambda = \hat{\lambda}$, and so the lemma's bound on the eigenvalue $\hat{\lambda}$ states that

$$\lambda \geq |\rho|. \tag{8.34}$$

Since our choice of basis does not affect the actual model behavior, but only how we express its equations in $\mathbb{R}^n$, we may therefore assume that the parameter constraints (8.33) and (8.34) are fulfilled, without sacrificing any generality of the model. We will translate (8.34) into a constraint involving λ and r_∞ instead later in this section.

Choice of $e_1, \dots, e_n$. Now let us complete our choice of the isotropic basis $\mathfrak{I}$, by choosing also our remaining basis vectors $e_1, \dots, e_n$. Our specific choice will be motivated by the goal of simplifying the shapes of both E and $\mathbf{v}$, by making use of our freedom described in Lemma 8.1 (ii).

Completing our isotropic basis by *any* set of vectors $e_1, \dots, e_n$ at first (as described by Lemma 8.1 (ii)), our previous choices of $e_\odot$ and $e_\times$ imply that the matrix $(\varepsilon^\mu{}_\nu)$ is of the form (8.20), with $\|\mathbf{v}\| = 1$ and $\mathbf{u} = \vec{0}$. According to the second part of Lemma 8.1 (ii), we can now modify our isotropic basis by choosing the vectors $e'_\odot, e'_\times, e'_1, \dots, e'_n$ given by the matrix equation

$$(e'_\odot, e'_\times, e'_1, \dots, e'_n) := (e_\odot, e_\times, e_1, \dots, e_n) P_+$$

instead, where $P_+ \in \mathbb{R}^{(n+2)\times(n+2)}$ is of the form

$$P_+ = \begin{pmatrix} 1 & 0 & -\ \vec{0}\ - \\ 0 & 1 & -\ \vec{0}\ - \\ | & | & \\ \vec{0} & \vec{0} & P \\ | & | & \end{pmatrix},$$

for any orthogonal matrix $P \in \mathbb{R}^{n \times n}$. Under this change of basis, the isotropic coordinates of a given vector $x = (x^\odot, x^\times, \vec{x}) \in \mathbb{T}$ transform as[5]

$$(x^\odot, x^\times, \vec{x})' = P_+^T (x^\odot, x^\times, \vec{x}),$$

and so simply writing ε instead of $(\varepsilon^\mu{}_\nu)$ from now on to ease our notation,[6] the coefficients of ε transform to

$$\varepsilon' = P_+^T \, \varepsilon \, P_+ = \begin{pmatrix} -\lambda & 0 & -\ -P^T\mathbf{u}\ - \\ 0 & \lambda & -\ -P^T\bar{\mathbf{v}}\ - \\ | & | & \\ P^T\bar{\mathbf{v}} & P^T\mathbf{u} & P^T E P \\ | & | & \end{pmatrix},$$

i.e., the model parameters that describe ε in the chosen basis transform as

$$\lambda' = \lambda, \qquad \mathbf{u}' = P^T\mathbf{u}, \qquad \bar{\mathbf{v}}' = P^T\bar{\mathbf{v}}, \qquad E' = P^T E P. \qquad (8.35)$$

In particular, our parameter constraints (8.30) and (8.33)–(8.34) remain the same for our modified basis (as expected, since we did not modify $e_\odot$ and $e_\times$).

Choosing $P := -A^T$, where A is the matrix given by Lemma 2.12, we have thus succeeded in completing our isotropic basis in such a way that E and $\bar{\mathbf{v}}$ have the specific forms of E^A and $\mathbf{v}^A$ listed in that lemma, except that the signs of the components of $\mathbf{v}^A$ are flipped. In other words, the matrix E and the vector $\mathbf{v}$ defined in (8.24) (which points into the direction of $-\bar{\mathbf{v}}$) are of the form

$$E = \begin{pmatrix} 0 & \eta_1 & & & \\ -\eta_1 & 0 & & \text{\Large 0} & \\ & & \ddots & & \\ & & & 0 & \eta_{\lfloor n/2 \rfloor} \\ \text{\Large 0} & & & -\eta_{\lfloor n/2 \rfloor} & 0 \end{pmatrix} \quad \text{or} \quad \begin{pmatrix} 0 & \eta_1 & & & & \\ -\eta_1 & 0 & & & \text{\Large 0} & \\ & & \ddots & & & \\ & & & 0 & \eta_{\lfloor n/2 \rfloor} & \\ & \text{\Large 0} & & -\eta_{\lfloor n/2 \rfloor} & 0 & \\ & & & & & 0 \end{pmatrix} \qquad (8.36)$$

---

[5]Despite our notation, the reader should interpret $(x^\odot, x^\times, \vec{x})$ as a column vector.

[6]We can afford to do so since we will no longer need to work with matrices like $(\varepsilon^{\mu\nu})$ from this point forward.

for even or odd values of n, respectively,[7] and

$$\mathbf{v} = \begin{cases} (v_1, 0, v_2, 0, \ldots, v_{\lceil n/2 \rceil}, 0) & \text{for even } n, \\ (v_1, 0, v_2, 0, \ldots, v_{\lceil n/2 \rceil}) & \text{for odd } n, \end{cases} \tag{8.37}$$

for some scalars

$$\eta_1 \geq \cdots \geq \eta_{\lfloor n/2 \rfloor} \geq 0 \qquad \text{and} \qquad v_1, \ldots, v_{\lceil n/2 \rceil} \geq 0. \tag{8.38}$$

We may therefore assume that E and $\mathbf{v}$ have these forms, without restricting the actual behavior of the model.

Closure of the state space under the zero-noise process. Finally, let us express our model assumption (7.113), which was introduced to guarantee that we have $\mathcal{U}_t(\mathcal{M}_0) \subseteq \mathcal{M}_0$ for $\forall t \geq 0$, in terms of the components of the matrix $\bar{\varepsilon}$ as well. By (8.14), (8.20), and (8.33) we have

$$\begin{aligned} \varepsilon^2 b_0 = \varepsilon^2 e_\odot = \varepsilon(-\lambda, 0, \bar{\mathbf{v}}) \\ &= \left(\lambda^2 - \langle \mathbf{u}, \bar{\mathbf{v}} \rangle, -\|\bar{\mathbf{v}}\|^2, -\lambda\bar{\mathbf{v}} + E\bar{\mathbf{v}} \right) \\ &= \left(\lambda^2, -\|\bar{\mathbf{v}}\|^2, -\lambda\bar{\mathbf{v}} + E\bar{\mathbf{v}} \right), \end{aligned}$$

and so by (8.8) and the antisymmetry of E thus

$$\begin{aligned} \langle \varepsilon^2 b_0, \varepsilon^2 b_0 \rangle &= -2\lambda^2\|\bar{\mathbf{v}}\|^2 + \|{-\lambda\bar{\mathbf{v}} + E\bar{\mathbf{v}}}\|^2 \\ &= -2\lambda^2\|\bar{\mathbf{v}}\|^2 + \lambda^2\|\bar{\mathbf{v}}\|^2 - 2\lambda\langle \bar{\mathbf{v}}, E\bar{\mathbf{v}} \rangle + \|E\bar{\mathbf{v}}\|^2 \\ &= -\lambda^2\|\bar{\mathbf{v}}\|^2 + \|E\bar{\mathbf{v}}\|^2. \end{aligned}$$

Using (8.31), our condition (7.122) can therefore be written as

$$0 \leq -\left(\langle \varepsilon^2 b_0, \varepsilon^2 b_0 \rangle + \zeta\rho^2 \right) = \lambda^2\|\bar{\mathbf{v}}\|^2 - \|E\bar{\mathbf{v}}\|^2 - \tfrac{4}{(n-2)^2}\rho^2.$$

To write this constraint in terms of r_∞ and $\mathbf{v}$ instead of ρ and $\bar{\mathbf{v}}$, we multiply by $\frac{(n-2)^3}{16}$ and use (8.23)–(8.24) and (8.32), which leads us to

$$\begin{aligned} 0 &\leq \tfrac{(n-2)^3}{16}\left(\lambda^2\|\bar{\mathbf{v}}\|^2 - \tfrac{4}{(n-2)^2}\rho^2 - \|E\bar{\mathbf{v}}\|^2 \right) \\ &= \tfrac{n-2}{4}\left(\lambda^2 - \rho^2 - \|E\mathbf{v}\|^2 \right) \\ &= \tfrac{n-2}{4}(\lambda + \rho)(\lambda - \rho) - \tfrac{n-2}{4}\|E\mathbf{v}\|^2 \\ &= \tfrac{n-2}{2}(\lambda + \rho) \cdot (\lambda - \tfrac{1}{2}(\lambda + \rho)) - \tfrac{n-2}{4}\|E\mathbf{v}\|^2 \\ &= r_\infty \cdot (\lambda - \tfrac{1}{n-2}r_\infty) - \tfrac{n-2}{4}\|E\mathbf{v}\|^2 \\ &= r_\infty\lambda - \tfrac{1}{n-2}r_\infty^2 - \tfrac{n-2}{4}\|E\mathbf{v}\|^2. \end{aligned} \tag{8.39}$$

---

[7]The matrices in (8.36) have (2×2)-matrices on the diagonal, with all other elements being zero; in other words, every other entry on the two off-diagonals vanishes. In particular, this means that the action on $\mathbb{R}^n$ of any matrix E of this form decouples into its actions on certain two-dimensional subspaces.

Note that in its version in the second line, this constraint in particular implies that $|\lambda| \geq |\rho|$, so that the only additional constraint in (8.34) is that

$$\lambda \geq 0, \tag{8.40}$$

and so we may replace (8.34) by the ρ-independent inequality (8.40) without weakening our collection of constraints.[8] Furthermore, observe that (8.39) and (8.40) together imply that

$$r_\infty \geq 0. \tag{8.41}$$

Summary.   There are two types of constraints on our model parameters: those that ensure that the model has all the desired properties, and those that are introduced to break the symmetries in the parametrization space. While there are many ways of achieving the latter, our specific choices have led us to the following (for now[9]) complete set of constraints:

> (i) $\mathbf{u} = \vec{0}$,
>
> (ii) $\mathbf{v}$ fulfills $\|\mathbf{v}\| = 1$ and is of the form (8.37), with only non-negative entries,
>
> (iii) E is of the form (8.36), with the values η_i satisfying the constraint in (8.38),
>
> (iv) $r_\infty \geq 0$,
>
> (v) $\lambda \geq 0$,
>
> (vi) $r_\infty \lambda - \frac{1}{n-2} r_\infty^2 - \frac{n-2}{4} \|E\mathbf{v}\|^2 \geq 0$.

The fact that $r_\infty \geq 0$ in fact follows from the constraints (v)–(vi) and is only listed for clarity. Note that given (iv), the constraints (v)–(vi) can equivalently be expressed as a stronger lower bound for λ (with an extra constraint if $r_\infty = 0$):

> (v)–(vi)'   • If $r_\infty = 0$ then $E\mathbf{v} = \vec{0}$.
>
> • $\lambda \geq \lambda_{\min} := \begin{cases} \dfrac{r_\infty}{n-2} + \dfrac{n-2}{4r_\infty} \|E\mathbf{v}\|^2 & \text{if } r_\infty > 0, \\ 0 & \text{if } r_\infty = 0. \end{cases}$

From this point forward we will assume that all these parameter constraints are fulfilled.

8.12   Evolution Operators

Our constraint $\mathbf{u} = \vec{0}$ (which was a result of our choice of the basis vector $e_\times$) and our constraint on the entries of E (which was a result of our choice of the

---

[8]Indeed, clearly $\lambda \geq |\rho|$ implies that $\lambda \geq 0$; conversely, if $|\lambda| \geq |\rho|$ and $\lambda \geq 0$ then $\lambda \geq |\rho|$.

[9]Additional constraints will be necessary to ensure the closure of the state space also under the full process including the noise. Those, however, are better derived in the $\mathbb{R}^n$-based setting, and so we introduced them only in Chapter 2, namely in (2.2), (2.5), (2.9b) and Lemma 2.4.

basis vectors $e_1, \ldots, e_n$) have an important advantage: They simplify the matrix in (8.20) associated to the operator ε to the form

$$
\varepsilon = \begin{pmatrix} -\lambda & 0 & - & \vec{0} & - \\ 0 & \lambda & - & -\bar{\mathbf{v}} & - \\ | & | & & & \\ \bar{\mathbf{v}} & \vec{0} & & E & \\ | & | & & & \end{pmatrix} \tag{8.42}
$$

(with E of the form given in (8.36)), which—as we will see in Lemma 8.11—will just be enough for us to compute the action of the operators $e^{t\varepsilon} : \mathbb{T} \to \mathbb{T}$ and thus of the operators $\mathcal{U}_t = e^{t(\varepsilon+\rho)}$ and $\mathcal{U}_t^* = e^{t(-\varepsilon+\rho)}$ explicitly, by computing the infinite sum of their defining exponential power series analytically. This in turn will make our formulas for the bond prices and forward rates more explicit, since the bond price formula (8.28) depends on the bond vectors $b_\tau = \mathcal{U}_{-\tau}^* e_\odot$, and since the forward rate formula described in Lemma 8.5 utilizes the operators $\mathcal{U}_t$ via (5.14) and (8.17).

Lemma 8.11. *For $\forall t \in \mathbb{R}$ we have*

$$
e^{t\varepsilon} = \begin{pmatrix} e^{-\lambda t} & 0 & - & \vec{0} & - \\ -\frac{1}{2} e^{-\lambda t} t^2 \| R_t \bar{\mathbf{v}} \|^2 & e^{\lambda t} & - & \left(-t e^{-tE} R_t \bar{\mathbf{v}} \right) & - \\ | & | & & & \\ e^{-\lambda t} t R_t \bar{\mathbf{v}} & \vec{0} & & e^{tE} & \\ | & | & & & \end{pmatrix}, \tag{8.43}
$$

where we defined the matrices

$$
R_t = \sum_{k=0}^{\infty} \frac{t^k}{(k+1)!} (E + \lambda I)^k. \tag{8.44}
$$

This allows us to compute the action of the evolution group operators $\mathcal{U}_t$ and $\mathcal{U}_t^*$ explicitly via the formulas

$$
\mathcal{U}_t = e^{\rho t} \cdot e^{t\varepsilon} \qquad and \qquad \mathcal{U}_t^* = e^{\rho t} \cdot e^{-t\varepsilon}. \tag{8.45}
$$

Proof. The formula (8.43) was originally derived by computing the first couple of powers of the matrix in (8.42), observing a general pattern, and then splitting the obtained exponential power series into solvable parts, eventually arriving at (8.43). However, once the formula has been found in this way, we can prove its validity more easily, by showing that the matrices on the right of (8.43)—let us call them G_t—satisfy the ODE $\partial_t G_t = G_t \varepsilon$ (where ε is given by (8.42)) and the initial condition $G_{t=0} = I$, just like $e^{t\varepsilon}$ does. The necessary calculations for this can be found in Appendix G.5.

The formulas in (8.45) follow directly from our definition (7.59). $\square$

8.13   Bond Prices, Part 2

As a first direct consequence of Lemma 8.11 we can make our bond price formula in Lemma 8.9 more explicit, by using (8.43) to derive explicit expressions for our bond vectors b_τ.

Lemma 8.12. *The bond dollar prices for any tenor $\tau \geq 0$, any process time $t \in \mathbb{R}$, and any state $\mathbf{x} \in M_0$ in the spot parametrization are (independently of t) given by*

$$B_{\$,t+\tau}(\mathbf{x},t) = \mathrm{e}^{-r_\infty \tau} \left\| \frac{1}{n-2}\tau \left\| R_{-\tau}\mathbf{v} \right\| \mathbf{x} + \frac{R_{-\tau}\mathbf{v}}{\left\| R_{-\tau}\mathbf{v} \right\|} \right\|^{2-n}. \qquad (8.46)$$

If $E + \lambda I$ is invertible (e.g., if $\lambda > 0$), then we have the representation

$$B_{\$,t+\tau}(\mathbf{x},t) = \begin{cases} \mathrm{e}^{-r_\infty\tau} \left\| \left\| \mathbf{c}_\tau \right\| \mathbf{x} - \frac{\mathbf{c}_\tau}{\left\| \mathbf{c}_\tau \right\|} \right\|^{2-n} & \text{for } \tau \in (0,T], \\ 1 & \text{for } \tau = 0, \end{cases} \qquad (8.47)$$

where the currency vectors $\mathbf{c}_\tau$, $\tau \geq 0$, are defined as

$$\mathbf{c}_\tau := \frac{1}{n-2}(E + \lambda I)^{-1}\big(\mathrm{e}^{-\tau(E+\lambda I)} - I\big)\mathbf{v}. \qquad (8.48)$$

Proof. By (7.91), (8.45), (8.14), and (8.43) the bond vectors b_τ are given by

$$b_\tau = \mathcal{U}_\tau^* b_0 = \mathrm{e}^{\rho\tau} \cdot \mathrm{e}^{-\tau\varepsilon} e_\odot = \mathrm{e}^{\rho\tau} \cdot \mathrm{e}^{-\tau\varepsilon}(1,0,\vec{0})$$

$$= \mathrm{e}^{\rho\tau} \cdot \big(\mathrm{e}^{\lambda\tau}, -\tfrac{1}{2}\mathrm{e}^{\lambda\tau}\tau^2\big\| R_{-\tau}\bar{\mathbf{v}} \big\|^2, \mathrm{e}^{\lambda\tau}(-\tau)R_{-\tau}\bar{\mathbf{v}}\big)$$

$$= \mathrm{e}^{(\rho+\lambda)\tau} \cdot \big(1, -\tfrac{1}{2}\tau^2\big\| R_{-\tau}\bar{\mathbf{v}} \big\|^2, -\tau R_{-\tau}\bar{\mathbf{v}}\big) \qquad (8.49)$$

for $\forall \tau \in \mathbb{R}$. Plugging this into our bond price formula in Lemma 8.9 and using our definitions (8.23)–(8.24), we therefore find that for $\forall \tau > 0$ we have

$$B_{\$,t+\tau}(\mathbf{x},t) = \big(-\tfrac{1}{2}b_\tau^\times \big\| \mathbf{x} - \vec{b}_\tau/b_\tau^\times \big\|^2\big)^{1-n/2}$$

$$= \left(\tfrac{1}{4}\mathrm{e}^{(\rho+\lambda)\tau}\tau^2 \big\| R_{-\tau}\bar{\mathbf{v}} \big\|^2 \left\| \mathbf{x} - \frac{2R_{-\tau}\bar{\mathbf{v}}}{\tau\big\| R_{-\tau}\bar{\mathbf{v}} \big\|^2} \right\|^2 \right)^{1-n/2}$$

$$= \mathrm{e}^{-(n/2-1)(\rho+\lambda)\tau} \left\| \tfrac{1}{2}\tau \big\| R_{-\tau}\bar{\mathbf{v}} \big\| \mathbf{x} - \frac{R_{-\tau}\bar{\mathbf{v}}}{\big\| R_{-\tau}\bar{\mathbf{v}} \big\|} \right\|^{2-n}$$

$$= \mathrm{e}^{-r_\infty\tau} \left\| \frac{1}{n-2}\tau \big\| R_{-\tau}\mathbf{v} \big\| \mathbf{x} + \frac{R_{-\tau}\mathbf{v}}{\big\| R_{-\tau}\mathbf{v} \big\|} \right\|^{2-n}.$$

Since for $\tau = 0$ this expression evaluates to 1 without any problems, just as the bond price formula in Lemma 8.9 does, we now no longer need to treat that case separately.

Finally, if $E + \lambda I$ is invertible (which Lemma 2.5 (i) showed to be the case if $\lambda > 0$) then the vectors $\mathbf{c}_\tau$ are well-defined and can be written as $\mathbf{c}_\tau = \frac{-\tau}{n-2}R_{-\tau}\mathbf{v}$, and plugging $R_{-\tau}\mathbf{v} = -\frac{n-2}{\tau}\mathbf{c}_\tau$ into (8.46) yields (8.47). $\square$

Note that although the exponent $2 - n$ in (8.46) is negative by (7.104), we do not have to worry about the norm potentially vanishing, since this formula was derived directly from our original bond price formula in Lemma 8.9, whose general validity we have already established. We can therefore *conclude* that the vector inside the norm does not vanish for any $\mathbf{x} \in M_0$.[10]

Also observe that (as a result of what we had said in Section 6.4.4) this formula indeed does not depend on the functional model parameter α that will control the size of the noise of our state process (X_t) (see (8.61) and (8.65)), i.e., the model is unspanned.

Corollary 8.1. *The functions* $B_{\$,T}(\mathbf{x}, t)$ *given by* (8.46) *fulfill*

$$B_{\$,T}(\mathbf{x}, t) = \mathbb{E}\Big[\exp\Big(-\int_t^T r(X_s)\,\mathrm{d}s\Big)\,\Big|\,X_t = \mathbf{x}\Big] \tag{8.50}$$

for $\forall (\mathbf{x}, t) \in M_0 \times [0, T]$.

Proof. As we had stated in Lemma 5.9 (i), it suffices to show that the functions $B_{\$,T}(\mathbf{x}, t)$ are continuous and bounded on $M_0 \times [0, T]$. This has been done in Lemma 2.9 (ii). $\qquad\square$

8.14 The Zero-Noise Process, Part 2

Next, we can combine our formula (8.43) with (8.16) to derive an explicit expression also for our evolution operators U_t in the flat parametrization. This in turn will then allow us to compute also the forward rate curve explicitly, as described in Lemma 8.5.

Lemma 8.13. *The evolution operators* $U_t \colon \bar{\mathbb{R}}^n \to \bar{\mathbb{R}}^n$, $t \in \mathbb{R}$, *are given by*

$$U_t(\mathbf{x}) = \frac{\mathrm{e}^{t(E+\lambda I)}\frac{\mathbf{x}}{\|\mathbf{x}\|^2} + \frac{1}{n-2}tR_t\mathbf{v}}{\left\|\mathrm{e}^{t(E+\lambda I)}\frac{\mathbf{x}}{\|\mathbf{x}\|^2} + \frac{1}{n-2}tR_t\mathbf{v}\right\|^2}\,, \tag{8.51}$$

where again we use the convention that $\frac{\vec{0}}{\|\vec{0}\|^2} = \infty$ *and* $\frac{\infty}{\|\infty\|^2} = \vec{0}$.

If $E + \lambda I$ *is invertible (e.g., if* $\lambda > 0$*) then this formula can be written as*

$$U_t(\mathbf{x}) = \frac{\mathrm{e}^{t(E+\lambda I)}\big(\frac{\mathbf{x}}{\|\mathbf{x}\|^2} - \mathbf{c}_t\big)}{\left\|\mathrm{e}^{t(E+\lambda I)}\big(\frac{\mathbf{x}}{\|\mathbf{x}\|^2} - \mathbf{c}_t\big)\right\|^2}\,. \tag{8.52}$$

Proof. Given any $\mathbf{x} \in \mathbb{R}^n$, let us abbreviate $x := (-\frac{1}{2}\|\mathbf{x}\|^2, 1, \mathbf{x})$ as usual. If $\mathbf{x} = \vec{0}$ then we have $x = (0, 1, \vec{0})$ and thus $\mathcal{U}_t x = \mathrm{e}^{\rho t}\mathrm{e}^{t\varepsilon}(0, 1, \vec{0}) = (0, \mathrm{e}^{(\rho+\lambda)t}, \vec{0})$ by Lemma 8.11, and so according to (8.16) we have $U_t(\mathbf{x}) = \vec{0}$, which coincides with the value returned by (8.51).

---

[10]Nevertheless, the interested reader can find an independent proof of this fact in Lemma 2.9 (ii).

Now assuming that $\mathbf{x} \neq \vec{0}$, by Lemma 8.11 the vector $y = (y^\circ, y^\times, \vec{y}) := \mathcal{U}_t x$ fulfills

$$\vec{y} = \mathrm{e}^{\rho t} \cdot \left(-\tfrac{1}{2}\|\mathbf{x}\|^2 \cdot \mathrm{e}^{-\lambda t} t R_t \bar{\mathbf{v}} + \mathrm{e}^{tE} \mathbf{x} \right),$$

$$y^\times = \mathrm{e}^{\rho t} \cdot \left[\left(-\tfrac{1}{2}\|\mathbf{x}\|^2 \right) \cdot \left(-\tfrac{1}{2}\mathrm{e}^{-\lambda t} t^2 \|R_t \bar{\mathbf{v}}\|^2 \right) + 1 \cdot \mathrm{e}^{\lambda t} + \left\langle \mathbf{x}, -t \mathrm{e}^{-tE} R_t \bar{\mathbf{v}} \right\rangle \right]$$

$$= \frac{\|\vec{y}\|^2}{\mathrm{e}^{(\rho-\lambda)t}\|\mathbf{x}\|^2} \, .$$

If $\vec{y} \neq \vec{0}$ (and thus $y^\times \neq 0$) then our original formula (8.16) together with (8.24) yields

$$U_t(\mathbf{x}) = \frac{\vec{y}}{y^\times} = \mathrm{e}^{(\rho-\lambda)t}\|\mathbf{x}\|^2 \cdot \frac{\mathrm{e}^{\rho t}\left(\mathrm{e}^{tE}\mathbf{x} - \tfrac{1}{2}\|\mathbf{x}\|^2 \mathrm{e}^{-\lambda t} t R_t \bar{\mathbf{v}}\right)}{\mathrm{e}^{2\rho t}\left\|\mathrm{e}^{tE}\mathbf{x} - \tfrac{1}{2}\|\mathbf{x}\|^2 \mathrm{e}^{-\lambda t} t R_t \bar{\mathbf{v}}\right\|^2}$$

$$= \frac{\mathrm{e}^{t(E+\lambda I)} \frac{\mathbf{x}}{\|\mathbf{x}\|^2} - \tfrac{1}{2} t R_t \bar{\mathbf{v}}}{\left\|\mathrm{e}^{t(E+\lambda I)} \frac{\mathbf{x}}{\|\mathbf{x}\|^2} - \tfrac{1}{2} t R_t \bar{\mathbf{v}}\right\|^2} = \frac{\mathrm{e}^{t(E+\lambda I)} \frac{\mathbf{x}}{\|\mathbf{x}\|^2} + \tfrac{1}{n-2} t R_t \mathbf{v}}{\left\|\mathrm{e}^{t(E+\lambda I)} \frac{\mathbf{x}}{\|\mathbf{x}\|^2} + \tfrac{1}{n-2} t R_t \mathbf{v}\right\|^2} \, ,$$

which is (8.51). (Note that in this case the vector in the numerator of (8.51) is $\mathrm{e}^{\lambda t}\|\mathbf{x}\|^{-2}\vec{y} \neq \vec{0}$.) If instead we have $\vec{y} = \vec{0}$ (and thus $y^\times = 0$) then according to (8.16) we have $U_t(\mathbf{x}) = \infty$, which is in agreement with (8.51) since the vector in the numerator of (8.51) is then $\mathrm{e}^{\lambda t}\|\mathbf{x}\|^{-2}\vec{y} = \vec{0}$.

Finally, if $\mathbf{x} = \infty$ then according to (8.17), Lemma 8.11, and (8.24) we have

$$U_t(\mathbf{x}) = \frac{(\mathcal{U}_t)^i{}_\odot}{(\mathcal{U}_t)^\times{}_\odot} = \frac{\mathrm{e}^{(\rho-\lambda)t} t R_t \bar{\mathbf{v}}}{-\tfrac{1}{2}\mathrm{e}^{(\rho-\lambda)t} t^2 \|R_t \bar{\mathbf{v}}\|^2} = \frac{-\tfrac{1}{2} t R_t \bar{\mathbf{v}}}{\left\|\tfrac{1}{2} t R_t \bar{\mathbf{v}}\right\|^2} = \frac{\tfrac{1}{n-2} t R_t \mathbf{v}}{\left\|\tfrac{1}{n-2} t R_t \mathbf{v}\right\|^2} \, ,$$

which is in agreement with (8.51) as well since in this case we have $\frac{\mathbf{x}}{\|\mathbf{x}\|^2} = \frac{\infty}{\|\infty\|^2} = \vec{0}$. This completes the proof of (8.51).

The alternative expression (8.52) then follows from the relation

$$\mathrm{e}^{t(E+\lambda I)} \mathbf{c}_t = \tfrac{1}{n-2}(E+\lambda I)^{-1}\left(I - \mathrm{e}^{t(E+\lambda I)}\right)\mathbf{v}$$

$$= -\tfrac{1}{n-2} \sum_{k=1}^\infty \tfrac{t^k}{k!}(E+\lambda I)^{k-1}\mathbf{v}$$

$$= -\tfrac{1}{n-2} \sum_{k=0}^\infty \tfrac{t^{k+1}}{(k+1)!}(E+\lambda I)^k \mathbf{v}$$

$$= -\tfrac{1}{n-2} t R_t \mathbf{v}. \qquad \square$$

Observe that $U_t(\vec{0}) = \vec{0}$ for $\forall t \in \mathbb{R}$, and that as expected we have $U_{t=0}(\mathbf{x}) = \mathbf{x}$ for $\forall \mathbf{x} \in \bar{\mathbb{R}}^n$. Further analyses on this formula (in particular, an independent proof of the group property of the maps U_t, and studies of their fixed points, their limiting behavior as $t \to \pm\infty$, and the resulting large-tenor behavior of the forward rate curves) have already been carried out in Section 2.8.

As a consequence of Lemma 2.13 in that section we now obtain the following result about the eigenvectors of ε.

Corollary 8.2. *For $\lambda > 0$ the two eigenvectors in $\mathbb{T}_0^\uparrow$ of the generator ε are $e_\times = (0, 1, \vec{0})$ (with eigenvalue λ) and $(-\frac{1}{2}, \|\hat{\mathbf{v}}\|^2, -\hat{\mathbf{v}})$ (with eigenvalue $-\lambda$), where $\hat{\mathbf{v}}$ is defined in (2.53);[11] for $\lambda = 0$ there is only one eigenvector in $\mathbb{T}_0^\uparrow$ (except for positive multiples), namely $e_\times$, with eigenvalue 0.*

In particular, this implies that the eigenvectors provided by Lemma 8.10 (ii) are unique even in the case $\hat{\lambda} = 0$.

Proof. As it was shown at the beginning of the proof of Lemma 8.10 (ii), every joint fixed point $\mathbf{x}$ of the maps U_t corresponds to a ray $h^{-1}(\mathbf{x}) \in \mathbb{P}_0$ of eigenvectors of ε. By (8.13) the fixed points given in Lemma 2.13 (ii) therefore lead us to the rays $h^{-1}(\vec{0}) = \pi((0, 1, \vec{0})) = \pi(e_\times)$ and (for $\lambda > 0$ only)

$$h^{-1}\left(\tfrac{-\hat{\mathbf{v}}}{\|\hat{\mathbf{v}}\|^2}\right) = \pi\left(\left(-\tfrac{1}{2}\left\|\tfrac{-\hat{\mathbf{v}}}{\|\hat{\mathbf{v}}\|^2}\right\|^2, 1, \tfrac{-\hat{\mathbf{v}}}{\|\hat{\mathbf{v}}\|^2}\right)\right) = \pi\left(\left(-\tfrac{1}{2}, \|\hat{\mathbf{v}}\|^2, -\hat{\mathbf{v}}\right)\right),$$

and thus to the eigenvectors stated in this corollary. The associated eigenvalues can be found by actually applying ε (given in (8.42)) to these vectors.

Conversely, every eigenvector $x \in \mathbb{T}_0$ of ε is an eigenvector of the maps $\mathcal{U}_t = e^{t(\varepsilon+\rho)}$, so that by (8.16) $\mathbf{x} = h(\pi(x))$ is a fixed point of the maps U_t. Lemma 2.13 (ii) thus also implies that these eigenvectors are in fact the only ones. □

As a final remark, recall that as a result of our choice to restrict E to the specific form (8.36), the matrices $(E + \lambda I)^{-1}$, e^{tE}, and R_t, and thus also the currency vectors $\mathbf{c}_\tau$ defined in (8.48) that occur repeatedly in our formulas here and in Section 2.8 can be computed analytically, without having to resort to any linear equation solver or to their original defining power series. Indeed, the necessary calculations decouple into those for the individual (2×2)-matrices of the form $\begin{pmatrix} 0 & \eta \\ -\eta & 0 \end{pmatrix}$ on the diagonal, and (for the last diagonal element if n is odd) for the (1×1)-matrix (0). See Section 2.7.2 for our explicit formulas.

8.15 The Kolmogorov Operator

We will now conclude our construction, by computing the formula for the classical Kolmogorov backward operator of the ACE model in the cash numeraire, based on the decomposition

$$\partial_t + D^{X,\$} = \nabla_t^{X,\$} + \tilde{D}^{X,\$} \tag{8.53}$$

given in (5.134a). We already know the deterministic part

$$\nabla_t^{X,\$} = \partial_t + \langle \mu, \nabla \rangle - r^X \tag{8.54}$$

from (5.136b), where μ and r^X are given by (8.26) (with $\mathbf{u} = \vec{0}$ by (8.33)) and by (8.22), respectively.

---

[11] Note that the fact that $e_\times$ is an eigenvector of ε is not coming as a surprise to us, as this has been our own modeling choice in Section 8.11; it is the formula for the second eigenvector that is interesting here.

In Lemma 8.15 we will now derive also a formula for the stochastic part $\tilde{D}^{X,\$}$, based on (5.138) and (6.56). (Recall that $\diamond$ is the counterpart in $\mathbb{R}^n$ of our operator $\diamond$ in $\mathbb{P}_0$ defined in Section 7.6.3.) To prepare, we quickly derive the following representation of the d'Alembert operator $\square$ (defined in (7.81)) on which our construction of $\diamond$ was based.

Lemma 8.14. *Given any isotropic basis* $\mathfrak{I}$*, the action of the d'Alembert operator on a function given in the form* $f(x^{\odot}, x^{\times}, \vec{x})$ *can be written as*

$$\square = 2\partial_{x^{\odot}}\partial_{x^{\times}} + \partial_{x^1}^2 + \cdots + \partial_{x^n}^2 \,. \tag{8.55}$$

Proof. See Appendix G.6. $\square$

Lemma 8.15. *For some specific choice of the gauge* m_0 *in* (6.56)*, we have*

$$\tilde{D}^{X,\$} = \alpha(\mathbf{x},t)\Delta = \alpha(\mathbf{x},t)\big(\partial_{\mathbf{x}^1}^2 + \cdots + \partial_{\mathbf{x}^n}^2\big), \tag{8.56}$$

where $\alpha \in \mathcal{F}_+(M_0 \times \mathbb{R})$ *is the functional model parameter introduced at the end of Section 6.4.3.*

Note that just as we had hoped for and in fact expected (recall the last paragraph of Section 6.6), the second-order part of this operator is indeed a positive function times the Laplacian.

Proof of Lemma 8.15. We compute the action of $\tilde{D}^{X,\$}$ on functions $f \in \mathcal{F}(M_0)$ by computing the action of the corresponding operator $\tilde{\mathcal{D}}^{X,\$}$ on the associated function $f^{\mathscr{M}} \in \mathcal{F}(\mathscr{M}_0)$ given by $f^{\mathscr{M}}(\mathbf{p}) := f(\mathbf{x}) = f(h(\mathbf{p}))$, as illustrated in the following commutative diagram:

$$
\begin{array}{ccc}
f^{\mathscr{M}} & \xrightarrow{\;\tilde{\mathcal{D}}^{X,\$}\;} & \tilde{\mathcal{D}}^{X,\$}f^{\mathscr{M}} \\
{\scriptstyle h}\downarrow & & \downarrow{\scriptstyle h} \\
f & \xrightarrow{\;\tilde{D}^{X,\$}\;} & \tilde{D}^{X,\$}f
\end{array}
\tag{8.57}
$$

By (5.138) and (6.56), the operator $\tilde{D}^{X,\$}$ is given by

$$\tilde{D}^{X,\$} = \tfrac{1}{\mathfrak{B}_0^Y} \circ \tilde{D}^{X,nf} \circ \mathfrak{B}_0^Y = \tfrac{1}{\mathfrak{B}_0^Y} \circ \big(\alpha \cdot \big(\tfrac{1}{m_0} \otimes \diamond\big)\big) \circ \mathfrak{B}_0^Y \,,$$

for some fixed bilinear form $m_0 \in c$ (which may choose at will later to simplify our result) and an arbitrary functional model parameter $\alpha \in \mathcal{F}_+(M_0)$ (which for time-inhomogeneous models we will allow to be time dependent at the end). The corresponding operator $\tilde{\mathcal{D}}^{X,\$}$ on the manifold $\mathscr{M}$ is therefore given by

$$\tilde{\mathcal{D}}^{X,\$} = \tfrac{1}{\mathcal{B}_0^Y} \circ \big(\alpha^{\mathscr{M}} \cdot \big(\tfrac{1}{\mathfrak{m}_0} \otimes \diamond\big)\big) \circ \mathcal{B}_0^Y \,,$$

for some fixed bilinear form $\mathfrak{m}_0 \in \mathfrak{c}$ and some function $\alpha^{\mathscr{M}} := \alpha \circ h \in \mathcal{F}_+(\mathscr{M}_0)$. Finally, rewriting this based on the representation $\diamond \colon \mathcal{S}(\mathscr{P}^{\otimes(-k)}) \to \mathcal{S}(\mathscr{P}^{\otimes(2-k)})$

of the diamond operator that we defined in Section 7.6 (where $k = 1 - \frac{n}{2}$), we have

$$\tilde{\mathcal{D}}^{X,\$} = \frac{1}{\mathcal{B}_0^Y} \circ \left(\alpha^{\mathscr{M}} \cdot \left(\frac{1}{\psi_{\mathbf{m}_0}} \otimes \Diamond \right) \right) \circ \mathcal{B}_0^Y$$

$$= \alpha^{\mathscr{M}} \cdot \left(\frac{1}{\psi_{\mathbf{m}_0} \otimes \mathcal{B}_0^Y} \circ \Diamond \circ \mathcal{B}_0^Y \right). \tag{8.58}$$

Now given any function $f(\mathbf{x})$, the associated function $f^{\mathscr{M}}(\mathbf{p})$ is given by

$$f^{\mathscr{M}}(\mathbf{p}) = f(h(\mathbf{p})) \overset{(8.12)}{=} f\left(\frac{\vec{x}}{x^\times} \right) \qquad \text{for any } x \in \mathbf{p} \setminus \{0\}.$$

Since we have defined our operator $\Diamond$ via its action on homogeneous functions (i.e., based on the duality between sections and homogeneous functions described in Section 7.5), computing the image of $f^{\mathscr{M}}$ under $\tilde{\mathcal{D}}^{X,\$}$ requires us to first translate $f^{\mathscr{M}}$ to its corresponding representation as a homogeneous function. As described at the end of Section 7.5.3, that function is the 0-homogeneous function defined on (a subset of) $\mathbb{T}_0^\uparrow$ that takes the constant value $f^{\mathscr{M}}(\mathbf{p})$ along each ray $\mathbf{p}_+$, i.e., we have the representation[12]

$$f^{\mathscr{M}} \quad \sim \quad x \mapsto f\left(\frac{\vec{x}}{x^\times} \right). \tag{8.59}$$

To apply the operator $\tilde{\mathcal{D}}^{X,nf}$ given in (8.58), we must begin by multiplying this function with the homogeneous function corresponding to the bond section $\mathcal{B}_0^Y$, i.e., with $[b_0](x) = \langle b_0, x \rangle^{1-n/2} = \langle e_\odot, x \rangle^{1-n/2} = (x^\times)^{1-n/2}$ by (7.88), (8.14), and (8.2), and we obtain

$$\mathcal{B}_0^Y \cdot f^{\mathscr{M}} \quad \sim \quad x \mapsto (x^\times)^{1-n/2} f\left(\frac{\vec{x}}{x^\times} \right). \tag{8.60}$$

Next, we need to apply the $\Diamond$-operator, which by its definition in Section 7.6.3 means that we have to apply the d'Alembert operator $\Box$ to any smooth k-homogeneous extension of the function on the right to some open neighborhood G of $\mathbb{T}_0^\uparrow \setminus \pi(e_0)$. Since as that extension we can choose the expression on the right of (8.60) itself, we therefore find that

$$\Diamond(\mathcal{B}_0^Y \cdot f^{\mathscr{M}}) \quad \sim \quad x \mapsto \Box\left[(x^\times)^{1-n/2} f\left(\frac{\vec{x}}{x^\times} \right) \right]$$

$$= \left(\partial_{x^1}^2 + \cdots + \partial_{x^n}^2 \right) \left[(x^\times)^{1-n/2} f\left(\frac{\vec{x}}{x^\times} \right) \right]$$

$$= (x^\times)^{-1-n/2} (\Delta f)\left(\frac{\vec{x}}{x^\times} \right),$$

where in the first step we used Lemma 8.14 and the fact that the expression on the right does not depend on $x^\odot$.

Now we need to divide by the homogeneous function associated to $\psi_{\mathbf{m}_0} \otimes \mathcal{B}_0^Y$, which by (7.76) is just the product of the two (-2)- and k-homogeneous functions that are associated to $\psi_{\mathbf{m}_0} \in \mathcal{S}_+(\mathscr{P}^{\otimes 2})$ and $\mathcal{B}_0^Y \in \mathcal{S}_+(\mathscr{P}^{\otimes(-k)})$, respectively. As we found above, the latter one is the function $x \mapsto (x^\times)^{1-n/2}$, and

---

[12]The fact that this function is undefined for $x^\times = 0$, i.e., for $x \in \pi(e_\odot)$, is okay since $\pi(e_\odot) = \pi(b_0) \notin \mathscr{M}_0$.

so if we choose $\psi_{\mathbf{m}_0}$ such that the former is the function $x \mapsto (x^\times)^{-2}$ (which is indeed (-2)-homogeneous and positive by (8.11)) then we have

$$\left(\frac{1}{\psi_{\mathbf{m}_0} \otimes \mathcal{B}_0^Y} \circ \Diamond \circ \mathcal{B}_0^Y \right) f^{\mathscr{M}} \quad \sim \quad x \mapsto (\Delta f)\left(\tfrac{\vec{x}}{x^\times}\right).$$

Finally, since just like in (8.59) the function $\alpha^{\mathscr{M}}$ is associated to the 0-homogeneous function $x \mapsto \alpha\left(\tfrac{\vec{x}}{x^\times}\right)$, we have

$$\tilde{\mathcal{D}}^{X,\$} f^{\mathscr{M}} = \alpha^{\mathscr{M}} \cdot \left(\frac{1}{\psi_{\mathbf{m}_0} \otimes \mathcal{B}_0^Y} \circ \Diamond \circ \mathcal{B}_0^Y \right) f^{\mathscr{M}} \quad \sim \quad x \mapsto (\alpha \cdot \Delta f)\left(\tfrac{\vec{x}}{x^\times}\right),$$

and so following the right arrow in the diagram (8.57) down to translate this result into $\mathbb{R}^n$, we obtain $\tilde{D}^{X,\$} f = \alpha \cdot \Delta f$. Now allowing also for time-dependent functions $\alpha \in \mathcal{F}_+(M_0 \times \mathbb{R})$, this is (8.56). □

Defining our functional noise parameter $\sigma \in \mathcal{F}_+(M_0 \times \mathbb{R})$ as

$$\sigma := \sqrt{2\alpha}, \tag{8.61}$$

which just like α we must ask to vanish sufficiently fast near ∂M_0 and to not grow too fast as $\mathbf{x} \to \infty$ (recall the end of Section 6.4.3), we can now combine all of our previous results to arrive at the final $\mathbb{R}^n$-based equations that describe the process dynamics of the ACE model under the risk-neutral measure.

Theorem 8.1. *The Kolmogorov backward operator of the ACE model in the cash numeraire (i.e., under the risk-neutral measure) is given by*

$$\partial_t + D^{X,\$} = \partial_t + \langle \mu(\mathbf{x}), \nabla \rangle + \tfrac{1}{2}\sigma^2(\mathbf{x}, t)\Delta - r^X(\mathbf{x}), \tag{8.62}$$

where $\quad \mu(\mathbf{x}) = (-\lambda I + E)\mathbf{x} + \tfrac{1}{n-2}\|\mathbf{x}\|^2 \mathbf{v} - \tfrac{2}{n-2}\langle \mathbf{v}, \mathbf{x}\rangle \mathbf{x} \tag{8.63}$

and $\quad r^X(\mathbf{x}) = r_\infty + \langle \mathbf{v}, \mathbf{x}\rangle, \tag{8.64}$

and so the dynamics of the ACE model's state process $(X_t)_{t \geq 0}$ under the risk-neutral measure (i.e., the measure that is used when writing the model in the form of a short rate model) is given by the SDE

$$\mathrm{d}X_t = \mu(X_t)\,\mathrm{d}t + \sigma(X_t, t)\,\mathrm{d}W_t, \qquad X_{t=0} = \mathbf{x}_0, \tag{8.65}$$

where $(W_t)_{t \geq 0}$ is an n-dimensional Brownian motion.

Proof. Combining our formulas in (8.53), (8.54), (8.56), and (8.61), we obtain (8.62), where the formulas for μ and r^X are given by (8.26) (with $\mathbf{u} = \vec{0}$ by (8.33)) and (8.22), respectively. Since (8.62) implies that

$$D^{X,t} \stackrel{(5.139a)}{=} D^{X,\$} + r^X \stackrel{(8.62)}{=} \langle \mu, \nabla \rangle + \tfrac{1}{2}\sigma^2\Delta,$$

the last statement of this theorem now follows from Lemma 5.9. □

8.16 Summary

This completes our parametrization of the ACE model in $\mathbb{R}^n$, providing us with an easily understandable description of the model that has been stripped of any remnants of its complicated construction on the manifold $\mathbb{P}_0$. Since all of its key equations and properties derived in this chapter have already been compiled in our "fast track to ACE" in Chapter 2, we will refrain from listing them here again in detail; instead, we will only give a brief summary that points the reader to the exact places in the preceding sections where they can be found.

State space dimension, model parameters. We have derived a unified $\mathbb{R}^n$-based description of the ACE model for all dimensions $n \geq 3$ (see (7.104)). For $n = 2$ no meaningful ACE model exists (since in this case all its short rates are zero by (8.64)), while the case $n = 1$ requires some special treatment to confine the process to its intended state space (recall our discussion at the end of Section 7.8.3), which we have decided to postpone to Section 2.9.

The model has four non-functional parameters $E \in \mathbb{R}^{n \times n}$, $\mathbf{v} \in \mathbb{R}^n$, and $\lambda, r_\infty \in \mathbb{R}$, subject to the constraints (8.19), (8.32), (8.39), and (8.40)—which also imply (8.41), and which can equivalently be rephrased as discussed at the end of Section 8.11—in addition to the initial state $\mathbf{x}_0$ in our state space $M_0 \subset \mathbb{R}^n$ given by (8.25), and the noise function $\sigma \in \mathcal{F}_+(M_0 \times [0, \infty))$, which must be chosen such that

σ vanishes fast enough near ∂M_0 and does not grow too fast as $\mathbf{x} \to \infty$.

(8.66)

(The constraint (8.66) is made rigorous in Lemma 2.4, which also requires the additional constraint (2.9b).) If σ is chosen to be time independent (i.e., as a function $\sigma(\mathbf{x})$) then the model will be time homogeneous; by allowing σ to depend on the process time t as well, one can sacrifice time homogeneity to improve the calibration to the swaption market.

Evolution equation, short rate function. Given any such set of parameters, the ACE model, written in the form of a short rate model, is given by an n-dimensional state process $(X_t)_{t \geq 0}$ and a short rate function $r^X \in \mathcal{F}_+(M_0)$, both of which are defined in Theorem 8.1. Since some of our parameter constraints were a translation of the condition (7.113) of Lemma 7.14, they guarantee that the drift points inwards at the boundary of M_0, and so the condition (8.66) ensures that the process $(X_t)_{t \geq 0}$ is indeed confined to M_0 almost surely.

By the very definition of the state space M_0, the short rate function r^X takes positive values on M_0, and so the ACE model simulates positive rates at all process times t almost surely. If one wants the model to impose a lower bound $r_0 \in \mathbb{R}$ other than 0, one can simply consider the modified short rate function $\tilde{r}(\mathbf{x}) := r(\mathbf{x}) + r_0$ instead, as discussed in the paragraph "Choice of the short rate lower bound" in Section 2.10.

Gauge fixing, matrix formulas.   The parameter constraints listed above are sufficient to give the model all its desired properties, and to ensure that our bond price and forward rate formulas below are valid. However, as is, there is still some redundancy in the model parameters, with different choices of $(E, \mathbf{v}, \mathbf{x}_0, \sigma)$ leading to the same exact short rate dynamics. We have removed these redundancies by imposing additional parameter constraints, namely by asking that E and $\mathbf{v}$ are of the forms (8.36)–(8.38) that originated in Lemma 2.12. As a result, we are left with a total of $2n + 1$ degrees of freedom (recall (2.45)) for our non-functional model parameters E, $\mathbf{v}$, r_∞, λ, and $\mathbf{x}_0$, in addition to the functional model parameter $\sigma(\mathbf{x}, t)$. This gives the model enough parametric freedom to calibrate well to both the discount curve and the swaption market.

While these redundancies could have been addressed in a variety of ways, our specific strategy was designed to allow us to derive explicit formulas for the matrices e^{tE}, $(E + \lambda I)^{-1}$, and R_τ that we encounter in our various formulas listed below (see Section 2.7.2).

Bond prices.   As stated in Corollary 8.1, the dollar value of the maturity-T bond, i.e.,

$$B_{\$,T}(\mathbf{x}, t) = \mathbb{E}\big[\exp\big(-\textstyle\int_t^T r(X_s)\,\mathrm{d}s\big) \,\big|\, X_t = \mathbf{x}\big], \qquad (8.67)$$

can for $\forall \mathbf{x} \in M_0$ and $\forall t \in [0, T]$ be computed via the explicit formula (8.46), where the matrices $R_\tau \in \mathbb{R}^{n \times n}$ are defined in (8.44). To prove this, we showed that (8.46) is a bounded and continuous solution of the Feynman–Kac PDE

$$\forall t \in [0, T)\ \forall \mathbf{x} \in M_0\colon\ \big(\partial_t + D^{X,\$}\big)B_{\$,T}(\mathbf{x}, t) = 0,$$
$$\forall \mathbf{x} \in M_0\colon\ B_{\$,T}(\mathbf{x}, T) = 1,$$

where $\partial_t + D^{X,\$}$ is the Kolmogorov backward operator in the cash numeraire given in Theorem 8.1; this sufficed since it is known that this PDE has only one such solution, namely the one given by the expression on the right of (8.67).

As one can see, the bond price formula (8.46) does not depend on the functional parameter $\sigma(\mathbf{x}, t)$, despite the fact that this parameter does enter the definition (8.67) of the bond prices $B_{\$,T}(\mathbf{x}, t)$ via the process (X_t) (see Section 6.4.4 and (8.61) for how this was achieved). This makes the ACE model unspanned with respect to $\sigma(\mathbf{x}, t)$, thus opening the door to more efficiently calibrating the model to the swaption market via bootstrapping by expiration as opposed to maturity.

Forward rates, zero-noise evolution.   As a result of the very definition of the spot parametrization in Section 5.2.1, the forward rate function

$$F(\mathbf{x}, t; s) := -\partial_s \log B_{\$,s}(\mathbf{x}, t),$$

where $\mathbf{x} \in M_0$ and $t \le s$, depends on t and s only via their difference $s - t$, i.e., we have $F(\mathbf{x}, t; s) = F_{\text{rel}}(\mathbf{x}; s - t)$ by (5.7); as a consequence—as we can see in (8.46)—the same holds true for the bond price function $B_{\$,s}(\mathbf{x}, t)$ itself as well.

As stated in Lemma 8.5, the relative forward rate function $F_{\text{rel}}(\mathbf{x}; \tau)$ can be computed via the formula (5.14), where the zero-noise evolution maps $U_t \colon M_0 \to M_0$ are defined as the solution of the ODE (5.10), with the drift vector field $\mu \colon \mathbb{R}^n \to \mathbb{R}^n$ given in (8.63). As a result of its origins on the compact manifold $\mathbb{P}_0$, this semigroup of maps can be extended to a proper group of maps $U_t \colon \bar{\mathbb{R}}^n \to \bar{\mathbb{R}}^n$ (where $\bar{\mathbb{R}}^n := \mathbb{R}^n \cup \{\infty\}$) whose actions are given by the explicit formula provided in Lemma 8.13. An independent proof of the group property of this family of maps can be found in Section 2.8.1.

As can easily be seen from (5.14), the forward rates of the ACE model are subject to the same lower bound as the short rate r^X. The limiting behavior of the maps U_t as $t \to \infty$ and the resulting large-tenor limiting behavior of the forward rate formula are explored in detail in Section 2.8.2.

Chapter 9

Conclusions

We have successfully designed a model with all of the desired features listed in Section 1.1: It is low-dimensional, complete, consistent, time homogeneous (if desired), and unspanned, with an explicit bond price formula, flexible model dynamics, and with short rates that are bounded below. A description of our final product—the $\mathbb{R}^n$-based version of the model and all its relevant formulas, along with independent proofs of their validity—and detailed discussions of its properties, of various aspects related to its calibration, and of possible model extensions can be found in **Part I: A Fast Track to ACE**. The features and the limitations of the ACE model are summarized in Section 2.12.

9.1 Recap: Model Derivation

Chapter 5: The Mathematical Framework. To derive our equations, we first devised a general interest rate modeling framework, based on two alternative parametrizations of the state variable (the spot parametrization $\mathbf{x}$ and the forward parametrization $\mathbf{y}$), on sections on the "wealth line bundle" that model the "numeraire-free" value of instruments or accounts, and on a group of "evolution operators" that link the bond sections of different maturities and that contain the information about the flow of the zero-noise limit of our state dynamics.

While the time homogeneity requirement was best described in the spot parametrization, the consistency requirement was more easily understood in the forward parametrization. The latter was expressed by way of a section-based variant of the Kolmogorov backward equation that all sections representing the value of self-financing accounts (and thus in particular our bond sections) have to satisfy. A decomposition of the Kolmogorov backward operator into its stochastic and its deterministic part was introduced to help us compute its representations in both parametrizations and in the various numeraires.

Chapter 6: Solution Strategy.   To facilitate the solution of the consistency requirement, we then introduced the (purely analytically motivated) scaling assumption, which asks that the Kolmogorov backward operator should commute with the evolution operators up to multiplicative factors. This reduced our consistency requirement from infinitely many equations (one Kolmogorov backward equation for each maturity) to only one equation for the maturity-zero bond section, at the cost of having to fulfill also the scaling assumption.

Looking for ways to satisfy this scaling assumption, we were guided by the insight that this assumption has two implications: (i) It reduces the construction of our section-based Kolmogorov operator to the one of a differential operator $\diamond$ (the "diamond operator") that leads into a different section space but that commutes with the evolution operators exactly, i.e., without any multiplicative factors. (ii) It implies that the action of our evolution operators on states in our state space must be conformal w.r.t. a conformal structure on the state space that is encoded in $\diamond$.

Arguing that we may—likely with no or only little loss of generality—choose to make this conformal structure the Euclidean one, the latter meant that the class of all possible evolution maps (i.e., the class of all conformal maps acting on $\mathbb{R}^n$) can best be understood by mapping our state space to the manifold $\mathbb{P}_0$ (the projective space of the set $\mathbb{T}_0$ of all isotropic points in the $(n{+}2)$-dimensional Minkowski space $\mathbb{T}$), where their counterparts are simply represented by the class of all pseudo-orthogonal operators acting on $\mathbb{T}$. We therefore decided to equivalently carry out our entire construction on $\mathbb{P}_0$, and only in the end to map our model back to $\mathbb{R}^n$.

Chapter 7: Solution.   *Definition of the state space manifold, the line bundle, and the conformal structure:* We then began the actual construction of our model by familiarizing the reader with the Minkowski space $\mathbb{T}$ and some relevant related spaces, objects, and techniques, such as the isotropic space $\mathbb{T}_0$, the projective space $\mathbb{P}_0$, pseudo-orthogonal operators, covariant and contravariant vectors, and the Einstein notation. We defined a signed line bundle $\mathscr{P}$ on $\mathbb{P}_0$ (with the plan to eventually use a power $\mathscr{P}^{\otimes(-k)}$ as our model's wealth line bundle), by simply considering for each state $\mathbf{p} \in \mathbb{P}_0$ the line $\mathbf{p}$ itself as the associated one-dimensional vector space, with the sign of a point $x \in \mathbf{p}$ given by the sign of its x^0-component. We also showed how bilinear forms on $\mathbb{P}_0$ can be represented by sections on $\mathscr{P}^{\otimes 2}$, and how the Minkowski inner product naturally induces a conformal structure $\mathfrak{c}$ on $\mathbb{P}_0$.

Definition of the evolution maps: We then demonstrated how every scalar multiple of a pseudo-orthogonal operator on $\mathbb{T}$ induces an action on our section spaces, and an action on $\mathbb{P}_0$ that is conformal w.r.t. $\mathfrak{c}$. Our group of evolution maps is therefore given by a group of such operators, which in turn is determined by its generator $\varepsilon + \rho$, where ε is an antisymmetric operator on $\mathbb{T}$, and where $\rho \in \mathbb{R}$. This generator is one of the model's parameters (along with the starting point $\mathbf{x}_0$ and the noise function $\sigma(\mathbf{x}, t)$ of the state process), equivalent to the parameters E, $\mathbf{v}$, λ, and r_∞ in our eventual $\mathbb{R}^n$-based version of the model.

Definition of the diamond operator: After showing that every section on $\mathscr{P}^{\otimes(-k)}$, $k \in \mathbb{Z}$, is naturally represented by a k-homogeneous function on $\mathbb{T}_0$ (which also allows us to even consider fractional exponents $k \in \mathbb{R}$), we defined the diamond operator $\diamondsuit$ (the counterpart of $\diamond$ in $\mathbb{P}_0$) as a differential operator acting on such functions. To do so, we recalled that pseudo-orthogonal operators are known to commute with the d'Alembert operator $\square$ (just like orthogonal operators in $\mathbb{R}^n$ commute with the Laplacian Δ); since we needed our evolution maps (which were derived from pseudo-orthogonal operators) to commute with $\diamondsuit$, we therefore decided to derive $\diamondsuit$ naturally from $\square$, by restricting the action of $\square$ from $\mathbb{T}$ to $\mathbb{T}_0$. To make this restriction unique, we found that the exponent k in our wealth line bundle $\mathscr{P}^{\otimes(-k)}$ must be chosen as $k = 1 - \frac{n}{2}$, where n is the dimension of $\mathbb{P}_0$ (and thus of our state space).

Definition of the bond sections, derivation of related functions: We then defined our bond section $\mathcal{B}_0^Y$ as the counterpart of the specific k-homogeneous function $[b_0](x) := \langle b_0, x \rangle^{1-n/2}$ (where b_0 is some fixed vector on the backward light cone $\mathbb{T}_0^{\downarrow}$), which is positive and harmonic, i.e., it satisfies the consistency condition $\diamondsuit[b_0] = 0$. By properly applying our evolution operators to this section, we then obtained the bond sections $\mathcal{B}_T^X$ for all maturities $T \in \mathbb{R}$, and thus (by taking their ratios) all bond exchange rate functions B_{T_1,T_2}^X and bond dollar price functions $B_{\$,T}^X$, and (by differentiating) the short rate function r^X.

Definition of the state space: We concluded our construction by defining the intended state space $\mathscr{M}_0 \subset \mathbb{P}_0$ of our state process as the subset of all states $\mathbf{p} \in \mathbb{P}_0$ with positive short rate $r^X(\mathbf{p})$. For $n \geq 3$, a constraint on the vector $\overset{\shortdownarrow}{b}_0^{\perp} := \varepsilon(\varepsilon - \rho)b_0$ ensures that the zero-noise process remains confined to this state space (i.e., that $\forall t \geq 0: \mathcal{U}_t(\mathscr{M}_0) \subset \mathscr{M}_0$), which in turn—together with our assumption that the noise vanishes sufficiently fast near $\partial \mathscr{M}_0$—will later imply that our eventual stochastic process $(X_t)_{t \geq 0}$ in $\mathbb{R}^n$ cannot escape its state space M_0, either. The dimension $n = 2$ is not considered, since in this dimension the short rate degenerates to $r^X \equiv 0$; the dimension $n = 1$ is treated separately in Section 2.9, using our eventual $\mathbb{R}^n$-based model description instead.

Chapter 8: Parametrization in $\mathbb{R}^n$. To map our model to $\mathbb{R}^n$, we first represented each vector in $\mathbb{T}$ by its coefficients w.r.t. an isotropic basis $\mathfrak{I}$, i.e., a basis whose first two vectors have inner product 1 and lie on the two opposing light cones $\mathbb{T}_0^{\downarrow}$ and $\mathbb{T}_0^{\uparrow}$, respectively, with all remaining basis vectors being pseudo-orthogonal. A simple map $h: \mathbb{P}_0 \to \mathbb{R}^n$ based on these coefficient vectors then mapped each state $\mathbf{p} \in \mathbb{P}_0$ down to a vector $\mathbf{x} \in \bar{\mathbb{R}}^n := \mathbb{R}^n \cup \{\infty\}$, in a way that maps our conformal structure $\mathfrak{c}$ on $\mathbb{P}_0$ to the Euclidean one on $\mathbb{R}^n$. Based on h we then consistently translated also all other objects of our model down to $\mathbb{R}^n$, choosing our isotropic basis (which has no impact on the actual model behavior) along the way to simplify our formulas. Finally, we read off the dynamics of our state process $(X_t)_{t \geq 0}$ from the $\mathbb{R}^n$-based Kolmogorov backward equation in the cash numeraire.

9.2   Restrictive Modeling Choices

Some modeling choices in our construction were motivated merely by numerical or analytical necessity, rather than by observed (or otherwise theoretically derived) properties of the current market prices or of the historical interest rate evolution. Such modeling decisions may restrict the generality of a model in unforeseen ways, and overcoming the issues they address in alternative ways may lead to a more general model.

To help the reader assess the generality of the ACE model, and to facilitate potential future generalizations, we will therefore now list some of the main technical assumptions that we had decided to make during the model's construction.

Unspanned volatility: Unspanned volatility was added to our list of desired model features to facilitate calibrating $\sigma(\mathbf{x}, t)$ via bootstrapping by expiration. However, leaving aside the issue of finding the optimal choice of σ *in practice*, there is no reason why unspanned models would describe the actual real-world interest rate evolution better than others.

Scaling assumption: Our way of finding a solution of the consistency condition (6.42) was to introduce the scaling assumption (6.43), which is solely of technical nature. There may be solutions that do not satisfy the scaling assumption.

Scalar-valued noise function: In Section 6.4.5 we have effectively decided to construct the ACE model with a scalar-valued (as opposed to matrix-valued) noise function $\sigma(\mathbf{x}, t)$, justified by the argument that any process with a matrix-valued noise function can at least locally be reparametrized to a process with a scalar-valued one. However, this might have led us to miss out on certain processes with matrix-valued noise functions that might not be reparametrizable in this way *globally*.

Choice of bond section: In (7.88) we have defined our bond section $\mathcal{B}_0^Y$ as one of the harmonic basis functions $[b_0]$ that we had found in Section 7.7.1. This choice led us to simple formulas for the short rate function and thus for the state space $\mathcal{M}_0$, and ultimately also to the simple constraint (7.110) to ensure that $\mathcal{M}_0$ is closed under the zero-noise flow.

Instead, we could have chosen to define $\mathcal{B}_0^Y$ as any linear combination of these harmonic basis functions (with positive coefficients), i.e., as sections of the form $\sum_i c_i [b^i]$ or $\int_0^1 c_\alpha [b^\alpha] \, d\alpha$. Overcoming the mathematical challenges resulting from such a generalization would lead to additional model parameters that can be used for fitting the discount curve.

9.3 Closing Statements

This concludes our derivation of the ACE equations. We recommend the reader to now go back to Part I and reread our "Fast Track To ACE," as now with our newly gained insight into the model's inner workings it will appear in a completely new light.

We would then like to encourage the reader to go ahead and actually implement the ACE model, to try out various calibration strategies, and to price some instruments of interest via backwards induction or Monte Carlo methods. Finally, we hope that the reader feels inspired by our successful use of the various mathematical techniques introduced in this book (and in particular the concept of numeraire-free section-based modeling), and to consider utilizing these newly-learned tools in their future modeling endeavors.

Appendix

Appendix A

Supporting Material for Chapter 1

A.1 Recombining HJM Models

To understand the notion of a recombining HJM model, let us take a look at the simplest one-dimensional example of a noise function of the form (1.7).

Example and definition. Consider the HJM model (1.5) with the deterministic noise process

$$\sigma_t(s) := \eta \, e^{-\lambda(s-t)}$$

for $0 \leq t \leq s$, where $\eta, \lambda > 0$ are model parameters. The HJM drift formula (1.6) then tells us that to obtain a consistent model, we must use the drift

$$\mu_t(s) = \sigma_t(s) \int_t^s \sigma_t(u) \, du = \tfrac{\eta^2}{\lambda}(e^{-\lambda(s-t)} - e^{-2\lambda(s-t)}).$$

The SDE (1.5) starting from the state $\mathbf{f}_0$ at time $t = 0$ is solved by

$$\begin{aligned}
\mathbf{f}_t(s) &= \mathbf{f}_0(s) + \int_0^t \mu_u(s) \, du + \int_0^t \sigma_u(s) \, dW_u \\
&= \mathbf{f}_0(s) + \tfrac{\eta^2}{\lambda^2}\big[\big(e^{-\lambda(s-t)} - e^{-\lambda s}\big) - \tfrac{1}{2}\big(e^{-2\lambda(s-t)} - e^{-2\lambda s}\big)\big] \\
&\qquad\qquad\qquad\qquad\qquad\qquad + \eta \int_0^t e^{-\lambda(s-u)} \, dW_u \,,
\end{aligned}$$

which can be written as

$$\mathbf{f}_t(s) = F(X_t, t; s) \tag{A.1}$$

for $0 \leq t \leq s$, where

$$F(\mathbf{x}, t; s) := \mathbf{f}_0(s) + \mathbf{x}_1 e^{-\lambda(s-t)} + \mathbf{x}_2 e^{-2\lambda(s-t)}, \tag{A.2}$$

$$X_t := \Big(\tfrac{\eta^2}{\lambda^2}(1 - e^{-\lambda t}) + \eta \int_0^t e^{-\lambda(t-u)} \, dW_u \,, \ -\tfrac{\eta^2}{2\lambda^2}(1 - e^{-2\lambda t})\Big). \tag{A.3}$$

This shows that although HJM processes may generally explore their full infinite-dimensional state space at any given time t, for this specific choice of $\sigma_t(s)$ the process remains confined to some two-dimensional affine subspace, parameterized by the two-dimensional process $(X_t)_{t\geq 0}$ via the function $F(\mathbf{x}, t; s)$. We say that an HJM model like this one whose forward rate process can be written in the form (A.1), for some function $F(\mathbf{x}, t; s)$ and some finite-dimensional process $(X_t)_{t\geq 0}$, *recombines*. This property is desirable since it reduces the model dynamics to the dynamics of a finite-dimensional process *without any loss of accuracy* (in contrast to the dimension reduction via brute-force discretization or the use of the BGM approach, both of which cannot accurately be undone via straightforward interpolation).

Observations. Note that the parameter η in our example does not enter our formula (A.2) for F, and thus our bond price formula

$$B_{\$,T}(\mathbf{x}, t) = \exp\left(-\int_t^T F(\mathbf{x}, t; s)\, \mathrm{d}s\right) \qquad (A.4)$$

in this parametrization. This remains true even when this calculation is generalized to a time-dependent parameter $\eta(t)$, and so in the finite-dimensional reformulation of this generalized model the functional parameter $\eta(t)$ is unspanned.

Furthermore, observe that we could in fact have chosen $(X_t)_{t\geq 0}$ as a *one-dimensional* process (by including its second component in the definition of F instead), but then F would no longer have been independent of η. Unlike completeness, the universally valid unspannedness of HJM models is not necessarily inherited by their finite-dimensional reformulations if they recombine; unspannedness of a parameter $\eta(t)$ only survives the dimension reduction if this parameter does not enter the formula for F.

Equivalence to the short rate approach. Our example suggests that there are two ways of constructing a model with all of our desired features listed in Section 1.1: One can either try to define a short rate model in such a way that one obtains an explicit bond price function that makes the model unspanned, or one can look for a recombining HJM model that preserves its unspannedness (just as the one in our example).

In fact, these two approaches are not as distinct as they may appear: Any recombining HJM model can be written as a short rate model since its short rate can be expressed as

$$r_t = \mathbf{f}_t(t) = F(X_t, t; t) =: r(X_t, t); \qquad (A.5)$$

on the other hand, any short rate model can be written in the form (A.1) of a recombining HJM model since its forward rate process can be expressed as

$$\mathbf{f}_t(s) = -\partial_s \log B_{\$,s}(X_t, t) =: F(X_t, t; s), \qquad (A.6)$$

with the function $B_{\$,s}(\mathbf{x}, t)$ given by (1.3).

The two approaches, which are both based on the same process $(X_t)_{t\geq0}$, have an explicit bond price function (given by (A.4)) if and only if the function F in the parametrization of the HJM model is known explicitly. Furthermore, since they share the same bond price function, any given model parameter is unspanned in one formulation if and only if it is unspanned in the other. Both approaches therefore lead to the same class of models, and so one may choose whichever one seems more tractable, without restricting the outcome a priori.

Our derivation of the ACE model in Chapters 5–8 follows neither of the two approaches; instead, our starting point lies right in between: We will construct a process $(X_t)_{t\geq0}$ and a function $F(\mathbf{x}, t; s)$ (or equivalently, a family of functions $B_{\$,T}(\mathbf{x}, t)$) that does not depend on our functional noise parameter, in such a way that all of our desired properties listed in Section 1.1 are fulfilled. Only at the very end is our model then rewritten in the more familiar form of a short rate model, as presented in Chapter 2.

Example (continued). To illustrate the equivalence of the two approaches, consider again the recombining HJM model from our example above. Following our instructions for converting this model into its short rate formulation $(r(X_t, t))_{t\geq0}$, by (A.5) and (A.2) we obtain the short rate function

$$r(\mathbf{x}, t) = F(\mathbf{x}, t; t) = \mathbf{f}_0(t) + \mathbf{x}_1 + \mathbf{x}_2, \tag{A.7}$$

while the two-dimensional process $(X_t)_{t\geq0}$ is simply the one given in (A.3). The bond price function $B_{\$,T}(\mathbf{x}_1, \mathbf{x}_2, t)$, obtained by plugging (A.2) into (A.4), does not depend on η, and so the short rate formulation inherits the unspannedness of this parameter from our original HJM model.

In fact, going one step further and plugging (A.3) into (A.7), we obtain the explicit formula

$$r_t = r(X_t, t) = \mathbf{f}_0(t) + \tfrac{\eta^2}{2\lambda^2}\left(1 - e^{-\lambda t}\right)^2 + \eta \int_0^t e^{-\lambda(t-u)}\,\mathrm{d}W_u$$

for the short rate process, which satisfies the SDE

$$\begin{aligned}
\mathrm{d}r_t &= \left[\mathbf{f}_0'(t) + \tfrac{\eta^2}{2\lambda^2}\left(2\lambda e^{-\lambda t} - 2\lambda e^{-2\lambda t}\right) - \eta\lambda \int_0^t e^{-\lambda(t-u)}\,\mathrm{d}W_u\right]\mathrm{d}t + \eta\,\mathrm{d}W_t \\
&= -\lambda\bigg[\left(\mathbf{f}_0(t) + \tfrac{\eta^2}{2\lambda^2}\left(1 - e^{-\lambda t}\right)^2 + \eta \int_0^t e^{-\lambda(t-u)}\,\mathrm{d}W_u\right) \\
&\qquad\qquad - \left(\mathbf{f}_0(t) + \tfrac{1}{\lambda}\mathbf{f}_0'(t) + \tfrac{\eta^2}{2\lambda^2}\left(1 - e^{-2\lambda t}\right)\right)\bigg]\mathrm{d}t + \eta\,\mathrm{d}W_t \\
&= -\lambda\big(r_t - \theta(t)\big)\,\mathrm{d}t + \eta\,\mathrm{d}W_t\,,
\end{aligned}$$

with $\theta(t) := \mathbf{f}_0(t) + \tfrac{1}{\lambda}\mathbf{f}_0'(t) + \tfrac{\eta^2}{2\lambda^2}\left(1 - e^{-2\lambda t}\right)$.

We therefore see that the model was in fact nothing but the HJM reformulation of the Hull–White short rate model with mean reversion rate λ, reversion target $\theta(t)$, and noise coefficient η.

To recover the standard Hull–White bond price function $B_{\$,T}(\mathbf{r}, t)$ (i.e., the one based on the short rate $\mathbf{r}$ as the state variable) from the function

$B_{\$,T}(\mathbf{x}_1, \mathbf{x}_2, t)$, we can successively substitute $\mathbf{x}_1 = \mathbf{r} - \mathbf{f}_0(t) - \mathbf{x}_2$ and then $\mathbf{x}_2 = -\frac{\eta^2}{2\lambda^2}(1 - e^{-2\lambda t})$. Note, however, that this formula will depend on η, and so in this standard one-dimensional form of the Hull–White model the parameter η—or in the generalized case $\eta(t)$—is *not* unspanned.

Appendix B

Proofs for Chapter 2

B.1 Mapping the State Space to a Ball

To prepare also for our proof of of Lemma 8.10 (ii) later in Section 8.11, we will prove a statement that is slightly more general than necessary for us at this point, at very little extra cost. For the purposes of Chapter 2, we will only need to consider the case $\mathbf{u} = \vec{0}$.

Lemma B.1. *For any fixed $\mathbf{a} \in \mathbb{R}^n \setminus \overline{M_0}$ the following statements hold:*

(i) The function

$$g \colon \mathbb{R}^n \setminus \{\mathbf{a}\} \to \mathbb{R}^n \setminus \{\vec{0}\}, \qquad g(\mathbf{x}) := \frac{\mathbf{x} - \mathbf{a}}{\|\mathbf{x} - \mathbf{a}\|^2},$$

is a bijection.

(ii) The image $g(M_0)$ is the open ball with radius $-\frac{1}{2r(\mathbf{a})} > 0$ and center $-\frac{1}{2r(\mathbf{a})}\mathbf{v}$. Furthermore, we have $g(\overline{M_0}) = \overline{g(M_0)} \setminus \{\vec{0}\}$, i.e., the image $g(\overline{M_0})$ is the closure of this ball with only the origin $\vec{0}$ removed from its boundary.

(iii) Let the generalized drift be defined as

$$\mu(\mathbf{x}) = \mathbf{u} + (-\lambda I + E)\mathbf{x} + \tfrac{1}{n-2}\|\mathbf{x}\|^2\mathbf{v} - \tfrac{2}{n-2}\langle \mathbf{v}, \mathbf{x}\rangle\mathbf{x} \tag{B.1}$$

for $\forall \mathbf{x} \in \mathbb{R}^n$, where $\mathbf{u} \in \mathbb{R}^n$ is an additional parameter. Then the image

$$\tilde{\mu} := (\nabla g \cdot \mu) \circ g^{-1} \tag{B.2}$$

of μ under g (i.e., the vector field whose flowlines are the images of the flowlines of μ) is given by

$$\tilde{\mu}(\mathbf{x}) = \tilde{\mathbf{u}} + (-\tilde{\lambda}I + \tilde{E})\mathbf{x} + \tfrac{1}{n-2}\|\mathbf{x}\|^2\tilde{\mathbf{v}} - \tfrac{2}{n-2}\langle \tilde{\mathbf{v}}, \mathbf{x}\rangle\mathbf{x}, \tag{B.3a}$$

where

$$\tilde{\lambda} := -\lambda - \tfrac{2}{n-2}\langle \mathbf{v}, \mathbf{a}\rangle, \qquad\qquad \tilde{\mathbf{u}} := \tfrac{1}{n-2}\mathbf{v}, \tag{B.3b}$$

$$\tilde{E} := E + \tfrac{2}{n-2}(\mathbf{v}\mathbf{a}^T - \mathbf{a}\mathbf{v}^T), \qquad\qquad \tilde{\mathbf{v}} := (n-2)\mu(\mathbf{a}). \tag{B.3c}$$

(iv) In particular, at $\mathbf{x} = \vec{0} \in \partial(g(M_0))$ *the drift* $\tilde{\mu}(\vec{0}) = \frac{1}{n-2}\mathbf{v}$ *given by* (B.3a) *points towards the center of the ball* $g(M_0)$, *and so no flowline of* $\tilde{\mu}$ *starting from* $g(M_0)$ *can exit* $g(M_0)$ *near* $\vec{0}$.

Proof. (i) The function g is a bijection since it maps into $\mathbb{R}^n \setminus \{\vec{0}\}$, and since its inverse $g^{-1}(\mathbf{x}) = \frac{\mathbf{x}}{\|\mathbf{x}\|^2} + \mathbf{a}$ is well defined on all of $\mathbb{R}^n \setminus \{\vec{0}\}$ and returns values in $\mathbb{R}^n \setminus \{\mathbf{a}\}$.

(ii) Using that $r(\mathbf{a}) < 0$ by (2.14a)–(2.14b), we see that the set $g(M_0)$ consists of all points $\tilde{\mathbf{x}} \in \mathbb{R}^n \setminus \{\vec{0}\}$ with $g^{-1}(\tilde{\mathbf{x}}) \in M_0$, i.e., with

$$0 < r_\infty + \langle \mathbf{v}, g^{-1}(\tilde{\mathbf{x}}) \rangle = r_\infty + \langle \mathbf{v}, \frac{\tilde{\mathbf{x}}}{\|\tilde{\mathbf{x}}\|^2} + \mathbf{a} \rangle$$

$$= r(\mathbf{a}) + \langle \mathbf{v}, \frac{\tilde{\mathbf{x}}}{\|\tilde{\mathbf{x}}\|^2} \rangle = \frac{r(\mathbf{a})}{\|\tilde{\mathbf{x}}\|^2} \left(\|\tilde{\mathbf{x}}\|^2 + \langle \frac{\mathbf{v}}{r(\mathbf{a})}, \tilde{\mathbf{x}} \rangle \right)$$

$$= \frac{r(\mathbf{a})}{\|\tilde{\mathbf{x}}\|^2} \left(\left\| \tilde{\mathbf{x}} + \frac{\mathbf{v}}{2r(\mathbf{a})} \right\|^2 - \frac{\|\mathbf{v}\|^2}{4r(\mathbf{a})^2} \right)$$

$$\Leftrightarrow \qquad \frac{\|\mathbf{v}\|^2}{4r(\mathbf{a})^2} > \left\| \tilde{\mathbf{x}} + \frac{\mathbf{v}}{2r(\mathbf{a})} \right\|^2,$$

and so $g(M_0)$ is an open ball with radius $\frac{\|\mathbf{v}\|}{2|r(\mathbf{a})|} = -\frac{1}{2r(\mathbf{a})}$ and center $-\frac{\mathbf{v}}{2r(\mathbf{a})}$. The statement about $g(\overline{M}_0)$ is shown analogously.

(iii) To understand the formula (B.2), observe that given any flowline $\mathbf{x}(t)$ of μ, i.e., any solution of the ODE $\frac{\mathrm{d}}{\mathrm{d}t}\mathbf{x}(t) = \mu(\mathbf{x}(t))$, according to the chain rule, its image $\tilde{\mathbf{x}}(t) := g(\mathbf{x}(t))$ under g satisfies the ODE

$$\frac{\mathrm{d}}{\mathrm{d}t}\tilde{\mathbf{x}}(t) = (\nabla g)(\mathbf{x}(t)) \cdot \frac{\mathrm{d}}{\mathrm{d}t}\mathbf{x}(t) = (\nabla g \cdot \mu)(\mathbf{x}(t)) = ((\nabla g \cdot \mu) \circ g^{-1})(\tilde{\mathbf{x}}(t)),$$

i.e., it is a flowline of the vector field $(\nabla g \cdot \mu) \circ g^{-1}$.

To compute $\tilde{\mu}$ explicitly, we write our map as $g = g_2 \circ g_1$, where $g_1(\mathbf{x}) := \mathbf{x} - \mathbf{a}$ and $g_2(\mathbf{x}) := \frac{\mathbf{x}}{\|\mathbf{x}\|^2}$, and carry out our calculation in two steps. The image of μ under g_1 is

$$\mu'(\mathbf{x}) = ((\nabla g_1 \cdot \mu) \circ g_1^{-1})(\mathbf{x}) = \mu(\mathbf{x} + \mathbf{a})$$

$$= \mathbf{u} + (-\lambda I + E)(\mathbf{x} + \mathbf{a}) + \frac{1}{n-2}\|\mathbf{x} + \mathbf{a}\|^2 \mathbf{v} - \frac{2}{n-2}\langle \mathbf{v}, \mathbf{x} + \mathbf{a} \rangle(\mathbf{x} + \mathbf{a})$$

$$= \mathbf{u}' + (-\lambda' I + E')\mathbf{x} + \frac{1}{n-2}\|\mathbf{x}\|^2 \mathbf{v}' - \frac{2}{n-2}\langle \mathbf{v}', \mathbf{x} \rangle \mathbf{x},$$

where $\lambda' := \lambda + \frac{2}{n-2}\langle \mathbf{v}, \mathbf{a} \rangle$, $E' := E + \frac{2}{n-2}(\mathbf{v}\mathbf{a}^T - \mathbf{a}\mathbf{v}^T)$, $\mathbf{v}' := \mathbf{v}$, and

$$\mathbf{u}' := \mathbf{u} + (-\lambda I + E)\mathbf{a} + \frac{1}{n-2}\|\mathbf{a}\|^2 \mathbf{v} - \frac{2}{n-2}\langle \mathbf{v}, \mathbf{a} \rangle \mathbf{a} = \mu(\mathbf{a}),$$

i.e., μ preserves its shape under this type of transformation. Note that E' is antisymmetric again (just like E, recall (2.4)).

The vector field $\tilde{\mu}$ is now the image of μ' under g_2. Abbreviating $\tilde{\mathbf{x}} := g_2(\mathbf{x})$, it is given by

$$\tilde{\mu}(\tilde{\mathbf{x}}) = ((\nabla g_2 \cdot \mu') \circ g_2^{-1})(\tilde{\mathbf{x}}) = (\nabla g_2 \cdot \mu')(\mathbf{x})$$

$$= \frac{1}{\|\mathbf{x}\|^2}\left(I - 2\frac{\mathbf{x}\mathbf{x}^T}{\|\mathbf{x}\|^2} \right)\mu'(\mathbf{x})$$

$$= \|\tilde{\mathbf{x}}\|^2 \Big(I - 2\tfrac{\tilde{\mathbf{x}}\tilde{\mathbf{x}}^T}{\|\tilde{\mathbf{x}}\|^2} \Big) \mu'\Big(\tfrac{\tilde{\mathbf{x}}}{\|\tilde{\mathbf{x}}\|^2} \Big)$$

$$= \|\tilde{\mathbf{x}}\|^2 \Big(I - 2\tfrac{\tilde{\mathbf{x}}\tilde{\mathbf{x}}^T}{\|\tilde{\mathbf{x}}\|^2} \Big) \Big[\mathbf{u}' + (-\lambda' I + E')\tfrac{\tilde{\mathbf{x}}}{\|\tilde{\mathbf{x}}\|^2}$$

$$\qquad\qquad + \tfrac{1}{n-2} \big\| \tfrac{\tilde{\mathbf{x}}}{\|\tilde{\mathbf{x}}\|^2} \big\|^2 \mathbf{v}' - \tfrac{2}{n-2} \big\langle \mathbf{v}', \tfrac{\tilde{\mathbf{x}}}{\|\tilde{\mathbf{x}}\|^2} \big\rangle \tfrac{\tilde{\mathbf{x}}}{\|\tilde{\mathbf{x}}\|^2} \Big]$$

$$= \Big(I - 2\tfrac{\tilde{\mathbf{x}}\tilde{\mathbf{x}}^T}{\|\tilde{\mathbf{x}}\|^2} \Big) \Big[\|\tilde{\mathbf{x}}\|^2 \mathbf{u}' + (-\lambda' I + E')\tilde{\mathbf{x}} + \tfrac{1}{n-2}\mathbf{v}' - \tfrac{2}{n-2}\tfrac{\langle \mathbf{v}', \tilde{\mathbf{x}} \rangle \tilde{\mathbf{x}}}{\|\tilde{\mathbf{x}}\|^2} \Big]$$

$$= \|\tilde{\mathbf{x}}\|^2 \mathbf{u}' - 2\langle \mathbf{u}', \tilde{\mathbf{x}} \rangle \tilde{\mathbf{x}} + (\lambda' I + E')\tilde{\mathbf{x}} + \tfrac{1}{n-2}\mathbf{v}'$$

$$= \tilde{\mathbf{u}} + (-\tilde{\lambda} I + \tilde{E})\tilde{\mathbf{x}} + \tfrac{1}{n-2}\|\tilde{\mathbf{x}}\|^2 \tilde{\mathbf{v}} - \tfrac{2}{n-2}\langle \tilde{\mathbf{v}}, \tilde{\mathbf{x}} \rangle \tilde{\mathbf{x}},$$

where $\tilde{\mathbf{u}} := \tfrac{1}{n-2}\mathbf{v}'$, $\tilde{\lambda} := -\lambda'$, $\tilde{E} := E'$, and $\tilde{\mathbf{v}} := (n-2)\mathbf{u}'$. (In the next to last step we used that E' is antisymmetric and therefore has the property in Lemma 2.2 (ii).) Plugging in our formulas for λ', E', $\mathbf{v}'$, and $\mathbf{u}'$ above now completes our proof of (B.3).

(iv) We have $\tilde{\mu}(\vec{0}) = \tilde{\mathbf{u}} = \tfrac{1}{n-2}\mathbf{v}$, which points towards the center $-\tfrac{1}{2r(\mathbf{a})}\mathbf{v}$ of the ball $g(M_0)$. $\qquad\square$

B.2   The Feller Condition On The Unit Ball

To prepare for our proof of Lemma 2.4 later in Section B.3, we will first prove a variant of the classical Feller condition that is interesting in its own right. The classical Feller condition states that the simple one-dimensional CIR model [15] given by the SDE

$$\mathrm{d}R_t = b(R_t)\,\mathrm{d}t + \sigma(R_t)\,\mathrm{d}W_t\,, \qquad R_{t=0} = \mathbf{r}_0,$$

where $b(\mathbf{r}) := -\nu(\mathbf{r} - \mathbf{r}_\infty)$ and $\sigma(\mathbf{r}) := \sigma_0\sqrt{\mathbf{r}}$ for some parameters $\mathbf{r}_0, \mathbf{r}_\infty, \nu, \sigma_0 > 0$, is guaranteed to have a unique strong solution that is confined to $(0, \infty)$ for $\forall t \geq 0$ if only we have

$$\tfrac{1}{2}\sigma_0^2 < b(0). \tag{B.4}$$

A generalization of this condition to the class of affine models is provided in [20].

The version we prove here applies to general SDEs with isotropic noise that are defined on the open unit ball B in $\mathbb{R}^n$. For such SDEs, it states that if the drift points inwards near ∂B, if the noise vanishes near ∂B at some specific rate that depends on the radial component of the drift (or faster), and if certain other technical conditions are met, then there exists a unique strong solution that is confined to B (i.e., that never reaches ∂B) for $\forall t \geq 0$. This condition, (B.6), is analogous to (B.4); however, note that it manages to avoid the need to evaluate the drift on the boundary ∂B itself, thus allowing also for the degenerate situation in which the drift itself vanishes on the boundary as well.[1]

---

[1]This will be important in our degenerate case $\delta = 0$ at the point $\mathbf{x} = \mathbf{x}_{\min}$.

Lemma B.2 (Feller condition on the unit ball). *Denoting the open unit ball in $\mathbb{R}^n$ by $B := \{\mathbf{y} \in \mathbb{R}^n \mid \|\mathbf{y}\| < 1\}$, let the three continuous functions $b\colon B \times [0,\infty) \to \mathbb{R}^n$, $\sigma\colon B \times [0,\infty) \to (0,\infty)$, and $A\colon B \times [0,\infty) \to \mathbb{R}^{n\times n}$ be given, where $A(\mathbf{y},t)$ is assumed to be an orthogonal matrix for $\forall \mathbf{y} \in B$ and $\forall t \geq 0$, and suppose that for some $T > 0$ the following conditions are satisfied: (i) All three functions of $(\mathbf{y},t)$ are locally Lipschitz continuous in $\mathbf{y}$ uniformly in $t \in [0,T]$,[2] (ii) we have*

$$\sup_{\substack{\mathbf{y}\in B \\ t\in[0,T]}} |\langle b(\mathbf{y},t), \mathbf{y}\rangle| < \infty, \tag{B.5}$$

and (iii) there $\exists \rho, \alpha \in (0,1)$ such that for $\forall \mathbf{y} \in \mathbb{R}^n$ with $1 - \rho < \|\mathbf{y}\| < 1$ and for $\forall t \in [0,T]$ we have

$$\sigma^2(\mathbf{y},t) \leq -2\alpha\langle b(\mathbf{y},t), \mathbf{y}\rangle(1 - \|\mathbf{y}\|). \tag{B.6}$$

Then for each $\mathbf{y}_0 \in B$ the SDE

$$dY_t = b(Y_t, t)\,dt + \sigma(Y_t, t)A(Y_t, t)\,dW_t, \qquad Y_{t=0} = \mathbf{y}_0 \tag{B.7}$$

has a unique strong solution $(Y_t)_{t\in[0,T]}$ that remains in B for $\forall t \in [0,T]$ almost surely.

If these conditions hold for $\forall T > 0$, then (B.7) has a unique strong solution $(Y_t)_{t\geq 0}$ that remains in B for $\forall t \geq 0$ almost surely.

Proof. To prove the first part, let $T > 0$ be fixed, and let $\rho \in (0,1)$ and $\alpha \in (0,1)$ be chosen such that (B.6) holds. By decreasing ρ if necessary, let us assume that ρ is so small that $\bar{\alpha} := \alpha/(1 - \frac{1}{2}\rho)$ is still in $(0,1)$. Then since for $\forall \mathbf{y} \in \mathbb{R}^n$ with $1 - \rho < \|\mathbf{y}\| < 1$ we have

$$2(1 - \|\mathbf{y}\|) = \frac{2(1 - \|\mathbf{y}\|^2)}{1 + \|\mathbf{y}\|} = \frac{2(1 - \|\mathbf{y}\|^2)}{2 - (1 - \|\mathbf{y}\|)} < \frac{2(1 - \|\mathbf{y}\|^2)}{2 - \rho} = \frac{1 - \|\mathbf{y}\|^2}{1 - \frac{1}{2}\rho}, \tag{B.8}$$

our condition (B.6) implies that for all such $\mathbf{y}$ and for $\forall t \in [0,T]$ we have

$$\sigma^2(\mathbf{y},t) \leq -\bar{\alpha}\langle b(\mathbf{y},t), \mathbf{y}\rangle(1 - \|\mathbf{y}\|^2). \tag{B.9}$$

Now given any $\varepsilon \in (0, 1 - \|\mathbf{y}_0\|^2)$, let us denote $B_\varepsilon := \{\mathbf{y} \in \mathbb{R}^n \mid \|\mathbf{y}\|^2 < 1 - \varepsilon\}$ (note that $\mathbf{y}_0 \in B_\varepsilon$), and let $b_\varepsilon\colon \mathbb{R}^n \times [0,T] \to \mathbb{R}^n$, $\sigma_\varepsilon\colon \mathbb{R}^n \times [0,T] \to (0,\infty)$, and $A_\varepsilon\colon \mathbb{R}^n \times [0,T] \to \mathbb{R}^{n\times n}$ be functions that coincide with b, σ, and A on $\bar{B}_\varepsilon \times [0,T]$, respectively, and that are bounded and *globally* Lipschitz continuous in $\mathbf{y}$ uniformly in $t \in [0,T]$.[3] Then according to the standard existence and uniqueness result in [32, Chapter 5.2] or [33, Chapter 5, Theorem 2.9], the SDE

$$dY_t^\varepsilon = b_\varepsilon(Y_t^\varepsilon, t)\,dt + \sigma_\varepsilon(Y_t^\varepsilon, t)A_\varepsilon(Y_t^\varepsilon, t)\,dW_t, \qquad Y_{t=0}^\varepsilon = \mathbf{y}_0 \tag{B.10}$$

---

[2] I.e., for $\forall \mathbf{y} \in B$ there $\exists \varepsilon, K > 0$ such that for $\forall \mathbf{y}_1, \mathbf{y}_2 \in B$ with $\|\mathbf{y}_1 - \mathbf{y}\| < \varepsilon$ and $\|\mathbf{y}_2 - \mathbf{y}\| < \varepsilon$ and for $\forall t \in [0,T]$ we have $\|b(\mathbf{y}_1,t) - b(\mathbf{y}_2,t)\| \leq K\|\mathbf{y}_1 - \mathbf{y}_2\|$, $|\sigma(\mathbf{y}_1,t) - \sigma(\mathbf{y}_2,t)| \leq K\|\mathbf{y}_1 - \mathbf{y}_2\|$, and $\|A(\mathbf{y}_1,t) - A(\mathbf{y}_2,t)\| \leq K\|\mathbf{y}_1 - \mathbf{y}_2\|$.

[3] One way of achieving this is to set $b_\varepsilon(\mathbf{y},t) := b\big((1 - \varepsilon)\frac{\mathbf{y}}{\|\mathbf{y}\|}, t\big)$ etc. for $\forall \mathbf{y} \in \mathbb{R}^n \setminus \bar{B}_\varepsilon$ and for $\forall t \in [0,T]$.

has a unique strong solution $(Y_t^\varepsilon)_{t\in[0,T]}$ (with values in all of $\mathbb{R}^n$).

Defining the stopping time $\tau_\varepsilon := \inf\{t\in[0,T]\,|\,Y_t^\varepsilon \notin B_\varepsilon\} \in (0,T]\cup\{\infty\}$, we have $\|Y_{\tau_\varepsilon\wedge T}^\varepsilon\|^2 \le 1-\varepsilon$, with equality holding if and only if $\tau_\varepsilon \le T$. Setting $\nu := \bar\alpha^{-1}-1 > 0$, by Chebyshev's inequality the events $Q_\varepsilon := \{\forall t\in[0,T]\colon Y_t^\varepsilon\in B_\varepsilon\} = \{\tau_\varepsilon=\infty\}$ (which increase with decreasing ε) therefore fulfill

$$\mathbb{P}(Q_\varepsilon^c) = \mathbb{P}(\tau_\varepsilon\le T) = \mathbb{P}\big(\|Y_{\tau_\varepsilon\wedge T}^\varepsilon\|^2 = 1-\varepsilon\big) = \mathbb{P}\big((1-\|Y_{\tau_\varepsilon\wedge T}^\varepsilon\|^2)^{-\nu} = \varepsilon^{-\nu}\big)$$
$$\le \varepsilon^\nu \mathbb{E}\big(1-\|Y_{\tau_\varepsilon\wedge T}^\varepsilon\|^2\big)^{-\nu}.$$

Now suppose that we can find an upper bound on $\mathbb{E}\big(1-\|Y_{\tau_\varepsilon\wedge T}^\varepsilon\|^2\big)^{-\nu}$ that does not depend on ε. Then we can conclude that $\lim_{\varepsilon\to 0}\mathbb{P}(Q_\varepsilon)=1$, and so choosing any sequence $(\varepsilon_k)_{k\in\mathbb{N}}$ that monotonically decreases to 0, the disjoint sets $C_k := Q_{\varepsilon_k}\setminus\bigcup_{i=1}^{k-1}Q_{\varepsilon_i}$ fulfill $\bigcup_{i=1}^k C_k = Q_{\varepsilon_k}$ and thus $\mathbb{P}\big(\bigcup_{i=1}^\infty C_k\big)=\lim_{\varepsilon\to 0}\mathbb{P}(Q_\varepsilon)=1$. This allows us to fully define a process $(Y_t)_{t\in[0,T]}$ as $Y_t(\omega) := Y_t^{\varepsilon_k}(\omega)$ for $\omega\in C_k$ (and arbitrarily on the remaining set $\big(\bigcup_{i=1}^\infty C_k\big)^c$, which is of measure zero).

By definition of the sets $C_k\subset Q_{\varepsilon_k}$ this process only takes values in $\bigcup_{k=1}^\infty B_{\varepsilon_k}=B$ almost surely. Furthermore, since the processes $(Y_t^{\varepsilon_k})_{t\in[0,T]}$ satisfy (B.10) and on C_k take values in B_{ε_k}, where b_{ε_k}, σ_{ε_k}, and A_{ε_k} coincide with b, σ, and A, respectively, we find that on each set C_k we have

$$Y_t = Y_t^{\varepsilon_k} = \mathbf{y}_0 + \int_0^t b_{\varepsilon_k}(Y_s^{\varepsilon_k},s)\,\mathrm{d}s + \int_0^t \sigma_{\varepsilon_k}(Y_s^{\varepsilon_k},s)A_{\varepsilon_k}(Y_s^{\varepsilon_k},s)\,\mathrm{d}W_s$$
$$= \mathbf{y}_0 + \int_0^t b(Y_s^{\varepsilon_k},s)\,\mathrm{d}s + \int_0^t \sigma(Y_s^{\varepsilon_k},s)A(Y_s^{\varepsilon_k},s)\,\mathrm{d}W_s$$
$$= \mathbf{y}_0 + \int_0^t b(Y_s,s)\,\mathrm{d}s + \int_0^t \sigma(Y_s,s)A(Y_s,s)\,\mathrm{d}W_s$$

for $\forall t\in[0,T]$, which shows that $(Y_t)_{t\in[0,T]}$ satisfies (B.7).

To show uniqueness, suppose that there were another solution $(\hat{Y}_t)_{t\in[0,T]}$ with $\mathbb{P}(\hat{Y}_\cdot\neq Y_\cdot)>0$. Then for sufficiently small $\varepsilon>0$ the event

$$G_\varepsilon := \big\{\hat{Y}_\cdot\neq Y_\cdot \text{ and } \forall t\in[0,T]\colon \hat{Y}_t,Y_t\in B_\varepsilon\big\}$$

would have positive probability as well (since $\lim_{\varepsilon\to 0}\mathbb{P}(G_\varepsilon)=\mathbb{P}(\hat{Y}_\cdot\neq Y_\cdot)>0$), and so at least one of the two events $G_\varepsilon^1 := \{Y_\cdot^\varepsilon\neq\hat{Y}_\cdot\}\cap G_\varepsilon$ and $G_\varepsilon^2 := \{Y_\cdot^\varepsilon\neq Y_\cdot\}\cap G_\varepsilon$ would have to have positive probability as well. Depending on which of the two events it is, we could then modify the process $(Y_t^\varepsilon)_{t\in[0,T]}$ to take the values of $(\hat{Y}_t)_{t\in[0,T]}$ on G_ε^1 or of $(Y_t)_{t\in[0,T]}$ on G_ε^2 instead, and obtain a different process that still solves (B.10), contradicting the uniqueness of $(Y_t^\varepsilon)_{t\in[0,T]}$. This shows that our constructed solution $(Y_t)_{t\in[0,T]}$ of (B.7) must be unique, thus concluding our proof of the first part of this lemma.

To complete the proof of that part, it therefore only remains to prove an upper bound on $\mathbb{E}\big(1-\|Y_{\tau_\varepsilon\wedge T}^\varepsilon\|^2\big)^{-\nu}$ that does not depend on ε. To find it, let us abbreviate $D_t^\varepsilon := 1 - \|Y_{\tau_\varepsilon \wedge t}^\varepsilon\|^2$ and $Z_t^\varepsilon := (D_t^\varepsilon)^{-\nu} = \left(1 - \|Y_{\tau_\varepsilon \wedge t}^\varepsilon\|^2\right)^{-\nu}$ for $t \in [0, T]$; in this notation, we need to find a bound on $\mathbb{E}Z_T^\varepsilon$.

Given any $t \in [0, T)$, on $\{\tau_\varepsilon \leq t\}$ we have $\mathrm{d}Z_t^\varepsilon = 0$, while on $\{\tau_\varepsilon > t\}$ by Itô's Lemma we have

$$\mathrm{d}Z_t^\varepsilon = -\nu(D_t^\varepsilon)^{-\nu-1}\langle -2Y_t^\varepsilon, \mathrm{d}Y_t^\varepsilon\rangle$$
$$+ \frac{1}{2}\Big\langle \mathrm{d}Y_t^\varepsilon, \Big[2\nu(D_t^c)^{-\nu-1}\cdot I + \nu(\nu+1)(D_t^\varepsilon)^{-\nu-2}\cdot 4Y_t^\varepsilon \otimes Y_t^\varepsilon\Big]\mathrm{d}Y_t^\varepsilon\Big\rangle$$
$$= 2\nu(D_t^\varepsilon)^{-\nu-1}\langle Y_t^\varepsilon, b_\varepsilon(Y_t^\varepsilon, t)\rangle\,\mathrm{d}t$$
$$+ \sigma_\varepsilon^2(Y_t^\varepsilon, t)\Big\langle \mathrm{d}W_t, A_\varepsilon(Y_t^\varepsilon, t)^T\Big[\nu(D_t^\varepsilon)^{-\nu-1}\cdot I$$
$$+ 2\nu(\nu+1)(D_t^\varepsilon)^{-\nu-2}\cdot Y_t^\varepsilon \otimes Y_t^\varepsilon\Big]A_\varepsilon(Y_t^\varepsilon, t)\,\mathrm{d}W_t\Big\rangle$$
$$+ 2\nu(D_t^\varepsilon)^{-\nu-1}\sigma_\varepsilon(Y_t^\varepsilon, t)\langle Y_t^\varepsilon, A_\varepsilon(Y_t^\varepsilon, t)\,\mathrm{d}W_t\rangle.$$

Applying Lemma 2.1 to the orthogonal matrix $A(Y_t^\varepsilon, t)^T$, the second of these three terms can be written as

$$\sigma_\varepsilon^2(Y_t^\varepsilon, t)\Big\langle \mathrm{d}W_t, \Big[\nu(D_t^\varepsilon)^{-\nu-1}\cdot I$$
$$+ 2\nu(\nu+1)(D_t^\varepsilon)^{-\nu-2}\cdot \left(A_\varepsilon(Y_t^\varepsilon, t)^T Y_t^\varepsilon\right)\otimes \left(A_\varepsilon(Y_t^\varepsilon, t)^T Y_t^\varepsilon\right)\Big]\mathrm{d}W_t\Big\rangle$$
$$= \sigma_\varepsilon^2(Y_t^\varepsilon, t)\cdot \mathrm{tr}[\ldots]\,\mathrm{d}t$$
$$= \sigma_\varepsilon^2(Y_t^\varepsilon, t)\Big[\nu(D_t^\varepsilon)^{-\nu-1}\cdot n + 2\nu(\nu+1)(D_t^\varepsilon)^{-\nu-2}\big\|A_\varepsilon(Y_t^\varepsilon, t)^T Y_t^\varepsilon\big\|^2\Big]\mathrm{d}t$$
$$= \nu Z_t^\varepsilon \sigma_\varepsilon^2(Y_t^\varepsilon, t)(D_t^\varepsilon)^{-1}\big(n + 2(\nu+1)(D_t^\varepsilon)^{-1}\|Y_t^\varepsilon\|^2\big)\,\mathrm{d}t.$$

Since on $\{\tau_\varepsilon > t\}$ we have $Y_t^\varepsilon \in B_\varepsilon$, which allows us to remove the index ε from b_ε, σ_ε, and A_ε, we therefore obtain

$$\mathrm{d}Z_t^\varepsilon = \nu Z_t^\varepsilon\Big[2(D_t^\varepsilon)^{-1}\langle Y_t^\varepsilon, b(Y_t^\varepsilon, t)\rangle$$
$$+ \sigma^2(Y_t^\varepsilon, t)(D_t^\varepsilon)^{-1}\big(n + 2(\nu+1)(D_t^\varepsilon)^{-1}\|Y_t^\varepsilon\|^2\big)\Big]\mathrm{d}t$$
$$+ 2\nu(D_t^\varepsilon)^{-\nu-1}\sigma(Y_t^\varepsilon, t)\big\langle A(Y_t^\varepsilon, t)^T Y_t^\varepsilon, \mathrm{d}W_t\big\rangle$$
$$= \nu Z_t^\varepsilon h(Y_t^\varepsilon, t)\,\mathrm{d}t + 2\nu(D_t^\varepsilon)^{-\nu-1}\sigma(Y_t^\varepsilon, t)\big\langle A(Y_t^\varepsilon, t)^T Y_t^\varepsilon, \mathrm{d}W_t\big\rangle,$$

where we define

$$h(\mathbf{y}, t) := \frac{2\langle \mathbf{y}, b(\mathbf{y}, t)\rangle + n\sigma^2(\mathbf{y}, t)}{1 - \|\mathbf{y}\|^2} + \frac{2(\nu+1)\sigma^2(\mathbf{y}, t)}{1 - \|\mathbf{y}\|^2}\cdot \frac{\|\mathbf{y}\|^2}{1 - \|\mathbf{y}\|^2}$$
$$= \frac{2\langle \mathbf{y}, b(\mathbf{y}, t)\rangle + n\sigma^2(\mathbf{y}, t)}{1 - \|\mathbf{y}\|^2} + \frac{2\bar\alpha^{-1}\sigma^2(\mathbf{y}, t)}{1 - \|\mathbf{y}\|^2}\cdot\left(\frac{1}{1 - \|\mathbf{y}\|^2} - 1\right)$$
$$= \frac{2}{1 - \|\mathbf{y}\|^2}\left(\langle \mathbf{y}, b(\mathbf{y}, t)\rangle + \frac{\bar\alpha^{-1}\sigma^2(\mathbf{y}, t)}{1 - \|\mathbf{y}\|^2}\right) + (n - 2\bar\alpha^{-1})\frac{\sigma^2(\mathbf{y}, t)}{1 - \|\mathbf{y}\|^2}.$$
$$\tag{B.11}$$

This shows that for $\forall t \in [0, T]$ we have

$$Z_t^\varepsilon = (1 - \|\mathbf{y}_0\|^2)^{-\nu} + \nu \int_0^{\tau_\varepsilon \wedge t} Z_s^\varepsilon h(Y_s^\varepsilon, s) \, \mathrm{d}s$$

$$+ 2\nu \int_0^{\tau_\varepsilon \wedge t} (D_s^\varepsilon)^{-\nu-1} \sigma(Y_s^\varepsilon, s) \langle A(Y_s^\varepsilon, s)^T Y_s^\varepsilon, \mathrm{d}W_s \rangle. \quad (B.12)$$

Since by the definition of τ_ε we have $Y_{\tau_\varepsilon \wedge t}^\varepsilon \in \bar{B}_\varepsilon$ and thus $Z_t^\varepsilon \in [1, \varepsilon^{-\nu}]$, and since as a consequence the first integral term in (B.12) is bounded as well (namely by $\nu T \varepsilon^{-\nu} \cdot \max_{\mathbf{y} \in \bar{B}_\varepsilon, \, s \in [0,T]} |h(\mathbf{y}, s)|$), the second integral term in (B.12) must be bounded, too, and so as a function of t it is in fact a martingale (and not just a local martingale). Its expected value must therefore vanish, and so we have

$$\mathbb{E} Z_t^\varepsilon = (1 - \|\mathbf{y}_0\|^2)^{-\nu} + \nu \, \mathbb{E} \int_0^{\tau_\varepsilon \wedge t} Z_s^\varepsilon h(Y_s^\varepsilon, s) \, \mathrm{d}s. \quad (B.13)$$

Now since for $1 - \rho < \|\mathbf{y}\| < 1$ and $t \in [0, T]$ our condition (B.9) ensures that

$$h(\mathbf{y}, t) \leq 0 + |n - 2\bar{\alpha}^{-1}| \cdot \bar{\alpha} \cdot \sup_{\substack{\mathbf{y} \in B \\ s \in [0,T]}} |\langle b(\mathbf{y}, s), \mathbf{y} \rangle| < \infty,$$

and since h is bounded also on the compact set $\{\mathbf{y} \in \mathbb{R}^n \mid \|\mathbf{y}\| \leq 1 - \rho\} \times [0, T]$ simply because of its continuity, we have

$$K := \sup_{\substack{\mathbf{y} \in B \\ s \in [0,T]}} h(\mathbf{y}, s) < \infty.$$

Our integral formula (B.13) therefore leads us to the estimate

$$\mathbb{E} Z_t^\varepsilon \leq (1 - \|\mathbf{y}_0\|^2)^{-\nu} + \nu K \, \mathbb{E} \int_0^{\tau_\varepsilon \wedge t} Z_s^\varepsilon \, \mathrm{d}s$$

$$\leq (1 - \|\mathbf{y}_0\|^2)^{-\nu} + \nu K \, \mathbb{E} \int_0^t Z_s^\varepsilon \, \mathrm{d}s$$

$$= (1 - \|\mathbf{y}_0\|^2)^{-\nu} + \nu K \int_0^t \mathbb{E} Z_s^\varepsilon \, \mathrm{d}s,$$

and so by applying Grönwall's inequality (e.g., [37, Theorem 1.2]) we find that

$$\mathbb{E} Z_t^\varepsilon \leq (1 - \|\mathbf{y}_0\|^2)^{-\nu} \cdot \mathrm{e}^{\nu K t}$$

for $\forall t \in [0, T]$. Setting $t := T$ now yields our desired ε-independent bound, completing the proof of the first part of this lemma.

The second part of this lemma is now an immediate consequence: If the conditions of this lemma hold for $\forall T > 0$, then by what we have proven so far, for each $T > 0$ the SDE (B.7) has a unique solution $(Y_t^T)_{t \in [0,T]}$ on $[0, T]$. The uniqueness then implies that any two processes $(Y_t^{T_1})_{t \in [0,T_1]}$ and $(Y_t^{T_2})_{t \in [0,T_2]}$

in this family of solutions must coincide on their joint domain $[0, T_1 \wedge T_2]$ almost surely. Therefore, if we define the process $(Y_t)_{t \geq 0}$ by setting $Y_{t=0} := \mathbf{y}_0$ and $Y_t := Y_t^{\lceil t \rceil}$ for $\forall t > 0$, we see that for each $T > 0$ it must coincide with $(Y_t^T)_{t \in [0,T]}$ on $[0, T]$ almost surely, which shows that it is a solution of (B.7).

To show uniqueness, suppose that there is another solution $(\tilde{Y}_t)_{t \geq 0}$. Then since its restriction to any interval $[0, T]$ is a solution on $[0, T]$, it must coincide with $(Y_t^T)_{t \in [0,T]}$ and thus with $(Y_t)_{t \geq 0}$ on $[0, T]$ almost surely, and so $(\tilde{Y}_t)_{t \geq 0}$ is equal to $(Y_t)_{t \geq 0}$ almost surely. $\qquad\square$

B.3   Proof Of Lemma 2.4

Proof of Lemma 2.4. We will only prove the first part of this lemma (i.e., the one focussing on a fixed interval $[0, T]$); the second part then follows analogously to our arguments used in the proof of Lemma B.2 above. To do so, let us fix some $T > 0$.

Again denoting the open unit ball in $\mathbb{R}^n$ by $B := \{\mathbf{y} \in \mathbb{R}^n \mid \|\mathbf{y}\| < 1\}$, we begin by defining a bijection $\bar{g} \colon M_0 \to B$. To do so, note that since at the point

$$\mathbf{a} := \mathbf{x}_{\min} - \tfrac{1}{2}\mathbf{v} = \tfrac{n-2}{2}E\mathbf{v} - (r_\infty + \tfrac{1}{2})\mathbf{v} \tag{B.14}$$

we have

$$r(\mathbf{a}) = r_\infty + \langle \mathbf{v}, \mathbf{a} \rangle = r_\infty + \tfrac{n-2}{2}\langle \mathbf{v}, E\mathbf{v} \rangle - (r_\infty + \tfrac{1}{2})\|\mathbf{v}\|^2$$
$$= r_\infty + 0 - (r_\infty + \tfrac{1}{2}) = -\tfrac{1}{2} < 0,$$

i.e., $\mathbf{a} \in \mathbb{R}^n \backslash \overline{M_0}$ by (2.14a)–(2.14b), according to Lemma B.1 (i)–(ii) the function $g(\mathbf{x}) := \frac{\mathbf{x} - \mathbf{a}}{\|\mathbf{x} - \mathbf{a}\|^2}$ maps M_0 bijectively onto the open ball with radius $-\frac{1}{2r(\mathbf{a})} = 1$ and center $-\frac{1}{2r(\mathbf{a})}\mathbf{v} = \mathbf{v}$. The function

$$\bar{g}(\mathbf{x}) := \frac{\mathbf{x} - \mathbf{a}}{\|\mathbf{x} - \mathbf{a}\|^2} - \mathbf{v}$$

therefore maps M_0 bijectively onto B, as desired.

This allows us to prove the first part of the desired statement by using Lemma B.2 to show that the SDE for the image process $Y_t := \bar{g}(X_t)$, obtained via Itô's Lemma, has a unique strong solution that almost surely remains confined to B for $\forall t \in [0, T]$. Indeed, given this solution $(Y_t)_{t \in [0,T]}$ of the image SDE, the process $X_t := \bar{g}^{-1}(Y_t)$ must be a solution of the original ACE SDE (2.10) that remains in $\bar{g}^{-1}(B) = M_0$ almost surely for $\forall t \in [0, T]$; furthermore, the solution of (2.10) must be unique since if there were two solutions then their images under $\bar{g}$ would be two solutions of the image SDE, contradicting Lemma B.2.

This shows that all we need to do is to verify that the drift and the noise function of the SDE for Y_t satisfy the conditions of the first part of Lemma B.2.

We therefore begin by computing the SDE for Y_t. Denoting the components of $\bar{g}$ and Y_t by $\bar{g}_i$ and Y_t^i, respectively, Itô's Lemma tells us that for $\forall i = 1, \ldots, n$ we have

$$\mathrm{d}Y_t^i = \left[\langle \nabla \bar{g}_i(X_t), \mu(X_t) \rangle + \frac{1}{2} \sigma^2(X_t, t) \sum_{j=1}^n \partial_j^2 \bar{g}_i(X_t) \right] \mathrm{d}t + \sigma(X_t, t) \langle \nabla \bar{g}_i(X_t), \mathrm{d}W_t \rangle. \tag{B.15}$$

The necessary derivatives of $\bar{g}$ are given by

$$\nabla \bar{g}(\mathbf{x}) = \frac{1}{\|\mathbf{x} - \mathbf{a}\|^2} \left(I - 2 \frac{(\mathbf{x} - \mathbf{a}) \otimes (\mathbf{x} - \mathbf{a})}{\|\mathbf{x} - \mathbf{a}\|^2} \right), \quad \text{i.e.,} \tag{B.16}$$

$$\partial_j \bar{g}_i(\mathbf{x}) = \frac{\delta_{ij}}{\|\mathbf{x} - \mathbf{a}\|^2} - \frac{2(\mathbf{x} - \mathbf{a})_i (\mathbf{x} - \mathbf{a})_j}{\|\mathbf{x} - \mathbf{a}\|^4},$$

$$\partial_j^2 \bar{g}_i(\mathbf{x}) = -\frac{2(\mathbf{x} - \mathbf{a})_j}{\|\mathbf{x} - \mathbf{a}\|^4} \delta_{ij} + \frac{8(\mathbf{x} - \mathbf{a})_i (\mathbf{x} - \mathbf{a})_j^2}{\|\mathbf{x} - \mathbf{a}\|^6} - \frac{2(\mathbf{x} - \mathbf{a})_i + 2\delta_{ij}(\mathbf{x} - \mathbf{a})_j}{\|\mathbf{x} - \mathbf{a}\|^4}$$

$$= -(2 + 4\delta_{ij}) \frac{(\mathbf{x} - \mathbf{a})_i}{\|\mathbf{x} - \mathbf{a}\|^4} + \frac{8(\mathbf{x} - \mathbf{a})_i (\mathbf{x} - \mathbf{a})_j^2}{\|\mathbf{x} - \mathbf{a}\|^6},$$

$$\sum_{j=1}^n \partial_j^2 \bar{g}_i(\mathbf{x}) = -(2n + 4) \frac{(\mathbf{x} - \mathbf{a})_i}{\|\mathbf{x} - \mathbf{a}\|^4} + 8 \frac{(\mathbf{x} - \mathbf{a})_i}{\|\mathbf{x} - \mathbf{a}\|^4} = (4 - 2n) \frac{(\mathbf{x} - \mathbf{a})_i}{\|\mathbf{x} - \mathbf{a}\|^4}.$$

Plugging them into (B.15) and combining the obtained SDEs for $i = 1, \ldots, n$ into a single vector-valued SDE, we arrive at

$$\mathrm{d}Y_t = \left[\nabla \bar{g}(X_t) \mu(X_t) + \frac{(2 - n)\sigma^2(X_t, t)}{\|X_t - \mathbf{a}\|^4} (X_t - \mathbf{a}) \right] \mathrm{d}t$$

$$+ \frac{\sigma(X_t, t)}{\|X_t - \mathbf{a}\|^2} \left(I - 2 \frac{(X_t - \mathbf{a}) \otimes (X_t - \mathbf{a})}{\|X_t - \mathbf{a}\|^2} \right) \mathrm{d}W_t$$

$$= \left(\tilde{\mu}_1(Y_t) + \tilde{\mu}_2(Y_t, t) \right) \mathrm{d}t + \tilde{\sigma}(Y_t, t) A(Y_t) \mathrm{d}W_t, \tag{B.17}$$

where with the abbreviation

$$\mathbf{y} := \bar{g}(\mathbf{x}) = \frac{\mathbf{x} - \mathbf{a}}{\|\mathbf{x} - \mathbf{a}\|^2} - \mathbf{v}, \quad \text{i.e.,} \quad \mathbf{x} := \bar{g}^{-1}(\mathbf{y}) = \frac{\mathbf{y} + \mathbf{v}}{\|\mathbf{y} + \mathbf{v}\|^2} + \mathbf{a}, \tag{B.18}$$

we define

$$\tilde{\sigma}(\mathbf{y}, t) := \frac{\sigma(\mathbf{x}, t)}{\|\mathbf{x} - \mathbf{a}\|^2}, \tag{B.19a}$$

$$A(\mathbf{y}) := I - 2 \frac{(\mathbf{x} - \mathbf{a}) \otimes (\mathbf{x} - \mathbf{a})}{\|\mathbf{x} - \mathbf{a}\|^2}, \tag{B.19b}$$

$$\tilde{\mu}_1(\mathbf{y}) := \nabla \bar{g}(\mathbf{x}) \mu(\mathbf{x}), \tag{B.19c}$$

$$\tilde{\mu}_2(\mathbf{y}, t) := \frac{(2 - n)\sigma^2(\mathbf{x}, t)}{\|\mathbf{x} - \mathbf{a}\|^4} (\mathbf{x} - \mathbf{a}). \tag{B.19d}$$

Now observe the following: (i) Since $\mathbf{a} \in \mathbb{R}^n \setminus \overline{M}_0$, all these functions are continuous on their respective domains B or $B \times [0, T]$, and in fact locally Lipschitz continuous in $\mathbf{x}$ and thus (because of the boundedness of $\nabla \bar{g}$ in (B.16) on $\overline{M}_0$) also in $\mathbf{y}$, in the case of $\tilde{\sigma}$ and $\tilde{\mu}_2$ uniformly in $t \in [0, T]$. (ii) For $\forall \mathbf{y} \in B$ the matrix $A(\mathbf{y})$ is orthogonal since $A(\mathbf{y}) A(\mathbf{y})^T = A(\mathbf{y}) A(\mathbf{y}) = I$. (iii) The drift $b(\mathbf{y}, t) := \tilde{\mu}_1(\mathbf{y}) + \tilde{\mu}_2(\mathbf{y}, t)$ of the image SDE (B.17) fulfills the boundedness condition (B.5) of Lemma B.2 (in fact, we even have $\sup_{\mathbf{y} \in B, t \in [0, T]} \|b(\mathbf{y}, t)\| < \infty$), since its first component

$$\tilde{\mu}_1 = (\nabla \bar{g} \cdot \mu) \circ \bar{g}^{-1} = (\nabla g \cdot \mu) \circ g^{-1} \circ (\,\cdot + \mathbf{v}) \tag{B.20}$$

has a continuous extension to $\bar{B}$ (in fact, to all of $\mathbb{R}^n$) by Lemma B.1 (iii), and since its second component $\tilde{\mu}_2$ is bounded as a result of our conditions (2.19)–(2.20).

Indeed, to see the boundedess of $\tilde{\mu}_2$, first note that by (B.19d) we have

$$\|\tilde{\mu}_2(\mathbf{y}, t)\| = \frac{(n-2)\sigma^2(\mathbf{x}, t)}{\|\mathbf{x} - \mathbf{a}\|^3} = \frac{(n-2)\sigma^2(\mathbf{x}, t)}{r(\mathbf{x})\|\mathbf{x}\|^2} \cdot \frac{r(\mathbf{x})\|\mathbf{x}\|^2}{\|\mathbf{x} - \mathbf{a}\|^3}, \tag{B.21}$$

so that by (2.19), (2.12), and (2.3) there $\exists R > 0$ such that for $\forall t \in [0, T]$ and $\forall \mathbf{x} \in M_0$ with $\|\mathbf{x}\| > R$ we have $\|\tilde{\mu}_2(\mathbf{x}, t)\| \le 2$. Starting from the first representation in (B.21) again, the condition (2.20) then allows us to obtain a bound also for $\forall \mathbf{x} \in M_0 \cap N$ with $\|\mathbf{x}\| \le R$. On the remaining compact set $\{\mathbf{x} \in M_0 \setminus N \mid \|\mathbf{x}\| \le R\} \times [0, T]$ (note that $M_0 \setminus N = \overline{M}_0 \setminus N$ is closed) $\tilde{\mu}_2$ is bounded simply because of its continuity.

To apply Lemma B.2 it would therefore suffice to show that there $\exists \rho, \alpha \in (0, 1)$ such that for $\forall \mathbf{y} \in \mathbb{R}^n$ with $1 - \rho < \|\mathbf{y}\| < 1$ and for $\forall t \in [0, T]$ we have

$$\tilde{\sigma}^2(\mathbf{y}, t) \le -2\alpha^{1/3} \langle \tilde{\mu}_1(\mathbf{y}) + \tilde{\mu}_2(\mathbf{y}, t), \mathbf{y} \rangle (1 - \|\mathbf{y}\|). \tag{B.22}$$

(We introduce the exponent $\frac{1}{3}$ because we will use the value $\alpha \in (0, 1)$ provided by our condition (2.20) and then verify the condition (B.6) of Lemma B.2 for $\tilde{\alpha} := \alpha^{1/3} \in (0, 1)$ instead.)

To prepare, first observe that our conditions (2.19)–(2.20) imply that there exist $\alpha \in (0, 1)$, $\varepsilon > 0$, an open set $N \supset \partial M_0$, and an $R_0 > 0$, such that for any given $R \ge R_0$ the estimate

$$\sup_{t \in [0, T]} \sigma^2(\mathbf{x}, t) \le 2\alpha r(\mathbf{x}) h_R(\mathbf{x}) \tag{B.23}$$

holds for $\forall \mathbf{x} \in M_0$ with $\mathbf{x} \in N$ or $\|\mathbf{x}\| > R$, where we define

$$h_R(\mathbf{x}) := \begin{cases} \frac{1}{n-2}\|\mathbf{x}\|^2 / \left(1 + 2(n-2)\frac{r(\mathbf{x})}{\|\mathbf{x}\|}\right) & \text{if } \|\mathbf{x}\| > R, \\ \delta + \frac{1}{n-2}\|\mathbf{x} - \mathbf{x}_{\min}\|^2 \\ \quad + \mathbb{1}_{\delta=0, \|\mathbf{x} - \mathbf{x}_{\min}\| < \varepsilon}\, r(\mathbf{x})\left(\frac{2}{n-2}r_\infty - \lambda\right) & \text{if } \|\mathbf{x}\| \le R. \end{cases} \tag{B.24}$$

In order to prove (B.22), we begin by claiming that it suffices to find some $R \geq R_0$ and some open set $\hat{N}$ with $\partial M_0 \subset \hat{N} \subset N$ such that for $\forall t \in [0, T]$ and $\forall \mathbf{x} \in M_0$ with $\mathbf{x} \in \hat{N}$ or $\|\mathbf{x}\| > R$ we have

$$-\|\mathbf{x} - \mathbf{a}\|^2 \langle \tilde{\mu}_1(\mathbf{y}) + \tilde{\mu}_2(\mathbf{y}, t), \mathbf{y} \rangle \geq \alpha^{2/3} h_R(\mathbf{x}), \qquad (\text{B.25})$$

where again we abbreviate $\mathbf{y} := \bar{g}(\mathbf{x})$. Indeed, once we have established the validity of (B.25), we can multiply it by $\alpha^{1/3}\|\mathbf{x} - \mathbf{a}\|^{-2}(1 + \|\mathbf{y}\|)^{-1}$ and by

$$
\begin{aligned}
1 - \|\mathbf{y}\|^2 &= 1 - \left\| \tfrac{\mathbf{x}-\mathbf{a}}{\|\mathbf{x}-\mathbf{a}\|^2} - \mathbf{v} \right\|^2 \\
&= 1 - \left(\|\mathbf{x} - \mathbf{a}\|^{-2} - 2\|\mathbf{x} - \mathbf{a}\|^{-2}\langle \mathbf{x} - \mathbf{a}, \mathbf{v} \rangle + \|\mathbf{v}\|^2 \right) \\
&= \|\mathbf{x} - \mathbf{a}\|^{-2}(-1 + 2\langle \mathbf{x} - \mathbf{a}, \mathbf{v} \rangle) \\
&= 2r(\mathbf{x})\|\mathbf{x} - \mathbf{a}\|^{-2}, \qquad (\text{B.26})
\end{aligned}
$$

where in the last step we used that

$$\langle \mathbf{x}-\mathbf{a}, \mathbf{v} \rangle = \left\langle \mathbf{x} - \tfrac{n-2}{2}E\mathbf{v} + (\tfrac{1}{2}+r_\infty)\mathbf{v}, \mathbf{v} \right\rangle = \langle \mathbf{x}, \mathbf{v} \rangle - 0 + \tfrac{1}{2} + r_\infty = r(\mathbf{x}) + \tfrac{1}{2}, \quad (\text{B.27})$$

to arrive at

$$
\begin{aligned}
-\alpha^{1/3}\langle \tilde{\mu}_1(\mathbf{y}) &+ \tilde{\mu}_2(\mathbf{y}, t), \mathbf{y} \rangle (1 - \|\mathbf{y}\|) \\
&\geq \quad \alpha h_R(\mathbf{x})\|\mathbf{x} - \mathbf{a}\|^{-2}(1 + \|\mathbf{y}\|)^{-1} \cdot 2r(\mathbf{x})\|\mathbf{x} - \mathbf{a}\|^{-2} \\
&\overset{(\text{B.23})}{\geq} \tfrac{1}{2}\sigma^2(\mathbf{x}, t)\|\mathbf{x} - \mathbf{a}\|^{-4} = \tfrac{1}{2}\tilde{\sigma}^2(\mathbf{y}, t),
\end{aligned}
$$

which is (B.22). This relation is shown to hold for $\forall t \in [0, T]$ and for all $\mathbf{y}$ in

$$\left\{ \mathbf{y} \in B \,\middle|\, \bar{g}^{-1}(\mathbf{y}) \in \hat{N} \text{ or } \|g^{-1}(\mathbf{y})\| > R \right\} = B \setminus \bar{g}\left(\left\{ \mathbf{x} \in M_0 \setminus \hat{N} \,\middle|\, \|\mathbf{x}\| \leq R \right\}\right),$$

and since the subtracted set on the right is a compact subset of B (note that $M_0 \setminus \hat{N} = \overline{M_0} \setminus \hat{N}$ is closed), this indeed includes $\forall \mathbf{y} \in \mathbb{R}^n$ with $1 - \rho < \|\mathbf{y}\| < 1$ for some sufficiently small $\rho \in (0, 1)$, as desired.

To now begin our proof of (B.25), we start by finding the right $R \geq R_0$ and focussing on the case $\|\mathbf{x}\| > R$. To do so, first note that by (B.20) and Lemma B.1 (iv) we have

$$\lim_{\mathbf{y} \to -\mathbf{v}} -\langle \tilde{\mu}_1(\mathbf{y}), \mathbf{y} \rangle = \left\langle \left((\nabla g \cdot \mu) \circ g^{-1} \right)(\vec{0}), \mathbf{v} \right\rangle = \left\langle \tfrac{1}{n-2}\mathbf{v}, \mathbf{v} \right\rangle = \tfrac{1}{n-2},$$

and that by (B.19d), (B.23)–(B.24), (2.12) and (2.3) we have

$$
\begin{aligned}
\limsup_{\mathbf{y} \to -\mathbf{v}} \|\tilde{\mu}_2(\mathbf{y}, t)\| \cdot {}&\left(\tfrac{\|\mathbf{x}\|}{r(\mathbf{x})} + 2(n-2) \right)\Big|_{\mathbf{x} = \bar{g}^{-1}(\mathbf{y})} \\
&= (n-2)\limsup_{\mathbf{x} \to \infty} \|\mathbf{x} - \mathbf{a}\|^{-3}\sigma^2(\mathbf{x}, t) \cdot \left(\tfrac{\|\mathbf{x}\|}{r(\mathbf{x})} + 2(n-2) \right) \\
&\leq 2\alpha(n-2)\limsup_{\mathbf{x} \to \infty} \|\mathbf{x} - \mathbf{a}\|^{-3}r(\mathbf{x})h_R(\mathbf{x}) \cdot \tfrac{\|\mathbf{x}\|}{r(\mathbf{x})}\left(1 + 2(n-2)\tfrac{r(\mathbf{x})}{\|\mathbf{x}\|} \right) \\
&\leq 2\alpha \limsup_{\mathbf{x} \to \infty} \|\mathbf{x} - \mathbf{a}\|^{-3}\|\mathbf{x}\|^3 = 2\alpha,
\end{aligned}
$$

where the last limit on the right is uniform in t. Since $\alpha < \alpha^{1/3} < 1$, there therefore $\exists \kappa > 0$ such that for $\forall t \in [0,T]$ and $\forall \mathbf{y} \in B$ with $\|\mathbf{y}+\mathbf{v}\| < \kappa$ we have

$$-\langle \tilde{\mu}_1(\mathbf{y}) + \tilde{\mu}_2(\mathbf{y},t), \mathbf{y} \rangle \geq \frac{\alpha^{1/3}}{n-2} - \frac{2\alpha^{1/3}}{\frac{\|\mathbf{x}\|}{r(\mathbf{x})} + 2(n-2)}$$

$$= \alpha^{1/3} \cdot \frac{\left(\frac{\|\mathbf{x}\|}{r(\mathbf{x})} + 2(n-2)\right) - 2 \cdot (n-2)}{(n-2) \cdot \left(\frac{\|\mathbf{x}\|}{r(\mathbf{x})} + 2(n-2)\right)}$$

$$= \frac{\alpha^{1/3}}{(n-2)\left(1 + 2(n-2)\frac{r(\mathbf{x})}{\|\mathbf{x}\|}\right)} = \frac{\alpha^{1/3} h_R(\mathbf{x})}{\|\mathbf{x}\|^2} > 0.$$

Choosing $R \geq R_0$ so large that for $\forall \mathbf{x} \in M_0$ with $\|\mathbf{x}\| > R$ we have $\frac{\|\mathbf{x}-\mathbf{a}\|}{\|\mathbf{x}\|} \geq \alpha^{1/6}$ and $\|\mathbf{y}+\mathbf{v}\| = \left\|\frac{\mathbf{x}-\mathbf{a}}{\|\mathbf{x}-\mathbf{a}\|^2}\right\| = \|\mathbf{x}-\mathbf{a}\|^{-1} < \kappa$, this shows that for $\forall t \in [0,T]$ and all such $\mathbf{x}$ we have

$$-\|\mathbf{x}-\mathbf{a}\|^2 \langle \tilde{\mu}_1(\mathbf{y}) + \tilde{\mu}_2(\mathbf{y},t), \mathbf{y} \rangle \geq \alpha^{1/3}\|\mathbf{x}\|^2 \cdot \frac{\alpha^{1/3} h_R(\mathbf{x})}{\|\mathbf{x}\|^2} = \alpha^{2/3} h_R(\mathbf{x}),$$

which is (B.25).

It therefore only remains to show (B.25) for $\forall t \in [0,T]$ and $\forall \mathbf{x} \in M_0$ with $\mathbf{x} \in \hat{N}$ and $\|\mathbf{x}\| \leq R$, for some conveniently chosen open set $\hat{N}$ with $\partial M_0 \subset \hat{N} \subset N$. We will choose $\hat{N}$ of the form

$$\hat{N}_\eta := \left\{\mathbf{x} \in \mathbb{R}^n \mid r(\mathbf{x}) \in (-\eta, \eta) \text{ and } \|\mathbf{x}\| < R+1\right\} \cup \left\{\mathbf{x} \in N \mid \|\mathbf{x}\| > R\right\},$$

which is open and fulfills $\hat{N}_\eta \supset \partial M_0$ for $\forall \eta > 0$ (recall (2.14b)), and which for sufficiently small $\eta > 0$ also fulfills $\hat{N}_\eta \subset N$ (since N is an open set that contains ∂M_0). Considering also (2.14a), this means that we need to show (B.25) for $\forall t \in [0,T]$ and $\forall \mathbf{x} \in M_0$ with $r(\mathbf{x}) \in (0,\eta)$ and $\|\mathbf{x}\| \leq R$.

To control $\tilde{\mu}_2$ first, we use (B.19d), the estimates $\|\mathbf{y}\| = \|\bar{g}(\mathbf{x})\| < 1$ and

$$\|\mathbf{x}-\mathbf{a}\| \geq \langle \mathbf{x}-\mathbf{a}, \mathbf{v} \rangle \stackrel{(B.27)}{=} r(\mathbf{x}) + \tfrac{1}{2} > \tfrac{1}{2},$$

and finally (B.23), to see that for $\eta \leq \frac{1}{4(n-2)\alpha}(\alpha^{1/3} - \alpha^{2/3})$ we have the estimate

$$\|\mathbf{x}-\mathbf{a}\|^2 \langle \tilde{\mu}_2(\mathbf{y},t), \mathbf{y} \rangle \leq \|\mathbf{x}-\mathbf{a}\|^2 |\langle \tilde{\mu}_2(\mathbf{y},t), \mathbf{y} \rangle|$$

$$= (n-2)\|\mathbf{x}-\mathbf{a}\|^{-2}\sigma^2(\mathbf{x},t)|\langle \mathbf{x}-\mathbf{a}, \mathbf{y} \rangle|$$

$$\leq (n-2)\|\mathbf{x}-\mathbf{a}\|^{-1}\sigma^2(\mathbf{x},t)$$

$$< 2(n-2)\sigma^2(\mathbf{x},t)$$

$$\leq 2(n-2) \cdot 2\alpha r(\mathbf{x}) h_R(\mathbf{x})$$

$$\leq 2(n-2) \cdot 2\alpha \eta h_R(\mathbf{x})$$

$$\leq (\alpha^{1/3} - \alpha^{2/3}) h_R(\mathbf{x}). \tag{B.28}$$

It therefore only remains that to show that for $\forall t \in [0, T]$ and $\forall \mathbf{x} \in M_0$ with $r(\mathbf{x}) \in (0, \eta)$ and $\|\mathbf{x}\| \leq R$ we also have

$$-\|\mathbf{x} - \mathbf{a}\|^2 \langle \tilde{\mu}_1(\mathbf{y}), \mathbf{y} \rangle \geq \alpha^{1/3} h_R(\mathbf{x}), \tag{B.29}$$

since subtracting (B.28) from (B.29) would then imply (B.25). Note that throughout our proof we may still further decrease η if necessary.

To show (B.29), we begin by evaluating its left-hand side: By (B.19c), (B.16), (B.18), (2.18), and (B.27) we have

$$
\begin{aligned}
-\|\mathbf{x} - \mathbf{a}\|^2 \langle \tilde{\mu}_1(\mathbf{y}), \mathbf{y} \rangle &= -\|\mathbf{x} - \mathbf{a}\|^2 \langle \nabla \bar{g}(\mathbf{x}) \mu(\mathbf{x}), \mathbf{y} \rangle \\
&= -\|\mathbf{x} - \mathbf{a}\|^2 \langle \mu(\mathbf{x}), \nabla \bar{g}(\mathbf{x})^T \mathbf{y} \rangle \\
&= -\langle \mu(\mathbf{x}), \left(I - 2 \tfrac{(\mathbf{x}-\mathbf{a}) \otimes (\mathbf{x}-\mathbf{a})}{\|\mathbf{x}-\mathbf{a}\|^2} \right) \left(\tfrac{\mathbf{x}-\mathbf{a}}{\|\mathbf{x}-\mathbf{a}\|^2} - \mathbf{v} \right) \rangle \\
&= -\langle \mu(\mathbf{x}), -\tfrac{\mathbf{x}-\mathbf{a}}{\|\mathbf{x}-\mathbf{a}\|^2} - \mathbf{v} + \tfrac{2\langle \mathbf{x}-\mathbf{a}, \mathbf{v} \rangle}{\|\mathbf{x}-\mathbf{a}\|^2}(\mathbf{x}-\mathbf{a}) \rangle \\
&= \langle \mu(\mathbf{x}), \mathbf{v} \rangle + \|\mathbf{x}-\mathbf{a}\|^{-2}(1 - 2\langle \mathbf{x}-\mathbf{a}, \mathbf{v} \rangle)\langle \mu(\mathbf{x}), \mathbf{x}-\mathbf{a} \rangle \\
&= \left[\delta + \tfrac{1}{n-2}\|\mathbf{x} - \mathbf{x}_{\min}\|^2 - r(\mathbf{x})(\lambda + \tfrac{2}{n-2}\langle \mathbf{v}, \mathbf{x} \rangle) \right] \\
&\qquad + \|\mathbf{x}-\mathbf{a}\|^{-2}(-2r(\mathbf{x}))\langle \mu(\mathbf{x}), \mathbf{x}-\mathbf{a} \rangle \\
&= \delta + \tfrac{1}{n-2}\|\mathbf{x} - \mathbf{x}_{\min}\|^2 + r(\mathbf{x})u(\mathbf{x}), \tag{B.30}
\end{aligned}
$$

where we abbreviate

$$u(\mathbf{x}) := -\lambda - \tfrac{2}{n-2}\langle \mathbf{v}, \mathbf{x} \rangle - 2\|\mathbf{x}-\mathbf{a}\|^{-2}\langle \mu(\mathbf{x}), \mathbf{x}-\mathbf{a} \rangle. \tag{B.31}$$

The inequality (B.29) is therefore equivalent to

$$\delta + \tfrac{1}{n-2}\|\mathbf{x} - \mathbf{x}_{\min}\|^2 + r(\mathbf{x})u(\mathbf{x}) \geq \alpha^{1/3} h_R(\mathbf{x}),$$

and thus, after plugging in (B.24) and rearranging terms, to

$$
r(\mathbf{x})\left[u(\mathbf{x}) - \mathbb{1}_{\delta=0, \|\mathbf{x}-\mathbf{x}_{\min}\|<\varepsilon}\, \alpha^{1/3}\left(\tfrac{2}{n-2} r_\infty - \lambda \right) \right] \\
+ (1 - \alpha^{1/3})\left(\delta + \tfrac{1}{n-2}\|\mathbf{x} - \mathbf{x}_{\min}\|^2 \right) \geq 0. \tag{B.32}
$$

We need to show this inequality for $\forall t \in [0, T]$ and $\forall \mathbf{x} \in M_0$ with $r(\mathbf{x}) \in (0, \eta)$ and $\|\mathbf{x}\| \leq R$.

To prove it for the case $\delta = 0$ first, observe that in this case by (B.31), (2.17), (B.14), Lemma 2.2 (ii), (2.3), (2.16), and finally (2.9b) we have

$$
\begin{aligned}
u(\mathbf{x}_{\min}) &= -\lambda - \tfrac{2}{n-2}\langle \mathbf{v}, \tfrac{n-2}{2}E\mathbf{v} - r_\infty \mathbf{v} \rangle - 2\|\tfrac{1}{2}\mathbf{v}\|^{-2}\langle \mu(\mathbf{x}_{\min}), \tfrac{1}{2}\mathbf{v} \rangle \\
&= -\lambda - 0 + \tfrac{2}{n-2}r_\infty - 2 \cdot 4 \cdot \tfrac{1}{2}\delta = \tfrac{2}{n-2}r_\infty - \lambda > 0;
\end{aligned}
$$

we can therefore find an $\hat{\varepsilon} \in (0, \varepsilon)$ such that for $\forall \mathbf{x} \in M_0$ with $\|\mathbf{x}-\mathbf{x}_{\min}\| < \hat{\varepsilon}$ we have $u(\mathbf{x}) \geq \alpha^{1/3}\left(\tfrac{2}{n-2}r_\infty - \lambda \right)$, which implies (B.32) for such $\mathbf{x}$. Furthermore, setting

$$L := \max\{ |u(\mathbf{x})| \mid \mathbf{x} \in \overline{M_0},\ r(\mathbf{x}) \in [0, 1],\ \|\mathbf{x}\| \leq R \}$$

and reducing η if necessary so that

$$\eta \leq \min\left\{1, \frac{(1-\alpha^{1/3})\frac{1}{n-2}\hat{\varepsilon}^2}{L + \alpha^{1/3}\left(\frac{2}{n-2}r_\infty - \lambda\right)}\right\},$$

we see that for $\forall \mathbf{x} \in M_0$ with $r(\mathbf{x}) \in (0,\eta)$, $\|\mathbf{x}\| \leq R$, and $\|\mathbf{x} - \mathbf{x}_{\min}\| \geq \hat{\varepsilon}$ we have

$$-r(\mathbf{x})\left[u(\mathbf{x}) - \mathbb{1}_{\delta=0,\,\|\mathbf{x}-\mathbf{x}_{\min}\|<\varepsilon}\,\alpha^{1/3}\left(\tfrac{2}{n-2}r_\infty - \lambda\right)\right]$$

$$\leq -r(\mathbf{x})\left[u(\mathbf{x}) - \alpha^{1/3}\left(\tfrac{2}{n-2}r_\infty - \lambda\right)\right]$$

$$\leq \eta\left(L + \alpha^{1/3}\left(\tfrac{2}{n-2}r_\infty - \lambda\right)\right)$$

$$\leq (1-\alpha^{1/3})\tfrac{1}{n-2}\hat{\varepsilon}^2$$

$$\leq (1-\alpha^{1/3}) \cdot \tfrac{1}{n-2}\|\mathbf{x} - \mathbf{x}_{\min}\|^2,$$

and so (B.32) holds also for those $\mathbf{x}$, completing the case $\delta = 0$.

Finally, if $\delta > 0$ then we can instead reduce η so much that $\eta \leq \min\{1, \frac{1}{L}(1-\alpha^{1/3})\delta\}$, which implies that for $\forall \mathbf{x} \in M_0$ with $r(\mathbf{x}) \in (0,\eta)$ and $\|\mathbf{x}\| \leq R$ we have

$$-r(\mathbf{x})u(\mathbf{x}) \leq \eta L \leq (1-\alpha^{1/3})\delta,$$

which also implies (B.32).

This completes our proof of (B.32), and thus of Lemma 2.4. $\square$

B.4   Proof of Lemma 2.8

Proof. For any given $0 \leq t < T$ and $\mathbf{x} \in M_0$, let $(X_s)_{s\in[t,T)}$ be the process satisfying the SDE (2.10)–(2.11), but starting from $X_t = \mathbf{x}$, and consider the processes $(Y_s)_{s\in[t,T]}$ and $(Z_s)_{s\in[t,T)}$ defined as

$$Y_s := \exp\left(-\int_t^s r(X_\theta)\,\mathrm{d}\theta\right) \qquad \text{and} \qquad Z_s := B_{\$,T}(X_s, s) \cdot Y_s.$$

By Itô's formula the increments of $(Z_s)_{s\in[t,T)}$ are of the form

$$\mathrm{d}Z_s = Y_s \cdot \left[\left(\partial_s + \langle\mu,\nabla\rangle + \tfrac{1}{2}\sigma^2\Delta - r\right)B_{\$,T}\right](X_s,s)\,\mathrm{d}s + (\dots)\,\mathrm{d}W_s,$$

and since by (2.35) the ds-term vanishes, this shows that $(Z_s)_{s\in[t,T)}$ is a local martingale. Since this process is bounded (since $B_{\$,T}$ is bounded and r is bounded below), it is therefore in fact a martingale (see for example [38, Corollary 2.2.6]), so that for $\forall\varepsilon \in (0, T-t)$ we have

$$B_{\$,T}(\mathbf{x},t) = Z_t = \mathbb{E}[Z_{T-\varepsilon}] = \mathbb{E}\left[B_{\$,T}(X_{T-\varepsilon}, T-\varepsilon) \cdot Y_{T-\varepsilon}\right].$$

Letting $\varepsilon \searrow 0$, applying the dominated convergence theorem, and using that $B_{\$,T}$ is continuous and by (2.35) takes the value 1 at $t = T$, we now obtain $B_{\$,T}(\mathbf{x},t) = \mathbb{E}Y_T$.[4] Since in this proof we have assumed that $X_t = \mathbf{x}$, this is (2.25). $\square$

---

[4]Defining Z_s only on $[t,T)$ and then taking the limit (instead of just choosing $\varepsilon = 0$ to begin with) relieves us from having to show that the PDE (2.35) is satisfied also at $t = T$.

B.5   Proof of Lemma 2.12

Proof. The proof will be carried out in two steps: In the first we construct an orthogonal matrix $A_1 \in \mathbb{R}^{n \times n}$ such that $E' := A_1^T E A_1$ has the desired form (2.41), with its entries η_j satisfying the constraint in (2.43), while the form of $\mathbf{v}' := A_1^T \mathbf{v}$ is not yet controlled. In a second step we will then construct another orthogonal matrix $A_2 \in \mathbb{R}^{n \times n}$ such that $E'' := A_2^T E' A_2 = E'$ and such that $\mathbf{v}'' := A_2^T \mathbf{v}'$ has the desired form (2.42), for some nonnegative entries v_j. The statement of this lemma thus holds for $A := (A_1 A_2)^T$.

To begin our construction of A_1, observe that the matrix $F := iE \in \mathbb{C}^{n \times n}$ is Hermitian, i.e., it is equal to its conjugate transpose:

$$F^* = (iE)^* = \bar{i}E^* = \bar{i}E^T = (-i)(-E) = iE = F.$$

By a standard result in linear algebra, this implies that all its eigenvalues $\eta_1, \ldots, \eta_k$ are real, that the associated eigenspaces $V_{\eta_1}, \ldots, V_{\eta_k}$ of F are orthogonal to each other, and that $\mathbb{C}^n = V_{\eta_1} \oplus \cdots \oplus V_{\eta_k}$. One can thus construct an orthogonal basis of $\mathbb{C}^n$ that diagonalizes F, by constructing orthogonal bases for each of the eigenspaces individually and then considering the union of all the k bases. In our specific situation where F is purely imaginary, we can equip these basis vectors with certain additional properties, as follows:

First, if $\eta = 0$ happens to be an eigenvalue of F then we can ensure that our basis of the eigenspace V_0 only consists of vectors in $\mathbb{R}^n$ (as opposed to $\mathbb{C}^n$). Indeed, to do so, we make use of the two observations that

$$
\begin{aligned}
\vec{u} \in V_0 \quad &\Rightarrow \quad F\vec{u} = \vec{0} \\
&\Rightarrow \quad F\Re(\vec{u}) = iE\Re(\vec{u}) = i\Re(E\vec{u}) = i\Im(iE\vec{u}) = i\Im(F\vec{u}) = \vec{0}, \\
& \qquad\quad F\Im(\vec{u}) = iE\Im(\vec{u}) = i\Im(E\vec{u}) = -i\Re(iE\vec{u}) = -i\Re(F\vec{u}) = \vec{0} \\
&\Rightarrow \quad \Re(\vec{u}), \Im(\vec{u}) \in V_0
\end{aligned}
$$

(where $\Re(\vec{z})$ and $\Im(\vec{z})$ denote the real and the imaginary part of a given vector $\vec{z} \in \mathbb{C}^n$, respectively), and that

$$\forall \vec{u} \in \mathbb{C}^n \; \forall \vec{u}' \in \mathbb{R}^n: \quad \vec{u} \perp \vec{u}' \quad \Rightarrow \quad \Re(\vec{u}), \Im(\vec{u}) \perp \vec{u}';$$

in the Gram–Schmidt algorithm these observations allow us to replace each newly constructed basis vector by either its real or its imaginary part (whichever one is non-zero).

Second, note that for any eigenvector $\vec{u}$ of F with eigenvalue η we have

$$F\bar{\vec{u}} = (-\overline{F})\bar{\vec{u}} = -\overline{F\vec{u}} = -\overline{\eta\vec{u}} = -\eta\bar{\vec{u}},$$

i.e., the complex conjugate $\bar{\vec{u}}$ is an eigenvector of F with eigenvalue $-\eta$; in other words, we have $\overline{V_\eta} \subseteq V_{-\eta}$ for every eigenvalue η of F. Replacing η by $-\eta$ and taking complex conjugates on both sides, this implies that $V_{-\eta} \subseteq \overline{V_\eta}$, and thus in fact $V_{-\eta} = \overline{V_\eta}$. As a consequence, given any eigenvalue $\eta > 0$ and any orthonormal basis $\{\vec{u}_1, \ldots, \vec{u}_{n_\eta}\}$ of the associated eigenspace V_η (obtained via the Gram–Schmidt algorithm), we can choose the set $\{\overline{\vec{u}}_1, \ldots, \overline{\vec{u}}_{n_\eta}\}$ as our orthonormal basis of the eigenspace $V_{-\eta}$.

In summary, we can construct an orthonormal basis of $\mathbb{C}^n$ consisting of only (i) pairs $(\vec{u}, \overline{\vec{u}})$ of eigenvectors of F with non-zero real eigenvalues $(\eta, -\eta)$ of opposing signs (with $\eta > 0$, and with the pairs sorted such that the values of η are weakly decreasing), and possibly in addition (ii) vectors in $\mathbb{R}^n$ that are eigenvectors of F with eigenvalue 0.

Once such an orthonormal basis of $\mathbb{C}^n$ is constructed, we can choose the columns of our matrix $A_1 \in \mathbb{R}^{n \times n}$ as follows: First we choose, for every eigenvector pair $(\vec{u}, \overline{\vec{u}})$ with eigenvalues $(\eta, -\eta)$, the vectors given by the (properly normalized) real and imaginary parts of u, i.e.,

$$\vec{w} := \tfrac{1}{\sqrt{2}}(\vec{u} + \overline{\vec{u}}) \qquad \text{and} \qquad \vec{w}' := \tfrac{1}{\sqrt{2}\,i}(\vec{u} - \overline{\vec{u}});$$

then, if 0 is an eigenvalue of F, we also add all of our basis vectors that are eigenvectors with eigenvalue 0 (as is). These are n vectors that are indeed all in $\mathbb{R}^n$, and they are still orthonormal, as one can see from the following calculations, which make repeated use of the fact that $\vec{u} \perp \overline{\vec{u}}$ in our specific case (since $\vec{u}$ and $\overline{\vec{u}}$ are both members of our original orthonormal basis):

$$\langle \vec{w}, \vec{w}' \rangle = \tfrac{1}{2i}\langle \vec{u} + \overline{\vec{u}}, \vec{u} - \overline{\vec{u}} \rangle = \tfrac{1}{2i}\big(\|\vec{u}\|^2 - \|\overline{\vec{u}}\|^2\big) = \tfrac{1}{2i}(1 - 1) = 0,$$
$$\|\vec{w}\|^2 = \tfrac{1}{2}\|\vec{u} + \overline{\vec{u}}\|^2 = \tfrac{1}{2}(\|\vec{u}\|^2 + \|\overline{\vec{u}}\|^2) = \tfrac{1}{2}(1 + 1) = 1,$$
$$\|\vec{w}'\|^2 = \tfrac{1}{2}\|\vec{u} - \overline{\vec{u}}\|^2 = \tfrac{1}{2}(\|\vec{u}\|^2 + \|\overline{\vec{u}}\|^2) = \tfrac{1}{2}(1 + 1) = 1.$$

(The remaining orthonormality requirements follow directly from the orthonormality of our original basis that diagonalizes F.)

Finally, observe that the vectors $\vec{w}$ and $\vec{w}'$ constructed above fulfill

$$E\vec{w} = \tfrac{1}{i}F\vec{w} = \tfrac{1}{\sqrt{2}\,i}(F\vec{u} + F\overline{\vec{u}}) = \tfrac{1}{\sqrt{2}\,i}(\eta\vec{u} - \eta\overline{\vec{u}}) = \eta\vec{w}',$$
$$E\vec{w}' = \tfrac{1}{i}F\vec{w}' = \tfrac{1}{\sqrt{2}\,i^2}(F\vec{u} - F\overline{\vec{u}}) = -\tfrac{1}{\sqrt{2}}(\eta\vec{u} + \eta\overline{\vec{u}}) = -\eta\vec{w},$$

and that every vector $\vec{u}$ with $F\vec{u} = \vec{0}$ fulfills also $E\vec{u} = \tfrac{1}{i}F\vec{u} = \vec{0}$. The matrix $E' = A_1^T E A_1$ thus has a (2×2)-matrix of the form $\left(\begin{smallmatrix} 0 & \eta \\ -\eta & 0 \end{smallmatrix}\right)$ on the diagonal for each pair $(\vec{u}, \overline{\vec{u}})$, with all remaining entries vanishing. Since the statement of our lemma allows the values η_i in (2.41) to vanish, this shows that E' is indeed of the form (2.41), with its entries η_j satisfying the constraint in (2.43).

Moving on to the construction of A_2, let us for $\forall (\alpha, \beta) \in \mathbb{R}^2$ define the matrix $C_{\alpha,\beta} \in \mathbb{R}^{2 \times 2}$ as

$$C_{\alpha,\beta} := \begin{cases} \dfrac{1}{\sqrt{\alpha^2 + \beta^2}} \begin{pmatrix} \alpha & -\beta \\ \beta & \alpha \end{pmatrix} & \text{for } (\alpha, \beta) \neq (0,0), \\[2ex] \begin{pmatrix} 1 & 0 \\ 0 & 1 \end{pmatrix} & \text{for } (\alpha, \beta) = (0,0). \end{cases}$$

Each of these matrices is orthogonal and fulfills

$$
C_{\alpha,\beta}^T \begin{pmatrix} 0 & 1 \\ -1 & 0 \end{pmatrix} C_{\alpha,\beta} = \frac{1}{\alpha^2 + \beta^2} \begin{pmatrix} \alpha & \beta \\ -\beta & \alpha \end{pmatrix} \begin{pmatrix} 0 & 1 \\ -1 & 0 \end{pmatrix} \begin{pmatrix} \alpha & -\beta \\ \beta & \alpha \end{pmatrix}
$$

$$
= \frac{1}{\alpha^2 + \beta^2} \begin{pmatrix} \alpha & \beta \\ -\beta & \alpha \end{pmatrix} \begin{pmatrix} \beta & \alpha \\ -\alpha & \beta \end{pmatrix}
$$

$$
= \begin{pmatrix} 0 & 1 \\ -1 & 0 \end{pmatrix},
$$

$$
C_{\alpha,\beta}^T \begin{pmatrix} \alpha \\ \beta \end{pmatrix} = \frac{1}{\sqrt{\alpha^2 + \beta^2}} \begin{pmatrix} \alpha & \beta \\ -\beta & \alpha \end{pmatrix} \begin{pmatrix} \alpha \\ \beta \end{pmatrix} = \begin{pmatrix} \sqrt{\alpha^2 + \beta^2} \\ 0 \end{pmatrix}
$$

for $(\alpha, \beta) \neq (0,0)$, and in fact for $(\alpha, \beta) = (0,0)$ these relations are obviously fulfilled as well. Denoting $\mathbf{v}' = (v_1, \ldots, v_n)$ and using the modified sign function

$$
\operatorname{sign}'(v) := \begin{cases} 1 & \text{if } v \geq 0, \\ -1 & \text{if } v < 0 \end{cases}
$$

for $\forall v \in \mathbb{R}$, the matrix A_2 defined as

$$
\begin{pmatrix} C_{v_1,v_2} & & \mathbf{0} \\ & \ddots & \\ \mathbf{0} & & C_{v_{n-1},v_n} \end{pmatrix} \quad \text{or} \quad \begin{pmatrix} C_{v_1,v_2} & & & \mathbf{0} \\ & \ddots & & \\ & & C_{v_{n-2},v_{n-1}} & \\ \mathbf{0} & & & \operatorname{sign}'(v_n) \end{pmatrix}
$$

for even and for odd n, respectively, is thus indeed orthogonal and fulfills $E'' = A_2^T E' A_2 = E'$ (recall that E' is of the form (2.41)) as well as

$$
\mathbf{v}'' = A_2^T \mathbf{v}' = \begin{cases} \left(\sqrt{v_1^2 + v_2^2}, 0, \ldots, \sqrt{v_{n-1}^2 + v_n^2}, 0\right) & \text{for even } n, \\ \left(\sqrt{v_1^2 + v_2^2}, 0, \ldots, \sqrt{v_{n-2}^2 + v_{n-1}^2}, 0, |v_n|\right) & \text{for odd } n \end{cases}
$$

(which is of the desired form (2.42), with only non-negative entries). This concludes our construction of A_2 and thus of $A = (A_1 A_2)^T$. $\qquad \square$

B.6 Proof of Lemma 2.15 (ii)

Proof. First let us assume that the conditions (a) and (b) hold for some $T > 0$. Using the map $\mathbf{y} := h(\mathbf{r}) := c(\mathbf{r} - \mathbf{r}_{\min}) - 1$, where $c := 2/(\mathbf{r}_{\max} - \mathbf{r}_{\min})$, to rescale the interval M_0 to $(-1, 1)$, the rescaled process $Y_t := h(R_t)$ follows the SDE

$$
\mathrm{d}Y_t = \tilde{\mu}(Y_t)\, \mathrm{d}t + \tilde{\sigma}(Y_t, t)\, \mathrm{d}W_t, \qquad Y_{t=0} = \mathbf{y}_0, \tag{B.33}
$$

where

$$
\tilde{\mu}(\mathbf{y}) := c\,\mu(h^{-1}(\mathbf{y})), \qquad \tilde{\sigma}(\mathbf{y}, t) := c\,\sigma(h^{-1}(\mathbf{y}), t), \qquad \text{and} \qquad \mathbf{y}_0 := h(\mathbf{r}_0).
$$

In order to apply Lemma B.2 to this process, we need to check its three conditions on these two continuous functions.

First, both $\tilde{\mu}(\mathbf{y})$ and $\tilde{\sigma}(\mathbf{y}, t)$ fulfill the Lipschitz condition (i) of Lemma B.2: $\tilde{\mu}$ by our definition (2.70) of μ, and $\tilde{\sigma}$ by our assumption (a) on σ. Second, since $\tilde{\mu}$ is continuous on $[-1, 1]$, also the boundedness condition (B.5) of Lemma B.2 holds. It therefore only remains to verify the inequality (B.6).

To do so, we begin by multiplying both sides of (2.72) by c to write it as

$$\limsup_{\mathbf{y} \searrow -1} \frac{\sup_{t \in [0,T]} \tilde{\sigma}^2(\mathbf{y}, t)}{2(1 + \mathbf{y})} < \tilde{\mu}(-1).$$

Choosing an $\alpha \in (0, 1)$ so large that this inequality still holds with the right-hand side replaced by $\alpha^2 \tilde{\mu}(-1)$, we therefore find that for $\forall \mathbf{y} \in (-1, 1)$ sufficiently close to -1 we have

$$\sup_{t \in [0,T]} \tilde{\sigma}^2(\mathbf{y}, t) \leq 2\alpha^2 \cdot \tilde{\mu}(-1) \cdot (1 + \mathbf{y}) \leq -2\alpha \cdot (\tilde{\mu}(\mathbf{y}) \cdot \mathbf{y}) \cdot (1 - |\mathbf{y}|),$$

which is (B.6).

In the case $\lambda > \mathbf{r}_{\max} - \mathbf{r}_\infty$ we can analogously derive this inequality for $\forall \mathbf{y} \in (-1, 1)$ sufficiently close to $+1$ from our condition (2.73a). In the case $\lambda = \mathbf{r}_{\max} - \mathbf{r}_\infty$ we can write $\mu(\mathbf{r}) = [\lambda + (\mathbf{r} - \mathbf{r}_{\max})](\mathbf{r} - \mathbf{r}_{\max})$, and so by (2.73b) we can similarly choose an $\alpha \in (0, 1)$ such that for $\forall \mathbf{r} \in M_0$ sufficiently close to $\mathbf{r}_{\max}$ we have

$$\sup_{t \in [0,T]} \sigma^2(\mathbf{r}, t) \leq 2\alpha^3 \lambda (\mathbf{r} - \mathbf{r}_{\max})^2 \leq 2\alpha^2 [\lambda + (\mathbf{r} - \mathbf{r}_{\max})] (\mathbf{r} - \mathbf{r}_{\max})^2$$
$$= 2\alpha^2 \mu(\mathbf{r}) (\mathbf{r} - \mathbf{r}_{\max}) = \tfrac{2}{c} \alpha^2 \mu(\mathbf{r}) (h(\mathbf{r}) - 1);$$

but this again implies that for $\forall \mathbf{y} \in (-1, 1)$ sufficiently close to $+1$ we have

$$\sup_{t \in [0,T]} \tilde{\sigma}^2(\mathbf{y}, t) \leq 2\alpha^2 \tilde{\mu}(\mathbf{y})(\mathbf{y} - 1) \leq -2\alpha \cdot (\tilde{\mu}(\mathbf{y}) \cdot \mathbf{y}) \cdot (1 - |\mathbf{y}|).$$

We have thus shown that in either case the condition (B.6) of Lemma B.2 indeed holds for all $\mathbf{y}$ sufficiently close to the boundary of $(-1, 1)$.

We can therefore invoke Lemma B.2 to conclude that for each starting point $\mathbf{y}_0 \in (-1, 1)$ the SDE (B.33) has a unique strong solution $(Y_t)_{t \in [0,T]}$ that almost surely remains in $(-1, 1)$ for $\forall t \in [0, T]$. This in turn means that for each starting point $\mathbf{r}_0 \in M_0$ the SDE (2.69) has a unique strong solution $(R_t)_{t \in [0,T]}$ that remains in M_0 for $\forall t \in [0, T]$.

If our conditions hold for $\forall T > 0$ then we can proceed analogously to apply the corresponding part of Lemma B.2 and thus obtain the desired statement about the process $(R_t)_{t \geq 0}$ defined for $\forall t \geq 0$. $\square$

B.7 Independent Proofs of (2.74)–(2.76)

Proof. First, we can prove the bond price formula (2.74) by confirming that it indeed solves the Feynman–Kac PDE, with the short rate given by $\mathbf{r}$:

$$\left[\partial_t + \tfrac{1}{2}\sigma^2(\mathbf{r},t)\partial_\mathbf{r}^2 + \mu(\mathbf{r})\partial_\mathbf{r}\right]B_{\$,T}(\mathbf{r},t)$$

$$= \left[\partial_t + \tfrac{1}{2}\sigma^2(\mathbf{r},t)\partial_\mathbf{r}^2 + \left(-\lambda(\mathbf{r}-\mathbf{r}_\infty) + (\mathbf{r}-\mathbf{r}_\infty)^2\right)\partial_\mathbf{r}\right]$$
$$\left[\left(1 - \tfrac{\mathbf{r}-\mathbf{r}_\infty}{\lambda}\right)\mathrm{e}^{-\mathbf{r}_\infty(T-t)} + \tfrac{\mathbf{r}-\mathbf{r}_\infty}{\lambda}\,\mathrm{e}^{-(\mathbf{r}_\infty+\lambda)(T-t)}\right]$$

$$= \left[\mathbf{r}_\infty\left(1 - \tfrac{\mathbf{r}-\mathbf{r}_\infty}{\lambda}\right) + 0 + \left(-\lambda(\mathbf{r}-\mathbf{r}_\infty) + (\mathbf{r}-\mathbf{r}_\infty)^2\right)\cdot\left(-\tfrac{1}{\lambda}\right)\right]\mathrm{e}^{-\mathbf{r}_\infty(T-t)}$$
$$+ \left[(\mathbf{r}_\infty+\lambda)\tfrac{\mathbf{r}-\mathbf{r}_\infty}{\lambda} + 0 + \left(-\lambda(\mathbf{r}-\mathbf{r}_\infty) + (\mathbf{r}-\mathbf{r}_\infty)^2\right)\cdot\tfrac{1}{\lambda}\right]\mathrm{e}^{-(\mathbf{r}_\infty+\lambda)(T-t)}$$

$$= \mathbf{r}\cdot\left(1 - \tfrac{\mathbf{r}-\mathbf{r}_\infty}{\lambda}\right)\mathrm{e}^{-\mathbf{r}_\infty(T-t)} + \mathbf{r}\cdot\tfrac{\mathbf{r}-\mathbf{r}_\infty}{\lambda}\,\mathrm{e}^{-(\mathbf{r}_\infty+\lambda)(T-t)}$$

$$= \mathbf{r}\cdot B_{\$,T}(\mathbf{r},t).$$

The validity of the boundary condition $B_{\$,T}(\mathbf{r},T) = 1$ for $\forall\mathbf{r}\in M_0$ is easy to see, and the uniform boundedness of the functions $B_{\$,T}$ on $M_0\times[0,T]$ simply holds because the expression in (2.74) is continuous on the compact set $\overline{M_0}\times[0,T]$.

The relative forward rate formula (2.75) can now be derived from the bond price formula via a simple differentiation, by combining (2.31) and (2.27) (for any arbitrary $t\geq 0$):

$$F_{\mathrm{rel}}(\mathbf{r};\tau) = F(\mathbf{r},t;t+\tau) = -\partial_\tau\log B_{\$,t+\tau}(\mathbf{r},t)$$

$$= -\frac{\partial_\tau B_{\$,t+\tau}(\mathbf{r},t)}{B_{\$,t+\tau}(\mathbf{r},t)}$$

$$= -\frac{-\mathbf{r}_\infty\left(1 - \tfrac{\mathbf{r}-\mathbf{r}_\infty}{\lambda}\right)\mathrm{e}^{-\mathbf{r}_\infty\tau} - (\mathbf{r}_\infty+\lambda)\tfrac{\mathbf{r}-\mathbf{r}_\infty}{\lambda}\,\mathrm{e}^{-(\mathbf{r}_\infty+\lambda)\tau}}{\left(1 - \tfrac{\mathbf{r}-\mathbf{r}_\infty}{\lambda}\right)\mathrm{e}^{-\mathbf{r}_\infty\tau} + \tfrac{\mathbf{r}-\mathbf{r}_\infty}{\lambda}\,\mathrm{e}^{-(\mathbf{r}_\infty+\lambda)\tau}}$$

$$= \mathbf{r}_\infty + \frac{(\mathbf{r}-\mathbf{r}_\infty)\mathrm{e}^{-(\mathbf{r}_\infty+\lambda)\tau}}{\left(1 - \tfrac{\mathbf{r}-\mathbf{r}_\infty}{\lambda}\right)\mathrm{e}^{-\mathbf{r}_\infty\tau} + \tfrac{\mathbf{r}-\mathbf{r}_\infty}{\lambda}\,\mathrm{e}^{-(\mathbf{r}_\infty+\lambda)\tau}}$$

$$= \mathbf{r}_\infty + \frac{\lambda(\mathbf{r}-\mathbf{r}_\infty)}{(\lambda+\mathbf{r}_\infty-\mathbf{r})\mathrm{e}^{\lambda\tau} + (\mathbf{r}-\mathbf{r}_\infty)}.$$

Finally, the following calculation shows that the maps U_t given by (2.75)–(2.76) indeed satisfy the zero-noise ODE:

$$\partial_t U_t(\mathbf{r}) = \partial_t\left[\mathbf{r}_\infty + \frac{\lambda(\mathbf{r}-\mathbf{r}_\infty)}{(\lambda+\mathbf{r}_\infty-\mathbf{r})\mathrm{e}^{\lambda t} + (\mathbf{r}-\mathbf{r}_\infty)}\right]$$

$$= -\frac{\lambda(\mathbf{r}-\mathbf{r}_\infty)\cdot\lambda(\lambda+\mathbf{r}_\infty-\mathbf{r})\mathrm{e}^{\lambda t}}{[(\lambda+\mathbf{r}_\infty-\mathbf{r})\mathrm{e}^{\lambda t} + (\mathbf{r}-\mathbf{r}_\infty)]^2}$$

$$= \frac{-\lambda^2(\mathbf{r}-\mathbf{r}_\infty)[(\lambda+\mathbf{r}_\infty-\mathbf{r})\mathrm{e}^{\lambda t} + (\mathbf{r}-\mathbf{r}_\infty)] + \lambda^2(\mathbf{r}-\mathbf{r}_\infty)^2}{[(\lambda+\mathbf{r}_\infty-\mathbf{r})\mathrm{e}^{\lambda t} + (\mathbf{r}-\mathbf{r}_\infty)]^2}$$

$$= -\lambda\cdot\frac{\lambda(\mathbf{r}-\mathbf{r}_\infty)}{(\lambda+\mathbf{r}_\infty-\mathbf{r})\mathrm{e}^{\lambda t} + (\mathbf{r}-\mathbf{r}_\infty)} + \left[\frac{\lambda(\mathbf{r}-\mathbf{r}_\infty)}{(\lambda+\mathbf{r}_\infty-\mathbf{r})\mathrm{e}^{\lambda t} + (\mathbf{r}-\mathbf{r}_\infty)}\right]^2$$

$$= -\lambda(U_t(\mathbf{r}) - \mathbf{r}_\infty) + (U_t(\mathbf{r}) - \mathbf{r}_\infty)^2$$
$$= \mu(U_t(\mathbf{r})). \qquad\qquad\qquad \square$$

B.8   Proof of Lemma 2.16 (ii)

Proof. The proof works analogously to the one of Lemma 2.15 (ii) in Appendix B.6. First let us assume that the conditions (2.81) and (2.82a) hold for some $T > 0$. Again using the map $\mathbf{y} := h(\mathbf{r}) := c(\mathbf{r} - \mathbf{r}_{\min}) - 1$, where $c := 2/(\mathbf{r}_{\max} - \mathbf{r}_{\min})$, to rescale the interval M_0 to $(-1, 1)$, the rescaled process $Y_t := h(R_t)$ follows the SDE

$$\mathrm{d}Y_t = \tilde{\mu}(Y_t, t)\,\mathrm{d}t + \tilde{\sigma}(Y_t, t)\,\mathrm{d}W_t, \qquad Y_{t=0} = \mathbf{y}_0,$$

where

$$\tilde{\mu}(\mathbf{y}, t) := c\,\mu(h^{-1}(\mathbf{y}), t), \qquad \tilde{\sigma}(\mathbf{y}, t) := c\,\sigma(h^{-1}(\mathbf{y}), t), \qquad \text{and} \qquad \mathbf{y}_0 := h(\mathbf{r}_0).$$

(Note that $\tilde{\mu}$ is now time-dependent.) In order to apply Lemma B.2 to the rescaled process, we need to check its three conditions on these two continuous functions.

First, both $\tilde{\mu}(\mathbf{y}, t)$ and $\tilde{\sigma}(\mathbf{y}, t)$ fulfill the Lipschitz condition (i) of Lemma B.2: $\tilde{\mu}$ by our definition (2.79) of μ and the continuity of $\theta(t)$ and $\lambda(t)$, and $\tilde{\sigma}$ by our assumption (a) on σ. Second, since $\tilde{\mu}$ is continuous on $[-1, 1] \times [0, T]$, also the boundedness condition (B.5) of Lemma B.2 holds. It therefore only remains to verify the inequality (B.6).

To prepare, first observe that with just a slight improvement of our arguments in our proof of part (i) we find that

$$\mu(\mathbf{r}_{\min}, t) > 0 \qquad \text{and} \qquad \mu(\mathbf{r}, t) < 0$$

for $\forall t \in [0, T]$ and $\forall \mathbf{r} \in \left(\max_{t\in[0,T]} \theta(t), \mathbf{r}_{\max}\right)$, i.e., that

$$\tilde{\mu}(-1, t) > 0 \qquad \text{and} \qquad \tilde{\mu}(\mathbf{y}, t) < 0$$

for $\forall t \in [0, T]$ and $\forall \mathbf{y} < 1$ sufficiently close to 1. Furthermore, since $\frac{\tilde{\mu}(\mathbf{y},t)}{\tilde{\mu}(-1,t)}$ is a polynomial in $\mathbf{y}$ whose coefficients are continuous functions of t, its $\mathbf{y}$-derivatives are continuous in $(\mathbf{y}, t)$, and so we have

$$\lim_{\mathbf{y}\searrow-1} \frac{\tilde{\mu}(\mathbf{y}, t)}{\tilde{\mu}(-1, t)} = 1 \qquad\qquad (\text{B.34})$$

uniformly in $t \in [0, T]$.

Now rewriting the condition (2.81) as

$$\limsup_{\mathbf{y}\searrow-1} \sup_{t\in[0,T]} \left(\frac{\tilde{\sigma}^2(\mathbf{y}, t)}{2(1 + \mathbf{y})} - \tilde{\mu}(-1, t) \right) < 0,$$

and choosing an $\alpha \in (0, 1)$ so large that this inequality holds with the right-hand side replaced by $-(1 - \alpha^2) \max_{t \in [0,T]} \tilde{\mu}(-1, t) < 0$, we see that for $\forall t \in [0, T]$ and for $\forall \mathbf{y} \in (-1, 1)$ sufficiently close to -1 we have

$$\frac{\tilde{\sigma}^2(\mathbf{y}, t)}{2(1 + \mathbf{y})} - \tilde{\mu}(-1, t) \leq -(1 - \alpha^2)\tilde{\mu}(-1, t)$$

and thus by (B.34)

$$\tilde{\sigma}^2(\mathbf{y}, t) \leq 2(1 + \mathbf{y}) \cdot \alpha^2 \tilde{\mu}(-1, t) \leq -2\alpha \cdot (\tilde{\mu}(\mathbf{y}, t) \cdot \mathbf{y}) \cdot (1 - |\mathbf{y}|),$$

which is (B.6).

Similarly, (2.82a) can be rewritten as

$$\limsup_{\mathbf{y} \nearrow 1} \; \sup_{t \in [0,T]} \frac{\tilde{\sigma}^2(\mathbf{y}, t)}{2(1 - \mathbf{y})(-\tilde{\mu}(\mathbf{y}, t))} < 1,$$

and so choosing $\alpha \in (0, 1)$ so large that this inequality holds with the right-hand side replaced by α^2, this means that for $\forall t \in [0, T]$ and for $\forall \mathbf{y} \in (-1, 1)$ sufficiently close to $+1$ we have

$$\tilde{\sigma}^2(\mathbf{y}, t) \leq \alpha^2 \cdot 2(1 - \mathbf{y})(-\tilde{\mu}(\mathbf{y}, t)) \leq -2\alpha \cdot (\tilde{\mu}(\mathbf{y}, t) \cdot \mathbf{y}) \cdot (1 - |\mathbf{y}|)$$

again.

We have thus shown that the condition (B.6) of Lemma B.2 indeed holds for all $\mathbf{y}$ sufficiently close to the boundary of $(-1, 1)$. We can therefore invoke Lemma B.2 and follow the arguments at the end of our proof of Lemma 2.15 (ii) in Appendix B.6 to arrive at the desired conclusion.

It now only remains to show that in the non-degenerate case where $\lambda(t) > \mathbf{r}_{\max} - \theta(t)$ for $\forall t \in [0, T]$ the condition (2.82a) is equivalent to (2.82b). In this case, $\mathbf{r}_{\max}$ cannot coincide with the right root $\theta(t) + \lambda(t)$ of $\mu(\cdot, t)$, and so we must have

$$\mu(\mathbf{r}_{\max}, t) < 0 \qquad \text{for} \qquad \forall t \in [0, T].$$

Denoting the left-hand side of the condition (2.82a) by $\alpha < 1$, this condition implies that

$$\limsup_{\mathbf{r} \nearrow \mathbf{r}_{\max}} \; \sup_{t \in [0,T]} \left(\frac{\sigma^2(\mathbf{r}, t)}{2(\mathbf{r}_{\max} - \mathbf{r})} + \mu(\mathbf{r}_{\max}, t) \right)$$

$$= \limsup_{\mathbf{r} \nearrow \mathbf{r}_{\max}} \; \sup_{t \in [0,T]} \left[\left(\frac{\sigma^2(\mathbf{r}, t)}{2(\mathbf{r}_{\max} - \mathbf{r})(-\mu(\mathbf{r}, t))} + \frac{\mu(\mathbf{r}_{\max}, t)}{-\mu(\mathbf{r}, t)} \right) \cdot (-\mu(\mathbf{r}, t)) \right]$$

$$\leq \left(\limsup_{\mathbf{r} \nearrow \mathbf{r}_{\max}} \; \sup_{t \in [0,T]} \frac{\sigma^2(\mathbf{r}, t)}{2(\mathbf{r}_{\max} - \mathbf{r})(-\mu(\mathbf{r}, t))} + \limsup_{\mathbf{r} \nearrow \mathbf{r}_{\max}} \; \sup_{t \in [0,T]} \frac{\mu(\mathbf{r}_{\max}, t)}{-\mu(\mathbf{r}, t)} \right)$$

$$\times \liminf_{\mathbf{r} \nearrow \mathbf{r}_{\max}} \; \inf_{t \in [0,T]} (-\mu(\mathbf{r}, t))$$

$$= (\alpha - 1) \cdot \min_{t \in [0,T]} (-\mu(\mathbf{r}_{\max}, t)) < 0,$$

i.e., that (2.82b) holds.[5]

Conversely, denoting the left-hand side of the condition (2.82b) by $\tilde{\alpha} < 0$, this condition implies that

$$
\begin{aligned}
\limsup_{\mathbf{r} \nearrow \mathbf{r}_{\max}} & \sup_{t \in [0,T]} \frac{\sigma^2(\mathbf{r}, t)}{2(\mathbf{r}_{\max} - \mathbf{r})(-\mu(\mathbf{r}, t))} \\
&= \limsup_{\mathbf{r} \nearrow \mathbf{r}_{\max}} \sup_{t \in [0,T]} \left[\left(\frac{\sigma^2(\mathbf{r}, t)}{2(\mathbf{r}_{\max} - \mathbf{r})} + \mu(\mathbf{r}_{\max}, t) \right) \cdot \frac{1}{-\mu(\mathbf{r}, t)} + \frac{\mu(\mathbf{r}_{\max}, t)}{\mu(\mathbf{r}, t)} \right] \\
&\leq \left[\limsup_{\mathbf{r} \nearrow \mathbf{r}_{\max}} \sup_{t \in [0,T]} \left(\frac{\sigma^2(\mathbf{r}, t)}{2(\mathbf{r}_{\max} - \mathbf{r})} + \mu(\mathbf{r}_{\max}, t) \right) \times \liminf_{\mathbf{r} \nearrow \mathbf{r}_{\max}} \inf_{t \in [0,T]} \frac{1}{-\mu(\mathbf{r}, t)} \right] \\
&\quad + \limsup_{\mathbf{r} \nearrow \mathbf{r}_{\max}} \sup_{t \in [0,T]} \frac{\mu(\mathbf{r}_{\max}, t)}{\mu(\mathbf{r}, t)} \\
&= \tilde{\alpha} \cdot \min_{t \in [0,T]} \frac{1}{-\mu(\mathbf{r}_{\max}, t)} + 1 < 1,
\end{aligned}
$$

i.e., that (2.82a) holds.

$\square$

B.9   An Independent Proof of (2.85)

Proof. For $\forall \mathbf{r} \in M_0$ and $\forall t \geq 0$ let $s \mapsto U_{t,s}(\mathbf{r})$, $s \geq t$, denote the solution of the ODE

$$
\partial_s U_{t,s}(\mathbf{r}) = \mu(U_{t,s}(\mathbf{r}), s), \qquad U_{t,t}(\mathbf{r}) = \mathbf{r}.
$$

Then for $\forall q \in [t, s)$ we have $U_{t,s}(\mathbf{r}) = U_{q,s}(U_{t,q}(\mathbf{r}))$ and thus

$$
\begin{aligned}
0 &= \partial_q U_{t,s}(\mathbf{r}) \\
&= \partial_q U_{q,s}(U_{t,q}(\mathbf{r})) + \partial_q U_{t,q}(\mathbf{r}) \cdot (\partial_{\mathbf{r}} U_{q,s})(U_{t,q}(\mathbf{r})) \\
&= \partial_q U_{q,s}(U_{t,q}(\mathbf{r})) + \mu(U_{t,q}(\mathbf{r}), q) \cdot (\partial_{\mathbf{r}} U_{q,s})(U_{t,q}(\mathbf{r})),
\end{aligned}
$$

which for $q := t$ reduces to

$$
(\partial_t + \mu(\mathbf{r}, t) \cdot \partial_{\mathbf{r}}) \, U_{t,s}(\mathbf{r}) = 0.
$$

As a result, the function

$$
\tilde{B}_{\$,T}(\mathbf{r}, t) := \exp\left(-\int_t^T U_{t,s}(\mathbf{r}) \, \mathrm{d}s \right) \tag{B.35}
$$

---

[5]Note that as the last step shows, for $\forall t \in [0, T]$ and for $\mathbf{r}$ sufficiently close to $\mathbf{r}_{\max}$, the two factors in the second line are negative and positive, respectively, which explains why we had to use the lim inf and the inf in the fourth line. The evaluations of the second term in the third line and of the term in the fourth line use the fact that near $\mathbf{r} = \mathbf{r}_{\max}$ the respective functions are locally Lipschitz continuous in $\mathbf{r}$ uniformly in $t \in [0, T]$.

fulfills $\tilde{B}_{\$,T}(\mathbf{r}, T) = 1$ and

$$
\begin{aligned}
\left(\partial_t + \mu\partial_\mathbf{r} - \mathbf{r}\right)\tilde{B}_{\$,T}(\mathbf{r}, t) &= \left[-\left(\partial_t + \mu\partial_\mathbf{r}\right)\int_t^T U_{t,s}(\mathbf{r})\,\mathrm{d}s - \mathbf{r}\right]\tilde{B}_{\$,T}(\mathbf{r}, t) \\
&= \left[U_{t,t}(\mathbf{r}) - \int_t^T \left(\partial_t + \mu\partial_\mathbf{r}\right)U_{t,s}(\mathbf{r})\,\mathrm{d}s - \mathbf{r}\right]\tilde{B}_{\$,T}(\mathbf{r}, t) \\
&= \left[\mathbf{r} - 0 - \mathbf{r}\right]\tilde{B}_{\$,T}(\mathbf{r}, t) = 0,
\end{aligned}
$$

i.e., it satisfies the Feynman–Kac PDE for $\sigma = 0$. Since the function $B_{\$,T}(\mathbf{r}, t)$ defined in (2.83)–(2.84) was shown in the proof of Lemma 2.16 to do the same, this implies that $\tilde{B}_{\$,T} = B_{\$,T}$, and thus by comparing (B.35) with the definition of F via (2.26) (with the role of $\mathbf{x}$ now taken over by our new state variable $\mathbf{r}$) that $F(\mathbf{r}, t; s) = U_{t,s}(\mathbf{r})$, which is what we wanted to prove. $\qquad\square$

Appendix C

Proofs for Chapter 3

C.1 Proof of Lemma 3.1

Proof. Comparing (3.4) with (3.5), we see that for the two expressions to coincide, we must define $\mathbb{P}_c$ such that the Radon–Nikodym derivative $\mathcal{E} := \frac{\mathrm{d}\mathbb{P}_c}{\mathrm{d}\mathbb{P}_{c_0}}$ is the process

$$\mathcal{E}_t = \frac{a(X_t, t)}{a(\mathbf{x}_0, 0)} \exp\left(\int_0^t \left(r^c(X_s) - r^{c_0}(X_s) \right) \mathrm{d}s \right). \tag{C.1}$$

This process is positive almost surely, but for it to indeed be a valid Radon–Nikodym derivative that one can use to define a new measure $\mathbb{P}_c$, it must be a martingale under $\mathbb{P}_{c_0}$. To show this, and to obtain a formula for the new SDE for $(X_t)_{t \geq 0}$ under $\mathbb{P}_c$, we will apply Girsanov's Theorem. This in turn requires us to write $\mathcal{E}_t$ as the Doléans-Dade exponential

$$\mathcal{E}_t = \exp\left(Z_t - \tfrac{1}{2} \langle Z \rangle_t \right) \tag{C.2}$$

for some local martingale $(Z_t)_{t \geq 0}$, where $\langle Z \rangle_t$ denotes the quadratic variation of $(Z_t)_{t \geq 0}$ on the interval $[0, t]$.

We claim that our desired process (C.1) is indeed of the form (C.2), for

$$Z_t := \int_0^t \langle Y_s, \mathrm{d}W_s \rangle, \qquad \text{where} \qquad Y_s := (\sigma \cdot \alpha^c)(X_s, s).$$

Note that Z_t is well defined since by Chebycheff's inequality and our assumption (3.10) we have

$$\exp\left(\frac{1}{2} \cdot \mathbb{E}_{c_0} \int_0^t \|Y_s\|^2 \, \mathrm{d}s \right) \leq \mathbb{E}_{c_0}\left[\exp\left(\frac{1}{2} \int_0^t \|Y_s\|^2 \, \mathrm{d}s \right) \right] < \infty$$

and thus $\mathbb{E}_{c_0} \int_0^t \|Y_s\|^2 \, \mathrm{d}s < \infty$.

To see that with this choice of Z_t the right-hand side in (C.2) indeed coincides with the one in (C.1), we abbreviate $a_t := a(X_t, t)$, $\nabla a_t := \nabla a(X_t, t)$, etc., and

289

use Itô's formula and (3.9) to show that

$$
\begin{aligned}
\mathrm{d}\log a_t &= \tfrac{1}{a_t}\mathrm{d}a_t - \tfrac{1}{2a_t^2}\mathrm{d}\langle a\rangle_t \\
&= \tfrac{1}{a_t}\Big(\big(\partial_t + \langle \mu_t, \nabla\rangle + \tfrac{1}{2}\sigma_t^2\Delta\big)a_t\,\mathrm{d}t + \sigma_t\langle \nabla a_t, \mathrm{d}W_t\rangle\Big) - \tfrac{1}{2a_t^2}\sigma_t^2\|\nabla a_t\|^2\,\mathrm{d}t \\
&= \big(r_t^{c_0} - r_t^c - \tfrac{\sigma_t^2}{2a_t^2}\|\nabla a_t\|^2\big)\,\mathrm{d}t + \tfrac{\sigma_t}{a_t}\langle \nabla a_t, \mathrm{d}W_t\rangle \\
&= \big(r_t^{c_0} - r_t^c - \tfrac{1}{2}\|Y_t\|^2\big)\,\mathrm{d}t + \langle Y_t, \mathrm{d}W_t\rangle.
\end{aligned}
$$

Integrating over $[0,t]$, we therefore find that

$$
\begin{aligned}
\log a_t - \log a_0 &= \int_0^t \big(r_s^{c_0} - r_s^c - \tfrac{1}{2}\|Y_s\|^2\big)\,\mathrm{d}s + \int_0^t \langle Y_s, \mathrm{d}W_s\rangle \\
&= \int_0^t \big(r_s^{c_0} - r_s^c\big)\,\mathrm{d}s - \tfrac{1}{2}\langle Z\rangle_t + Z_t\,,
\end{aligned}
$$

and rearranging terms and exponentiating yields

$$
\exp\big(Z_t - \tfrac{1}{2}\langle Z\rangle_t\big) = \frac{a(X_t,t)}{a(\mathbf{x}_0,0)}\exp\left(\int_0^t \big(r^c(X_s) - r^{c_0}(X_s)\big)\,\mathrm{d}s\right),
$$

as desired.

We have therefore managed to write our intended Radon–Nikodym derivative (C.1) in the form (C.2) required by Girsanov's Theorem. Furthermore, since $(Z_t)_{t\geq 0}$ is a local martingale, so is the Doléans-Dade exponential process $(\mathcal{E}_t)_{t\geq 0}$, and since by our assumption (3.10) the Novikoff condition $\mathbb{E}_{c_0}\exp\big(\tfrac{1}{2}\int_0^t \|Y_s\|^2\,\mathrm{d}s\big) < \infty$ is fulfilled, $(\mathcal{E}_t)_{t\geq 0}$ is in fact a proper martingale.

This allows us to apply Girsanov's Theorem and conclude that there exists a new measure $\mathbb{P}_c$ on our probability space such that the Radon–Nikodym derivative $\frac{\mathrm{d}\mathbb{P}_c}{\mathrm{d}\mathbb{P}_{c_0}}$ is given by the process $(\mathcal{E}_t)_{t\geq 0}$, which means that the expressions in (3.4) and (3.5) indeed coincide for any asset A, as intended.

For the second statement of this lemma, note that under this new measure $\mathbb{P}_c$ the process $(W_t)_{t\geq 0}$ is no longer a Brownian motion. However, according to a corollary of Girsanov's theorem, denoting by $\langle W, Z\rangle_t$ the quadratic covariation between the processes $(W_t)_{t\geq 0}$ and $(Z_t)_{t\geq 0}$, the process

$$
\tilde{W}_t^c := W_t - \langle W, Z\rangle_t = W_t - \int_0^t Y_s\,\mathrm{d}s = W_t - \int_0^t (\sigma\cdot\alpha^c)(X_s,s)\,\mathrm{d}s
$$

is a Brownian motion under $\mathbb{P}_c$, and we have

$$
\begin{aligned}
\mathrm{d}X_t &= \mu(X_t)\,\mathrm{d}t + \sigma(X_t,t)\,\mathrm{d}W_t \\
&= \mu(X_t)\,\mathrm{d}t + \sigma(X_t,t)\big[\mathrm{d}\tilde{W}_t + (\sigma\cdot\alpha^c)(X_t,t)\,\mathrm{d}t\big] \\
&= \big[\mu(X_t) + (\sigma^2\cdot\alpha^c)(X_t,t)\big]\,\mathrm{d}t + \sigma(X_t,t)\,\mathrm{d}\tilde{W}_t \\
&= \mu^c(X_t,t)\,\mathrm{d}t + \sigma(X_t,t)\,\mathrm{d}\tilde{W}_t^c\,.
\end{aligned}
$$

This concludes our proof.   □

C.2   Proof of Lemma 3.3

Proof. The SDE for $(X_t)_{t\geq 0}$ under $\mathbb{P}_c$ is given by the SDE (3.11a)–(3.11b) in Lemma 3.1. Applying Itô's formula and abbreviating $a := a^{c/c_0}$, the SDE driving the process $\hat{X}_t^c := G_c(X_t)$ under $\mathbb{P}_c$ is therefore given by

$$
\begin{aligned}
\mathrm{d}(\hat{X}_t^c)^i &= (\nabla G_c^i)\,\mathrm{d}X_t + \tfrac{1}{2}\nabla\nabla G_c^i : \mathrm{d}\langle X\rangle_t \\
&= \left[(\nabla G_c^i)\mu^c + \tfrac{1}{2}\sigma^2\Delta G_c^i\right]\mathrm{d}t + \sigma\nabla G_c^i\,\mathrm{d}\tilde{W}_t^c \\
&= \left[(\nabla G_c^i)\mu + \tfrac{1}{2}\sigma^2\left(\tfrac{2}{a}\nabla G_c^i(\nabla a)^T + \Delta G_c^i\right)\right]\mathrm{d}t + \sigma\nabla G_c^i\,\mathrm{d}\tilde{W}_t^c \\
&= \left[(\nabla G_c^i)\mu + \tfrac{1}{2a}\sigma^2\Delta(a\cdot G_c^i))\right]\mathrm{d}t + \sigma\nabla G_c^i\,\mathrm{d}\tilde{W}_t^c
\end{aligned}
\tag{C.3}
$$

for $\forall i = 1,\dots,n$, where all functions are evaluated at X_t or (X_t,t), the colon denotes summation over all matrix components, $\langle X\rangle_t$ is the (matrix-valued) quadratic covariation, and where in the last step we used the relation

$$
\begin{aligned}
\Delta(a\cdot G_c^i) &= (\Delta a)\cdot G_c^i + 2\nabla G_c^i(\nabla a)^T + a\cdot(\Delta G_c^i) \\
&= 0 + 2\nabla G_c^i(\nabla a)^T + a\cdot(\Delta G_c^i)
\end{aligned}
\tag{C.4}
$$

(obtained by applying the product rule twice and using (3.13)). We can now check off the statements (i)–(iii):

(i) If the noise term in (C.3) can be written as

$$
\sigma(X_t,t)\nabla G_c(X_t)\,\mathrm{d}\tilde{W}_t^c = \hat{\sigma}^c(\hat{X}_t^c,t)\mathrm{d}\hat{W}_t^c,
\tag{C.5}
$$

for some scalar function $\hat{\sigma}^c(\hat{\mathbf{x}},t)$ and some Brownian motion $(\hat{W}_t^c)_{t\geq 0}$, then multiplying the ith with the jth component for all pairs (i,j),[1] we find that we must have

$$
\sigma^2(X_t,t)(\nabla G_c(X_t))^T\nabla G_c(X_t)\,\mathrm{d}t = (\hat{\sigma}^c)^2(\hat{X}_t^c,t)\,I\,\mathrm{d}t.
$$

Since ∇G_c is assumed to be an invertible matrix (since G_c is a diffeomorphism), this shows that wherever σ is non-zero, $\hat{\sigma}^c$ must be non-zero as well, and that (3.21) must hold in these regions of M_0, with

$$
s_c(\mathbf{x}) = \frac{\hat{\sigma}^c(G_c(\mathbf{x}),t)}{\sigma(\mathbf{x},t)}\,.
$$

(In particular, we can conclude that the ratio on the right does not depend on t.) Considering any choice of σ that is positive on all of M_0, we can therefore conclude that (3.21) must hold on all of M_0.

---

[1]Technically more accurately: We integrate both sides of (C.5) from 0 to t, for each side compute the quadratic covariation process of the ith and the jth component of the obtained process, and then differentiate with respect to t.

To see the reverse direction, suppose that (3.21) holds for some function $s_c > 0$, and that we are given any (not necessarily strictly positive) functional parameter $\sigma(\mathbf{x}, t)$. Then if we define $\hat{W}_t^c$ as

$$\hat{W}_t^c := \int_0^t s_c(X_s)^{-1} \nabla G_c(X_s) \, d\tilde{W}_s^c, \qquad (C.6)$$

we have

$$\int_0^t s_c(X_s) \sigma(X_s, s) \, d\hat{W}_s^c = \int_0^t \sigma(X_s, s) \nabla G_c(X_s) \, d\tilde{W}_s^c$$

i.e., (C.5) holds with

$$\hat{\sigma}^c(\hat{\mathbf{x}}, t) := \left[s_c(\mathbf{x}) \cdot \sigma(\mathbf{x}, t) \right]_{\mathbf{x} = G_c^{-1}(\hat{\mathbf{x}})}. \qquad (C.7)$$

Since because of (3.21) the covariation processes of the components of $(\hat{W}_t^c)_{t \geq 0}$ are given by

$$\left\langle \hat{W}^{c,i}, \hat{W}^{c,j} \right\rangle_t = \int_0^t s_c(X_s)^{-2} \left[(\nabla G_c(X_s))^T \nabla G_c(X_s) \right]_{ij} ds = \int_0^t \delta_{ij} \, dt = \delta_{ij} t,$$

Lévy's Theorem [38, Section 3, Exercise 4.1] tells us that $(\hat{W}_t^c)_{t \geq 0}$ is a Brownian motion.

(ii) This part immediately follows from (C.3).

(iii) The formulas (3.23a)–(3.23d) now follow from the arguments above, by combining (C.3), (3.22), (C.5), (C.7), and (C.6). The second statement then follows from the arguments preceding this lemma.   □

C.3   Proof of Lemma 3.4 (iv)

Proof. The smoothness is clear for the case $\mathbf{c} = \vec{0}$ on all of $\mathbb{R}^n$ by (ii.1), and for the case $\mathbf{c} \neq \vec{0}$ on $\mathbb{R}^n \setminus \left\{ \frac{\mathbf{c}}{\|\mathbf{c}\|^2}, \vec{0} \right\}$ by (3.40). The smoothness for the case $\mathbf{c} \neq \vec{0}$ near $\mathbf{x} = \vec{0}$ can be seen from the Taylor expansion

$$H_{\mathbf{c}}(\mathbf{x}) = \frac{\frac{\mathbf{x}}{\|\mathbf{x}\|^2} - \mathbf{c}}{\left\| \frac{\mathbf{x}}{\|\mathbf{x}\|^2} - \mathbf{c} \right\|^2} = \frac{\mathbf{x} - \|\mathbf{x}\|^2 \mathbf{c}}{\left\| \frac{\mathbf{x}}{\|\mathbf{x}\|} - \|\mathbf{x}\| \mathbf{c} \right\|^2} = \frac{\mathbf{x} - \|\mathbf{x}\|^2 \mathbf{c}}{1 - \left(2\langle \mathbf{x}, \mathbf{c} \rangle - \|\mathbf{x}\|^2 \|\mathbf{c}\|^2 \right)}$$

$$= \left(\mathbf{x} - \|\mathbf{x}\|^2 \mathbf{c} \right) \sum_{k=0}^{\infty} \left(2\langle \mathbf{x}, \mathbf{c} \rangle - \|\mathbf{x}\|^2 \|\mathbf{c}\|^2 \right)^k \qquad (C.8)$$

$$= \mathbf{x} + O(\|\mathbf{x}\|^2), \qquad (C.9)$$

which holds in a neighborhood of the origin. (The calculation leading to (C.8) does not work at the point $\mathbf{x} = \vec{0}$ itself, but clearly the formula (C.8) continues to hold when $\mathbf{x} = \vec{0}$ is included.)

To obtain (3.45), let us begin with the case $\mathbf{c} \neq \vec{0}$. The gradient at $\mathbf{x} = \vec{0}$ follows right from (C.9). To compute the gradient at all other points, we can utilize the functions $g_{\mathbf{a}}(\mathbf{x}) := (\mathbf{x}-\mathbf{a})/\|\mathbf{x}-\mathbf{a}\|^2$ that we had defined in Lemma B.1 in Appendix B.1 to write $H_{\mathbf{c}} = g_{\mathbf{c}} \circ g_{\vec{0}}$, and then use that

$$\nabla g_{\mathbf{a}}(\mathbf{x}) = \frac{1}{\|\mathbf{x}-\mathbf{a}\|^2}\left(I - 2\frac{(\mathbf{x}-\mathbf{a}) \otimes (\mathbf{x}-\mathbf{a})}{\|\mathbf{x}-\mathbf{a}\|^2}\right)$$

for $\mathbf{x} \neq \mathbf{a}$ to find that

$$\nabla H_{\mathbf{c}}(\mathbf{x}) = (\nabla g_{\mathbf{c}})(g_{\vec{0}}(\mathbf{x})) \cdot \nabla g_{\vec{0}}(\mathbf{x})$$

$$= \frac{1}{\left\|\frac{\mathbf{x}}{\|\mathbf{x}\|^2} - \mathbf{c}\right\|^2}\left(I - 2\frac{\left(\frac{\mathbf{x}}{\|\mathbf{x}\|^2} - \mathbf{c}\right) \otimes \left(\frac{\mathbf{x}}{\|\mathbf{x}\|^2} - \mathbf{c}\right)}{\left\|\frac{\mathbf{x}}{\|\mathbf{x}\|^2} - \mathbf{c}\right\|^2}\right) \cdot \frac{1}{\|\mathbf{x}\|^2}\left(I - 2\frac{\mathbf{x} \otimes \mathbf{x}}{\|\mathbf{x}\|^2}\right)$$

$$= \frac{1}{\left\|\|\mathbf{c}\|\mathbf{x} - \frac{\mathbf{c}}{\|\mathbf{c}\|}\right\|^2} U_{H_{\mathbf{c}}(\mathbf{x})} U_{\mathbf{x}}$$

for $\mathbf{x} \in \mathbb{R}^n \setminus \left\{\frac{\mathbf{c}}{\|\mathbf{c}\|^2}, \vec{0}\right\}$, where in the last step we used the relation (3.43).

In the case $\mathbf{c} = \vec{0}$ we have $H_{\mathbf{c}}(\mathbf{x}) = \mathbf{x}$ by part (ii.1) and thus $\nabla H_{\mathbf{c}}(\mathbf{x}) = I$ for $\forall \mathbf{x} \in \mathbb{R}^n$, and so it remains to show that the formula in (3.45) indeed reduces to I for $\mathbf{c} = \vec{0}$ and $\mathbf{x} \neq \vec{0}$. This in turn is simple: The prefactor $\left\|\|\vec{0}\|\mathbf{x} - \frac{\vec{0}}{\|\vec{0}\|}\right\|$ equals 1 by our convention (3.44), and we have $U_{H(\mathbf{x})} U_{\mathbf{x}} = I$ because of part (ii.1) and

$$U_{\mathbf{y}} U_{\mathbf{y}} = \left(I - 2\frac{\mathbf{y} \otimes \mathbf{y}}{\|\mathbf{y}\|^2}\right)\left(I - 2\frac{\mathbf{y} \otimes \mathbf{y}}{\|\mathbf{y}\|^2}\right) = I - 4\frac{\mathbf{y} \otimes \mathbf{y}}{\|\mathbf{y}\|^2} + 4\frac{\mathbf{y} \otimes \mathbf{y}}{\|\mathbf{y}\|^2} \cdot \frac{\mathbf{y} \otimes \mathbf{y}}{\|\mathbf{y}\|^2} = I.$$

Finally, the symmetry of $U_{\mathbf{y}}$ is obvious, and the orthogonality follows from the other two properties: $U_{\mathbf{y}}^T U_{\mathbf{y}} = U_{\mathbf{y}} U_{\mathbf{y}} = I$. $\qquad\square$

C.4   Proof of Lemma 3.5

Proof. (i) Each set M^c is open because it is the pre-image of the open half space $\{\mathbf{x} \in \mathbb{R}^n \mid r_\infty^c + \langle \mathbf{v}^c, \mathbf{x}\rangle > 0\}$ under the continuous function $H_{\mathbf{c}^c}$, it is connected because it is the image of that half space under the continuous function $H_{-\mathbf{c}^c}$, and we have $\vec{0} \in M^c$ because

$$r^c(\vec{0}) \stackrel{(3.50)}{=} r_\infty^c + \langle \mathbf{v}^c, G_c(\vec{0})\rangle \stackrel{(3.48)}{=} r_\infty^c + \langle \mathbf{v}^c, H_{\mathbf{c}^c}(\vec{0})\rangle \stackrel{(3.42)}{=} r_\infty^c + \langle \mathbf{v}^c, \vec{0}\rangle \stackrel{(3.27)}{=} r_\infty^c > 0.$$

Finally, as the intersection of finitely many sets M_c, M_0 must have all these properties as well.

(ii) If $\mathbf{c}^c = \vec{0}$ then we have $G_c = id$ by Lemma 3.4 (ii.1), so that the definition (3.50) reduces to $r^c(\mathbf{x}) = r_\infty^c + \langle \mathbf{v}^c, \mathbf{x}\rangle$, which impies that the set M^c defined in (3.52b) is indeed given by the half space (3.54).

Now suppse that $\mathbf{c}^c \neq \vec{0}$. Then for $\forall \mathbf{x} \in \mathbb{R}^n \setminus \left\{\vec{0}, \frac{\mathbf{c}^c}{\|\mathbf{c}^c\|^2}\right\}$ we have $r^c(\mathbf{x}) > 0$ if and only if

$$0 < \|\mathbf{x}\|^2 \left\|\frac{\mathbf{x}}{\|\mathbf{x}\|^2} - \mathbf{c}^c\right\|^2 r^c(\mathbf{x})$$

$$= r_\infty^c \|\mathbf{x}\|^2 \left\|\frac{\mathbf{x}}{\|\mathbf{x}\|^2} - \mathbf{c}^c\right\|^2 + \left\langle \mathbf{v}^c, \mathbf{x} - \mathbf{c}^c\|\mathbf{x}\|^2\right\rangle$$

$$= r_\infty^c \big(1 - 2\langle \mathbf{x}, \mathbf{c}^c\rangle + \|\mathbf{c}^c\|^2\|\mathbf{x}\|^2\big) + \langle \mathbf{v}^c, \mathbf{x}\rangle - \langle \mathbf{v}^c, \mathbf{c}^c\rangle\|\mathbf{x}\|^2$$
$$= \big(r_\infty^c\|\mathbf{c}^c\|^2 - \langle \mathbf{v}^c, \mathbf{c}^c\rangle\big)\|\mathbf{x}\|^2 + \langle \mathbf{v}^c - 2r_\infty^c\mathbf{c}^c, \mathbf{x}\rangle + r_\infty^c \qquad (C.10)$$
$$= r^c(\infty)\|\mathbf{x}\|^2 + \langle \mathbf{v}^c - 2r_\infty^c\mathbf{c}^c, \mathbf{x}\rangle + r_\infty^c. \qquad (C.11)$$

Since for $\mathbf{x} = \vec{0}$ this statement holds true as well (because $\vec{0} \in M^c$ by part (i) and because $r_\infty^c > 0$ by (3.27)), and since the function r^c is not defined at the point $\frac{\mathbf{c}^c}{\|\mathbf{c}^c\|^2}$, the set M^c is therefore given by all points $\mathbf{x} \in \mathbb{R}^n$ for which (C.11) is positive.

This shows that if $r_\infty^c(\infty) = 0$ then M^c is indeed given by (3.54), which is a half space because $\mathbf{v}^c - 2r_\infty^c\mathbf{c}^c \neq \vec{0}$ (indeed, if we had $\mathbf{v}^c = 2r_\infty^c\mathbf{c}^c$ then (3.53) would yield $r^c(\infty) = -r_\infty^c < 0$).

(iii) In all other $c \in C$, the prefactor of $\|\mathbf{x}\|^2$ in (C.10) is non-zero, which allows us to complete the square in (C.10) and show that M^c consists of all those points for which

$$0 < \big(r_\infty^c\|\mathbf{c}^c\|^2 - \langle \mathbf{v}^c, \mathbf{c}^c\rangle\big)\Bigg[\bigg\|\mathbf{x} + \frac{\mathbf{v}^c - 2r_\infty^c\mathbf{c}^c}{2\big(r_\infty^c\|\mathbf{c}^c\|^2 - \langle \mathbf{v}^c, \mathbf{c}^c\rangle\big)}\bigg\|^2$$
$$+ \frac{4r_\infty^c\big(r_\infty^c\|\mathbf{c}^c\|^2 - \langle \mathbf{v}^c, \mathbf{c}^c\rangle\big) - \|\mathbf{v}^c - 2r_\infty^c\mathbf{c}^c\|^2}{4\big(r_\infty^c\|\mathbf{c}^c\|^2 - \langle \mathbf{v}^c, \mathbf{c}^c\rangle\big)^2}\Bigg]$$
$$= \|\mathbf{c}^c\|^2\,r^c(\infty)\Bigg[\bigg\|\mathbf{x} - \frac{r_\infty^c\mathbf{c}^c - \frac{1}{2}\mathbf{v}^c}{\|\mathbf{c}^c\|^2\,r^c(\infty)}\bigg\|^2 - \bigg(\frac{\frac{1}{2}\|\mathbf{v}^c\|}{\|\mathbf{c}^c\|^2\,r^c(\infty)}\bigg)^2\Bigg],$$

which leads to the statement in (iii).

(iv) To see that $\frac{\mathbf{c}^c}{\|\mathbf{c}^c\|^2} \in \partial M^c$, first note that this certainly holds if $\mathbf{c}^c = \vec{0}$, since in the case of part (ii) it is understood that $\infty \in \partial M_0$. In the other case covered by part (ii), namely if $r^c(\infty) = 0$, this holds because

$$\big\langle \mathbf{v}^c - 2r_\infty^c\mathbf{c}^c, \tfrac{\mathbf{c}^c}{\|\mathbf{c}^c\|^2}\big\rangle + r_\infty^c = \big\langle \mathbf{v}^c, \tfrac{\mathbf{c}^c}{\|\mathbf{c}^c\|^2}\big\rangle - r_\infty^c = -r^c(\infty) = 0.$$

In the cases of part (iii), we need to compute the distance of $\frac{\mathbf{c}^c}{\|\mathbf{c}^c\|^2}$ to the center of the ball M_0 given in (3.55):

$$\bigg\|\frac{\mathbf{c}^c}{\|\mathbf{c}^c\|^2} - \frac{r_\infty^c\mathbf{c}^c - \frac{1}{2}\mathbf{v}^c}{\|\mathbf{c}^c\|^2\,r^c(\infty)}\bigg\| = \bigg\|\frac{(r^c(\infty) - r_\infty^c)\mathbf{c}^c + \frac{1}{2}\mathbf{v}^c}{\|\mathbf{c}^c\|^2\,r^c(\infty)}\bigg\| = \bigg\|\frac{\langle \mathbf{v}^c, \frac{\mathbf{c}^c}{\|\mathbf{c}^c\|^2}\rangle\mathbf{c}^c + \frac{1}{2}\mathbf{v}^c}{\|\mathbf{c}^c\|^2\,r^c(\infty)}\bigg\|$$
$$= \bigg\|\frac{\frac{1}{2}\big(I - 2\frac{\mathbf{c}^c \otimes \mathbf{c}^c}{\|\mathbf{c}^c\|^2}\big)\mathbf{v}^c}{\|\mathbf{c}^c\|^2\,r^c(\infty)}\bigg\|.$$

Since the matrix $I - 2\frac{\mathbf{c}^c \otimes \mathbf{c}^c}{\|\mathbf{c}^c\|^2}$ is orthogonal (in fact, it is symmetric, and applying it twice yields the identity), this distance is equal to the ball's radius, concluding our proof that $\frac{\mathbf{c}^c}{\|\mathbf{c}^c\|^2} \in \partial M^c$. In particular, since by part (i) the set M^c is open, this implies that $\frac{\mathbf{c}^c}{\|\mathbf{c}^c\|^2} \notin M^c$ and thus $\frac{\mathbf{c}^c}{\|\mathbf{c}^c\|^2} \notin M_0$ by (3.52b), and so by (3.48) and Lemma 3.4 (i) we have

$$G_c(\mathbf{x}) = H_{\mathbf{c}^c}(\mathbf{x}) \neq \infty.$$

To see the second statement of (iv), suppose we are given any (w.l.o.g. connected[2]) neighborhood of $\frac{\mathbf{c}^c}{\|\mathbf{c}^c\|^2}$. Then for sufficiently small (positive or negative) values $\eta \neq 0$, the points $\mathbf{x}_\eta := \frac{\mathbf{c}^c + \eta \mathbf{v}^c}{\|\mathbf{c}^c + \eta \mathbf{v}^c\|^2}$ are in the given neighborhood (note that this holds no matter whether $\mathbf{c}^c = \vec{0}$ or $\mathbf{c}^c \neq \vec{0}$). These points fulfill $\frac{\mathbf{x}_\eta}{\|\mathbf{x}_\eta\|^2} = \mathbf{c}^c + \eta \mathbf{v}^c$ and thus

$$G_c(\mathbf{x}_\eta) = \frac{(\mathbf{c}^c + \eta \mathbf{v}^c) - \mathbf{c}^c}{\|(\mathbf{c}^c + \eta \mathbf{v}^c) - \mathbf{c}^c\|^2} = \frac{\mathbf{v}^c}{\eta \|\mathbf{v}^c\|^2},$$

and therefore $r^c(\mathbf{x}_\eta) = r^c_\infty + \langle \mathbf{v}^c, G_c(\mathbf{x}_\eta) \rangle = r^c_\infty + \frac{1}{\eta}$. This shows that r^c takes every sufficiently large (positive or negative) real value in the given neighborhood, and since this neighborhood, when punctured at $\frac{\mathbf{c}^c}{\|\mathbf{c}^c\|^2}$, is a connected set on which r^c is continuous, r^c must therefore in fact take *all* real values on it.

The last statement of part (iv) follows simply from the continuity of r^c on $\bar{\mathbb{R}}^n \setminus \left\{ \frac{\mathbf{c}^c}{\|\mathbf{c}^c\|^2} \right\}$. □

C.5   Proof of Lemma 3.6

Proof of (3.12). By (3.51), (3.49), and (3.44) we have

$$a^{c/c_0}(\mathbf{x}, t) = \frac{a^c}{a^{c_0}} \cdot e^{(r^{c_0}_\infty - r^c_\infty)t} \cdot \frac{1}{\left\| \|\mathbf{c}^c\| \mathbf{x} - \frac{\mathbf{c}^c}{\|\mathbf{c}^c\|} \right\|^{n-2}} \cdot \tag{C.12}$$

Checking the condition (3.12), i.e.,

$$\left(\partial_t + \langle \mu(\mathbf{x}), \nabla \rangle \right) a^{c/c_0}(\mathbf{x}, t) = (r^{c_0}(\mathbf{x}) - r^c(\mathbf{x})) a^{c/c_0}(\mathbf{x}, t), \tag{C.13}$$

requires the computation of two derivatives: The first one yields

$$\partial_t a^{c/c_0}(\mathbf{x}, t) = (r^{c_0}_\infty - r^c_\infty) a^{c/c_0}(\mathbf{x}, t), \tag{C.14}$$

giving us the constant terms of the two rate functions on the right of (C.13). The computation of the second derivative in (C.13) boils down to understanding the expression

$$\left\langle \mu(\mathbf{x}), \nabla \left\| \|\mathbf{c}^c\| \mathbf{x} - \frac{\mathbf{c}^c}{\|\mathbf{c}^c\|} \right\|^{-(n-2)} \right\rangle$$

$$= -(n-2) \left\| \|\mathbf{c}^c\| \mathbf{x} - \frac{\mathbf{c}^c}{\|\mathbf{c}^c\|} \right\|^{-n} \|\mathbf{c}^c\| \left\langle \mu(\mathbf{x}), \|\mathbf{c}^c\| \mathbf{x} - \frac{\mathbf{c}^c}{\|\mathbf{c}^c\|} \right\rangle.$$

---

[2]If the given neighborhood is not connected, we can always consider an open connected subset containing $\frac{\mathbf{c}^c}{\|\mathbf{c}^c\|^2}$ instead.

The last two factors of the right-hand side can be written as

$$\|\mathbf{c}^c\|\big\langle \mu(\mathbf{x}), \|\mathbf{c}^c\|\mathbf{x} - \tfrac{\mathbf{c}^c}{\|\mathbf{c}^c\|}\big\rangle$$

$$= \big\langle (-\lambda I + E)\mathbf{x} + \tfrac{1}{n-2}\|\mathbf{x}\|^2\mathbf{v}^{c_0} - \tfrac{2}{n-2}\langle\mathbf{v}^{c_0},\mathbf{x}\rangle\mathbf{x}, \|\mathbf{c}^c\|^2\mathbf{x} - \mathbf{c}^c\big\rangle$$

$$= -\lambda\|\mathbf{c}^c\|^2\|\mathbf{x}\|^2 + \big\langle(\lambda I + E)\mathbf{c}^c,\mathbf{x}\big\rangle + \tfrac{1}{n-2}\|\mathbf{c}^c\|^2\|\mathbf{x}\|^2\langle\mathbf{v}^{c_0},\mathbf{x}\rangle$$

$$\quad - \tfrac{1}{n-2}\|\mathbf{x}\|^2\langle\mathbf{c}^c,\mathbf{v}^{c_0}\rangle - \tfrac{2}{n-2}\|\mathbf{c}^c\|^2\|\mathbf{x}\|^2\langle\mathbf{v}^{c_0},\mathbf{x}\rangle + \tfrac{2}{n-2}\langle\mathbf{v}^{c_0},\mathbf{x}\rangle\langle\mathbf{c}^c,\mathbf{x}\rangle$$

$$= \big\langle\tfrac{1}{n-2}\mathbf{v}^{c_0} + (\lambda I + E)\mathbf{c}^c, \mathbf{x} - \|\mathbf{x}\|^2\mathbf{c}^c\big\rangle$$

$$\qquad\qquad\qquad - \tfrac{1}{n-2}\langle\mathbf{v}^{c_0},\mathbf{x}\rangle\big(\|\mathbf{c}^c\|^2\|\mathbf{x}\|^2 - 2\langle\mathbf{c}^c,\mathbf{x}\rangle + 1\big)$$

$$= \big\langle\tfrac{1}{n-2}\mathbf{v}^c, \mathbf{x} - \|\mathbf{x}\|^2\mathbf{c}^c\big\rangle - \tfrac{1}{n-2}\langle\mathbf{v}^{c_0},\mathbf{x}\rangle\big\|\|\mathbf{c}^c\|\mathbf{x} - \tfrac{\mathbf{c}^c}{\|\mathbf{c}^c\|}\big\|^2,$$

where we repeatedly used the antisymmetry of E, and where in the last step we used our definition (3.47) of $\mathbf{c}^c$ to modify the first expression. Plugging this into the preceding relation, we therefore find that

$$\Big\langle \mu(\mathbf{x}), \nabla\big\|\|\mathbf{c}^c\|\mathbf{x} - \tfrac{\mathbf{c}^c}{\|\mathbf{c}^c\|}\big\|^{-(n-2)}\Big\rangle$$

$$= -\big\|\|\mathbf{c}^c\|\mathbf{x} - \tfrac{\mathbf{c}^c}{\|\mathbf{c}^c\|}\big\|^{-(n-2)}\bigg(\Big\langle\mathbf{v}^c, \tfrac{\mathbf{x} - \|\mathbf{x}\|^2\mathbf{c}^c}{\big\|\|\mathbf{c}^c\|\mathbf{x} - \tfrac{\mathbf{c}^c}{\|\mathbf{c}^c\|}\big\|^2}\Big\rangle - \langle\mathbf{v}^{c_0},\mathbf{x}\rangle\bigg)$$

$$\overset{(3.43)}{=} -\big\|\|\mathbf{c}^c\|\mathbf{x} - \tfrac{\mathbf{c}^c}{\|\mathbf{c}^c\|}\big\|^{-(n-2)}\bigg(\Big\langle\mathbf{v}^c, \tfrac{\tfrac{\mathbf{x}}{\|\mathbf{x}\|^2} - \mathbf{c}^c}{\big\|\tfrac{\mathbf{x}}{\|\mathbf{x}\|^2} - \mathbf{c}^c\big\|^2}\Big\rangle - \langle\mathbf{v}^{c_0},\mathbf{x}\rangle\bigg)$$

$$\overset{(3.40)}{=} -\big\|\|\mathbf{c}^c\|\mathbf{x} - \tfrac{\mathbf{c}^c}{\|\mathbf{c}^c\|}\big\|^{-(n-2)}\Big(\langle\mathbf{v}^c, H_{\mathbf{c}^c}(\mathbf{x})\rangle - \langle\mathbf{v}^{c_0},\mathbf{x}\rangle\Big)$$

$$\overset{(3.48),(3.49)}{=} -\big\|\|\mathbf{c}^c\|\mathbf{x} - \tfrac{\mathbf{c}^c}{\|\mathbf{c}^c\|}\big\|^{-(n-2)}\Big(\langle\mathbf{v}^c, G_c(\mathbf{x})\rangle - \langle\mathbf{v}^{c_0}, G_{c_0}(\mathbf{x})\rangle\Big)$$

$$\overset{(3.50)}{=} -\big\|\|\mathbf{c}^c\|\mathbf{x} - \tfrac{\mathbf{c}^c}{\|\mathbf{c}^c\|}\big\|^{-(n-2)}\big((r^c(\mathbf{x}) - r^c_\infty) - (r^{c_0}(\mathbf{x}) - r^{c_0}_\infty)\big).$$

Multiplying both sides with $\tfrac{a^c}{a^{c_0}}\cdot e^{(r^{c_0}_\infty - r^c_\infty)t}$ and comparing the result with (C.12), this shows that

$$\big\langle\mu(\mathbf{x}), \nabla a^{c/c_0}(\mathbf{x},t)\big\rangle = a^{c/c_0}(\mathbf{x},t)\Big[\big(r^{c_0}(\mathbf{x}) - r^c(\mathbf{x})\big) - \big(r^{c_0}_\infty - r^c_\infty\big)\Big],$$

and adding (C.14) finally yields the desired relation (C.13), i.e., (3.12).   □

Proof of (3.13). By (C.12), the condition (3.13) asks that

$$\Delta\left[\frac{1}{\big\|\|\mathbf{c}\|\mathbf{x} - \tfrac{\mathbf{c}}{\|\mathbf{c}\|}\big\|^{n-2}}\right] = 0 \tag{C.15}$$

for $\mathbf{c} = \mathbf{c}^c$. Since the function in the square brackets is just a shifted and rescaled version of $\|\mathbf{x}\|^{2-n}$, the validity of (C.15) for $\forall\mathbf{c}\in\mathbb{R}^n$ follows from the fact that $\Delta\|\mathbf{x}\|^{2-n} = 0$, which we have already shown in our proof of Lemma 2.10.   □

Proof of (3.17). The validity of the triangle relation (3.17) follows right from our definition of the exchange rate functions (3.51). □

Proof of (3.21). To check the validity of the conformality condition (3.21), first note that in the case $\mathbf{c}^c \neq \vec{0}$ Lemma 3.4 (iv) implies that for $\forall \mathbf{x} \in \mathbb{R}^n \setminus \left\{ \frac{\mathbf{c}^c}{\|\mathbf{c}^c\|^2}, \vec{0} \right\}$ we have

$$(\nabla G_c(\mathbf{x}))(\nabla G_c(\mathbf{x}))^T = \frac{1}{\left\| \|\mathbf{c}^c\| \mathbf{x} - \frac{\mathbf{c}^c}{\|\mathbf{c}^c\|} \right\|^4} \cdot U_{G_c(\mathbf{x})} U_{\mathbf{x}} \cdot U_{\mathbf{x}}^T U_{G_c(\mathbf{x})}^T$$

$$= \frac{1}{\left\| \|\mathbf{c}^c\| \mathbf{x} - \frac{\mathbf{c}^c}{\|\mathbf{c}^c\|} \right\|^4} \cdot I, \tag{C.16}$$

and that for $\mathbf{x} = \vec{0}$ (C.16) holds as well because of the second case in (3.45). For $\mathbf{c}^c = \vec{0}$ we have $G_c = id$ by Lemma 3.4 (ii.1) and thus $\nabla G_c(\mathbf{x}) = I$, and so by our convention (3.44) the relation (C.16) holds in this case as well. This shows that (3.21) is indeed fulfilled for any $\mathbf{c}^c \in \mathbb{R}^n$, with

$$s_c(\mathbf{x}) = \left\| \|\mathbf{c}^c\| \mathbf{x} - \frac{\mathbf{c}^c}{\|\mathbf{c}^c\|} \right\|^{-2}. \tag{C.17}$$

□

Proof of (3.22). To see (3.22), we compute the gradient with respect to $\mathbf{c}$ of both sides of the relation (C.15):

$$\vec{0} = \Delta \nabla_{\mathbf{c}} \left[\frac{1}{\left\| \|\mathbf{c}\| \mathbf{x} - \frac{\mathbf{c}}{\|\mathbf{c}\|} \right\|^{n-2}} \right]$$

$$= -(n-2) \cdot \Delta \left[\frac{\left(\|\mathbf{c}\| \mathbf{x} - \frac{\mathbf{c}}{\|\mathbf{c}\|} \right)^T \nabla_{\mathbf{c}} \left(\|\mathbf{c}\| \mathbf{x} - \frac{\mathbf{c}}{\|\mathbf{c}\|} \right)}{\left\| \|\mathbf{c}\| \mathbf{x} - \frac{\mathbf{c}}{\|\mathbf{c}\|} \right\|^n} \right]$$

$$= -(n-2) \cdot \Delta \left[\frac{\left(\|\mathbf{c}\| \mathbf{x} - \frac{\mathbf{c}}{\|\mathbf{c}\|} \right)^T \left[\mathbf{x} \otimes \frac{\mathbf{c}}{\|\mathbf{c}\|} - \frac{1}{\|\mathbf{c}\|} \left(I - \frac{\mathbf{c}}{\|\mathbf{c}\|} \otimes \frac{\mathbf{c}}{\|\mathbf{c}\|} \right) \right]}{\left\| \|\mathbf{c}\| \mathbf{x} - \frac{\mathbf{c}}{\|\mathbf{c}\|} \right\|^n} \right]$$

$$= -(n-2) \cdot \Delta \left[\frac{\|\mathbf{x}\|^2 \mathbf{c} - \mathbf{x} + \left\langle \mathbf{x}, \frac{\mathbf{c}}{\|\mathbf{c}\|^2} \right\rangle \mathbf{c} - \left\langle \frac{\mathbf{c}}{\|\mathbf{c}\|^2}, \mathbf{x} \right\rangle \mathbf{c} + \vec{0}}{\left\| \|\mathbf{c}\| \mathbf{x} - \frac{\mathbf{c}}{\|\mathbf{c}\|} \right\|^n} \right]$$

$$= -(n-2) \cdot \Delta \left[\frac{\|\mathbf{x}\|^2 \mathbf{c} - \mathbf{x}}{\left\| \|\mathbf{c}\| \mathbf{x} - \frac{\mathbf{c}}{\|\mathbf{c}\|} \right\|^n} \right]$$

$$= (n-2) \cdot \Delta \left[\frac{1}{\left\| \|\mathbf{c}\| \mathbf{x} - \frac{\mathbf{c}}{\|\mathbf{c}\|} \right\|^{n-2}} \cdot \frac{\frac{\mathbf{x}}{\|\mathbf{x}\|^2} - \mathbf{c}}{\frac{1}{\|\mathbf{x}\|^2} \left\| \|\mathbf{c}\| \mathbf{x} - \frac{\mathbf{c}}{\|\mathbf{c}\|} \right\|^2} \right]$$

$$\overset{(3.43)}{=} (n-2) \cdot \Delta \left[\frac{1}{\left\| \|\mathbf{c}\| \mathbf{x} - \frac{\mathbf{c}}{\|\mathbf{c}\|} \right\|^{n-2}} \cdot \frac{\frac{\mathbf{x}}{\|\mathbf{x}\|^2} - \mathbf{c}}{\left\| \frac{\mathbf{x}}{\|\mathbf{x}\|^2} - \mathbf{c} \right\|^2} \right].$$

Now setting $\mathbf{c} = \mathbf{c}^c$, multiplying by $\frac{1}{n-2} \cdot \frac{a^c}{a^{c_0}} \cdot e^{(r_\infty^{c_0} - r_\infty^c)t}$, and comparing the result with (C.12), we finally arrive at

$$
0 = \Delta \left[\frac{a^c}{a^{c_0}} \cdot e^{(r_\infty^{c_0} - r_\infty^c)t} \cdot \frac{1}{\left\| \|\mathbf{c}^c\| \mathbf{x} - \frac{\mathbf{c}^c}{\|\mathbf{c}^c\|} \right\|^{n-2}} \cdot \frac{\frac{\mathbf{x}}{\|\mathbf{x}\|^2} - \mathbf{c}^c}{\left\| \frac{\mathbf{x}}{\|\mathbf{x}\|^2} - \mathbf{c}^c \right\|^2} \right]
$$
$$
= \Delta \big(a^{c/c_0}(\mathbf{x}, t) \cdot G_c(\mathbf{x}) \big). \qquad \qquad \square
$$

C.6   Proof of Lemma 3.7

Proof. Beginning with the drift $\hat{\mu}^c$ of the SDE for $(\hat{X}_t^c)_{t \geq 0}$, which by (3.23b) is simply the image[3] of μ under G_c, we utilize the decomposition $G_c = g_{\mathbf{c}^c} \circ g_{\vec{0}}$ that we already used in our proof of Lemma 3.4 (iv), where $g_{\mathbf{a}}(\mathbf{x}) := (\mathbf{x} - \mathbf{a})/\|\mathbf{x} - \mathbf{a}\|^2$; this allows us to compute $\hat{\mu}$ by first computing its image under $g_{\vec{0}}$ and then the image of the resulting vector field under $g_{\mathbf{c}^c}$. Luckily, the images of the ACE drift under maps of the form $g_{\mathbf{a}}$ are well understood, as Lemma B.1 (iii) tells us how these maps transform the generalized ACE drift (B.1) with parameters $(\lambda, E, \mathbf{u}, \mathbf{v})$. Applying the transformation rules (B.3b)–(B.3c) twice (first with $\mathbf{a} = \vec{0}$ and the with $\mathbf{a} = \mathbf{c}^c$), we therefore find that the ACE drift (where $\mathbf{u} = \vec{0}$) transforms as follows:

$$
(\lambda, E, \vec{0}, \mathbf{v}^{c_0}) \xrightarrow{g_{\vec{0}}} \left(-\lambda, E, \tfrac{1}{n-2}\mathbf{v}^{c_0}, \vec{0}\right) \xrightarrow{g_{\mathbf{c}^c}} \left(\lambda, E, \vec{0}, \mathbf{v}^{c_0} + (n-2)(\lambda I + E)\mathbf{c}^c\right).
$$

Since by (3.47) the final parameter set is in fact equal to $\left(\lambda, E, \vec{0}, \mathbf{v}^c\right)$, we have thus found that $\hat{\mu}$ is the ACE drift vector field again, but with parameters λ, E, and $\mathbf{v}^c$. This proves (3.56b).

Next, we compute the SDE's noise function $\hat{\sigma}^c$ using the formula (3.23c). Plugging in (C.17), recalling (3.48), and then using Lemma 3.4 (ii)–(iii), we obtain

$$
\begin{aligned}
\hat{\sigma}^c(\hat{\mathbf{x}}, t) &= \big[s_c(\mathbf{x}) \cdot \sigma(\mathbf{x}, t) \big]_{\mathbf{x} = G_c^{-1}(\hat{\mathbf{x}})} \\
&= \left[\left\| \|\mathbf{c}^c\| \mathbf{x} - \tfrac{\mathbf{c}^c}{\|\mathbf{c}^c\|} \right\|^{-2} \right]_{\mathbf{x} = G_c^{-1}(\hat{\mathbf{x}})} \cdot \sigma\big(G_c^{-1}(\hat{\mathbf{x}}), t\big) \\
&= \left[\frac{\|H_{\mathbf{c}^c}(\mathbf{x})\|^2}{\|\mathbf{x}\|^2} \right]_{\mathbf{x} = H_{\mathbf{c}^c}^{-1}(\hat{\mathbf{x}})} \cdot \sigma\big(G_c^{-1}(\hat{\mathbf{x}}), t\big) \\
&= \frac{\|\hat{\mathbf{x}}\|^2}{\left\| H_{\mathbf{c}^c}^{-1}(\hat{\mathbf{x}}) \right\|^2} \cdot \sigma\big(G_c^{-1}(\hat{\mathbf{x}}), t\big) \qquad\qquad\qquad\text{(C.18)} \\
&= \frac{\|\hat{\mathbf{x}}\|^2}{\left\| H_{-\mathbf{c}^c}(\hat{\mathbf{x}}) \right\|^2} \cdot \sigma\big(G_c^{-1}(\hat{\mathbf{x}}), t\big) \\
&= \left\| \|\mathbf{c}^c\| \hat{\mathbf{x}} + \tfrac{\mathbf{c}^c}{\|\mathbf{c}^c\|} \right\|^2 \cdot \sigma\big(G_c^{-1}(\hat{\mathbf{x}}), t\big),
\end{aligned}
$$

---

[3]I.e., the flowlines of $\hat{\mu}$ are the images of the flowlines of μ under G_c.

which proves (3.56c). (Note that while this calculation only works for $\mathbf{x} \neq \vec{0}$, the expressions in lines 2 and 6 coincide also for $\mathbf{x} = \vec{0}$, since by (3.48) and (3.42) we have $G_c(\vec{0}) = \vec{0}$.)

Finally, let us derive the formula for the Brownian motion increments $\mathrm{d}\hat{W}_t^c$. By plugging (C.17) and the first case of (3.45) into (3.23d) and then using (3.11c), whenever $X_t \neq \vec{0}$ we obtain

$$
\begin{aligned}
\mathrm{d}\hat{W}_t^c &= s_c(X_t)^{-1} \cdot \nabla G_c(X_t)\, \mathrm{d}\tilde{W}_t^c \\
&= \left\| \|\mathbf{c}^c\|X_t - \tfrac{\mathbf{c}^c}{\|\mathbf{c}^c\|} \right\|^{-2} \cdot \left\| \|\mathbf{c}^c\|X_t - \tfrac{\mathbf{c}^c}{\|\mathbf{c}^c\|} \right\|^{-2} U_{H_{\mathbf{c}^c}(X_t)} U_{X_t}\, \mathrm{d}\tilde{W}_t^c \\
&= U_{H_{\mathbf{c}^c}(X_t)} U_{X_t}\, \mathrm{d}\tilde{W}_t^c \\
&= U_{H_{\mathbf{c}^c}(X_t)} U_{X_t} \big(\mathrm{d}W_t - (\sigma \cdot \alpha^c)(X_t, t)\, \mathrm{d}t \big),
\end{aligned}
$$

and since by (C.12) we have

$$
\begin{aligned}
\alpha^c(\mathbf{x}, t) &= \frac{(\nabla a^{c/c_0}(\mathbf{x}, t))^T}{a^{c/c_0}(\mathbf{x}, t)} = -(n-2)\|\mathbf{c}^c\| \frac{\|\mathbf{c}^c\|\mathbf{x} - \tfrac{\mathbf{c}^c}{\|\mathbf{c}^c\|}}{\left\| \|\mathbf{c}^c\|\mathbf{x} - \tfrac{\mathbf{c}^c}{\|\mathbf{c}^c\|} \right\|^2} \\
&= (n-2) \frac{\tfrac{\mathbf{c}^c}{\|\mathbf{c}^c\|^2} - \mathbf{x}}{\left\| \tfrac{\mathbf{c}^c}{\|\mathbf{c}^c\|^2} - \mathbf{x} \right\|^2}, \quad\quad\quad\quad (\mathrm{C}.19)
\end{aligned}
$$

this is the first case of (3.57). Whenever $X_t = \vec{0}$, we can use the second case in (3.45) and the fact that (C.17) and (C.19) yield $s_c(\vec{0}) = 1$ and $\alpha^c(\vec{0}, t) = (n-2)\mathbf{c}^c$, respectively, to obtain

$$
\begin{aligned}
\mathrm{d}\hat{W}_t^c &= s_c(X_t)^{-1} \cdot \nabla G_c(X_t)\, \mathrm{d}\tilde{W}_t^c = 1 \cdot I\, \mathrm{d}\tilde{W}_t^c \\
&= \mathrm{d}W_t - (\sigma \cdot \alpha^c)(X_t, t)\, \mathrm{d}t = \mathrm{d}W_t - \sigma(\vec{0}, t) \cdot (n-2)\mathbf{c}^c\, \mathrm{d}t,
\end{aligned}
$$

which is the second case of (3.57).

Interesting (but for us meaningless) side note: Observe the remarkable reappearance of the expression defining G_c in the ratio in (C.19), but with $\mathbf{x}$ and $\mathbf{c}^c$ interchanged. $\quad\square$

C.7 Proof of Lemma 3.8

Proof. Using Lemma 3.4 (ii), the images of the short rate functions under $G_{\bar{c}_0}$ can be written as

$$
\begin{aligned}
\hat{r}_{\bar{c}_0}^c(\hat{\mathbf{x}}) &= r^c\big(G_{\bar{c}_0}^{-1}(\hat{\mathbf{x}})\big) \overset{(3.20)}{=} r_\infty^c + \big\langle \mathbf{v}^c, G_c\big(G_{\bar{c}_0}^{-1}(\hat{\mathbf{x}})\big)\big\rangle \\
&= r_\infty^c + \big\langle \mathbf{v}^c, H_{\mathbf{c}^c}\big(H_{-\mathbf{c}^{\bar{c}_0}}(\hat{\mathbf{x}})\big)\big\rangle = r_\infty^c + \big\langle \mathbf{v}^c, H_{\mathbf{c}^c - \mathbf{c}^{\bar{c}_0}}(\hat{\mathbf{x}})\big\rangle,
\end{aligned}
$$

and since by (3.47) and (3.62) we have

$$
\begin{aligned}
\mathbf{c}^c - \mathbf{c}^{\bar{c}_0} &= \tfrac{1}{n-2}(\lambda I + E)^{-1}[(\mathbf{v}^c - \mathbf{v}^{c_0}) - (\mathbf{v}^{c_0} - \mathbf{v}^{c_0})] \\
&= \tfrac{1}{n-2}(\lambda I + E)^{-1}(\mathbf{v}^c - \mathbf{v}^{\bar{c}_0}) \\
&= \hat{\mathbf{c}}^c,
\end{aligned}
$$

together with (3.63) we arrive at

$$\hat{r}_{\bar{c}_0}^c(\hat{\mathbf{x}}) = r_\infty^c + \left\langle \mathbf{v}^c, H_{\hat{\mathbf{c}}^c}(\hat{\mathbf{x}}) \right\rangle = r_\infty^c + \left\langle \mathbf{v}^c, \hat{G}_c(\hat{\mathbf{x}}) \right\rangle,$$

which is (3.60).

To compute the images of the exchange rate functions under $G_{\bar{c}_0}$, first note that by (3.43) we can rewrite our exchange rate formula (3.51) for $\forall \mathbf{x} \neq \vec{0}$ as

$$a^{c_1/c_2}(\mathbf{x}, t) = \frac{a^{c_1} \cdot \mathrm{e}^{-r_\infty^{c_1} t} \cdot \|H_{\mathbf{c}^{c_1}}(\mathbf{x})\|^{n-2}}{a^{c_2} \cdot \mathrm{e}^{-r_\infty^{c_2} t} \cdot \|H_{\mathbf{c}^{c_2}}(\mathbf{x})\|^{n-2}} . \qquad (C.20)$$

This allows us to write

$$\begin{aligned}
\hat{a}_{\bar{c}_0}^{c_1/c_2}(\hat{\mathbf{x}}, t) &= a^{c_1/c_2}\left(G_{\bar{c}_0}^{-1}(\hat{\mathbf{x}}), t\right) \\
&= a^{c_1/c_2}\left(H_{-\mathbf{c}^{\bar{c}_0}}(\hat{\mathbf{x}}), t\right) \\
&= \frac{a^{c_1} \cdot \mathrm{e}^{-r_\infty^{c_1} t} \cdot \|H_{\mathbf{c}^{c_1}}(H_{-\mathbf{c}^{\bar{c}_0}}(\hat{\mathbf{x}}))\|^{n-2}}{a^{c_2} \cdot \mathrm{e}^{-r_\infty^{c_2} t} \cdot \|H_{\mathbf{c}^{c_2}}(H_{-\mathbf{c}^{\bar{c}_0}}(\hat{\mathbf{x}}))\|^{n-2}} \\
&= \frac{a^{c_1} \cdot \mathrm{e}^{-r_\infty^{c_1} t} \cdot \|H_{\mathbf{c}^{c_1}-\mathbf{c}^{\bar{c}_0}}(\hat{\mathbf{x}})\|^{n-2}}{a^{c_2} \cdot \mathrm{e}^{-r_\infty^{c_2} t} \cdot \|H_{\mathbf{c}^{c_2}-\mathbf{c}^{\bar{c}_0}}(\hat{\mathbf{x}})\|^{n-2}} \\
&= \frac{a^{c_1} \cdot \mathrm{e}^{-r_\infty^{c_1} t} \cdot \|H_{\hat{\mathbf{c}}^{c_1}}(\hat{\mathbf{x}})\|^{n-2}}{a^{c_2} \cdot \mathrm{e}^{-r_\infty^{c_2} t} \cdot \|H_{\hat{\mathbf{c}}^{c_2}}(\hat{\mathbf{x}})\|^{n-2}} ,
\end{aligned}$$

and now reversing the transformation leading to (C.20) again leads us to (3.60). For $\mathbf{x} = \vec{0}$ the definition of $\hat{a}_{\bar{c}_0}^{c_1/c_2}$ yields

$$\hat{a}_{\bar{c}_0}^{c_1/c_2}(\vec{0}, t) = a^{c_1/c_2}\left(G_{\bar{c}_0}^{-1}(\vec{0}), t\right) = a^{c_1/c_2}(\vec{0}, t) = \tfrac{a^{c_1}}{a^{c_2}} \mathrm{e}^{(r_\infty^{c_2} - r_\infty^{c_1})t},$$

which coincides with the value returned by (3.60) as well. $\square$

C.8   Proof of Lemma 3.10

To prepare for our proof of Lemma 3.10, we begin by generalizing Lemma 2.3 (i) so that it allows for the generalized parameter sets of the single-currency ACE model discussed in Section 2.7.1.

Lemma C.1. *In a generalized single-currency ACE model (i.e., one that does not necessarily satisfy the constraint $\|\mathbf{v}\| = 1$, as discussed in the last paragraph in Section 2.7.1), we have for $\forall \mathbf{x} \in \partial M_0$*

$$\langle \nabla r(\mathbf{x}), \mu(\mathbf{x}) \rangle = \tfrac{1}{n-2}\|\mathbf{v}\|^2 \|\mathbf{x} - \mathbf{x}_{\min}\|^2 + \delta, \qquad (C.21)$$

$$\text{where} \quad \mathbf{x}_{\min} := \tfrac{n-2}{2} \tfrac{E\mathbf{v}}{\|\mathbf{v}\|^2} - r_\infty \tfrac{\mathbf{v}}{\|\mathbf{v}\|^2} \in \partial M_0, \qquad (C.22)$$

and where δ is given by (2.48).

Proof. There are two ways of proving this generalization of Lemma 2.3 (i): Either one applies Lemma 2.3 (i) to the equivalent model given by the transformation (2.46), which does fulfill this constraint, or one simply generalizes the proof of Lemma 2.3 (i) itself. Since the latter is more straight forward and yields the additional more general formula (C.23), we will choose that approach.

First note that we have $\mathbf{x}_{\min} \in \partial M_0$ according to (2.14b), since by (2.12), (2.4), and Lemma 2.2 (ii) we have

$$r(\mathbf{x}_{\min}) = r_\infty + \left\langle \mathbf{v}, \tfrac{n-2}{2} \tfrac{E\mathbf{v}}{\|\mathbf{v}\|^2} - r_\infty \tfrac{\mathbf{v}}{\|\mathbf{v}\|^2} \right\rangle = r_\infty + 0 - r_\infty \cdot 1 = 0.$$

Using (2.11), (2.12), (2.4), (2.48), Lemma 2.2 (ii), and (C.22), we now find that

$$\langle \nabla r(\mathbf{x}), \mu(\mathbf{x}) \rangle + r(\mathbf{x})\big(\lambda + \tfrac{2}{n-2}\langle \mathbf{v}, \mathbf{x} \rangle\big)$$
$$= \big\langle \mathbf{v}, (-\lambda I + E)\mathbf{x} + \tfrac{1}{n-2}\|\mathbf{x}\|^2 \mathbf{v} - \tfrac{2}{n-2}\langle \mathbf{v}, \mathbf{x} \rangle \mathbf{x} \big\rangle + r(\mathbf{x})\big(\lambda + \tfrac{2}{n-2}\langle \mathbf{v}, \mathbf{x} \rangle\big)$$
$$= (r(\mathbf{x}) - \langle \mathbf{v}, \mathbf{x} \rangle)\big(\lambda + \tfrac{2}{n-2}\langle \mathbf{v}, \mathbf{x} \rangle\big) + \langle \mathbf{v}, E\mathbf{x} \rangle + \tfrac{1}{n-2}\|\mathbf{x}\|^2 \|\mathbf{v}\|^2$$
$$= r_\infty\big(\lambda + \tfrac{2}{n-2}\langle \mathbf{v}, \mathbf{x} \rangle\big) - \langle E\mathbf{v}, \mathbf{x} \rangle + \tfrac{1}{n-2}\|\mathbf{x}\|^2 \|\mathbf{v}\|^2$$
$$= \delta + \tfrac{1}{n-2}\big(r_\infty^2 + \tfrac{(n-2)^2}{4}\tfrac{\|E\mathbf{v}\|^2}{\|\mathbf{v}\|^2}\big) - \tfrac{2}{n-2}\big\langle -r_\infty \mathbf{v} + \tfrac{n-2}{2}E\mathbf{v}, \mathbf{x} \big\rangle + \tfrac{1}{n-2}\|\mathbf{x}\|^2 \|\mathbf{v}\|^2$$
$$= \delta + \tfrac{1}{n-2}\big\| -r_\infty \tfrac{\mathbf{v}}{\|\mathbf{v}\|} + \tfrac{n-2}{2}\tfrac{E\mathbf{v}}{\|\mathbf{v}\|} \big\|^2 - \tfrac{2}{n-2}\big\langle -r_\infty \mathbf{v} + \tfrac{n-2}{2}E\mathbf{v}, \mathbf{x} \big\rangle + \tfrac{1}{n-2}\|\mathbf{x}\|^2 \|\mathbf{v}\|^2$$
$$= \delta + \tfrac{1}{n-2}\|\mathbf{v}\|^2 \|\mathbf{x}_{\min}\|^2 - \tfrac{2}{n-2}\|\mathbf{v}\|^2 \langle \mathbf{x}_{\min}, \mathbf{x} \rangle + \tfrac{1}{n-2}\|\mathbf{x}\|^2 \|\mathbf{v}\|^2$$
$$= \delta + \tfrac{1}{n-2}\|\mathbf{v}\|^2 \|\mathbf{x}_{\min} - \mathbf{x}\|^2.$$

This shows that for $\forall \mathbf{x} \in \mathbb{R}^n$ we have

$$\langle \nabla r(\mathbf{x}), \mu(\mathbf{x}) \rangle = \delta + \tfrac{1}{n-2}\|\mathbf{v}\|^2 \|\mathbf{x} - \mathbf{x}_{\min}\|^2 - r(\mathbf{x})\big(\lambda + \tfrac{2}{n-2}\langle \mathbf{v}, \mathbf{x} \rangle\big), \qquad \text{(C.23)}$$

which by (2.14b) implies (C.21). □

Proof of Lemma 3.10. (i) Using the notation (3.23b), we have

$$\langle \nabla r^c(\mathbf{x}), \mu(\mathbf{x}) \rangle = \langle \nabla G_c(\mathbf{x})^T \mathbf{v}^c, \mu(\mathbf{x}) \rangle = \langle \mathbf{v}^c, \nabla G_c(\mathbf{x}) \mu(\mathbf{x}) \rangle = \langle \mathbf{v}^c, \hat{\mu}^c(G_c(\mathbf{x})) \rangle.$$

By Lemma 3.3 $\hat{\mu}^c$ is the drift of the single-currency ACE model for the parameters λ, E, and $\mathbf{v}^c$, and so we can now apply Lemma C.1. (Note we cannot use Lemma 2.3 (i) directly, since in the multi-currency model we do not enforce the constraint $\|\mathbf{v}^c\| = 1$.) Plugging the result of that lemma into the right-hand side above, we arrive at (3.67)–(3.68).

(ii) The first statement implies the second because of the definition (3.52b) of M_0, which also implies that $\partial M_0 \subset \bigcup_{c \in C} \partial M^c$.

To see that the first statement holds, note that as a particular consequence of (3.67), we have $\langle \nabla r^c(\mathbf{x}), \mu(\mathbf{x}) \rangle \geq \delta_c \geq \delta > 0$ for $\forall \mathbf{x} \in \partial M^c \setminus \{ \tfrac{\mathbf{c}^c}{\|\mathbf{c}^c\|^2} \}$. For all flowlines starting from any of those points $\mathbf{x}$, this statement therefore follows from the fact that by (3.52b) $\nabla r^c(\mathbf{x})$ is an inward-pointing normal vector to ∂M^c at $\mathbf{x}$.

To show it also for flowlines starting at $\mathbf{x} = \frac{\mathbf{c}^c}{\|\mathbf{c}^c\|^2}$ (if $\mathbf{c}^c \neq \vec{0}$), we need to compute $\langle \mathbf{n}, \mu(\frac{\mathbf{c}^c}{\|\mathbf{c}^c\|^2}) \rangle$, where $\mathbf{n} := \lim_{\mathbf{x} \to \mathbf{c}^c/\|\mathbf{c}^c\|^2} \frac{\nabla r^c(\mathbf{x})}{\|\nabla r^c(\mathbf{x})\|}$ is the normalized inward-pointing normal vector to ∂M^c at $\frac{\mathbf{c}^c}{\|\mathbf{c}^c\|^2}$ (recall that by Lemma 3.5 (ii)–(iii) the boundary of M^c is smooth). Since by Lemma 3.4 (iii)–(iv) we have

$$\|\nabla r^c(\mathbf{x})\| = \left\| \nabla G_c(\mathbf{x})^T \mathbf{v}^c \right\| = \frac{\|\mathbf{v}^c\|}{\left\| \|\mathbf{c}^c\| \mathbf{x} - \frac{\mathbf{c}^c}{\|\mathbf{c}^c\|} \right\|^2} = \frac{\|\mathbf{v}^c\| \|G_c(\mathbf{x})\|^2}{\|\mathbf{x}\|^2}$$

for $\forall \mathbf{x} \neq \frac{\mathbf{c}^c}{\|\mathbf{c}^c\|^2}$, dividing both sides of (3.67) by $\|\nabla r^c(\mathbf{x})\|$, letting $\mathbf{x} \to \frac{\mathbf{c}^c}{\|\mathbf{c}^c\|^2}$, and recalling that by (3.42) we have $G_c(\frac{\mathbf{c}^c}{\|\mathbf{c}^c\|^2}) = \infty$ yields

$$\langle \mathbf{n}, \mu(\tfrac{\mathbf{c}^c}{\|\mathbf{c}^c\|^2}) \rangle = \lim_{\mathbf{x} \to \frac{\mathbf{c}^c}{\|\mathbf{c}^c\|^2}} \frac{\langle \nabla r^c(\mathbf{x}), \mu(\mathbf{x}) \rangle}{\|\nabla r^c(\mathbf{x})\|}$$

$$= \lim_{\mathbf{x} \to \frac{\mathbf{c}^c}{\|\mathbf{c}^c\|^2}} \frac{\frac{1}{n-2}\|\mathbf{v}^c\|^2 \|G_c(\mathbf{x}) - \mathbf{x}_{\min}^c\|^2 + \delta_c}{\|\mathbf{v}^c\| \|G_c(\mathbf{x})\|^2 / \|\mathbf{x}\|^2}$$

$$= \frac{1}{n-2}\|\mathbf{v}^c\| \left\| \frac{\mathbf{c}^c}{\|\mathbf{c}^c\|^2} \right\|^2 = \frac{1}{n-2} \frac{\|\mathbf{v}^c\|}{\|\mathbf{c}^c\|^2} > 0,$$

which proves the validity of the first statement also for this case.

(iii) The proof of this part is analogous to the one of Lemma 2.3 (iii). $\square$

Appendix D

Proofs for Chapter 5

D.1   Proof of Lemma 5.4

Proof. Uniqueness: Suppose some extension $(\hat{U}_s)_{s \in \mathbb{R}}$ fulfills (5.74) and thus (5.73), and let $s \in \mathbb{R}$ and $\psi \in \mathcal{S}(\mathcal{V})$. Then abbreviating $\psi_T := \psi|_{M_T}$ for any $T \in \mathbb{R}$, we have

$$(\hat{U}_s \psi)\big|_{M_{T-s}} \stackrel{(5.68)}{=} \hat{U}_s \psi_T = \hat{U}_s\big(\tfrac{\psi_T}{\mathfrak{B}_T} \cdot \mathfrak{B}_T\big)$$

$$\stackrel{(5.71)}{=} \big(\bar{U}_s \tfrac{\psi_T}{\mathfrak{B}_T}\big) \cdot \hat{U}_s \mathfrak{B}_T \stackrel{(5.73)}{=} \big(\bar{U}_s \tfrac{\psi_T}{\mathfrak{B}_T}\big) \cdot \mathfrak{B}_{T-s}$$

$$= \big[\mathfrak{B}_{T-s} \circ \bar{U}_s \circ \tfrac{1}{\mathfrak{B}_T}\big](\psi|_{M_T}). \tag{D.1}$$

This uniquely determines the values of $\hat{U}_s \psi$ on M_{T-s}, and since $T \in \mathbb{R}$ was arbitrary, by (5.32) in fact on all of M.

Existence: We would now like to use our insight from (D.1) to *define*

$$\forall \psi \in \mathcal{S}(\mathcal{V}): \quad (\hat{U}_s \psi)(\mathbf{y}) := \big(\big[\mathfrak{B}_{T-s} \circ \bar{U}_s \circ \tfrac{1}{\mathfrak{B}_T}\big](\psi|_{M_T})\big)(\mathbf{y}) \tag{D.2}$$

for $\forall \mathbf{y} \in M$, where $T \in \mathbb{R}$ is arbitrarily chosen such that $\mathbf{y} \in M_{T-s}$. To do so, however, we would need to show that the right-hand side of (D.2) does not depend on the choice of T. Therefore let $T, T' \in \mathbb{R}$ be large enough that $\mathbf{y} \in M_{T-s} \cap M_{T'-s}$. Then we have

$$\big(\big[\mathfrak{B}_{T-s} \circ \bar{U}_s \circ \tfrac{1}{\mathfrak{B}_T}\big](\psi|_{M_T})\big)(\mathbf{y}) = \big(\mathfrak{B}_{T-s} \cdot \big(\bar{U}_s \tfrac{\psi_T}{\mathfrak{B}_T}\big)\big)(\mathbf{y})$$

$$= \big(\mathfrak{B}_{T-s} \cdot \big(\bar{U}_s\big(B_{T,T'} \cdot \tfrac{\psi_{T'}}{\mathfrak{B}_{T'}}\big)\big)\big)(\mathbf{y})$$

$$\stackrel{(5.58)}{=} \big(\mathfrak{B}_{T-s} \cdot \big(\bar{U}_s B_{T,T'}\big) \cdot \big(\bar{U}_s \tfrac{\psi_{T'}}{\mathfrak{B}_{T'}}\big)\big)(\mathbf{y})$$

$$\stackrel{(5.64)}{=} \big(\mathfrak{B}_{T-s} \cdot B_{T-s,T'-s} \cdot \big(\bar{U}_s \tfrac{\psi_{T'}}{\mathfrak{B}_{T'}}\big)\big)(\mathbf{y})$$

$$= \big(\mathfrak{B}_{T'-s} \cdot \big(\bar{U}_s \tfrac{\psi_{T'}}{\mathfrak{B}_{T'}}\big)\big)(\mathbf{y})$$

$$= \big(\big[\mathfrak{B}_{T'-s} \circ \bar{U}_s \circ \tfrac{1}{\mathfrak{B}_{T'}}\big](\psi|_{M_{T'}})\big)(\mathbf{y}).$$

This shows that $\hat{U}_s\psi$ is well defined via (D.2) for any $\psi \in \mathcal{S}(\mathcal{V})$. For sections defined on a smaller domain we define

$$\forall T \in \mathbb{R} \; \forall \psi \in \mathcal{S}(\mathcal{V}_T): \quad \hat{U}_s\psi := \left[\mathfrak{B}_{T-s} \circ \bar{U}_s \circ \tfrac{1}{\mathfrak{B}_T} \right]\psi, \tag{D.3}$$

which fulfills our consistency requirements (5.67)–(5.68).

We immediately find that these operators fulfill (5.70), since the sections $\mathfrak{B}_T^Y$ are assumed to be positive and since the operators $\bar{U}_s$ fulfill (5.59).

To see that the operators $\hat{U}_s$ defined by (D.2) and (D.3) form a group, let $s_1, s_2, T \in \mathbb{R}$, and first let $\psi \in \mathcal{S}(\mathcal{V}_T)$. Then we have

$$\begin{aligned}
\hat{U}_{s_1}(\hat{U}_{s_2}\psi) &= \left[\mathfrak{B}_{(T-s_2)-s_1} \circ \bar{U}_{s_1} \circ \tfrac{1}{\mathfrak{B}_{T-s_2}} \right]\left[\mathfrak{B}_{T-s_2} \circ \bar{U}_{s_2} \circ \tfrac{1}{\mathfrak{B}_T} \right]\psi \\
&= \left[\mathfrak{B}_{T-(s_1+s_2)} \circ \bar{U}_{s_1+s_2} \circ \tfrac{1}{\mathfrak{B}_T} \right]\psi = \hat{U}_{s_1+s_2}\psi.
\end{aligned} \tag{D.4}$$

The group property for $\psi \in \mathcal{S}(\mathcal{V})$ then follows from (D.4) and repeated application of (5.68): For $\forall \psi \in \mathcal{S}(\mathcal{V})$ and $\forall T \in \mathbb{R}$ we have

$$\begin{aligned}
(\hat{U}_{s_1}(\hat{U}_{s_2}\psi))|_{M_{T-(s_1+s_2)}} &= (\hat{U}_{s_1}(\hat{U}_{s_2}\psi))|_{M_{(T-s_2)-s_1}} = \hat{U}_{s_1}((\hat{U}_{s_2}\psi)|_{M_{T-s_2}}) \\
&= \hat{U}_{s_1}(\hat{U}_{s_2}(\psi|_{M_T})) = \hat{U}_{s_1+s_2}(\psi|_{M_T}) \\
&= (\hat{U}_{s_1+s_2}\psi)|_{M_{T-(s_1+s_2)}},
\end{aligned}$$

and since T was arbitrary, we find that $\hat{U}_{s_1}(\hat{U}_{s_2}\psi) = \hat{U}_{s_1+s_2}\psi$ on all of M.

To see that the property (5.71) holds, first let $s, T \in \mathbb{R}$, $\psi \in \mathcal{S}(\mathcal{V}_T)$, and $f \in \mathcal{F}(M_T)$. Then using (5.58) we find that

$$\begin{aligned}
\hat{U}_s(f\psi) &= \left[\mathfrak{B}_{T-s} \circ \bar{U}_s \circ \tfrac{1}{\mathfrak{B}_T} \right](f\psi) = \mathfrak{B}_{T-s} \cdot \left(\bar{U}_s\left(\tfrac{f\psi}{\mathfrak{B}_T} \right) \right) = \mathfrak{B}_{T-s} \cdot \left(\bar{U}_s\left(f \cdot \tfrac{\psi}{\mathfrak{B}_T} \right) \right) \\
&= \mathfrak{B}_{T-s} \cdot (\bar{U}_s f) \cdot \left(\bar{U}_s \tfrac{\psi}{\mathfrak{B}_T} \right)) = (\bar{U}_s f) \cdot \left[\mathfrak{B}_{T-s} \circ \bar{U}_s \circ \tfrac{1}{\mathfrak{B}_T} \right]\psi \\
&= (\bar{U}_s f) \cdot (\hat{U}_s\psi).
\end{aligned}$$

For $\psi \in \mathcal{S}(\mathcal{V})$ and $f \in \mathcal{F}(M)$, this relation now follows by repeated application of (5.56) and (5.68): For $\forall \psi \in \mathcal{S}(\mathcal{V})$ and $\forall T \in \mathbb{R}$ we have

$$\begin{aligned}
(\hat{U}_s(f\psi))\big|_{M_{T-s}} &= \hat{U}_s(f|_{M_T}\psi|_{M_T}) = (\bar{U}_s(f|_{M_T})) \cdot (\hat{U}_s(\psi|_{M_T})) \\
&= ((\bar{U}_s f)|_{M_{T-s}}) \cdot ((\hat{U}_s\psi)|_{M_{T-s}}) = ((\bar{U}_s f) \cdot (\hat{U}_s\psi))\big|_{M_{T-s}},
\end{aligned}$$

and since $T \in \mathbb{R}$ was arbitrary, we can conclude that $\hat{U}_s(f\psi) = (\bar{U}_s f) \cdot (\hat{U}_s\psi)$ holds on all of M.

Finally, the property (5.74) holds because by definition (D.3) we have

$$\begin{aligned}
\hat{U}_T \mathfrak{B}_T &= \left[\mathfrak{B}_{T-T} \circ \bar{U}_T \circ \tfrac{1}{\mathfrak{B}_T} \right]\mathfrak{B}_T = \mathfrak{B}_0 \cdot \left(\bar{U}_T \tfrac{\mathfrak{B}_T}{\mathfrak{B}_T} \right) \\
&= \mathfrak{B}_0 \cdot (\bar{U}_T 1) = \mathfrak{B}_0 \cdot 1 = \mathfrak{B}_0
\end{aligned}$$

for $\forall T \in \mathbb{R}$, where we denoted by 1 the constant functions taking the value 1 on M_T and M_0, respectively. $\square$

D.2 Proof of Lemma 5.7

Proof.

$$((\partial_t + D^{Z^{\Delta t}})f)(\mathbf{z}, t)$$
$$= \lim_{\varepsilon \searrow 0} \tfrac{1}{\varepsilon} \mathbb{E}\big[f(Z^{\Delta t}_{t+\varepsilon}, t+\varepsilon) - f(\mathbf{z}, t) \,\big|\, Z^{\Delta t}_t = \mathbf{z}\big]$$
$$= \lim_{\varepsilon \searrow 0} \tfrac{1}{\varepsilon} \mathbb{E}\big[f(Z_{t+\varepsilon-\Delta t}, t+\varepsilon) - f(\mathbf{z}, t) \,\big|\, Z_{t-\Delta t} = \mathbf{z}\big]$$
$$= \lim_{\varepsilon \searrow 0} \tfrac{1}{\varepsilon} \mathbb{E}\big[(S_{\Delta t}f)(Z_{(t-\Delta t)+\varepsilon}, (t-\Delta t)+\varepsilon) - (S_{\Delta t}f)(\mathbf{z}, t-\Delta t) \,\big|\, Z_{t-\Delta t} = \mathbf{z}\big]$$
$$= \big((\partial_t + D^Z)(S_{\Delta t}f)\big)(\mathbf{z}, t-\Delta t)$$
$$= \big((S_{-\Delta t} \circ (\partial_t + D^Z) \circ S_{\Delta t})f\big)(\mathbf{z}, t)$$

This shows (5.85), and Equation (5.86) now simply follows from the fact that $S_{-\Delta t} \circ \partial_t \circ S_{\Delta t} = \partial_t$. $\qquad\square$

D.3 Proof of Lemma 5.8

Proof. (i) Let $f_Y \in \mathcal{F}(M_{Y,T})$ and $(\mathbf{y}, t) \in M_{Y,T}$. Then setting $f_X := \bar{U}f_Y$ $\in \mathcal{F}(M_{X,T})$ and $(\mathbf{x}, t) := U(\mathbf{y}, t)$, and using (5.42), we have

$$\big((\partial_t + D^{Y,T})f_Y\big)(\mathbf{y}, t) = \lim_{\Delta t \searrow 0} \tfrac{1}{\Delta t} \mathbb{E}[f_Y(Y_{t+\Delta t}, t+\Delta t) - f_Y(\mathbf{y}, t)|Y_t = \mathbf{y}]$$
$$= \lim_{\Delta t \searrow 0} \tfrac{1}{\Delta t} \mathbb{E}[f_X(X_{t+\Delta t}, t+\Delta t) - f_X(\mathbf{x}, t)|X_t = \mathbf{x}]$$
$$= \big((\partial_t + D^{X,T})f_X\big)(\mathbf{x}, t)$$
$$= \big(\bar{U}^{-1}(\partial_t + D^{X,T})f_X\big)(\mathbf{y}, t)$$
$$= \big(\bar{U}^{-1}(\partial_t + D^{X,T})\bar{U}f_Y\big)(\mathbf{y}, t).$$

(ii) Let $f_X \in \mathcal{F}(M_{X,T_1 \wedge T_2})$ and $\forall (\mathbf{x}, t) \in M_{X,T_1 \wedge T_2}$. Then we have

$$\big((\partial_t + D^{X,T_2})f_X\big)(\mathbf{x}, t)$$
$$= \lim_{\Delta t \searrow 0} \tfrac{1}{\Delta t} \mathbb{E}^{T_2}\big[f_X(X_{t+\Delta t}, t+\Delta t) - f_X(\mathbf{x}, t) \,\big|\, X_t = \mathbf{x}\big]$$
$$= \lim_{\Delta t \searrow 0} \tfrac{1}{\Delta t} \mathbb{E}^{T_1}\Bigg[\frac{B^X_{T_1,T_2}(X_{T_1 \wedge T_2}, T_1 \wedge T_2)}{B^X_{T_1,T_2}(\mathbf{x}, t)}$$
$$\times \big(f_X(X_{t+\Delta t}, t+\Delta t) - f_X(\mathbf{x}, t)\big) \,\bigg|\, X_t = \mathbf{x}\Bigg]$$
$$= \tfrac{1}{B^X_{T_1,T_2}(\mathbf{x}, t)} \lim_{\Delta t \searrow 0} \tfrac{1}{\Delta t} \mathbb{E}^{T_1}\big[(B^X_{T_1,T_2} \cdot f_X)(X_{t+\Delta t}, t+\Delta t)$$
$$- (B^X_{T_1,T_2} \cdot f_X)(\mathbf{x}, t) \,\big|\, X_t = \mathbf{x}\big]$$
$$= \Big(B^X_{T_2,T_1}(\partial_t + D^{X,T_1})(B^X_{T_1,T_2} \cdot f_X)\Big)(\mathbf{x}, t),$$

which is (5.88a). (In the third step we used the fact that $(B^X_{T_1,T_2}(X_t, t))_{t \le T_1 \wedge T_2}$ is a martingale under the T_1-measure.) Equation (5.88b) is shown analogously. $\quad\square$

D.4   Proof of (5.112)

Proof. Since for $\forall f \in \mathcal{F}(M_{Y,T})$ we have

$$(\hat{U} \circ \mathfrak{B}_T^Y)f = \hat{U}(f \cdot \mathfrak{B}_T^Y) \overset{(5.51)}{=} (\bar{U}f) \cdot (\hat{U}\mathfrak{B}_T^Y) \overset{(5.52)}{=} (\bar{U}f) \cdot \mathfrak{B}_T^X = (\mathfrak{B}_T^X \circ \bar{U})f,$$

the relation

$$\hat{U} \circ \mathfrak{B}_T^Y = \mathfrak{B}_T^X \circ \bar{U} \tag{D.5}$$

holds. We can therefore conclude that

$$
\begin{aligned}
\tilde{D}^{X,T} \quad &= \quad \bar{U} \circ \tilde{D}^{Y,T} \circ \bar{U}^{-1} \\
&\overset{(5.104)}{=} \bar{U} \circ \frac{1}{\mathfrak{B}_T^Y} \circ \tilde{D}^{Y,nf} \circ \mathfrak{B}_T^Y \circ \bar{U}^{-1} \\
&\overset{(5.110)}{=} \bar{U} \circ \frac{1}{\mathfrak{B}_T^Y} \circ \hat{U}^{-1} \circ \tilde{D}^{X,nf} \circ \hat{U} \circ \mathfrak{B}_T^Y \circ \bar{U}^{-1} \\
&= \quad \left(\hat{U} \circ \mathfrak{B}_T^Y \circ \bar{U}^{-1}\right)^{-1} \circ \tilde{D}^{X,nf} \circ \left(\hat{U} \circ \mathfrak{B}_T^Y \circ \bar{U}^{-1}\right) \\
&\overset{(D.5)}{=} \frac{1}{\mathfrak{B}_T^X} \circ \tilde{D}^{X,nf} \circ \mathfrak{B}_T^X . \qquad \qquad \square
\end{aligned}
$$

D.5   Proof of (5.116)

Proof. For $\forall \psi \in \mathcal{S}(\mathcal{V}_X)$ we have

$$
\begin{aligned}
\left(\nabla_t^{X,nf}\psi\right)(\cdot,t) &\overset{(5.114)}{=} \left((\hat{U} \circ \nabla_t^{Y,nf} \circ \hat{U}^{-1})\psi\right)(\cdot,t) \\
&\overset{(5.106)}{=} \left((\hat{U} \circ \partial_t \circ \hat{U}^{-1})\psi\right)(\cdot,t) \\
&\overset{(5.65)}{=} \hat{U}_t\big(((\partial_t \circ \hat{U}^{-1})\psi)(\cdot,t)\big) \\
&\overset{(5.65)}{=} \hat{U}_t\big(\partial_t(\hat{U}_{-t}\psi(\cdot,t))\big) \\
&= \quad \hat{U}_t\big((\partial_t\hat{U}_{-t}) + \hat{U}_{-t}\partial_t\big)\psi(\cdot,t) \\
&= \quad \hat{U}_t\big((\partial_s\hat{U}_{-(t+s)})\big|_{s=0} + \hat{U}_{-t}\partial_t\big)\psi(\cdot,t) \\
&\overset{(5.69)}{=} \big(\partial_s\hat{U}_{-s}\big|_{s=0} + \partial_t\big)\psi(\cdot,t) \\
&\overset{(5.117)}{=} \big(-\nabla_{\hat{U}_t} + \partial_t\big)\psi(\cdot,t) . \qquad \qquad \square
\end{aligned}
$$

D.6   Proofs of (5.132)–(5.133)

Proof. First observe that analogously to $(D.5)$ one can derive the relation

$$\hat{U} \circ \mathfrak{B}_t^Y = \mathfrak{B}_t^X \circ \bar{U}, \tag{D.6}$$

and that taking the inverses on both sides leads us to the relation

$$\frac{1}{\mathfrak{B}_t^Y} \circ \hat{U}^{-1} = \bar{U}^{-1} \circ \frac{1}{\mathfrak{B}_t^X} . \tag{D.7}$$

We therefore find that

$$\partial_t + D^{Y,\$} \overset{(5.130)}{=} \frac{1}{\mathfrak{B}_t^Y} \circ \left(\nabla_t^{Y,nf} + \tilde{D}^{Y,nf}\right) \circ \mathfrak{B}_t^Y$$

$$\overset{(5.110),(5.114)}{=} \frac{1}{\mathfrak{B}_t^Y} \circ \hat{U}^{-1} \circ \left(\nabla_t^{X,nf} + \tilde{D}^{X,nf}\right) \circ \hat{U} \circ \mathfrak{B}_t^Y$$

$$\overset{(D.7),(D.6)}{=} \bar{U}^{-1} \circ \frac{1}{\mathfrak{B}_t^X} \circ \left(\nabla_t^{X,nf} + \tilde{D}^{X,nf}\right) \circ \mathfrak{B}_t^X \circ \bar{U}$$

$$\overset{(5.127)}{=} \bar{U}^{-1} \circ \left(\partial_t + \tilde{D}^{X,\$}\right) \circ \bar{U},$$

which is (5.132). The proof the relation (5.133a) is much simpler:

$$\partial_t + D^{X,T} \overset{(5.120)}{=} \frac{1}{\mathfrak{B}_T^X} \circ \left(\nabla_t^{X,nf} + \tilde{D}^{X,nf}\right) \circ \mathfrak{B}_T^X$$

$$\overset{(5.127)}{=} \frac{1}{\mathfrak{B}_T^X} \circ \mathfrak{B}_t^X \circ \left(\nabla_t^{X,\$} + \tilde{D}^{X,\$}\right) \circ \frac{1}{\mathfrak{B}_t^X} \circ \mathfrak{B}_T^X$$

$$= B_{T,t}^X \circ \left(\nabla_t^{X,\$} + \tilde{D}^{X,\$}\right) \circ B_{t,T}^X.$$

The relation (5.133b) is shown analogously. $\qquad\square$

D.7 Proof of (5.136b)

Proof. Let $f \in \mathcal{F}(M_X)$. Then we have for $\forall (\mathbf{y}, t) \in M_Y$

$$((\partial_t \circ \bar{U}^{-1})f)(\mathbf{y},t) \overset{(5.45b)}{=} \frac{\mathrm{d}}{\mathrm{d}t} f(U_t(\mathbf{y}),t) \overset{(5.24)}{=} ((\partial_t + \langle \mu, \nabla \rangle)f)(U_t(\mathbf{y}),t)$$

$$\overset{(5.45a)}{\Rightarrow} (\bar{U} \circ \partial_t \circ \bar{U}^{-1})f = (\partial_t + \langle \mu, \nabla \rangle)f.$$

The second part can be shown as follows:

$$(\bar{U} \circ r^Y \circ \bar{U}^{-1})f = \bar{U}(r^Y \cdot (\bar{U}^{-1}f)) \overset{(5.48a)}{=} (\bar{U}r^Y) \cdot (\bar{U}(\bar{U}^{-1}f)) \overset{(5.47)}{=} r^X \cdot f.$$

$\qquad\square$

D.8 Proof of Lemma 5.10

Proof. The equivalence of (i) and (ii) follows directly from Lemma 5.7, which says that the left-hand side of (ii) is the generator of the delayed process (X_t) under the T-measure, and from the fact the a process's measure is uniquely determined by its generator.

Next, each of the statements (ii)–(iv) are equivalent to their analogues (ii')–(iv') with the generators replaced by the corresponding Kolmogorov operators, since the missing pieces—∂_t for the generators in (ii) and (iii), and $\nabla_t^{X,nf} = \partial_t - \nabla_{\hat{U}}$ (recall (5.116)) for the one in (iv)—commute with $S_{\Delta t}$.

In particular, this allows us to prove the equivalence of (ii)–(iv) for these analogous statements (ii')–(iv') with the Kolmogorov operators instead.

(ii') $\Leftrightarrow$ (iv'): First observe that by (5.40) we have

$$(S_{-\Delta t}\mathfrak{B}_T^X)(\mathbf{x},t) = \mathfrak{B}_T^X(\mathbf{x}, t - \Delta t) = \mathfrak{B}_{T-(t-\Delta t)}^Y(\mathbf{x}) = \mathfrak{B}_{(T+\Delta t)-t}^Y(\mathbf{x})$$

$$= \mathfrak{B}_{T+\Delta t}^X(\mathbf{x},t)$$

for $\forall T, \Delta t \in \mathbb{R}$, and thus $S_{-\Delta t} \circ \mathfrak{B}_T^X = \mathfrak{B}_{T+\Delta t}^X \circ S_{-\Delta t}$. Furthermore, taking the inverse on both sides, we also find that $\frac{1}{\mathfrak{B}_T^X} \circ S_{\Delta t} = S_{\Delta t} \circ \frac{1}{\mathfrak{B}_{T+\Delta t}^X}$.

With this, we now see that (ii') is equivalent to asking that

$$\mathfrak{B}_{T+\Delta t}^X \circ (\partial_t + D^{X,T+\Delta t}) \circ \frac{1}{\mathfrak{B}_{T+\Delta t}} = \mathfrak{B}_{T+\Delta t}^X \circ S_{-\Delta t} \circ (\partial_t + D^{X,T}) \circ S_{\Delta t} \circ \frac{1}{\mathfrak{B}_{T+\Delta t}^X}$$
$$= S_{-\Delta t} \circ \mathfrak{B}_T^X \circ (\partial_t + D^{X,T}) \circ \frac{1}{\mathfrak{B}_T^X} \circ S_{\Delta t}$$

for $\forall T, \Delta t \in \mathbb{R}$, or written differently using (5.94a), that

$$\mathfrak{B}_{T+\Delta t}^X \circ (\nabla_t^X + \tilde{D}^{X,T+\Delta t}) \circ \frac{1}{\mathfrak{B}_{T+\Delta t}} = S_{-\Delta t} \circ \mathfrak{B}_T^X \circ (\nabla_t^X + \tilde{D}^{X,T}) \circ \frac{1}{\mathfrak{B}_T^X} \circ S_{\Delta t}$$

for $\forall T, \Delta t \in \mathbb{R}$. Using (5.111) and (5.115), we now see that this is equivalent to asking that

$$\nabla_t^{X,nf} + \tilde{D}^{X,nf} = S_{-\Delta t} \circ (\nabla_t^{X,nf} + \tilde{D}^{X,nf}) \circ S_{\Delta t} \tag{D.8}$$

for $\forall \Delta t \in \mathbb{R}$, which is (iv').

(iii') $\Leftrightarrow$ (iv'): If we now multiply (D.8) from the left and right by $\frac{1}{\mathfrak{B}_t^X}$ and $\mathfrak{B}_t^X$, respectively, and then use the fact that (as mentioned in (5.131)) the section $\mathfrak{B}_t^X = \mathfrak{B}_0^Y$ does not depend on t and therefore as a multiplication operator commutes with $S_{\Delta t}$, we obtain the equivalent statement

$$\frac{1}{\mathfrak{B}_t^X} \circ (\nabla_t^{X,nf} + \tilde{D}^{X,nf}) \circ \mathfrak{B}_t^X = \frac{1}{\mathfrak{B}_t^X} \circ S_{-\Delta t} \circ (\nabla_t^{X,nf} + \tilde{D}^{X,nf}) \circ S_{\Delta t} \circ \mathfrak{B}_t^X$$
$$= S_{-\Delta t} \circ \frac{1}{\mathfrak{B}_t^X} \circ (\nabla_t^{X,nf} + \tilde{D}^{X,nf}) \circ \mathfrak{B}_t^X \circ S_{\Delta t}.$$

By our definition (5.127), this in turn can be rewritten as

$$\partial_t + D^{X,\$} = S_{-\Delta t} \circ (\nabla_t^{X,\$} + \tilde{D}^{X,\$}) \circ S_{\Delta t},$$

which is (iii'). $\square$

Appendix E

Proofs for Chapter 6

E.1 Proof of Lemma 6.1

Proof. "$\Rightarrow$": If $\frac{1}{\psi_2} \circ \tilde{D}^{X,nf} \circ \psi_1$ is a kth-order differential operator on M_0 for some $\psi_1, \psi_2 \in \mathcal{S}_\varnothing(\mathcal{V}_0)$, then

$$\tilde{D}^{X,T} \overset{(5.112)}{=} \frac{1}{\mathfrak{B}_T^X} \circ \tilde{D}^{X,nf} \circ \mathfrak{B}_T^X = \frac{\psi_2}{\mathfrak{B}_T^X} \circ \left(\frac{1}{\psi_2} \circ \tilde{D}^{X,nf} \circ \psi_1 \right) \circ \frac{\mathfrak{B}_T^X}{\psi_1}$$

is a kth-order differential operator for $\forall T \in \mathbb{R}$, and thus by (5.99b) also $D^{X,T}$.

"$\Leftarrow$": If $D^{X,T}$ is a kth-order differential operator for $\forall T \in \mathbb{R}$ then in particular also the operator

$$D^{X,0} - \langle \mu, \nabla \rangle \overset{(5.99b)}{=} \tilde{D}^{X,0} \overset{(5.112)}{=} \frac{1}{\mathfrak{B}_0^X} \circ \tilde{D}^{X,nf} \circ \mathfrak{B}_0^X.$$

According to Definition 6.1, with $\psi_1 := \psi_2 := \mathfrak{B}_0^X \big|_{t=0}$, this shows that $\tilde{D}^{X,nf}$ is a kth-order differential operator on M_0. $\qquad \square$

E.2 Proof of Lemma 6.4

Proof. Let $\psi \in \mathcal{S}_\varnothing(\mathcal{V})$ be such that $\frac{1}{\psi} \circ D_{\text{sec}} \circ \psi$ is a kth-order differential operator, and let $\psi' := \hat{U}_s \psi$. Then we have

$$\frac{1}{\psi'} \circ (U_s D_{\text{sec}}) \circ \psi' = \frac{1}{\hat{U}_s \psi} \circ \hat{U}_s \circ D_{\text{sec}} \circ \hat{U}_{-s} \circ (\hat{U}_s \psi)$$
$$= \bar{U}_s \circ \left(\frac{1}{\psi} \circ D_{\text{sec}} \circ \psi \right) \circ \bar{U}_{-s}, \qquad (\text{E.1})$$

and so $\frac{1}{\psi'} \circ (U_s D_{\text{sec}}) \circ \psi'$ and thus $U_s D_{\text{sec}}$ is a kth-order differential operator as well. The reverse direction is shown in the same way. $\qquad \square$

E.3   Proof of Lemma 6.5

Proof. Taking a closer look at (E.1), we see that by Definition 6.2 the second-order coefficient matrix of its left-hand side is A_{U_sD}, and so this matrix-valued function can be obtained by computing the second-order coefficient matrix of the operator on the right of (E.1). Clearly, this only requires us to keep track of the second-order derivative terms of the operator in the parentheses on the right, whose coefficients are denoted by A_D. Writing $\cong$ whenever two expressions have the same second-order derivative terms, this shows that

$$\sum_{k,l}(A_{U_sD})_{k,l}\partial_{\mathbf{x}_k}\partial_{\mathbf{x}_l} \cong \bar{U}_s \circ \Big(\sum_{i,j}(A_D)_{i,j}\partial_{\mathbf{x}_i}\partial_{\mathbf{x}_j}\Big) \circ \bar{U}_{-s}. \qquad (\text{E.2})$$

To extract the second-order coefficients from the operator on the right, now consider any $f \in C^2(M,\mathbb{R})$ and $(\mathbf{x},t) \in M \times \mathbb{R}$. Then we have

$$\partial_{\mathbf{x}_j}(\bar{U}_{-s}f)(\mathbf{x}) = \partial_{\mathbf{x}_j}[f(U_s(\mathbf{x}))]$$
$$= \sum_l(\partial_l f)(U_s(\mathbf{x})) \cdot (\nabla U_s(\mathbf{x}))_{lj}\,,$$
$$\partial_{\mathbf{x}_i}\partial_{\mathbf{x}_j}(\bar{U}_{-s}f)(\mathbf{x}) \cong \sum_{k,l}(\partial_k\partial_l f)(U_s(\mathbf{x})) \cdot (\nabla U_s(\mathbf{x}))_{ki} \cdot (\nabla U_s(\mathbf{x}))_{lj}\,,$$

and thus

$$\sum_{i,j}(A_D)_{ij}(\mathbf{x})\partial_{\mathbf{x}_i}\partial_{\mathbf{x}_j}(\bar{U}_{-s}f)(\mathbf{x})$$
$$\cong \sum_{i,j,k,l}(A_D)_{ij}(\mathbf{x}) \cdot (\partial_k\partial_l f)(U_s(\mathbf{x})) \cdot (\nabla U_s(\mathbf{x}))_{ki} \cdot (\nabla U_s(\mathbf{x}))_{lj}$$
$$\cong \sum_{k,l}\big[\nabla U_s(\mathbf{x})A_D(\mathbf{x})(\nabla U_s(\mathbf{x}))^T\big]_{kl} \cdot (\partial_k\partial_l f)(U_s(\mathbf{x})).$$

Combining this with (E.2), we therefore have

$$\sum_{k,l}(A_{U_sD}(\mathbf{x}))_{k,l}(\partial_k\partial_l f)(\mathbf{x})$$
$$\cong \left[\Big(\bar{U}_s \circ \Big(\sum_{i,j}(A_D)_{i,j}\partial_{\mathbf{x}_i}\partial_{\mathbf{x}_j}\Big) \circ \bar{U}_{-s}\Big)f\right](\mathbf{x})$$
$$\cong \bar{U}_s \sum_{k,l}\big[\nabla U_s(\mathbf{x})A_D(\mathbf{x})(\nabla U_s(\mathbf{x}))^T\big]_{kl} \cdot (\partial_k\partial_l f)(U_s(\mathbf{x}))$$
$$= \sum_{k,l}\big[\nabla U_s(\mathbf{y})A_D(\mathbf{y})(\nabla U_s(\mathbf{y}))^T\big]_{kl} \cdot (\partial_k\partial_l f)(\mathbf{x}),$$

where $\mathbf{y} := U_s(\mathbf{x})$. This proves that

$$A_{U_sD}(\mathbf{x}) = \nabla U_s(\mathbf{y})\,A_D(\mathbf{y})\,\nabla U_s(\mathbf{y})^T,$$

and thus the statement of Lemma 6.5. $\square$

E.4 Proof of Lemma 6.6

Proof.

$$
\begin{aligned}
\langle \mathrm{d}\mathbf{x}_1, \mathrm{d}\mathbf{x}_2 \rangle_{\mathbf{x}}^{U_s m_A} & \overset{(6.31)}{=} \langle \nabla U_{-s}(\mathbf{x})\,\mathrm{d}\mathbf{x}_1, \nabla U_{-s}(\mathbf{x})\,\mathrm{d}\mathbf{x}_2 \rangle_{U_{-s}(\mathbf{x})}^{m_A} \\
& \overset{(6.39)}{=} \langle \nabla U_{-s}(\mathbf{x})\,\mathrm{d}\mathbf{x}_1, A(U_{-s}(\mathbf{x}))^{-1}\,\nabla U_{-s}(\mathbf{x})\,\mathrm{d}\mathbf{x}_2 \rangle \\
& = \langle \mathrm{d}\mathbf{x}_1, \nabla U_{-s}(\mathbf{x})^T\,A(U_{-s}(\mathbf{x}))^{-1}\,\nabla U_{-s}(\mathbf{x})\,\mathrm{d}\mathbf{x}_2 \rangle \\
& = \langle \mathrm{d}\mathbf{x}_1, \left[\nabla U_{-s}(\mathbf{x})^{-1}\,A(U_{-s}(\mathbf{x}))\,(\nabla U_{-s}(\mathbf{x})^{-1})^T \right]^{-1} \mathrm{d}\mathbf{x}_2 \rangle \\
& \overset{(6.33)}{=} \langle \mathrm{d}\mathbf{x}_1, \left[\nabla U_s(\mathbf{y})\,A(\mathbf{y})\,\nabla U_s(\mathbf{y})^T \right]^{-1} \mathrm{d}\mathbf{x}_2 \rangle \\
& \overset{(6.36)}{=} \langle \mathrm{d}\mathbf{x}_1, (U_s A)(\mathbf{x})^{-1}\,\mathrm{d}\mathbf{x}_2 \rangle \\
& \overset{(6.39)}{=} \langle \mathrm{d}\mathbf{x}_1, \mathrm{d}\mathbf{x}_2 \rangle_{\mathbf{x}}^{m_{U_s A}} \qquad\qquad \square
\end{aligned}
$$

E.5 Proof of Lemma 6.8

Proof. This equivalence follows from the relation

$$
\frac{1}{\psi_1'} \circ (\psi \otimes D) \circ \psi_1 = \frac{1}{\frac{1}{\psi} \otimes \psi_1'} \circ D \circ \psi_1 \tag{E.3}
$$

for any $\psi_1, \psi_1' \in \mathcal{S}_\varnothing(\mathcal{V}_1)$: If $\psi \otimes D$ is a kth-order differential operator then the left-hand side of (E.3) is a kth-order differential operator for some $\psi_1, \psi_1' \in \mathcal{S}_\varnothing(\mathcal{V}_1)$, and since $\frac{1}{\psi} \otimes \psi_1' \in \mathcal{S}(\mathcal{V}_2)$, we can therefore conclude that D is a kth-order differential operator. The reverse direction can be proven in the same way, since every section $\psi_2 \in \mathcal{S}_\varnothing(\mathcal{V}_2)$ can be written as $\psi_2 = \frac{1}{\psi} \otimes \psi_1'$ for some $\psi_1' \in \mathcal{S}_\varnothing(\mathcal{V}_1)$ (take $\psi_1' := \psi \otimes \psi_2$). $\qquad\qquad \square$

.

Appendix F

Proofs for Chapter 7

F.1   Proof of Lemma 7.1

Proof. (i) As usual let us write $x = (x_0, \vec{x})$ and $y = (y_0, \vec{y})$. Since x is isotropic, we have $0 = \langle x, x \rangle = -x_0^2 + \|\vec{x}\|^2$ and thus $x_0 = \text{sign}(x_0)\|\vec{x}\|$. Similarly, we have $y_0 = \text{sign}(y_0)\|\vec{y}\|$, and thus

$$\langle x, y \rangle = -x_0 y_0 + \langle \vec{x}, \vec{y} \rangle = -\text{sign}(x_0 y_0)\|\vec{x}\|\|\vec{y}\| + \langle \vec{x}, \vec{y} \rangle.$$

Therefore, if x and y are from opposite cones then we have

$$\langle x, y \rangle = \|\vec{x}\|\|\vec{y}\| + \langle \vec{x}, \vec{y} \rangle \geq 0,$$

with equality if and only if $\vec{x} = a\vec{y}$ for some $a < 0$. Since this condition also implies that $x_0 = \text{sign}(x_0)\|\vec{x}\| = -\text{sign}(y_0)\|a\vec{y}\| = a\,\text{sign}(y_0)\|\vec{y}\| = ay_0$ and thus $x = ay$, this is just the case in which $x \parallel y$.

If instead x and y are from the same cone then we have

$$\langle x, y \rangle = -\|\vec{x}\|\|\vec{y}\| + \langle \vec{x}, \vec{y} \rangle \leq 0,$$

with equality if and only if $\vec{x} = a\vec{y}$ for some $a > 0$, which again can analogously be shown to be the case in which $x \parallel y$.

(ii) If $y \in \mathbb{T}^\downarrow \cup \mathbb{T}^\uparrow$ then we have $0 > \langle y, y \rangle = -y_0^2 + \|\vec{y}\|^2$ and therefore $y_0 = \text{sign}(y_0)(\|\vec{y}\| + \delta)$ for some $\delta > 0$, which leads us to the representation

$$\langle x, y \rangle = -\text{sign}(x_0 y_0)\|\vec{x}\|(\|\vec{y}\| + \delta) + \langle \vec{x}, \vec{y} \rangle.$$

Therefore, if $x \in \mathbb{T}_0^\uparrow$ and $y \in \mathbb{T}^\downarrow$ then we have

$$\langle x, y \rangle = \|\vec{x}\|(\|\vec{y}\| + \delta) + \langle \vec{x}, \vec{y} \rangle > 0,$$

and similarly, if $x \in \mathbb{T}_0^\uparrow$ and $y \in \mathbb{T}^\uparrow$ then we have $\langle x, y \rangle < 0$.

(iii) For any given $x \in \mathbb{T}_0^\downarrow$, this part now follows by applying part (ii) to $-x \in \mathbb{T}_0^\uparrow$ instead of x. $\qquad\square$

F.2   Proof of Lemma 7.7 (Part 2)

Proof. Let us now show the necessity of the given conditions on ε and $\hat{\varepsilon}$.

(i) For the operators $\mathrm{e}^{s\varepsilon}$ to be pseudo-orthogonal, i.e.,

$$\forall s \in \mathbb{R} \ \forall x, y \in \mathbb{T}: \ \langle \mathrm{e}^{s\varepsilon} x, \mathrm{e}^{s\varepsilon} y \rangle = \langle x, y \rangle$$

by (7.34), we necessarily need to have

$$0 = \partial_s \langle \mathrm{e}^{s\varepsilon} x, \mathrm{e}^{s\varepsilon} y \rangle \big|_{s=0} = \langle \varepsilon x, y \rangle + \langle x, \varepsilon y \rangle = \langle (\varepsilon + \varepsilon^*) x, y \rangle$$

for $\forall x, y \in \mathbb{T}$, and thus $\varepsilon + \varepsilon^* = 0$, i.e., ε must be antisymmetric.

(ii) Now suppose that $\mathrm{e}^{s\bar{\varepsilon}} \in \hat{O}(\mathbb{T})$ for $\forall s \in \mathbb{R}$, i.e., that $\mathrm{e}^{s\bar{\varepsilon}} = \nu_s \hat{A}_s$ for some $\nu_s > 0$ and $\hat{A}_s \in O^\uparrow(\mathbb{T})$. First observe that $(\nu_s^2)_{s\in\mathbb{R}}$ inherits the group property from $(\mathrm{e}^{s\bar{\varepsilon}})_{s\in\mathbb{R}}$. Indeed, for $\forall x \in \mathbb{T}$ we have

$$\langle \mathrm{e}^{s\bar{\varepsilon}} x, \mathrm{e}^{s\bar{\varepsilon}} x \rangle = \langle \nu_s \hat{A}_s x, \nu_s \hat{A}_s x \rangle \overset{(7.33)}{=} \nu_s^2 \langle x, x \rangle \,,$$

and so picking any $x \in \mathbb{T}^\uparrow$ (so that $\langle x, x \rangle \neq 0$ by (7.4a)), we have

$$\begin{aligned}
\nu_{s_1+s_2}^2 &= \frac{\langle \mathrm{e}^{(s_1+s_2)\bar{\varepsilon}} x, \mathrm{e}^{(s_1+s_2)\bar{\varepsilon}} x \rangle}{\langle x, x \rangle} = \frac{\langle \mathrm{e}^{s_1\bar{\varepsilon}} \mathrm{e}^{s_2\bar{\varepsilon}} x, \mathrm{e}^{s_1\bar{\varepsilon}} \mathrm{e}^{s_2\bar{\varepsilon}} x \rangle}{\langle x, x \rangle} \\
&= \frac{\langle \mathrm{e}^{s_1\bar{\varepsilon}} \mathrm{e}^{s_2\bar{\varepsilon}} x, \mathrm{e}^{s_1\bar{\varepsilon}} \mathrm{e}^{s_2\bar{\varepsilon}} x \rangle}{\langle \mathrm{e}^{s_2\bar{\varepsilon}} x, \mathrm{e}^{s_2\bar{\varepsilon}} x \rangle} \times \frac{\langle \mathrm{e}^{s_2\bar{\varepsilon}} x, \mathrm{e}^{s_2\bar{\varepsilon}} x \rangle}{\langle x, x \rangle} = \nu_{s_1}^2 \nu_{s_2}^2 \,.
\end{aligned}$$

(Note that $\langle \mathrm{e}^{s_2\bar{\varepsilon}} x, \mathrm{e}^{s_2\bar{\varepsilon}} x \rangle \neq 0$ since by Lemma 7.4 we have $\mathrm{e}^{s_2\bar{\varepsilon}} x \in \mathbb{T}^\uparrow$.)

We therefore have $\nu_s^2 = \mathrm{e}^{2\rho s}$ for some $\rho \in \mathbb{R}$ and $\forall s \in \mathbb{R}$, and thus $\hat{A}_s = \frac{1}{\nu_s} \mathrm{e}^{s\bar{\varepsilon}} = \mathrm{e}^{s(-\rho+\bar{\varepsilon})}$. This representation shows that the operators $\hat{A}_s$ form a group as well, and so we can apply part (i) to show that $\hat{A}_s = \mathrm{e}^{s\varepsilon}$ for some antisymmetric operator ε. We can thus conclude that $\mathrm{e}^{s\bar{\varepsilon}} = \nu_s \hat{A}_s = \mathrm{e}^{\rho s} \mathrm{e}^{s\varepsilon} = \mathrm{e}^{s(\varepsilon+\rho)}$, and computing the derivative at $s = 0$ on both sides, we find that $\bar{\varepsilon} = \varepsilon + \rho$. $\square$

F.3   Proof of (7.77)

Proof. Let $x \in \mathbb{T}_0^\uparrow$. First note that the functions defined in (7.68) fulfill

$$\lambda_{A^{-1}x}(A^{-1}y) = \frac{A^{-1}x}{A^{-1}y} = \frac{A^{-1}\left(\frac{x}{y} \cdot y\right)}{A^{-1}y} = \frac{x}{y} = \lambda_x(y),$$

so that by (7.67) we have

$$g_{\pi(x)}(x) = \lambda_x = \lambda_{A^{-1}x} \circ A^{-1} = g_{\pi(A^{-1}x)}(A^{-1}x) \circ A^{-1}. \tag{F.1}$$

Given any section $\psi \in \mathcal{S}(\mathscr{P})$, we therefore have

$$
\begin{aligned}
\widehat{A\psi}(x) &\overset{(7.72)}{=} \big(g_{\pi(x)}\big((A\psi)(\pi(x))\big)\big)(x) \\
&\overset{(7.47)}{=} \big(g_{\pi(x)}\big(A \cdot \psi(A^{-1}\pi(x))\big)\big)(x) \\
&\overset{(F.1)}{=} \big(g_{\pi(A^{-1}x)}\big(\psi(A^{-1}\pi(x))\big)\big)(A^{-1}x) \\
&\overset{(7.42)}{=} \big(g_{\pi(A^{-1}x)}\big(\psi(\pi(A^{-1}x))\big)\big)(A^{-1}x) \\
&\overset{(7.72)}{=} \hat{\psi}(A^{-1}x).
\end{aligned}
$$

This shows the desired property (7.77) for sections $\psi \in \mathcal{S}(\mathscr{P})$, and the analogous statement for sections $\psi^* \in \mathcal{S}(\mathscr{P}^{\otimes(-1)})$ follows analogously from (7.48). For sections $\psi \in \mathcal{S}(\mathscr{P}^{\otimes k})$, $k \in \mathbb{Z} \setminus \{0\}$, it then follows from (7.49a)–(7.49b) and (7.76). The case $k = 0$, i.e., the one for functions $f \in \mathcal{F}(\mathbb{P}_0)$ can best be seen directly:

$$
\widehat{Af}(x) \overset{(7.73)}{=} (Af)(\pi(x)) \overset{(7.43)}{=} f(A^{-1}\pi(x)) \overset{(7.42)}{=} f(\pi(A^{-1}x)) \overset{(7.73)}{=} \hat{f}(A^{-1}x).
$$

$\square$

F.4 Proof of Lemma 7.9

Proof. Let $A \in \hat{O}(\mathbb{T})$, and let $\nu := |A| > 0$ and $\hat{A} \in O^\uparrow(\mathbb{T})$ be such that $A = \nu\hat{A}$. Then by (7.35) we have $A^{-1} = \nu^{-1}\hat{A}^{-1} = \nu^{-1}\hat{A}^*$, and thus

$$
\partial_i(A^{-1}x)^j = \partial_i(A^{-1})^j{}_l x^l = (\nu^{-1}\hat{A}^*)^j{}_l \partial_i x^l \overset{(7.32)}{=} \nu^{-1}\hat{A}_l{}^j \delta_i^l = \nu^{-1}\hat{A}_i{}^j
$$
$$
\Rightarrow \quad \partial_i(f(A^{-1}x)) = (\partial_j f)(A^{-1}x) \cdot \partial_i(A^{-1}x)^j = (\partial_j f)(A^{-1}x) \cdot \nu^{-1}\hat{A}_i{}^j,
$$

and raising the index i on both sides, exchanging the vertical positions of the indices j (recall (7.30)), and then renaming j to l, we also have

$$
\partial^i(f(A^{-1}x)) = (\partial^l f)(A^{-1}x) \cdot \nu^{-1}\hat{A}^i{}_l.
$$

Combining this with (7.35), which in index notation can be written as

$$
\hat{A}_i{}^j \hat{A}^i{}_l = (\hat{A}^*)^j{}_i \hat{A}^i{}_l = (\hat{A}^*\hat{A})^j{}_l = I^j{}_l = \delta_l^j,
$$

we therefore obtain

$$
\begin{aligned}
(\square(Af))(x) = \partial^i \partial_i(f(A^{-1}x)) &= \nu^{-2}\hat{A}^i{}_l \hat{A}_i{}^j (\partial^l \partial_j f)(A^{-1}x) \\
&= \nu^{-2}\delta_l^j (\partial^l \partial_j f)(A^{-1}x) = \nu^{-2}(\partial^j \partial_j f)(A^{-1}x) \\
&= \nu^{-2}(\square f)(A^{-1}x) = |A|^{-2} \cdot (A\square f)(x).
\end{aligned}
$$

$\square$

F.5   Proof of Lemma 7.13

Proof. We have

$$
\begin{aligned}
\mathscr{M}_t &\overset{(5.28)}{=} \mathcal{U}_{-t}(\mathscr{M}_0) \\
&\overset{(7.106)}{=} \mathcal{U}_{-t}\big(\pi\big(\{x \in \mathbb{T}_0^\uparrow \,|\, \langle \dot{b}_0, x\rangle > 0\}\big)\big) \\
&\overset{(7.42)}{=} \pi\big(\mathcal{U}_{-t}\big(\{x \in \mathbb{T}_0^\uparrow \,|\, \langle \dot{b}_0, x\rangle > 0\}\big)\big) \\
&\overset{\text{L.7.4}}{=} \pi\big(\{y \in \mathbb{T}_0^\uparrow \,|\, \langle \dot{b}_0, \mathcal{U}_t y\rangle > 0\}\big) \\
&= \pi\big(\{y \in \mathbb{T}_0^\uparrow \,|\, \langle \mathcal{U}_t^* \dot{b}_0, y\rangle > 0\}\big) \\
&\overset{(7.97a)}{=} \pi\big(\{y \in \mathbb{T}_0^\uparrow \,|\, \langle \dot{b}_t, y\rangle > 0\}\big),
\end{aligned}
$$

where in the fourth step we substituted $y := \mathcal{U}_{-t}x$. In particular, since as in (7.100) one can show that $\langle \dot{b}_t, b_t\rangle = 0$, this implies that

$$
\pi(b_t) \in \pi\big(\{y \in \mathbb{T}_0^\uparrow \,|\, \langle \dot{b}_t, y\rangle = 0\}\big) = \partial \mathscr{M}_t\,. \qquad \square
$$

F.6   Proof of Lemma 7.14

Proof. Let us now treat the remaining case $\ddot{b}_0^\perp \in \mathbb{T}_0^\downarrow$, which by (7.112) and (7.108) in particular implies that $\pi(\ddot{b}_0^\perp) \in \partial \mathscr{M}_0$.

In this case the last inequality in (7.119) only holds for $\forall \mathbf{p} \in \partial \mathscr{M}_0 \setminus \{\pi(\ddot{b}_0^\perp)\}$. Therefore, if for any $\hat{\mathbf{p}} \in \mathscr{M}_0$ the path $\{\mathcal{U}_t(\hat{\mathbf{p}}) \,|\, t \geq 0\}$ would lead out of $\mathscr{M}_0$, then our previously used argument would only suffice to conclude that its first exit point from $\mathscr{M}_0$ can only be $\pi(\ddot{b}_0^\perp)$; while this may well be the case, we could then try to obtain a contradiction by constructing a nearby point $\hat{\mathbf{p}}' \in \mathscr{M}_0$ whose associated first exit point is different from $\pi(\ddot{b}_0^\perp)$. This would prove (7.114), and thus that the sets $\mathscr{M}_t$ are increasing, also for this case.

To construct $\hat{\mathbf{p}}'$, it suffices to show that

$$
\mathscr{M}_0 \setminus \mathcal{U}_{-\hat{t}}(\overline{\mathscr{M}_0}) \neq \varnothing, \tag{F.2}
$$

where $\hat{t} > 0$ denotes the first exit time of the path emanating from the given $\hat{\mathbf{p}}$ (which implies that $\mathcal{U}_{\hat{t}}(\hat{\mathbf{p}}) = \pi(\ddot{b}_0^\perp)$). Once (F.2) is shown, since this set is an open subset of $\mathbb{P}_0$ (which is of dimension $n \geq 3$), it must contain a point $\hat{\mathbf{p}}'$ that does not lie on the path $\{\mathcal{U}_{-t}(\pi(\ddot{b}_0^\perp)) \,|\, t \in [0, \hat{t}]\} \subset \mathbb{P}_0$. For any such point, the path $\{\mathcal{U}_t(\hat{\mathbf{p}}') \,|\, t \in [0, \hat{t}]\}$ would lead from $\hat{\mathbf{p}}' \in \mathscr{M}_0$ to $\mathcal{U}_{\hat{t}}(\hat{\mathbf{p}}') \in \mathbb{P}_0 \setminus \overline{\mathscr{M}_0}$ without passing $\pi(\ddot{b}_0^\perp)$, and so in particular its exit point from $\mathscr{M}_0$ must be different from $\pi(\ddot{b}_0^\perp)$.

To show (F.2), we need to construct an $x \in \mathbb{T}_0^\uparrow$ with

$$
\langle \dot{b}_0, x\rangle > 0 \qquad \text{and} \qquad \langle \dot{b}_0, \mathcal{U}_{\hat{t}}x\rangle < 0, \tag{F.3}
$$

as this would show that $\pi(x) \in \mathscr{M}_0 \setminus \mathcal{U}_{-\hat{t}}(\overline{\mathscr{M}_0})$. In order to construct such an x, let $\hat{x} \in \hat{\mathbf{p}}_+$, let us write $\hat{x} =: (x^0, \vec{x})$ and $\dot{b}_{\hat{t}} =: (y^0, \vec{y})$, let $\eta > 0$ be so small that

$x^0 - \eta y^0 > 0$ (note that $x^0 > 0$ since $\hat{x} \in \hat{\mathbf{p}}_+ \subset \mathbb{T}_0^\uparrow$), and let us define

$$c_\eta := \sqrt{(x^0 - \eta y^0)^2 + \eta^2 \langle \dot{b}_{\hat{t}}, \dot{b}_{\hat{t}} \rangle} - (x^0 - \eta y^0)$$

$$\text{and} \quad x_\eta := \hat{x} - \eta \dot{b}_{\hat{t}} + (c_\eta, \vec{0}). \tag{F.4}$$

(Note that by (7.97a), (7.41), and (7.39) our assumption (7.102) implies that in fact we have

$$\forall t \in \mathbb{R}: \quad \langle \dot{b}_t, \dot{b}_t \rangle > 0, \tag{F.5}$$

so that c_η is well defined and positive.) Then since by definition of $\hat{t}$ and by (7.108) and (7.97a) we have

$$0 = \langle \dot{b}_0, \mathcal{U}_{\hat{t}} \hat{x} \rangle = \langle \mathcal{U}_{\hat{t}}^* \dot{b}_0, \hat{x} \rangle = \langle \dot{b}_{\hat{t}}, \hat{x} \rangle, \tag{F.6}$$

and since $\langle \hat{x}, \hat{x} \rangle = 0$, our definition (F.4) implies that

$$\langle x_\eta, x_\eta \rangle = \eta^2 \langle \dot{b}_{\hat{t}}, \dot{b}_{\hat{t}} \rangle - 2 c_\eta (x^0 - \eta y^0) - c_\eta^2 = 0.$$

Since by (F.4) we have $x_\eta^0 = x_0 - \eta y^0 + c_\eta = \sqrt{\ldots} > 0$, this shows that $x_\eta \in \mathbb{T}_0^\uparrow$. Furthermore, (7.97a), (F.4)–(F.6), and the fact that $c_\eta = O(\eta^2)$ imply that

$$\langle \dot{b}_0, \mathcal{U}_{\hat{t}} x_\eta \rangle = \langle \mathcal{U}_{\hat{t}}^* \dot{b}_0, x_\eta \rangle = \langle \dot{b}_{\hat{t}}, x_\eta \rangle = 0 - \eta \langle \dot{b}_{\hat{t}}, \dot{b}_{\hat{t}} \rangle - y^0 c_\eta < 0$$

for sufficiently small $\eta > 0$. Finally, since $\langle \dot{b}_0, \hat{x} \rangle > 0$ (recall that $\pi(\hat{x}) = \hat{\mathbf{p}} \in \mathcal{M}_0$) and since $\lim_{\eta \searrow 0} x_\eta = \hat{x}$, we can choose $\eta > 0$ so small that in addition we have $\langle \dot{b}_0, x_\eta \rangle > 0$. The vector $x := x_\eta$ then has both of our desired properties in (F.3). This concludes the proof of (7.114) also in this case, which as before also implies that the sets $\mathcal{M}_t$ are increasing.

To prove also (7.116) for this case $\ddot{b}_0^\perp \in \mathbb{T}_0^\downarrow$, we need to refine our arguments of the first case: Since (7.120b) now only holds for $\mathbf{p} \in \partial \mathcal{M}_0 \setminus \{\pi(\ddot{b}_0^\perp)\}$, we can only conclude that $(\partial \mathcal{M}_0 \setminus \{\pi(\ddot{b}_0^\perp)\}) \cap \partial \mathcal{M}_t = \varnothing$, i.e., $\partial \mathcal{M}_0 \cap \partial \mathcal{M}_t \subseteq \{\pi(\ddot{b}_0^\perp)\}$, for $\forall t > 0$, and applying $\mathcal{U}_t$ to both sides, we find that

$$\forall t > 0: \quad \partial \mathcal{M}_{-t} \cap \partial \mathcal{M}_0 \subseteq \{\mathcal{U}_t(\pi(\ddot{b}_0^\perp))\}. \tag{F.7}$$

Now observe that since the sets $\mathcal{M}_t$ are increasing, the sets $\partial \mathcal{M}_{-t} \cap \partial \mathcal{M}_0$, are decreasing in t for $t > 0$. Therefore, the only way in which some set $\partial \mathcal{M}_{-t} \cap \partial \mathcal{M}_0$ can be non-empty is that all the intersections $\partial \mathcal{M}_{-t'} \cap \partial \mathcal{M}_0$, $t' \in (0, t]$, are non-empty as well and thus contain the same single point. But this means that $\mathcal{U}_{t'}(\pi(\ddot{b}_0^\perp))$ has to be the same point $\pi(\ddot{b}_0^\perp)$ for $\forall t' \in [0, t]$, so that $\mathcal{U}_{t'} \ddot{b}_0^\perp = u_{t'} \ddot{b}_0^\perp$ for $\forall t' \in [0, t]$ and some values $u_{t'} \in \mathbb{R}$. Differentiating with respect to t' at $t' = 0$, we can then conclude that $(\rho + \varepsilon) \ddot{b}_0^\perp = \dot{u}_0 \ddot{b}_0^\perp$, i.e., $\ddot{b}_0^\perp$ must be an eigenvector of ε (with eigenvalue $\dot{u}_0 - \rho$). In short, we have found that if $\partial \mathcal{M}_{-t} \cap \partial \mathcal{M}_0$ is non-empty for any $t > 0$ then we must have $\partial \mathcal{M}_{-t} \cap \partial \mathcal{M}_0 = \{\pi(\ddot{b}_0^\perp)\}$, and $\ddot{b}_0^\perp$ must be an eigenvector of ε.

The reverse also holds: If $\ddot{b}_0^\perp$ is an eigenvector of ε with some eigenvalue ν then we have for $\forall t \in \mathbb{R}$

$$\mathcal{U}_{-t}\ddot{b}_0^\perp = e^{-(\rho+\nu)t}\ddot{b}_0^\perp$$
$$\Rightarrow \quad \mathcal{U}_{-t}(\pi(\ddot{b}_0^\perp)) = \pi(\ddot{b}_0^\perp) \in \partial\mathcal{M}_0$$
$$\Rightarrow \quad \pi(\ddot{b}_0^\perp) \in \mathcal{U}_t(\partial\mathcal{M}_0) = \partial\mathcal{M}_{-t}$$
$$\Rightarrow \quad \pi(\ddot{b}_0^\perp) \in \partial\mathcal{M}_{-t} \cap \mathcal{M}_0$$
$$\overset{(\text{F.7})}{\Rightarrow} \quad \partial\mathcal{M}_{-t} \cap \mathcal{M}_0 = \left\{\pi(\ddot{b}_0^\perp)\right\}.$$

We have thus shown that for $\forall t > 0$ we have

$$\partial\mathcal{M}_{-t} \cap \mathcal{M}_0 = \begin{cases} \left\{\pi(\ddot{b}_0^\perp)\right\} & \text{if } \ddot{b}_0^\perp \text{ is an eigenvector of } \varepsilon, \\ \varnothing & \text{else.} \end{cases}$$

Finally, given any two values $t_2 > t_1$, we can now set $t := t_2 - t_1$ and apply $\mathcal{U}_{-t_2}$ to both sides (which leaves the right-hand side unchanged) to prove the remaining cases of (7.116). $\qquad\square$

F.7 Proof of (7.121)

Proof. Abbreviating $\bar{r}^X := r^X/(\frac{n}{2}-1)$, by (7.99) we have

$$\partial_t \bar{r}^X(\mathcal{U}_t(\mathbf{p}))\big|_{t=0} = \partial_t\left(\rho - \frac{\langle\varepsilon b_0, \mathcal{U}_t x\rangle}{\langle b_0, \mathcal{U}_t x\rangle}\right)\bigg|_{t=0}$$
$$= -\frac{\langle\varepsilon b_0, (\varepsilon+\rho)x\rangle\langle b_0, x\rangle - \langle\varepsilon b_0, x\rangle\langle b_0, (\varepsilon+\rho)x\rangle}{\langle b_0, x\rangle^2}$$
$$= \frac{\langle(\varepsilon-\rho)\varepsilon b_0, x\rangle}{\langle b_0, x\rangle} - \frac{\langle\varepsilon b_0, x\rangle\langle(\varepsilon-\rho)b_0, x\rangle}{\langle b_0, x\rangle^2}$$
$$= \frac{\langle\ddot{b}_0^\perp, x\rangle}{\langle b_0, x\rangle} - \frac{\langle\varepsilon b_0, x\rangle}{\langle b_0, x\rangle}\left(\frac{\langle\varepsilon b_0, x\rangle}{\langle b_0, x\rangle} - \rho\right)$$
$$= \frac{\langle\ddot{b}_0^\perp, x\rangle}{\langle b_0, x\rangle} - \left(-\bar{r}^X(\mathbf{p}) + \rho\right)\left(-\bar{r}^X(\mathbf{p})\right)$$
$$= \frac{\langle\ddot{b}_0^\perp, x\rangle}{\langle b_0, x\rangle} - \bar{r}^X(\mathbf{p})\left(\bar{r}^X(\mathbf{p}) - \rho\right).$$

On $\partial\mathcal{M}_0$ we have $\bar{r}^X = 0$, so that the second term vanishes, and so multiplying by $(\frac{n}{2}-1)$ yields (7.121). $\qquad\square$

Appendix G

Proofs for Chapter 8

G.1 Proof of Lemma 8.2

Proof. (i) Since $\mathbb{T}_0^\uparrow$ is a connected set, its image $f(\mathbb{T}_0^\uparrow)$ under the continuous map $f \colon \mathbb{T}_0^\uparrow \to \mathbb{R}$ defined as $f(x) := x^\times - x^\odot$ is connected as well, i.e., it must be an interval. Since $e^\times \in \mathbb{T}_0^\uparrow$, we have $1 = 1 - 0 = f(e^\times) \in f(\mathbb{T}_0^\uparrow)$. Therefore, if (8.10a) were wrong and there existed a point $x \in \mathbb{T}_0^\uparrow$ with $f(x) \le 0$ then there would also be a point $x \in \mathbb{T}_0^\uparrow$ with $f(x) = 0$, i.e., with $x^\times = x^\odot$. However, by (8.8) that point would fulfill $0 = \langle x, x \rangle = 2x^\odot x^\times + \|\vec{x}\|^2 = 2(x^\times)^2 + \|\vec{x}\|^2$, so that $\vec{x} = 0$ and $0 = x^\times = x^\odot$, i.e., $x = 0$. But $0 \notin \mathbb{T}_0^\uparrow$, and so (8.10a) must hold. Equation (8.10b) can be shown analogously, using that $e_\odot \in \mathbb{T}_0^\downarrow$.

(ii) If $x^\times = 0$ then by (8.8) we also have $0 = \langle x, x \rangle = \|\vec{x}\|^2$ and thus $x \in \pi(e_\odot)$. If $x^\times > 0$ then $0 = \langle x, x \rangle = 2x^\odot x^\times + \|\vec{x}\|^2$ implies that $x^\odot = -\frac{1}{2x^\times}\|\vec{x}\|^2 \le 0 < x^\times$ and thus $x \in \mathbb{T}_0^\uparrow$ by (8.10a), and analogously $x^\times < 0$ implies that $x \in \mathbb{T}_0^\downarrow$. Since for $\forall x \in \pi(e_\odot)$ we trivially have $x^\times = 0$, this completes the proof. $\qquad\square$

G.2 Proof of Lemma 8.4

Proof. Computing the exponentials of the matrix in (8.20), we find the following:

$$((\mathcal{U}_t^{\lambda,0,\vec{0},\vec{0}})^\mu{}_\nu) = \begin{pmatrix} e^{-\lambda t} & 0 & - & \vec{0} & - \\ 0 & e^{\lambda t} & - & \vec{0} & - \\ | & | & & & \\ \vec{0} & \vec{0} & & I & \\ | & | & & & \end{pmatrix}$$

$$((\mathcal{U}_t^{0,E,\vec{0},\vec{0}})^\mu{}_\nu) = \begin{pmatrix} 1 & 0 & - & \vec{0} & - \\ 0 & 1 & - & \vec{0} & - \\ | & | & & & \\ \vec{0} & \vec{0} & & e^{tE} & \\ | & | & & & \end{pmatrix}$$

$$((\mathcal{U}_t^{0,0,\mathbf{u},\vec{0}})^\mu{}_\nu) = \begin{pmatrix} 1 & -\frac{1}{2}t^2\|\mathbf{u}\|^2 & - & -t\mathbf{u} & - \\ 0 & 1 & - & \vec{0} & - \\ | & | & & & \\ \vec{0} & t\mathbf{u} & & I & \\ | & | & & & \end{pmatrix}$$

$$((\mathcal{U}_t^{0,0,\vec{0},\bar{\mathbf{v}}})^\mu{}_\nu) = \begin{pmatrix} 1 & 0 & - & \vec{0} & - \\ -\frac{1}{2}t^2\|\bar{\mathbf{v}}\|^2 & 1 & - & -t\bar{\mathbf{v}} & - \\ | & | & & & \\ t\bar{\mathbf{v}} & \vec{0} & & I & \\ | & | & & & \end{pmatrix}$$

Now plugging these matrices into (8.17), we obtain

$$U_t^{\lambda,0,\vec{0},\vec{0}}(\mathbf{x}) = \frac{\vec{0} + \vec{0} + I\mathbf{x}}{0 + \mathrm{e}^{\lambda t} + 0} = \mathrm{e}^{-\lambda t}\mathbf{x}\,,$$

$$U_t^{0,E,\vec{0},\vec{0}}(\mathbf{x}) = \frac{\vec{0} + \vec{0} + \mathrm{e}^{tE}\mathbf{x}}{0 + 1 + 0} = \mathrm{e}^{tE}\mathbf{x}\,,$$

$$U_t^{0,0,\mathbf{u},\vec{0}}(\mathbf{x}) = \frac{\vec{0} + t\mathbf{u} + I\mathbf{x}}{0 + 1 + 0} = \mathbf{x} + t\mathbf{u}\,,$$

$$U_t^{0,0,\vec{0},\bar{\mathbf{v}}}(\mathbf{x}) = \frac{t\bar{\mathbf{v}} \cdot (-\frac{1}{2}\|\mathbf{x}\|^2) + \vec{0} + I\mathbf{x}}{(-\frac{1}{2}t^2\|\bar{\mathbf{v}}\|^2) \cdot (-\frac{1}{2}\|\mathbf{x}\|^2) + 1 + \langle -t\bar{\mathbf{v}}, \mathbf{x}\rangle} = \frac{\frac{\mathbf{x}}{\|\mathbf{x}\|^2} - \frac{1}{2}t\bar{\mathbf{v}}}{\left\|\frac{\mathbf{x}}{\|\mathbf{x}\|^2} - \frac{1}{2}t\bar{\mathbf{v}}\right\|^2}\,.$$

The very last transformation is usually only valid for $\mathbf{x} \neq \vec{0}$; however, for $\mathbf{x} = \vec{0}$ the preceding expression returns $\vec{0}$, just like the last expression does if we interpret $\frac{\vec{0}}{\|\vec{0}\|^2} = \infty$ and $\frac{\infty}{\|\infty\|^2} = \vec{0}$. For $\mathbf{x} = \frac{2}{t\|\bar{\mathbf{v}}\|^2}\bar{\mathbf{v}}$ the denominator on the left vanishes, so that we have $U_t(\mathbf{x}) = \infty$ according to what is said below (8.17); since in that case the last expression evaluates to $\frac{\vec{0}}{\|\vec{0}\|^2} = \infty$ as well, this shows that this expression is indeed valid for $\forall \mathbf{x} \in \bar{\mathbb{R}}^n$. □

G.3   Proof of Lemma 8.8

Proof. Starting from (8.7), we can use the fact that by Lemma 8.2 (ii) we have $x^\times \neq 0$ and $y^\times \neq 0$, and then the relation (8.8) and finally our notation (8.12) (for both x and y) to obtain

$$\begin{aligned} \langle x, y\rangle &= x^\odot y^\times + x^\times y^\odot + \langle \vec{x}, \vec{y}\rangle_{\mathbb{R}^n} \\ &= x^\times y^\times \left(\frac{x^\odot x^\times}{(x^\times)^2} + \frac{y^\odot y^\times}{(y^\times)^2} + \frac{\langle \vec{x}, \vec{y}\rangle}{x^\times y^\times}\right) \\ &= x^\times y^\times \left(\frac{-\frac{1}{2}\|\vec{x}\|^2}{(x^\times)^2} + \frac{-\frac{1}{2}\|\vec{y}\|^2}{(y^\times)^2} + \frac{\langle \vec{x}, \vec{y}\rangle}{x^\times y^\times}\right) \\ &= x^\times y^\times \left(-\frac{1}{2}\|\mathbf{x}\|^2 - \frac{1}{2}\|\mathbf{y}\|^2 + \langle \mathbf{x}, \mathbf{y}\rangle\right) \\ &= -\frac{1}{2}x^\times y^\times \|\mathbf{x} - \mathbf{y}\|^2. \end{aligned}$$

□

G.4   Proof of Lemma 8.10 (ii)

Proof. We begin by claiming that it suffices to only prove the existence of an eigenvector $x \in \mathbb{T}_0^{\uparrow}$ with $\pi(x) \in \overline{\mathcal{M}}_0$, since the bound of its eigenvalue $\hat{\lambda}$ and the existence of the other eigenvector $x' \in \mathbb{T}_0^{\uparrow}$ with $\pi(x') \in \mathbb{P}_0 \setminus \mathcal{M}_0$ and with eigenvalue $-\hat{\lambda}$ will then follow.

To see this, first observe that by (7.99) and the antisymmetry of ε we have

$$r^X(\pi(x)) = \left(\tfrac{n}{2} - 1\right)\left(\rho - \frac{\langle \varepsilon b_0, x \rangle}{\langle b_0, x \rangle}\right) = \left(\tfrac{n}{2} - 1\right)\left(\rho + \frac{\langle b_0, \varepsilon x \rangle}{\langle b_0, x \rangle}\right)$$
$$= \left(\tfrac{n}{2} - 1\right)(\rho + \hat{\lambda}), \tag{G.1}$$

so that by (7.105) the location $\pi(x) \in \overline{\mathcal{M}}_0$ of the first eigenvector just says that

$$\hat{\lambda} \geq -\rho. \tag{G.2}$$

Second, observe that the constraints (7.102) and (7.113), i.e.,

$$\langle \varepsilon e_{\odot}, \varepsilon e_{\odot} \rangle = 1 \quad \text{and} \quad \langle \varepsilon^2 e_{\odot}, \varepsilon^2 e_{\odot} \rangle + \rho^2 \leq 0, \tag{G.3}$$

continue to be satisfied if one replaces ρ by $-\rho$. Therefore, once a proof for the existence of an eigenvector $x \in \mathbb{T}_0^{\uparrow}$ with $\hat{\lambda} \geq -\rho$ is provided, we can apply it also to the model with the modified parameter $\rho' := -\rho$ and obtain an eigenvector of ε with eigenvalue $\hat{\lambda} \geq -\rho' = \rho$. Combining this with the result for the unmodified model, we would thus be guaranteed the existence of an eigenvector $x \in \mathbb{T}_0^{\uparrow}$ of ε with eigenvalue

$$\hat{\lambda} \geq \max\{-\rho, \rho\} = |\rho|. \tag{G.4}$$

In a second step we can then apply this entire statement also to the model with the modified parameter $\varepsilon' := -\varepsilon$ (which also still satisfies the constraints in (G.3)), and in this way prove the existence of an eigenvector $x' \in \mathbb{T}_0^{\uparrow}$ of ε' with an eigenvalue $\hat{\lambda}' \geq |\rho'| = |\rho|$, i.e., of an eigenvector of ε with eigenvalue

$$-\hat{\lambda}' \leq -|\rho|. \tag{G.5}$$

Now if our two eigenvectors x and x' of ε are not collinear, then by part (i.2) of this lemma their eigenvalues $\hat{\lambda}$ and $-\hat{\lambda}'$ must fulfill $\hat{\lambda} = -(-\hat{\lambda}') = \hat{\lambda}'$, as claimed. If instead x and x' are collinear, then their eigenvalues $\hat{\lambda}$ and $-\hat{\lambda}'$ must coincide, and so by (G.4)–(G.5) we have

$$|\rho| \leq \hat{\lambda} = -\hat{\lambda}' \leq -|\rho|,$$

which implies that $\rho = \hat{\lambda} = \hat{\lambda}' = 0$, and thus that $\hat{\lambda}' = -\hat{\lambda}$ again.

Finally, since we have shown that $\hat{\lambda} \geq |\rho| \geq -\rho$ and $-\hat{\lambda}' \leq -|\rho| \leq -\rho$, by (G.1) we have $r^X(\pi(x)) \geq 0$ and analogously $r^X(\pi(x')) \leq 0$, which means that $\pi(x) \in \overline{\mathcal{M}}_0$ and $\pi(x') \in \mathbb{P}_0 \setminus \mathcal{M}_0$ as claimed.

This leaves us with the task of showing the existence of an eigenvector x with $\pi(x) \in \overline{\mathscr{M}_0}$. To begin, note that it suffices to find a $\mathbf{p} \in \overline{\mathscr{M}_0}$ with $\mathcal{U}_t(\mathbf{p}) = \mathbf{p}$ for $\forall t \in \mathbb{R}$, i.e., with $\mathcal{U}_t x = \eta(t)x$ for $\forall x \in \mathbf{p}$ and some function $\eta \in \mathcal{F}_+(\mathbb{R})$, as differentiating both sides at $t = 0$ would then imply that $(\varepsilon + \rho)x = \dot{\eta}(0)x$, so that these vectors x (for $x \neq 0$) are eigenvectors of ε with eigenvalue $\dot{\eta}(0) - \rho$.

To find such a $\mathbf{p}$, it suffices to find an $\mathbf{x} \in \overline{M_0}$ with $U_t(\mathbf{x}) = \mathbf{x}$ for $\forall t \in \mathbb{R}$, as $\mathbf{p} := h^{-1}(\mathbf{x}) \in \overline{\mathscr{M}_0}$ would then have the desired property $\mathcal{U}_t(\mathbf{p}) = h^{-1}(U_t(\mathbf{x})) = h^{-1}(\mathbf{x}) = \mathbf{p}$ for $\forall t \in \mathbb{R}$.

Finally, since the function $t \mapsto U_t(\mathbf{x})$ is the solution of the ODE (5.10), this is equivalent to finding an $\mathbf{x} \in \mathbb{R}^n$ with $\mu(\mathbf{x}) = \vec{0}$.

Now if $\overline{M_0}$ were a *compact* convex subset of $\mathbb{R}^n$ and if μ would be defined and C^1 on a neighborhood of $\overline{M_0}$, then the existence of such an $\mathbf{x}$ would follow from the Brouwer Fixed Point Theorem.[1] Indeed, since by (5.9) the functions U_t map M_0 into itself for $\forall t > 0$, they map $\overline{M_0}$ into itself as well, and so the Brouwer Fixed Point Theorem could then conclude that $\forall t > 0 \; \exists \mathbf{x}_t \in \overline{M_0}: \; U_t(\mathbf{x}_t) = \mathbf{x}_t$. The compactness of $\overline{M_0}$ would then ensure the existence of a sequence $t_k \searrow 0$ and an $\mathbf{x} \in \overline{M_0}$ such that $\lim_{k \to \infty} \mathbf{x}_{t_k} = \mathbf{x}$, and it would allow us to estimate

$$
\begin{aligned}
\left\| U_{t_k}(\mathbf{x}) - \mathbf{x} \right\| &= \left\| \left(U_{t_k}(\mathbf{x}_{t_k}) - \mathbf{x}_{t_k} \right) + \left(U_{t_k}(\mathbf{x}) - U_{t_k}(\mathbf{x}_{t_k}) \right) - (\mathbf{x} - \mathbf{x}_{t_k}) \right\| \\
&= \left\| \vec{0} + \int_0^{t_k} \partial_s \left(U_s(\mathbf{x}) - U_s(\mathbf{x}_{t_k}) \right) ds \right\| \\
&= \left\| \int_0^{t_k} \left(\mu(U_s(\mathbf{x})) - \mu(U_s(\mathbf{x}_{t_k})) \right) ds \right\| \\
&= \left\| \int_0^{t_k} \int_0^1 \partial_{s'} \mu \left(U_s(\mathbf{x}_{t_k} + s'(\mathbf{x} - \mathbf{x}_{t_k})) \right) ds' \, ds \right\| \\
&= \left\| \int_0^{t_k} \int_0^1 \nabla \mu \left(U_s(\mathbf{x}_{t_k} + s'(\mathbf{x} - \mathbf{x}_{t_k})) \right) \right. \\
&\qquad\qquad \left. \times (\nabla U_s)(\mathbf{x}_{t_k} + s'(\mathbf{x} - \mathbf{x}_{t_k}))(\mathbf{x} - \mathbf{x}_{t_k}) \, ds' \, ds \right\| \\
&\leq t_k \cdot \max_{\mathbf{y} \in \overline{M_0}} \| \nabla \mu(\mathbf{y}) \| \cdot \max_{\substack{s \in [0, t_k] \\ \mathbf{y} \in \overline{M_0}}} \| \nabla U_s(\mathbf{y}) \| \cdot \| \mathbf{x} - \mathbf{x}_{t_k} \|.
\end{aligned}
$$

Now dividing by t_k and taking the limit $k \to \infty$, the right-hand side would go to zero, while the left-hand side goes to $\| \mu(\mathbf{x}) \|$, proving that $\mu(\mathbf{x}) = \vec{0}$.

To finish our proof, it thus suffices to find a diffeomorphism $g \colon \overline{M_0} \to g(\overline{M_0}) \subset \mathbb{R}^n$ such that $g(M_0)$ is a bounded convex subset of $\mathbb{R}^n$, and such that the associated image $\tilde{\mu} := (\nabla g \cdot \mu) \circ g^{-1}$ of μ (i.e., the vector field whose flowlines are the images of the flowlines of μ) has a C^1 continuation $\tilde{\mu}^+$ to $\overline{g(M_0)} \supsetneq g(\overline{M_0})$.[2] This would allow us to apply the above argument to $\tilde{\mu}^+$ instead of μ and to the associated evolution maps $\tilde{U}_t$ (which on $g(\overline{M_0})$ coincide

---

[1] The Brouwer Fixed Point Theorem states that any continuous function $f(\mathbf{x})$ mapping a compact convex subset of $\mathbb{R}^n$ into itself has a fixed point, i.e., a point $\mathbf{x} \in \mathbb{R}^n$ with $f(\mathbf{x}) = \mathbf{x}$.

[2] Note that $\overline{g(M_0)} \supsetneq g(\overline{M_0})$, since $g(\overline{M_0})$ is missing the point that could be interpreted as $g(\infty)$.

with $g \circ U_t \circ g^{-1}$), proving the existence of a point $\tilde{\mathbf{x}} \in \overline{g(M_0)}$ with $\tilde{\mu}^+(\tilde{\mathbf{x}}) = \vec{0}$. If we can then show that $\tilde{\mathbf{x}} \in g(\overline{M_0})$, then the (consequently well-defined) point $\mathbf{x} := g^{-1}(\tilde{\mathbf{x}}) \in \overline{M_0}$ would fulfill

$$\mu(\mathbf{x}) = \big((\nabla g)^{-1} \cdot (\tilde{\mu} \circ g)\big)(\mathbf{x}) = (\nabla g)^{-1}(\mathbf{x}) \cdot \tilde{\mu}(\tilde{\mathbf{x}}) = (\nabla g)^{-1}(\mathbf{x}) \cdot \tilde{\mu}^+(\tilde{\mathbf{x}}) = \vec{0}.$$

Luckily, we have already constructed such a map in Lemma B.1: The image $g(M_0)$ under that map is an open ball and therefore bounded and convex, and the expression (B.3a) for $\tilde{\mu}$ is in fact smooth on all of $\mathbb{R}^n$ and therefore naturally defines a C^1 continuation $\tilde{\mu}^+$ of $\tilde{\mu}$ to $\overline{g(M_0)}$. Our arguments above therefore guarantee the existence of a point $\tilde{\mathbf{x}} \in \overline{g(M_0)}$ with $\tilde{\mu}^+(\tilde{\mathbf{x}}) = \vec{0}$, and so it only remains to show that in fact we have $\tilde{\mathbf{x}} \in g(\overline{M_0})$. Since $\overline{g(M_0)} \setminus g(\overline{M_0}) = \{\vec{0}\}$ by Lemma B.1 (ii), this means that we only need to show that $\tilde{\mathbf{x}} \neq \vec{0}$. But this follows from the fact that

$$\tilde{\mu}^+(\vec{0}) \overset{\text{(B.3a)}}{=} \tilde{\mathbf{u}} \overset{\text{(B.3b)}}{=} \frac{1}{n-2}\mathbf{v} \overset{\text{(8.32)}}{\neq} \vec{0} = \tilde{\mu}^+(\tilde{\mathbf{x}}).$$

This concludes the proof of the existence of the eigenvector x, and thus the proof of this lemma. $\square$

G.5   Proof of Lemma 8.11

Proof. Denoting the right-hand side of (8.43) by G_t, we immediately see that $G_{t=0} = I$, and we obtain

$$G_t \varepsilon = \begin{pmatrix} \mathrm{e}^{-\lambda t} & 0 & -\ \vec{0}\ - \\ -\frac{1}{2}\mathrm{e}^{-\lambda t}t^2\|R_t\bar{\mathbf{v}}\|^2 & \mathrm{e}^{\lambda t} & -\ (-t\mathrm{e}^{-tE}R_t\bar{\mathbf{v}})\ - \\ \begin{array}{c} | \\ \mathrm{e}^{-\lambda t}t R_t\bar{\mathbf{v}} \\ | \end{array} & \begin{array}{c} | \\ \vec{0} \\ | \end{array} & \mathrm{e}^{tE} \end{pmatrix} \begin{pmatrix} -\lambda & 0 & -\ \vec{0}\ - \\ 0 & \lambda & -\ -\bar{\mathbf{v}}\ - \\ \begin{array}{c} | \\ \mathbf{v} \\ | \end{array} & \begin{array}{c} | \\ \vec{0} \\ | \end{array} & E \end{pmatrix}$$

$$= \begin{pmatrix} -\lambda\mathrm{e}^{-\lambda t} & 0 & -\ \vec{0}\ - \\ \frac{1}{2}\lambda\mathrm{e}^{-\lambda t}t^2\|R_t\bar{\mathbf{v}}\|^2 - t\langle \mathrm{e}^{-tE}R_t\bar{\mathbf{v}}, \bar{\mathbf{v}}\rangle & \lambda\mathrm{e}^{\lambda t} & -\ (-\mathrm{e}^{\lambda t}\bar{\mathbf{v}} + tE\mathrm{e}^{-tE}R_t\bar{\mathbf{v}})\ - \\ \begin{array}{c} | \\ -\lambda\mathrm{e}^{-\lambda t}t R_t\bar{\mathbf{v}} + \mathrm{e}^{tE}\bar{\mathbf{v}} \\ | \end{array} & \begin{array}{c} | \\ \vec{0} \\ | \end{array} & E\mathrm{e}^{tE} \end{pmatrix}.$$

(Note that since the entry in the third column of the second row of G_t is a *row* vector, we had to multiply it by the transpose $E^T = -E$, which explains how the minus sign in the second term disappeared.)

Six entries of this matrix can immediately be seen to indeed coincide with the corresponding entries of $\partial_t G_t$ since they are either zero or simple exponential expressions. The only non-trivial calculations are those for the first and the third column in the second row, and the one for the first column in the third row; however, using the relations $\partial_t(tR_t) = e^{t(E+\lambda I)}$ and $(e^{tE})^T = e^{tE^T} = e^{-tE}$ quickly leads to the desired expressions:

$$\partial_t \|tR_t\bar{\mathbf{v}}\|^2 = 2\langle tR_t\bar{\mathbf{v}}, e^{t(E+\lambda I)}\bar{\mathbf{v}}\rangle = 2te^{\lambda t}\langle e^{-tE}R_t\bar{\mathbf{v}}, \bar{\mathbf{v}}\rangle$$

$$\Rightarrow \quad \partial_t\left(-\tfrac{1}{2}e^{-\lambda t}t^2\|R_t\bar{\mathbf{v}}\|^2\right) = \tfrac{1}{2}\lambda e^{-\lambda t}t^2\|R_t\bar{\mathbf{v}}\|^2 - \tfrac{1}{2}e^{-\lambda t}\partial_t\|tR_t\bar{\mathbf{v}}\|^2$$

$$= \tfrac{1}{2}\lambda e^{-\lambda t}t^2\|R_t\bar{\mathbf{v}}\|^2 - t\langle e^{-tE}R_t\bar{\mathbf{v}}, \bar{\mathbf{v}}\rangle,$$

$$\partial_t(-te^{-tE}R_t\bar{\mathbf{v}}) = tEe^{-tE}R_t\bar{\mathbf{v}} - e^{-tE}e^{t(E+\lambda I)}\bar{\mathbf{v}}$$

$$= -e^{\lambda t}\bar{\mathbf{v}} + tEe^{-tE}R_t\bar{\mathbf{v}},$$

$$\partial_t(e^{-\lambda t}tR_t\bar{\mathbf{v}}) = e^{-\lambda t}\left(-\lambda tR_t + e^{t(E+\lambda I)}\right)\bar{\mathbf{v}}$$

$$= -\lambda e^{-\lambda t}tR_t\bar{\mathbf{v}} + e^{tE}\bar{\mathbf{v}}. \qquad \square$$

G.6   Proof of Lemma 8.14

Proof. For the purpose of this proof let us denote the vectors of the standard basis of $\mathbb{T}$ by e_ν, and let us denote those of $\mathfrak{I}$ by $\tilde{e}_\nu$. Furthermore, as described in Sections 7.3.2 and 8.1, let $Q_{\mu\nu} := \langle e_\mu, e_\nu\rangle$ and $\tilde{Q}_{\mu\nu} := \langle \tilde{e}_\mu, \tilde{e}_\nu\rangle$, and let $Q^{\mu\nu}$ and $\tilde{Q}^{\mu\nu}$ be the components of the inverse matrices $(Q_{\mu\nu})^{-1}$ and $(\tilde{Q}_{\mu\nu})^{-1}$, respectively. Finally, let us denote the coefficients of a given vector $x \in \mathbb{T}$ with respect to these two bases by x^ν and $\tilde{x}^\nu$, respectively, i.e., we have

$$x^\nu e_\nu = x = \tilde{x}^\nu \tilde{e}_\nu, \tag{G.6}$$

where from the context it is clear that on the left ν traverses $0, \ldots, m$ and on the right it traverses $\odot, \times, 1, \ldots, n$.

Multiplying (G.6) by $Q^{\eta\mu}e_\eta$, we obtain

$$x^\mu = x^\nu \delta^\mu_\nu \overset{(7.28)}{=} x^\nu Q_{\nu\eta}Q^{\eta\mu} = x^\nu\langle e_\nu, e_\eta\rangle Q^{\eta\mu} = \langle x, e_\eta\rangle Q^{\eta\mu}$$

$$\Rightarrow \quad x = x^\mu e_\mu = Q^{\eta\mu}\langle x, e_\eta\rangle e_\mu \tag{G.7}$$

for $\forall x \in \mathbb{T}$. Analogously, we find that

$$\tilde{x}^\mu = \cdots = \tilde{x}^\nu\langle \tilde{e}_\nu, \tilde{e}_\eta\rangle \tilde{Q}^{\eta\mu} \overset{(G.6)}{=} x^\nu\langle e_\nu, \tilde{e}_\eta\rangle \tilde{Q}^{\eta\mu},$$

so that by the chain rule of differentiation we have

$$\partial_{x^\nu} = \langle e_\nu, \tilde{e}_\eta\rangle\tilde{Q}^{\eta\mu}\partial_{\tilde{x}^\mu}. \tag{G.8}$$

Plugging this into the definition (7.81) of the d'Alembert operator, and then applying (G.7) to $x = \tilde{e}_{\eta'}$, we therefore find that

$$\square \overset{(7.81)}{=} \partial_\nu\partial^\nu = Q^{\nu\nu'}\partial_{x^\nu}\partial_{x^{\nu'}}$$

$$\overset{(G.8)}{=} Q^{\nu\nu'}\langle e_\nu, \tilde{e}_\eta\rangle\tilde{Q}^{\eta\mu}\partial_{\tilde{x}^\mu}\langle e_{\nu'}, \tilde{e}_{\eta'}\rangle\tilde{Q}^{\eta'\mu'}\partial_{\tilde{x}^{\mu'}}$$

$$= \tilde{Q}^{\eta\mu}\tilde{Q}^{\eta'\mu'}\langle \tilde{e}_\eta, Q^{\nu'\nu}\langle \tilde{e}_{\eta'}, e_{\nu'}\rangle e_\nu\rangle\partial_{\tilde{x}^\mu}\partial_{\tilde{x}^{\mu'}}$$

$$\overset{(G.7)}{=} \quad \tilde{Q}^{\eta\mu}\tilde{Q}^{\eta'\mu'}\langle \tilde{e}_\eta, \tilde{e}_{\eta'}\rangle \partial_{\tilde{x}^\mu}\partial_{\tilde{x}^{\mu'}}$$

$$= \quad \tilde{Q}^{\eta\mu}\tilde{Q}^{\eta'\mu'}\tilde{Q}_{\eta\eta'}\partial_{\tilde{x}^\mu}\partial_{\tilde{x}^{\mu'}}$$

$$\overset{(7.28)}{=} \quad \tilde{Q}^{\eta\mu}\delta^{\mu'}_\eta \partial_{\tilde{x}^\mu}\partial_{\tilde{x}^{\mu'}}$$

$$= \quad \tilde{Q}^{\mu'\mu}\partial_{\tilde{x}^\mu}\partial_{\tilde{x}^{\mu'}}$$

$$\overset{(8.2)}{=} \quad 2\partial_{\tilde{x}^\odot}\partial_{\tilde{x}^\times} + \partial^2_{\tilde{x}^1} + \cdots + \partial^2_{\tilde{x}^n}.$$

Renaming $\tilde{x}^\nu$ back to x^ν, this is (8.55).   □

Bibliography

[1] G. Pelts, "Unspanned Volatility in Non-Affine Short Rate Models and Conformal Symmetry." Working paper, https://www.researchgate.net/publication/292963562_Unspanned_Volatility_in_Non-Affine_Short_Rate_Models_and_Conformal_Symmetry, 2012.

[2] Wikipedia, "Interest Rate Cap and Floor – Wikipedia, The Free Encyclopedia," 2017. [online; accessed 05/18/2018].

[3] Wikipedia, "Swaption – Wikipedia, The Free Encyclopedia," 2017. [online; accessed 05/18/2018].

[4] Wikipedia, "Swap – Wikipedia, The Free Encyclopedia," 2017. [online; accessed 05/18/2018].

[5] P. Collin-Dufresne and R. S. Goldstein, "Do Bonds Span the Fixed Income Markets? Theory and Evidence for Unspanned Stochastic Volatility," *The Journal of Finance*, vol. 57, no. 4, pp. 1685–1730, 2002.

[6] J. Casassus, P. Collin-Dufresne, and B. Goldstein, "Unspanned stochastic volatility and fixed income derivatives pricing," *Journal of Banking & Finance*, vol. 29, no. 11, pp. 2723–2749, 2005.

[7] R. Bikbov and M. Chernov, "Unspanned stochastic volatility in affine models: Evidence from eurodollar futures and options," *Management Science*, vol. 55, no. 8, pp. 1292–1305, 2009.

[8] X. Gabaix, "Linearity-Generating Processes: A Modelling Tool Yielding Closed Forms for Asset Prices." Working paper, https://scholar.harvard.edu/files/xgabaix/files/linearity-generating_processes_modeling_tool.pdf, 2009.

[9] D. Brigo and F. Mercurio, *Interest rate models – theory and practice: With smile, inflation and credit.* Springer Science & Business Media, 2007.

[10] P. Veronesi, *Fixed income securities: Valuation, risk, and risk management.* John Wiley & Sons, 2010.

[11] L. B. G. Andersen and V. V. Piterbarg, *Interest Rate Modeling. Volume 1: Foundations And Vanilla Models.* Atlantic Financial Press, 2010.

[12] L. B. G. Andersen and V. V. Piterbarg, *Interest Rate Modeling. Volume 2: Term Structure Models*. Atlantic Financial Press, 2010.

[13] M. Musiela and M. Rutkowski, *Martingale Methods in Financial Modeling*, vol. 36 of *Stochastic Modeling and Applied Probability*. Springer-Verlag Berlin Heidelberg, second ed., 2005.

[14] J. Hull and A. White, "Pricing Interest-Rate-Derivative Securities," *Review of financial studies*, vol. 3, no. 4, pp. 573–592, 1990.

[15] J. C. Cox, J. E. Ingersoll, and S. A. Ross, "A Theory of the Term Structure of Interest Rates," *Econometrica*, vol. 53, no. 2, pp. 385–407, 1985.

[16] F. Black, E. Derman, and W. Toy, "A one-factor model of interest rates and its application to treasury bond options," *Financial Analysts Journal*, vol. 46, no. 1, pp. 33–39, 1990.

[17] F. Black and P. Karasinski, "Bond and option pricing when short rates are lognormal," *Financial Analysts Journal*, vol. 47, no. 4, pp. 52–59, 1991.

[18] I. Lekkos, "A critique of factor analysis of interest rates," *The Journal of Derivatives*, vol. 8, no. 1, pp. 72–83, 2000.

[19] B. N. Golub and L. M. Tilman, "Measuring Yield Curve Risk Using Principal Components, Analysis, Value, At Risk, And Key Rate Durations," *The Journal of Portfolio Management*, vol. 23, no. 4, pp. 72–84, 1997.

[20] D. Duffie and R. Kan, "A Yield-Factor Model of Interest Rates," *Mathematical Finance*, vol. 6, no. 4, pp. 379–406, 1996.

[21] Q. Dai and K. J. Singleton, "Specification Analysis of Affine Term Structure Models," *The Journal of Finance*, vol. 55, no. 5, pp. 1943–1978, 2000.

[22] D. Heath, R. Jarrow, and A. Morton, "Bond pricing and the term structure of interest rates: A new methodology for contingent claims valuation," *Econometrica: Journal of the Econometric Society*, pp. 77–105, 1992.

[23] A. Brace, D. Gatarek, and M. Musiela, "The market model of interest rate dynamics," *Mathematical Finance*, vol. 7, no. 2, pp. 127–155, 1997.

[24] Wikipedia, "Spread Option – Wikipedia, The Free Encyclopedia," 2017. [online; accessed 05/18/2018].

[25] O. Cheyette, "Markov Representation of the Heath–Jarrow–Morton Model," *working paper*, 1992–96.

[26] O. Cheyette, "Term structure dynamics and mortgage valuation," *The Journal of Fixed Income*, vol. 1, no. 4, pp. 28–41, 1992.

[27] O. Cheyette, "Interest Rate Models," 2002.

[28] I. Beyna, *Interest Rate Derivatives: Valuation, Calibration and Sensitivity Analysis*, vol. 666 of *Lecture Notes in Economics and Mathematical Systems*. Springer-Verlag, 2013.

[29] P. Ritchken and L. Sankarasubramanian, "Volatility Structures of Forward Rates and the Dynamics of the Term Structure," *Mathematical Finance*, vol. 5, no. 1, pp. 55–72, 1995.

[30] P. S. Hagan and D. E. Woodward, "Markov interest rate models," *Applied Mathematical Finance*, vol. 6, no. 4, pp. 233–260, 1999.

[31] G. Pelts, "Unspanned Stochastic Volatility & Conformal Symmetry." Conference paper for "Global Derivatives, Trading & Risk Management 2016," Budapest, Hungary, `https://www.researchgate.net/publication/303810671_Unspanned_Stochastic_Volatility_Conformal_Symmetry`, May 2016.

[32] B. Øksendal, *Stochastic Differential Equations – An Introduction with Applications*. Springer-Verlag Berlin Heidelberg, sixth ed., 2003.

[33] I. Karatzas and S. E. Shreve, *Brownian Motion and Stochastic Calculus*, vol. 113 of *Graduate Texts in Mathematics*. Springer-Verlag New York, second ed., 1998.

[34] V. A. Kholodnyi, *Beliefs-preferences gauge symmetry group and replication of contingent claims in a general market environment*. IES Press, 1998.

[35] V. A. Kholodnyi, "Valuation and dynamic replication of contingent claims in the framework of the beliefs-preferences gauge symmetry," *The European Physical Journal B – Condensed Matter and Complex Systems*, vol. 27, no. 2, pp. 229–238, 2002.

[36] V. A. Kholodnyi, "Beliefs-Preferences Gauge Symmetry and Dynamic Replication of Contingent Claims in a General Market Environment," *Journal of the Dynamics of Continuous, Discrete, and Impulsive Systems B*, vol. 10, no. 1, pp. 81–94, 2003.

[37] F. Verhulst, *Nonlinear Differential Equations and Dynamical Systems*. Universitext, Springer Berlin Heidelberg, second ed., 2006.

[38] R. Durrett, *Stochastic calculus: A practical introduction*. CRC Press, 1996.

Index

Made in the USA
Monee, IL
14 December 2024

3c1bb41d-c840-4f83-b103-7167e90dd082R01